Applied Game Theory

Institut für Höhere Studien − Institute for Advanced Studies
IHS-Studies No. 1

Applied Game Theory

Proceedings of a Conference
at the Institute for Advanced Studies, Vienna,
June 13−16, 1978

Edited by
S.J. Brams, A. Schotter, and G. Schwödiauer

Springer-Verlag Berlin Heidelberg GmbH
1979

CIP-Kurztitelaufnahme der Deutschen Bibliothek

Applied game theory / ed. by S.J. Brams . . . —
Würzburg, Wien : Physica-Verlag, 1979. —

NE: Brams, S.J. [Hrsg.]

ISBN 978-3-7908-0208-5   ISBN 978-3-662-41501-6 (eBook)
DOI 10.1007/978-3-662-41501-6

© Springer-Verlag Berlin Heidelberg 1979
Originally published by Physica-Verlag, Rudolf Liebing GmbH + Co., Würzburg 1979.

Preface

The present volume is the outcome of a conference organized at the Institute for Advanced Studies in Vienna which aimed at presenting and reviewing latest applications of the theory of games to real-world problems and phenomena.

Game theory is as such not a specific empirically testable and falsifiable theory. Rather it is a mathematical framework suitable for modeling the decision-making problems that typically arise in a multi-person context. In this respect, the theory of games resembles other mathematical disciplines like geometry, analytical mechanics, or probability theory which also gain empirical content and meaning only when applied to the study of processes observed in the real world. Three aspects of the theory can be distinguished (W. Feller): Its formal logical content, its intuitive background, and its applications. The intuitive background of game theory has originally been provided by parlor games of strategy and the analogies that can be drawn between them and situations of conflict and cooperation in the military, economic, political, etc., spheres. Intuition, however, develops with the development of formal theory and the progress made in its application. Formal game theory deals with the several conceivable mathematical descriptions of a game, e.g., its extensive or dynamical form, its normal or strategical form, and its characteristic-function or coalitional form, giving precise meaning to concepts like move, information, utility payoff, strategy, coalition structure, etc. But above all, the formal theory deals with the definition, the existence, and the computation of "solutions". It has often been claimed that game theory assumes rational behavior on the part of the players. However, this is correct only in so far as the theory presupposes the players' ability and willingness to act in accordance with their consistently formed preferences (usually represented by utility functions) – a postulate which is as such insufficient to predict and/or precribe the decisions eventually resulting in situations characterized by the interaction of several agents with generally conflicting goals. Thus, in going beyond the desiderata of individual rationality in isolated decision making but without introducing any collective valuations, the various game-theoretic solution concepts offer properties that a strategy combination and/or a utility distribution (or sets of such outcomes) ought to possess in order to be acceptable to the participants in a game. The multitude of solution concepts proposed and studied in the game-theoretic literature has frequently been a cause of frustration and confusion among empirical researchers. A good deal of this multiplicity can be attributed to the underlying assumptions about the conditions of communication and the extent to which the players are able to commit themselves irrevocably to a course of action (non-cooperative versus cooperative solutions); within the realm of cooperative game theory the diversity of solution concepts is due, on the one hand, to differing assumptions about the feasibility of side payments and utility transfers as well as restrictions imposed on coalition formations, on the other hand to the essentially static, equilibrium character of the propounded theories in which the details of the dynamics of bargaining are more or less suppressed. There is however still another, in some sense more profound, reason for the abundance of solution concepts in n-person game theory: The extension of the concept of rationality from the level of the individual to that of the group or society is no trivial task, in fact, the difficulties encountered in the formulation of game-theoretic solution concepts demonstrate that there is no straightforward generalization of the concept of rational behavior of the isolated decision maker to situations involving more than a single individual. This intellectual experience in itself provides an important insight and should make people think who are sometimes inclined to speak too easily about societal rationality.

To empirical scientists, philosophers, organizers and the like, formal game theory provides a tool kit for structuring, modeling and solving their specific problems. In applications, the use which can be made of the abstract mathematical models is not predetermined by their original motivations or any preconceived ideas about their interpretation. The same analytical concept may be fruitfully applied to quite different real-world situations — exemplified by the use which is made of the concept of value of an n-person game in the papers by Schotter, Nevison, Grofman, Hart, Lemaire, Allen, Owen and Charnes contained in this volume. On the other hand, different models may be employed to describe the same empirical phenomenon: A firm, e.g., may in one case be treated as a single player in a normal-form game, in another case, in order to stress the role of information, as a player made up of individual agents in a game in extensive form, in still another case as a coalition of players; the non-cooperative solution concept of equilibrium point, to give another example, may be used to study the allocation of resources in electoral campaigns (as in the papers by Aldrich and Lake), to analyze pricing strategies in oligopolistic markets (Friedman's paper), or the implications of international monetary policies (Hamada), it may be however also employed to solve distributional problems arising in a cooperative context (papers by Young and Kalai), to explain the observed behavioral regularities in combat between male crabs (Hyatt), or to give a rational interpretation to the resolution of biblical conflicts between God and man (Brams).

It is also not a priori clear what is meant by an "application" of game theory. The successful practical applications of mathematical methods in statistics, industrial engineering and operations research as well as the paradigmatic analogies to parlor games like chess or poker, where we are indeed interested in learning how to win, have supported the view that the most promising applications of game theory would consist in giving advice to individual decision makers in business, politics or war how to "play the game" most advantageously. However, though decision makers can learn a lot from a description of their situation in game-theoretic terms, game-theoretic solution concepts are of little if any value in advising single actors in game-like situations as to how they ought to behave optimally. Only for two-person zero-sum games, which are rarely adequate for analyzing socio-economic, political, or even military problems (see, however, the interesting applications of this model to two-party political contests by Aldrich and Lake and to the deception problem by Axelrod), there exist strategies for each player which may be considered optimal for the individual actor independently of the course of action followed by his adversary. For all other types of games, the various solution concepts would be of practical prescriptive value only if it were possible to address all the decision makers involved simultaneously – a condition seldom realized in real-life situations. But there are other potential addressees of game-theoretically supported advice: Policy makers who are in the position to affect the rules of the game will be interested in knowing how changes in the parameters under their control are likely to influence the eventual outcome of the game. Thus, in policy-making problems where the task essentially consists in controlling systems of interacting decision makers by designing appropriate rules, game-theoretical analysis may prove useful to the rule makers provided that the respective situation can be modeled in a way that the chosen solution concept, e.g. the non-cooperative equilibrium, may be regarded as a reasonably accurate predictor.

Apart from practical policy applications of game theory there are what may be called "theoretical applications". They contribute to the elucidation and extension of theoretical notions developed in the various empirical sciences. Economic theory has in this respect a long tradition of applying game-theoretical models and solutions – in fact, notions like the non-cooperative equilibrium, imputations, and the core had essentially been known to economists long before the creation of game theory proper. For sociology and, in particular, political science who only comparatively recently have become engaged in the construction of rigorous theoretical models, the mathematical framework provided by game theory becomes more and more indispensable. Even in cases where the theory of games is used neither for prescriptive nor predictive purposes it proves highly useful in the design of experiments and the interpretation of their results (as indicated, e.g., in the papers by Rapoport/Kahan and McKelvey/Ordeshook). The recognition that in actual decision-making situations simulated by means of experimental gaming techniques the behavior of the players does not live up to the rationality requirements set by game theoretical solution concepts does not render these concepts superfluous – quite to the contrary, "irrational" or "sub-rational" behavior can reasonably only be judged by using the standards of rationality provided by game theory as a yardstick. It must, however, also be emphasized that the application of a mathematical theory is no one-way street. Mathematical theories develop through the interaction with real-world problems and it is to be hoped that the demands on game theory from the empirical sciences will grow in the future.

The editors wish to acknowledge the financial and organizational support given to the International Conference on Applied Game Theory and the publication of its proceedings by the Institute for Advanced Studies, Vienna, and the New York University Center for Applied Economics. They also wish to thank all the contributors and participants. A very special note of appreciation is due Mrs. Hannelore Loos for the superb job she did in the organization of the conference and her tireless, skillful and effective management of the practical tasks involved in the publication of the proceedings.

S.J. Brams, A. Schotter, and G. Schwödiauer

Contents

Models of Control and Confrontation

Power Analysis

Applied Game Theory. 1979 © Physica-Verlag, Wuerzburg/Germany

Power and Negotiation[1])

By *P. Levine*, Paris[2]), and *J.-P. Ponssard*, Paris[3])

Abstract: This paper is concerned with the formalization of the notion of power in the context of two person negotiations. The notion of power is defined as follows: the amount of power of A over B is related to the level of achievement of B's objectives in the interaction $A-B$. As for the origin of power, it is taken as related to the strategies available to both parties in their interaction (such as rewards, punishments, etc.).

The development of these ideas relies mainly on the concepts of game theory but specific factors drawn from the psychological and psychosociological literature, such as the notion of representation, are explicitly introduced in the model.

Besides the formalization, the paper includes a detailed discussion of two examples enhancing the role of representation in power analysis. An axiomatic treatment of the model is briefly reported in the appendix.

Introduction

The subject of this paper is the formalization of the notion of power in the context of two person negotiations. The formalization relies mainly on the concepts of game theory but specific psychological factors which appear essential in the analysis are also introduced.

The approach which is taken here is both methodological and operational in the sense that it provides a theoretic discussion on the notion of power as well as a modelization of negotiation which allows the derivation of power indices.

The notion of power is defined as follows: the amount of power of A over B is related to the level of achievement of B's objectives in the interaction $A-B$. As for the origin of power, it is taken as related to the strategies available to both parties in their interaction (such as rewards, punishments, etc.). This paper is concerned with the formalization of these ideas, as well as with the connections with similar approaches initiated by *Dahl* [1957] and followed by many others such as *Blau* [1964], *Crozier* [1963], *Hickson* et al. [1974].

It is not the first time that the concepts of game theory are used in connection with power. The most remarkable development is certainly associated with the works of *Shapley/Shubik* [1954] in the context of politics [see *Brams* for recent references]. *Harsanyi* [1962] also worked on the subject, formalizing *Dahl*'s approach through a

---

[1]) This note greatly benefited from the comments of P. Mayer, J.L. Peaucelle and J. Ullmo. A French version of this note appeared in Revue Française ole Sociologie **2**, 1979.

[2]) *Pierre Levine*, Laboratoire d'Econométrie, Université de Paris VI, F –Paris.

[3]) *Jean-Pierre Ponssard*, Centre de Recherche en Gestion, Ecole Polytechnique, Paris Ve, France.

specific solution concept: the Nash solution to the bargaining problem.

The present development may be considered as different from these approaches since it does not try to use a well defined game solution concept to show its interest in terms of power analysis, but attempts to operate in the opposite direction. As such it is directly concerned with the formalization of sociological and psychosociological developments on the notion of power. These developments recognize the facts that "players" develop subjective representation of reality which includes bounded rationality and inconsistent beliefs, and follow "linear" lines of thought which by far are less sophisticated than "circular stability arguments" used in game theory (on the role of representation see *Abric* et al. [1967], *Apfelbaum* [1967], on "linear" lines of thoughts, see for instance *Selten* [1972]). The modelization of negotiation which is tempted here includes at least some ideas related to the notion of representation, this is done in section 1. Two detailed examples are presented in section 2, they give an opportunity to confront the analysis with intuitive ideas on power but also show the important role of representation in the modelization. In the appendix, an axiomatic treatment of the model is briefly reported.

1. The Concept of Negotiation with Representation as a Framework for Power Analysis

A negotiation between several actors could in principle be formalized using the game concepts of: players, strategy sets and utility functions [von *Neumann/Morgenstern*]. But, whenever one tries to use operationally these concepts to analyse a real situation, frustrating difficulties arise. The limits of the game being played are quite fuzzy, incomplete information on the player's strategy sets and utility functions is the general rule, social standards which confer legitimacy to the arguments being used play a complex role in the negotiation[4]), rules in some subgame may be strategies in a supergame, etc.

The present development is characterized by a two level approach to operationally analyse such situations in terms of power:

— at the first level, the interaction between the actors is limited to the specific and immediate issues of the interaction and complete information is assumed;
— at the second level, the long term implications, the externalities, the psychological environment including the actors' own perception are introduced as representation parameters playing a significant role in the first level model.

This will be made more precise in the context of two person negotiation.

---

[4]) Many real negotiations appears as largely structured by some rules used as a sort of "constitution" that gives objectivity to the arguments being used even if the actors keep some limited freedom (for a development of these considerations, see *Charvet/Levine/Ponssard* [1977], *Charvet/Ponssard* [1977]).

1.1 The Negotiation Domain at the First Level

At this level, we shall use the *Nash* [1951] formulation which starts from a physical transaction set (such as in a commercial agreement between a buyer and a seller, or the possible values of some parameters in a contractual agreement which may include several terms) to end up as a negotiation domain using the players' utility functions. This domain is usually assumed to be convex and the point agreed upon under conditions of complete information will certainly be efficient, that is Pareto optimal[5]). The unfamiliar reader is refered to *Luce/Raiffa* [1957] for a detailed presentation of the Nash framework.

Note that at this level no assumption is made as regards what may occur in case the negotiation does not end by an agreement, for instance no status quo is specified.

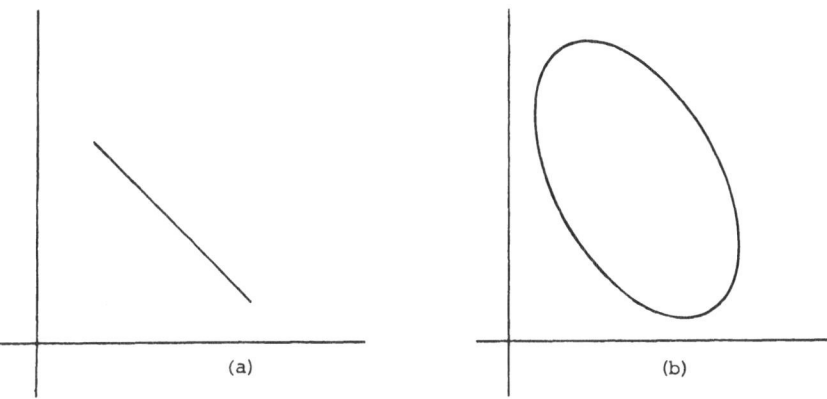

(a)   (b)

Fig. 1

1.2 The Players' Objective as Representation Parameters and the Implication in Terms of Power

This development is concerned with power. In our formulation power will be related to the players' objectives. Let us start by an example to illustrate how we see this relationship.

Take a simple interaction such as two players playing a Nim game: it is simple enough so that the winner may be objectively determined using the rules of the game and backward induction. The players' utility function may reflect the satisfaction of winning or losing but if their perceptions are correct, and it may certainly be in that case, their objectives cannot differ from the theoretic solution and they may be qualified as being identical. We shall say that in this case there is no power problem between the players. Suppose however that the interaction becomes more complicated so that

---

[5]) See figure 1 for a graphical presentation of the negotiation domain, figure 1a) is called a zero-sum situation corresponding usually to a one term negotiation such as the price of a second hand car; figure 1b) is called a non zero-sum situation and this will be the general framework in case of multi-term negotiation.

theoretic solution cannot be derived because of mathematical complexities or because the game concept use to derive it is subject to criticism (such as the non-cooperative *Nash* concept) then it is not unlikely that the players' objectives will be different. Moreover, many subjective considerations including perceptions of social standards and psychosociological aspects will play a significant role to make each player's objective appear to him as something he feels he deserves. It is not something he selects but something he has a right to get in the negotiation domain. It will take the form of two points, one for each player. The power relationship which is associated with a specific outcome of the negotiation is defined using these two points as reference points. Loosely speaking, the power relationship will depend on the distance of the final outcome with respect to the players' own objectives. This is now made more precise.

Let A and B stand for the two players. Denote by M_A and M_B their respective objectives whereas m_A and m_B denote what the other will get if A or B respectively achieve his own objectives. Finally let y_A and y_B stand for the outcome of the negotiation. The power of A over B, noted $\Pi_{A/B}$, and the power of B over A, noted as $\Pi_{B/A}$, are defined as (see figure 2):

$$\Pi_{A/B} = \frac{M_B - y_B}{M_B - m_B} \qquad \Pi_{B/A} = \frac{M_A - y_A}{M_A - m_A}.$$

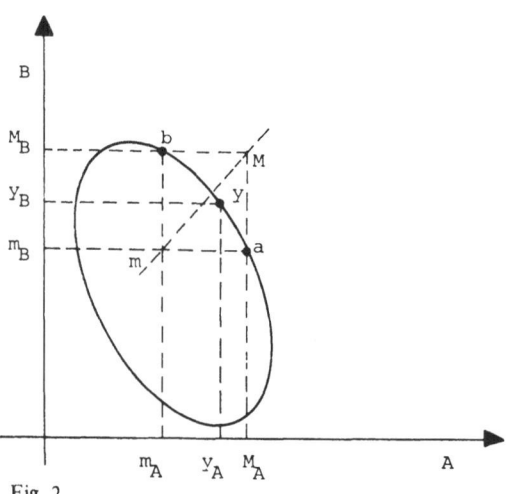

Fig. 2

Let us interpret, for instance, the index $\Pi_{A/B}$. The numerator measures from B's point of view, the distance from his objective M_B and the outcome of the negotiation y_B. The denominator measures also from B's point of view, the distance between M_B and the utility level he would get had he to submit totally to the objective of A, that is m_B. In other words, we assume that the players perceive the power relationship through an intercomparison of utilities using the players' objectives as two reference points. If the objectives coincide there is no power relationship. But, when they are different the outcome of the negotiation gives rise to a power relationship which is perceived as more or less favourable to A or B.

Note also that the power relationship is balanced only for the points on the line mM in which $m = (m_A, m_B)$ and $M = (M_A, M_B)$ (Figure 2). Thus, this line divides the negotiation domain in two regions, one which contains the point $a = (M_A, m_A)$ and in which $\Pi_{A/B} > \Pi_{B/A}$ and the other which contains $b = (M_B, m_B)$ and for which $\Pi_{B/A} > \Pi_{A/B}$.

Comparing the present development with some classics on the subject such as those of *Blau* [1964], *Crozier* [1963], *Dahl* [1957], it may be seen that it integrates the two well-known aspects of power. First, power is the ability to impose one's own will (the achievement of one's own objectives: $\Pi_{A/B}$ increases as y goes to a); second, power is the capacity to affect the behaviour of the other player (the non achievement of the other's own objectives: $\Pi_{A/B}$ increases as y goes away from b on the Pareto frontier). Moreover, the two indices $\Pi_{A/B}$ and $\Pi_{B/A}$ give an indication on who is at his advantage in terms of the exchange that takes place at the negotiation table, an idea which is developed at length by *Crozier*.

1.3 The Players' Threat Levels as Representation Parameters and the Implication in Terms of Strength

The power indices as defined in the previous section may be used to decribe the outcome of a negotiation in terms of power. However, it may be interesting to relate the power relationship which takes place to a theory as regards its origin. It is the purpose of concepts such as strengths, weaknesses, threats, or ability to cope with uncertainty, centrality, and others [see for instance *Hickson* et al.] to develop such a theory. Two strength indices will be proposed to formalize these ideas.

Broadly speaking strength is related to the strategy sets available to each player respectively in case of a breaking down of the negotiation. These strategy sets are part of the environment and no attempt will be made to describe them fully. Instead, it is assumed that they induce a representation for the players in terms of "threat levels" that is two utility levels, one for each player, under which they would rather break the negotiation. These levels will be denoted $(\bar{m}_A, \bar{m}_B)$ respectively. In special cases, it may correspond to a statu quo point, in others to reservation prices; at any rate, it is taken as given and known to each player. Then, the strength of A over B is defined as

$$P_{A/B} = \frac{M_B - \bar{m}_B}{M_B - m_B}$$

and similarly for the strength of B over A,

$$P_{B/A} = \frac{M_A - \bar{m}_A}{M_A - m_A}.$$

The strength index measures the maximal deviation that a player may be forced to concede because of his interaction with the other player. This deviation is taken relative to the same reference points as those used to define power indices. It is useful to summarize the two indices by the strength ratio $\rho_{A/B} = P_{A/B} / P_{B/A}$ so that we shall say that the strength ratio is in favor of A if $\rho_{A/B} \geqslant 1$ and vice versa (note that $\rho_{A/B}$ is invariant through linear transformations of the utility functions).

The graphical interpretation of these definitions is straightforward. Let $\overline{m} = (\overline{m}_A, \overline{m}_B)$, if a and $\overline{m}$ are on the same side of the line mM, the strength ratio is in favour of A; if not, it is in favour of B.

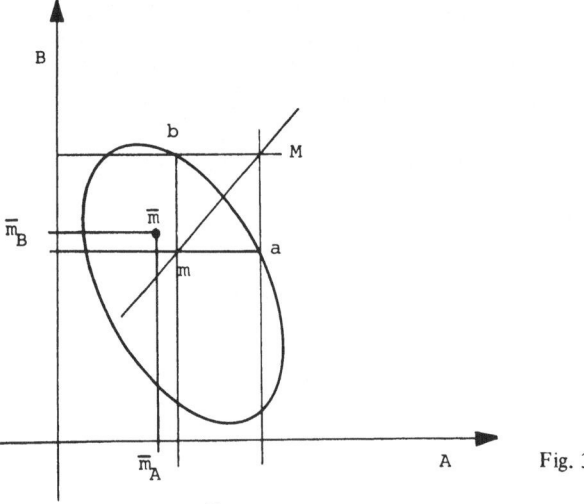

Fig. 3

In fact, the point $\overline{m}$ may fall in any one of the four quadrants from the point m and each one of these situations deserved some comments:

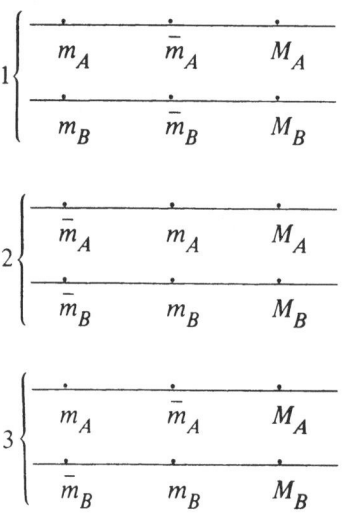

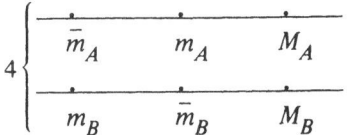

In the first case, the situation appears as very conflictual since by threatening the other player each player can obtain more than if he were to submit to the other player's will. In the second case the reverse is true. Note that in these two cases the strength ratio cannot be determined at the qualitative view of the figure. On the contrary, in case 3 it can be seen immediately that the situation is in favor of A whereas in case 4 it is in favor of B. In these latter cases, the situation is asymmetric: one player can obtain more in submitting to the other player's will than he thinks he can get outside and this is not true for the other player.

From an outsider point of view, there is something "inconsistent" in cases 1, 3 and 4 as regards objectives and threat levels taken altogether. The two players' representations of the environment are in contradiction. Still such cases are certainly interesting to study even if they would not be considered as relevant in a normative approach.

1.4 Concluding the Formalization: Strength as a Basis for Power

We would like to conclude this formalization by putting together the two notions that have been developed so far. This will be done by simply postulating the following equality:

$$\frac{\Pi_{A/B}}{\Pi_{B/A}} = \rho_{A/B}.$$

Graphically, this formalization selects as the outcome of the negotiation the point S: the intersection of the Pareto Frontier with the line $\bar{m}M$ (see figure 4).

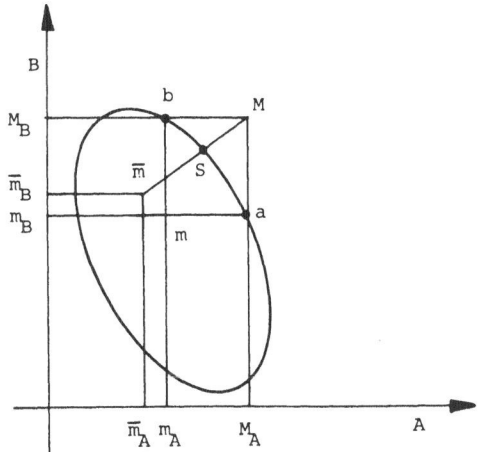

Fig. 4

This equality may indeed be derived formally from a general system of axioms concerning the negotiation procedure. This is done in the appendix.

But, this formalization was developed as an attempt to use some of the game concepts to analyse real situations and so we rather propose to evaluate the model by the inferences that can be drawn from the analysis of some typical situations.

2. Two Examples

2.1 Power and the Manipulation of the Rules of the Game

This first example is an illustration of the classical "Battle of the Sexes" scenario [*Luce/Raiffa*]. Simple as it is, it is quite useful to confront analysis and intuition. In particular, we whall discuss the advantage of commitment in such a context [*Schelling*] which results in an apparant paradox in terms of power analysis as noted and discussed at length by *Crozier/Friedberg* [1977].

Let us briefly recall the scenario. A man, player A, and a woman, player B, each have two choices for an evening's entertainment: a prize fight or a ballet. The man prefers the prize fight and the woman the ballet; however, to both it is more important that they go out together than each see the preferred entertainement. The associated game is summarized by the following matrix.

Our analysis shall be organized in three parts: first the basic case in which threats are credible and objectives are associated with the most preferred outcome; second, one of the player, say B, is "partially alienated" in his objectives and sets for himself lower objectives than his most preferred outcome; finally, we shall consider the possibility of commitment.

A \ B	Prize Fight	Ballet
Prize Fight	y_B / 1	0 / 0
Ballet	0 / 0	1 / y_A

$0 < y_A < 1$

$0 < y_B < 1$

Clearly, coordination on the prize fight or the ballet is preferable to the outcome in which each player goes his own way.

Case 1: Basic case

The reference points will be taken as follows:

(i)   each player's objective is his most preferred outcome then
$$a = (M_A, m_B) = (1, y_B),$$
$$b = (m_A, M_B) = (y_A, 1)$$

(ii) threats are credible which means that it cannot be exclude that each player goes his own way
$$m = (\bar{m}_A, \bar{m}_B) = (0, 0).$$

The analysis of the situation is graphically summarized figure 5. It corresponds to a type 2 situation (Cf. 1.3).

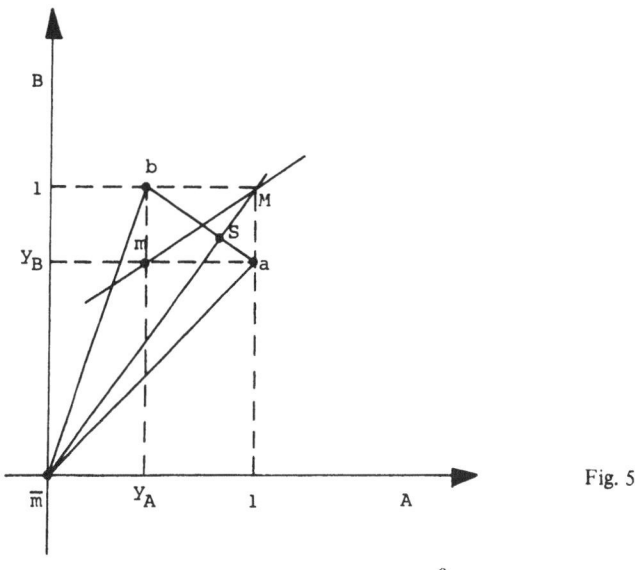

Fig. 5

It is easily seen figure 5 that the strength ratio is in favor of A if and only if $y_A < y_B$ (more precisely: $\rho_{A/B} = (1 - y_A) / (1 - y_B)$).

This result may be interpreted in several ways which make intuitive sense: the strength ratio is in favor of: (i) the player who will lose less by going his way as compared to submit to the other player's will $(y_A - 0 < y_B - 0)$ or (ii) the player who will lose more by submitting to the other's will as compared to impose his own will $(1 - y_A > 1 - y_B)$ [another way of saying this last sentence may be that the strength ratio is in favour of the player who cares the most on which one of the two entertainements will be selected by the couple].

Case 2: B is partially alienated

The basic case shall be modified in the following way: for cultural or other reasons player B's objective is not his most preferred outcome but something less "ambitious". Namely, B's objective is now taken as an intermediary point between the most preferred outcome for B and the most preferred outcome for A:

$b' = (m'_A, M'_B)$ where

$$M'_B = 1(1-\alpha) + y_B \alpha = 1 - \alpha(1-y_B),$$

$$m'_A = y_A(1-\alpha) + 1\alpha = y_A + \alpha(1-y_A).$$

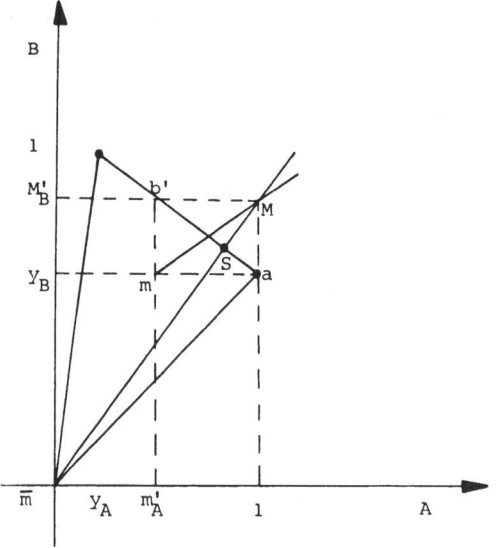

Fig. 6

It follows that:

$$\rho'_{A/B} = \frac{1-\alpha(1-y_B)}{(1-y_B)(1-\alpha)} \frac{(1-y_A)(1-\alpha)}{1} = \frac{1-y_A}{1-y_B}[1-\alpha(1-y_B)]$$

$$= \rho_{A/B}(1-\alpha(1-y_B)).$$

Then $\rho'_{A/B} < \rho_{A/B}$ as soon as $\alpha > 0$ which means that the strength ratio in case 2 is more favorable to player B than it is in case 1. On the other hand, player A's utility at the solution is higher in case 2 than in case 1 (see figure 6 for a graphical check and foot-note for the exact value[6]).

As a result, it appears that by having lower objectives, player B gives an opportunity to player A to obtain a higher utility level but makes the power relationship at the negociated outcome less favorable to player A. This is a general result[7] on "partial alienation" as defined here (alienation is taken as having a "low" objective).

---

[6] $S_A(\alpha) = (1-y_A y_B)/(2-y_A-y_B-\alpha(1-y_A)(1-y_B))$ so that as α increases from 0 to 1 (b' moving from b to a), $S_A(\alpha)$ increases from $S_A(0)$ to M_A.

[7] Indeed, since $\rho_{A/B} = \dfrac{M_B - \bar{m}_B}{M_B - m_B} \dfrac{M_A - m_A}{M_A - \bar{m}_A}$, the differentiation in b gives

Though it is intuitive that for an actor the higher his objectives the more vulnerable he is in a negotiation and so the less favorable to him is the strength ratio, to some extent the application of this result in this context may appear as leading to a paradox.

Given that if player B is partially alienated, player A enjoys a higher level of utility in the negotiated outcome, it seems natural to conclude that he has more power. This conclusion is in contradiction with the present analysis in which power, and strength, is subjective and depends on the representations of the individuals (summarized here in the objectives and the threats). Now, it is possible to say that this conclusion would be valid for an observer who brings along with him his own representations. In this case, if they are egalitarian, it follows that B's alienation gives more "objective" power to A, in terms of the observer's representations.

Case 3: The use of prior commitment by player B

Schelling [1960] points out the advantage of prior commitment in several examples such as this one. In this game, it is advantageous to B to commit himself to go to the ballet (i.e. restrict his strategy set to a unique strategy). By doing so, it is intuitive that he will change an uncertain bargaining process to one in which his strength comes from the very fact that he has no choice (in the scenario, he may do so by buying the tickets beforehands and, in that way, spend all the money available for entertainment). This seems contrary to the idea that strength is associated with freedom of choice and ability to be unpredictable. Indeed, *Crozier/Friedberg* [1977] recognizing the paradox go into a somewhat complicated and unconvincing argument to get rid of the problem. We would like to show here that our formalization avoids this problem and at the same time provides insights into the situation.

Consider the modified game, the one resulting from B's commitment:

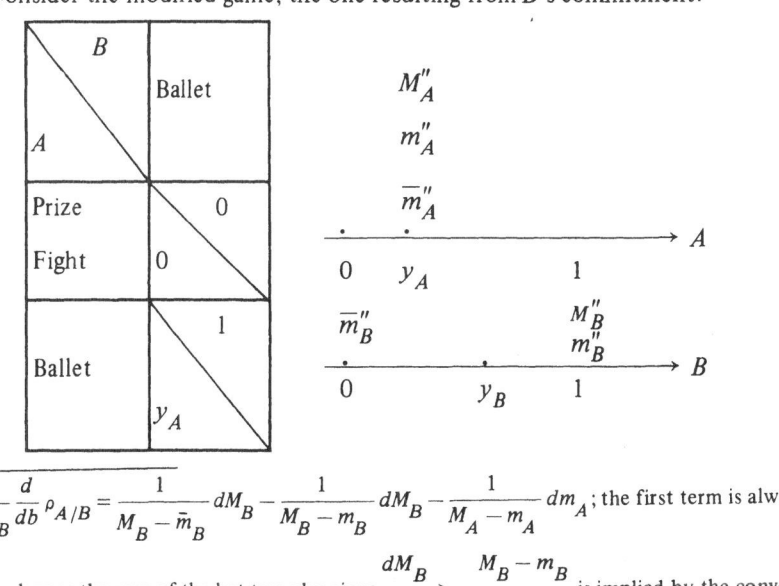

$$\frac{1}{\rho_{A/B}}\frac{d}{db}\rho_{A/B} = \frac{1}{M_B - \bar{m}_B}dM_B - \frac{1}{M_B - m_B}dM_B - \frac{1}{M_A - m_A}dm_A \text{; the first term is always nega-}$$

tive whereas the sum of the last two also since $\dfrac{dM_B}{dm_A} > -\dfrac{M_B - m_B}{M_A - m_A}$ is implied by the convexity

between a and b of the negotiation domain.

At the formal level, the notion of power is not defined in this new game since both players have the same objective (recall that $0 \leqslant y_A \leqslant 1$)

$$(M''_A, m''_B) = (m''_B, M''_B) = (y_A, 1).$$

Then, this point is also the solution of the negotiation. Comparing with case 1, we may say that by committing himself B modifies the rules of the game to his own advantage in spite of the fact that by doing so he gives up his freedom of choice. It is the coincidence of the players' objectives in the new game which makes irrelevant the analysis in terms of power and strength. However, if y_A were negative, then a power analysis would be meaningfull and the strength ratio would be in complete favor of A as is always the case in a game in which B has only a unique strategy. It follows that commitment may give rise to unstable analysis in terms of power. For instance, if A discovers that B is manipulating the rules of the game, the frustration of having been played upon may be so important that a modification of his own utility function could result. This is very critical for B since player A has the solution of the new game under his own will.

2.2 Power and the Ability to Cope with Uncertainty

We want to study the interaction between two departments within the same firm: the marketing department and the head department. Assume that the production process is well mastered whereas major uncertainties remain about the market of the firm. For simplicity, assume that the market may turn out to be either "good" or "bad". We shall say that the marketing department is the only one that can cope with this uncertainty because he is the only one that has the ability to make a diagnosis of the market and eventually take appropriate actions to change a bad market into a good market. Moreover, the head department cannot evaluate the seriousness of the marketing department's diagnosis. This last part of the assumption simply means that the head department cannot take over the responsibilities of the marketing department, he has not the "know-how". If this were a complete description of the situation, it would intuitively follow that the marketing department has a lot of strength in his relationship with the head department and a formal model would not add much to the understanding of the situation.

Suppose however that the head department also controls a major source of uncertainty concerning the allocation of bonuses among the various departments. The head department is the only one to have access to the detailed financial results of the firm both internally (costs) and externally (prices). His private information gives the head department much freedom on the allocation process. The strength ration between the two departments will certainly be affected by the existence of this other source of uncertainty. Is it possible to say anything more using the formal model developed in this paper?

We shall try to pursue the analysis and in particular to compare two modes of allocation of the bonuses: the first is incentive-oriented, it depends on the outcome of the market which is controlled by the marketing department, the second one if independent of this outcome.

Case 1: Incentive allocation system

The situation is summarized in the following table in which one assumes that the good market provides a marginal gain of 1 unit over the bad one. Given this unit, α and β are interpreted as follows:

α is the maximal bonus that the marketing department can possibly obtain $(0 \leqslant \alpha \leqslant 1)$.

β is the cost personnaly incurred to the marketing department for taking appropriate actions $(0 \leqslant \beta \leqslant \alpha)$.

The two departments' utility functions are computed accordingly.

Marketing A B Head	Take actions	Do nothing
Give max. bonus	$\alpha - \beta$ / $1 - \alpha$	0 / 0
Give no bonus	$-\beta$ / 1	0 / 0

Incentive allocation system – case 1

The graphical analysis is summarized figure 7. The reference points are selected as follows:

— the objective of the head department is to give no bonus and have the marketing department take actions; the objective of the marketing department is to take actions and receive the maximal bonus:

$(M_A, m_B) = (1, -\beta)$,

$(M_B, m_A) = (\alpha - \beta, 1 - \alpha)$.

— it cannot be excluded that nobody does anything:

$\bar{m} = (\bar{m}_A, \bar{m}_B) = (0, 0)$.

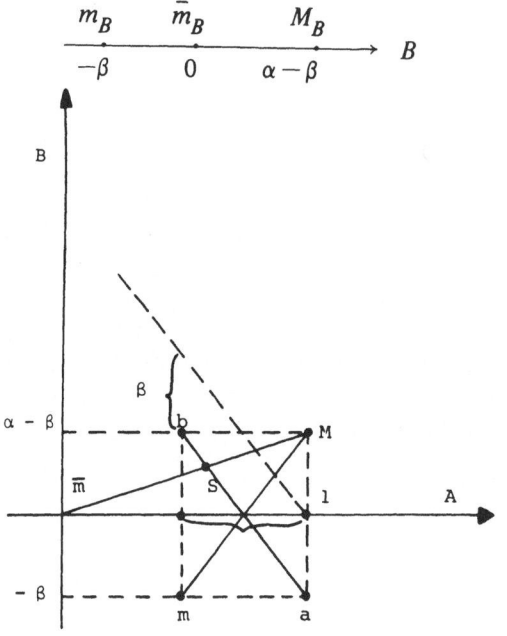

Fig. 7

Note that the resulting situation is inconsistent and the strength ratio is in favour of the marketing department whatever the value of the parameters (type 4 situation, Cf. 1.3). Numerically we have $\rho_{A/B} = \alpha - \beta$ (recall that $0 \leqslant \beta \leqslant \alpha \leqslant 1$). It also follows that if the maximal bonus, α, is increased then the strength ratio increases in favour of the head department but his own share in the negotiated outcome decreases[8]).

Head A	Marketing B	Take actions	Do nothing
Give max. bonus		$\alpha - \beta$ $1 - \alpha$	$-\alpha$ α
Give no bonus		$-\beta$ 1	0 0

Independent allocation system — case 2

---

[8]) See footnote 7)

Case 2: Independent allocation system

It is assumed here that the level of the bonus is independent, at least in its principle, from the outcome of the market.

If the two departments do not modify their respective objectives, it directly follows that the analysis of the situation also remains unmodified. But, we may certainly assume that the marketing departement will take a different perspective at the situation and it seems reasonable to assign as his new objective to do nothing and still obtain the maximal bonus:

$$(M_B, m_A) = (\alpha, -\alpha).$$

The situation may then be summarized figure 8.

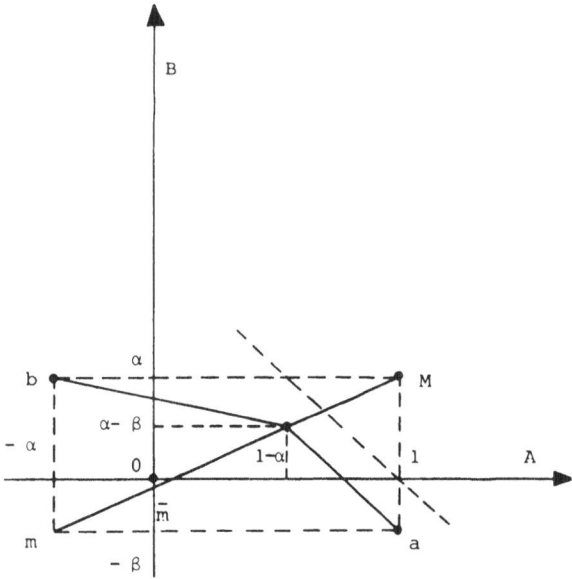

Fig. 8

Under these conditions the situation is inconsistent and of type 1 (Cf. 1.3). It is not possible to say who has the strength ratio in his favour independently of the parameters. In fact, it may be shown that:

$$\rho_{A/B} = \alpha\,(1 + \alpha)/(\alpha + \beta)$$

so that the strength ratio is in favour of the head department if $\alpha > \sqrt{\beta}\,(0 \leqslant \beta \leqslant \alpha \leqslant 1)$. and in favor of the marketing department if $\beta < \alpha < \sqrt{\beta}$. It may be noted that the negotiated outcome is particularly simple in this case since it corresponds to the point $(1 - \alpha, \alpha - \beta)$ which means that the head department gives away the maximal bonus whereas the marketing department takes actions. Comparing with case 1, one sees there that at the negotiated outcome the head department gives away only part of the bonus. As for the strength ratio, for α et β remaining fixed, the incentive system appears more favorable to the marketing department than the independent system precisely because its objective are lower in the first case than in the second one.

In summary, the effect of the internal reward system on the objective of the department which is to be controlled appears in this model as the key variable to determine the strength ratio. On the contrary the question whether it is incentive oriented is not.

Though this example is hypothetical, it seems worth-while to note that the formal model developed in this paper enhances the significance of the representations which go along with the control system, most often unnoticed.

Appendix

The purpose of this appendix is to give an axiomatic justification to the equality (as defined in the text):

$$\frac{\Pi_{A/B}}{\Pi_{B/A}} = \rho_{A/B}.$$

At this end we shall adopt the classical approach in game theory which consists in defining axioms concerning the outcome of any negotiation [see Nash, 1951; Kalai/ Smorodinsky].

First, let us give some definitions and notations. We shall call negotiation the data $(D, a, b, \bar{m})$ where

– D is a convex compact subset of R^2
– a (resp. b) is A's (resp. B's) objectives,
– $\bar{m}$ is the threat level of D. It satisfies $\bar{m}_A \leqslant M_A$ and $\bar{m}_B \leqslant M_B$.

Then, we shall call solution function, the function f which assigns to each negotiation $(D, a, b, \bar{m})$ its outcome $f(D, a, b, \bar{m}) \in D$.

Let us consider the two following axioms:

(H_1) $\bar{m} \leqslant f(D, a, b, \bar{m})$

(H_2) If $(D, a, b, \bar{m})$ and $(D, a, b, \bar{m}')$ are two negotiations for which the strength ratio $\rho_{A/B}$ is identical then
$$f(D, a, b, \bar{m}) = f(D, a, b, \bar{m}').$$

Proposition 1: If f satisfies (H_1) and (H_2) then $f(D, a, b, \bar{m})$ is the unique Pareto point of D such that:

$$\frac{\Pi_{A/B}}{\Pi_{A/B}} = \rho_{A/B}.$$

Proof: Let $(D, a, b, \bar{m})$ be a negotiation and consider the unique Pareto point $\bar{m}'$ of D such that:

$$\frac{M_A - \bar{m}'_A}{M_B - \bar{m}'_B} = \frac{M_A - m_A}{M_B - \bar{m}_B}.$$

The strength ratio $\rho'_{A/B}$ with respect to $(D, a, b, \bar{m}')$ is still $\rho_{A/B}$. But, by (H_1), $f(D, a, b, \bar{m}') = \bar{m}'$. From (H_2) it follows that $f(D, a, b, \bar{m}) = \bar{m}'$. Remark that $\bar{m}'$ satisfies by definition

$$\frac{\Pi_{A/B}}{\Pi_{B/A}} = \rho_{A/B}$$

which completes the proof.

One can also obtain the same solution function with axioms not using the strength ratio concept. Let us list the following:

(H_3) $f(D, a, b, \bar{m})$ is a Pareto point of D.
(H_4) If D is symmetrical, if $M_A = M_B$, $m_A = m_B$ and if $\bar{m}_A = \bar{m}_B$ then
$$f_A(D, a, b, \bar{m}) = f_B(D, a, b, \bar{m}).$$
(H_5) Let be two negotiations $(D, a, b, \bar{m})$ and $(D', a', b', \bar{m}')$ satisfying
    - $D \subset D'$
    - $M_A = M'_A$ et $M_B = M'_B$.
Then $f(D, a, b, \bar{m}) \leqslant f(D', a', b', \bar{m})$.
(H_6) If T is the transformation of R^2:
$$(x, y) \rightarrow (\alpha_1 x + \beta_1, \alpha_2 y + \beta_2) \text{ with } \alpha_1 \text{ and } \alpha_2 > 0$$
then $f(T(D), T(a), T(b), T(\bar{m})) = T[f(D, a, b, \bar{m})]$.

Proposition 2: If f satisfies (H_3) up to (H_6) then $f(D, a, b, \bar{m})$ is the unique Pareto point of D, such that:

$$\frac{\Pi_{A/B}}{\Pi_{B/A}} = \rho_{A/B}.$$

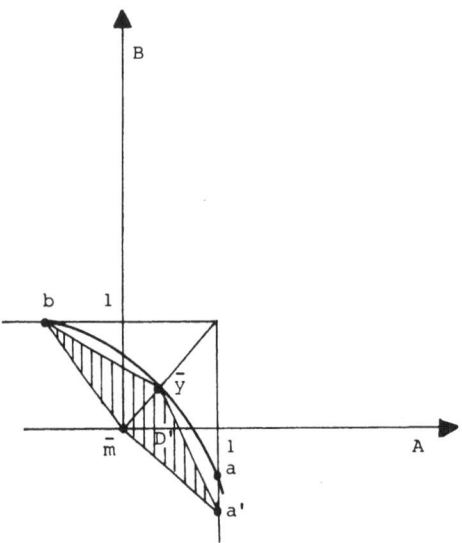

Fig. 9

Proof: Normalize the negotiation $(D, a, b, \bar{m})$ by taking $\bar{m} = 0$ and $M_A = M_B = 1$. (This is always possible by (H_6)).

Then consider the negotiation $(D', a', b, \bar{m})$ of figure 9. This negotiation is symmetrical thus $f(D', a', b, \bar{m}) = \bar{y}$ by (H_3) and (H_4). By using (H_5) it comes:

$$f(D, a, b, \bar{m}) \geqslant \bar{y}$$

and so:

$$f(D, a, b, \bar{m}) = \bar{y}.$$

Note that by construction, at the point $\bar{y}$,

$$\frac{\Pi_{A/B}}{\Pi_{B/A}} = \rho_{A/B}.$$

Remarks: Axioms (H_3) up to (H_6) generalize the axioms stated by *Kalai/Smorodinsky* [1974]. Moreover they are compatible with the axiom of independence of irrelevant alternatives:

(H_7) Let be two negotiations $(D, a, b, \bar{m})$ and $(D', a, b, \bar{m})$ such that:

  – $D \subset D'$
  – $f(D, a, b, \bar{m}) \in D'$

Then $f(D, a, b, \bar{m}) = f(D', a, b, \bar{m})$.

Finally, let us note that (H_7) is used by *Roth* [1977] in the special case where a (resp. b) is the point of D which corresponds to the maximum of utility for A (resp. B).

References

Abric, J.C., C. Faucheux, S. Moscovici and *M. Plon*: Rôle de l'image du partenaire sur la coopération en situation de jeu. Psychologie française **12** (4), 1967.

Apfelbaum, E.: Représentation du partenaire et interaction à propos d'un dilemme du prisonnier. Psychologie française, **12** (4), 1967, 287–295.

Blau, P.: Exchange and Power in Social Life. New York 1964.

Brams, S.J.: Game theory and politics. New York 1975.

Charvet, B., P. Levine and *J.P. Ponssard*: Analyse de la négociation: le cas des achats industriels de l'Administration. Congrès AFCET, Novembre 1977.

Charvet, B., and *J.P. Ponssard*: Simulations d'une négociation d'avenant en matière d'achats de travaux publics. Note de recherche, Centre de Recherche en Gestion, Décembre 1977.

Chazel, F.: Pouvoir, cause et force. Revue française de Sociologie **XV**, 1974.

Crozier, M.: Le phénomène bureaucratique. Paris 1963.

–: La société bloquée. Paris 1970.

Crozier, M., and *E. Friedberg*: L'acteur et le système. Paris 1977.

Dahl, R.: The concept of Power. Behavioral Science **2**, 1957.

Harsanyi, J.: Measurement of Social Power, Opportunity Costs, and the Theory of two-person Bargaining Games. Behavioral Science **7**, 1962.

Hickson, D.J., C.R. Hining, C.A. Lee, R.E. Schneck and *J.M. Pennings*: A Strategic Contingencies Theory of Intraorganizational Power. Adm. Sience Quarterly, 1974.

Kalai, E., and *M. Smorodinsky*: Other solutions to Nash Bargaining Problem. Econometrica, 1975.

Littlechild, L.S.: An entrepeneurial theory of games. Communication à International Symposium on Extremal Methods, Austin, Texas, Sept. 1977.

Luce, D., and *H. Raiffa*: Games and Decision. New York 1957.

March, J.G.: An introduction to the Theory and Measurement of Influence. American Political Science Review, 1955.

March, J.G., and *H.A. Simon*: Organizations. New York 1958.

Mayer, P.: Analyse organisationnelle d'un système social restreint: le cas d'une unité d'exploitation dans une société de service, Publication ADSSA, 1974.

Nash, J.: The Bargaining Problem. Econometrica, 1951.

–: Two Person Cooperative Games. Econometrica, 1953.

Von Neumann, J., and *O. Morgenstern*: The Theory of Games and Economic Behavior. Princeton 1947.

Peaucelle, J.L.: Théorie des Jeux et Sociologie des Organisations. Sociologie du Travail **II**, 1969, 22–43.

Roth, A.: Independance of Irrelevant Alternatives. Journal of Economic Theory **16**, 1977.

Schelling, T.C.: The Strategy of Conflict. Harvard 1960.

Selten, R.: Equal share Analysis of characteristic Function Experiments. Contributions to experimental Economics. Ed. by H. Sauermann. Volum III. Tübingen 1972.

Shapley, L., and *M. Shubik*: A Method for evaluating the Distribution of Power in a Committee System. American Political Science Review, Sept. 1954.

Tietz, R., and *H.J. Weber*: On the nature of the Bargaining Process. Contributions to experimental Economics. Volum III. Ed. by H. Sauermann. Tübingen 1972.

Applied Game Theory. 1979 © Physica-Verlag, Wuerzburg/Germany

Exploitable Surplus in N-Person Games

By *H.P. Young*, Laxenburg[1])

Abstract: Any cooperative n-person game with transferable utility has a noncooperative mode in which the players sell out of their positions to an external market of entrepreneurial organizing agents. Assuming a market of price takers, this game of competitive self-valuation always has an equilibrium price solution. Every core imputation in the original game constitutes a set of equilibrium prices. If there is no core the entrepreneurs can exploit the coalitions for a profit, i.e., they realize a positive rent for their organizing function. Application is made to determining fair wages to labor, and finding equilibrium prices for legislators selling their votes.

In this paper we describe a new approach to the valuation of n-person cooperative games with transferable utility. The idea is that values are determined competitively by creating a "market" for the players (or for the players' positions). Specifically, if value in the game is transferable, then outside entrepreneurs will view potential combinations of players as a source of potential profits. In such an environment any proposed valuation of the players will be seen as a set of prices by the entrepreneurs, who can acquire control of coalitions by paying these prices or more. It is natural then to ask whether a given valuation is in equilibrium, i.e. whether, given the others' prices, a player could charge more (or less) and do better.

The conclusion is that, in the face of profit maximizing price-takers, an equilibrium in pure strategies always exists in which every player gets what he asks. These valuations are characterized by certain marginal conditions on the subsets and hence are called "marginal values." It turns out that every core imputation is a marginal value. On the other hand if the core does not exist the players will not be able to divide the whole value of the game and the entrepreneurs realize a "rent" from their contribution as organizers. In other words, the nonexistence of the core means that in a sense the players can be "exploited" due their inability to cooperate. We now define these ideas more precisely and illustrate with two applications: the 'fair wage' problem, and 'political bribery'.

It is useful to think of a cooperative game with transferable utility as a *production process*. The players $\{1, 2, \ldots, n\} = N$ are the *factors*, and their joint payoff is what they can produce. Then, the *production function* is simply the characteristic function of the game, v. We make the following assumptions on v:

---

[1]) *H.P. Young*, International Institute for Applied Systems Analysis, Laxenburg, Austria.

Free disposal:

$$v(S) \geqslant 0 \quad \text{for all } S \subseteq N \text{ and } v(\phi) = 0; \tag{1}$$

Joint production:

$$v(S \cup T) \geqslant v(S) + v(T) \quad \text{whenever } S \cap T = \phi. \tag{2}$$

Conversely, given any production function satisfying (1) and (2) on factor set N, v may be interpreted as a game by supposing that each factor i is *represented* by some agent who is a player. For the present we assume that distinct factors are identified with distinct players. However, it is also possible within this framework to treat the case where a player simultaneously represents several different factors (see the fair wage problem below).

Now suppose that there is a market of outside agents or *entrepreneurs* who are potential buyers: their role is to buy up sets of factors and cause them to produce effectively. The problem is to determine what constitutes a fair wage or *value* for the individual factors.

We propose the following answer. Let each player (i.e., factor representative) announce what he thinks he is worth: thus, each i quotes a price $p_i \geqslant 0$. Now let the potential buyers arrive. Each of them perceives the same production function, v, and has an unlimited budget. We suppose that they arrive in some order and take the prices as given. The first buyer in line will then buy some set that maximizes his potential *profit*, $v(S) - \sum_S p_i$. Typically there will only be one such maximum profit set; however, in case of ties a specific tie-breaking rule must be used. We say that the tie-breaking rule is *efficient* if whenever T^* is the set of factors bought at prices p then $v(T^*) \geqslant v(T)$ for all maximum profit sets T.

Now define the *sell-out game* as follows: for strategies $p = (p_1, p_2, \ldots, p_n)$ the *payoff* to i is

$$\varphi_i(p) = \begin{cases} p_i & \text{if } i \text{ is bought,} \\ v(i) & \text{otherwise.} \end{cases} \tag{3}$$

A vector $\bar{p}$ is a *strong equilibrium* for this game if no collection of players can simultaneously change their strategies and all do better (assuming the others hold fast). It may then be shown [*Young*, 1978d]:

For any efficient tie-breaking rule, a strong equilibrium in pure strategies always exists. Moreover, there is always a strong equilibrium $\bar{p}$ in which each player receives what he asks. These $\bar{p}$ are characterized by the condition:

> *N is a maximum profit set and no factor is contained in every maximum profit set.* (4)

This is equivalent to the following *marginal conditions*, $v(N) - v(N - S)$ being the marginal value of S:

For every S, $\sum_S \bar{p}_i \leqslant v(N) - v(N - S)$ and for every i there is some S

containing i for which $\sum_S \bar{p}_i = v(N) - v(N - S)$. (5)

Any $\bar{p}$ satisfying (5) (equivalently, (4)) is called a *marginal value* of the game v.

A simple example will illustrate these ideas. Three laborers may be organized in different combinations to produce a divisible output. The outputs of the different combinations are shown below, where the larger combinations exhibit the advantages of a division of labor, and not all laborers are equally skilled.

$$v(\phi) = 0$$

$v(1) = 6$	$v(1, 2) = 27$
$v(2) = 7$	$v(1, 3) = 29$
$v(3) = 8$	$v(2, 3) = 32$

$$v(1, 2, 3) = 40 .$$

There is a unique vector $\bar{p}$ satisfying conditions (4) and (5), namely $\bar{p}_1 = 8$, $\bar{p}_2 = 11, \bar{p}_3 = 13$. These are the wages (in units of output) that one might expect to see if the laborers are unable to organize to produce by themselves, and if there are outside entrepreneurs who compete for control.

Notice that each laborer's wage is greater than the amount he can produce in isolation, as it should be. But the sum of all wages is less than the total output, meaning that the entrepreneur realizes a profit of eight units. At prices $\bar{p}$ there are several combinations of factors that are equally profitable: each of the sets $\{1, 2\}, \{1, 3\}$, and $\{1, 2, 3\}$ would yield a profit of 8 units to an organizer. For equilibrium to hold, the tie must be broken efficiently, i.e., by employing the set with highest output, namely $\{1, 2, 3\}$. An explanatory mechanism for this outcome is to imagine that each of the laborers shades his asking price by a small amount ϵ; then $\{1, 2, 3\}$ is the *unique* most profitable set. Thus an efficient tie-breaking rule has the property that it exhibits continuous behavior of the outcome as the equilibrium is approached from below.

There is an important relation between the class of marginal values and the core. In fact, every imputation in the core is a marginal value. To see this, consider the conditions for a core imputation: $\sum_S p_i \geqslant v(S)$ for all $S \subset N$ and $\sum_N p_i = v(N)$. This says that no set is profitable and the set N yields zero profit. Thus condition (4) is satisfied, since the *empty* set is also a maximum profit set in this case.

If the core is empty, however, then there are no strong equilibrium prices that permit the players to divide the whole value of the game. In this situation an outside entrepreneur will always be able to realize a surplus. This fact is illustrated in the following application.

A Fair Wage Problem

Let $1, 2, \ldots, n$ designate laborers who are available for hire by entrepreneurs. The laborers have different skills, and each combination $S \subseteq N$ has a potential productive value $v(S)$ (in, say, units of output). We assume that joint production is possible, e.g., is not prevented by exogenous fixed factors of production.

Instead of trying to undercut each other, suppose the laborers form a union to set their wages jointly. Then the union representative has the problem of finding a *wage structure* $w_1, \ldots, w_n$ that maximizes the return to labor. The employer has the problem of hiring a set of laborers that will maximize his profits. If there is only one potential employer and the union is in a position to call a general strike then this is a bargaining problem. However, suppose instead that there are other potential employers, and that the union does not feel itself strong enough internally to risk calling a general strike. (This may very well be the case if v has no core). The employers can then be expected to act as price takers: faced with a set of wage demands $w_1, w_2, \ldots, w_n$ they employ some combination S yielding maximum profits and walk away from the rest. On the other hand the union representative must face the possibility that if wages are set too high, some laborers will go unemployed. The real wage of such unemployed laborers will then be whatever they are paid by the union as unemployment compensation. Moreover this compensation must come out of the other workers' wages. Hence the *real* wage structure $w_1, w_2, \ldots, w_n$ is only sustainable if all are employed at these wages, that is, only if N is a maximum profit set at wages $\underset{\sim}{w}$.

The union representative therefore solves the problem

$$\max_{N} \Sigma \, w_i \qquad (6)$$

subject to

$$v\,(N) - \underset{N}{\Sigma} \, w_i \geqslant v\,(S) - \underset{S}{\Sigma} \, w_i \qquad \text{for all } S \subset N.$$

An optimal solution $\underset{\sim}{w}^*$ to (6) always exists. By definition, N is a maximum profit set under $\underset{\sim}{w}^*$. Moreover, if some factor i were in *every* maximum profit set, then w_i^* could be increased and N would still be a maximum profit set, a contradiction. Therefore every optimal solution satisfies conditions (4) and (5), hence is a marginal value for v. These are called the *core marginal values* for v.

A core marginal value $\underset{\sim}{w}^*$ represents a wage structure that yields the highest total return to the factors, and the least profit to the entrepreneurs. This profit, $\pi^* = v\,(N) - \underset{N}{\Sigma} \, w_i^*$ is called the *exploitable surplus* of the game v. A positive exploitable surplus exists if and only if v has no core. If v has a core then the set of core marginal values equals the core.

The meaning of (6) becomes clearer if we re-write it as follows:

$$\min \pi \qquad (7)$$

subject to

$$\underset{S}{\Sigma} \, w_i \geqslant v\,(S) - \pi \qquad \text{all } S \subset N$$

$$\underset{N}{\Sigma} \, w_i = v\,(N) - \pi \,.$$

This says that the exploitable surplus represents the least amount that must be skimmed off the value of *all* coalitions for the core to first appear, and the core marginal values are precisely the imputations in the core of the game that is "left over." While this notion bears a certain formal similarity to the "least core," the values it gives, and their interpretation, are quite different.

The existence of an exploitable surplus was predicated on the assumption that the union did not consider a general strike as a viable option. If this *were* an option, then it would appear that they could ask for any wages such that $\sum_N w_i = v(N)$ and, because of competition among the entrepreneurs, they will all be assured of employment. However this argument is only plausible if $\underset{\sim}{w}$ is in the core. If $\underset{\sim}{w}$ is not in the core, then for some S $v(S) - \sum_S w_i > 0$. But then an entrepreneur could bid away S by offering them higher wages and still make a profit, and the strike could collapse. Thus if the core does not exist, a strike is vulnerable and one can expect to observe exploitable surplus for the entrepreneur and a core market value for the wage structure. On the other hand, if the core does exist, the core market values coincide with the core.

Political Bribery

In the sell-out game (3), it was assumed in the definition of the payoff function that player i gets $v(i)$ – the amount he can "produce by himself" – even if he is not bought. However, this hypothesis overlooks two points. The first is the possibility that $v(i)$ does not represent value that i can obtain acting alone, but rather, is value that i's actions have to someone else. The second point is, that i may incur an opportunity cost by selling out; that is, there may be an inherent value to i in *not* selling out which is different from $v(i)$. Both of these situations require an appropriate modification of the payoff function (3), and both arise in the following model of political bribery.

A legislature may be thought of as a production process in which the legislators are the factors, voting is the process, and legislation the output. This output is valuable, – not generally to the legislators themselves – but to outside *interest groups* having a stake in the legislation. Moreover it is not too far-fetched to say that there exist entrepreneurs who might try to organize the factors to produce in a certain way – namely, lobbyists representing these interest groups.

Suppose a lobbyist proposes a special-interest bill having potential value M, and to pass it he will need to bribe a winning coalition of the legislature. The production function for this "legislative game" is easily given:

$$v(S) = \begin{cases} M \text{ if } S \text{ is a winning coalition} \\ 0 \text{ if } S \text{ is a losing coalition.} \end{cases}$$

Notice that value in this game does not accrue directly to the legislators. However, even though $v(i)$ in such a game is typically zero, the opportunity cost to i of selling out may well be positive, since selling oneself may involve certain risks or perhaps even pangs of conscience.

Let p_i^0 represent the opportunity cost to legislator i of selling out, that is, the minimum price needed to get him to go along with the bill. If the legislators all have equal votes and are arranged in increasing order of p_i^0, then we have a monotone increasing "supply curve" for votes as shown in Figure 1.

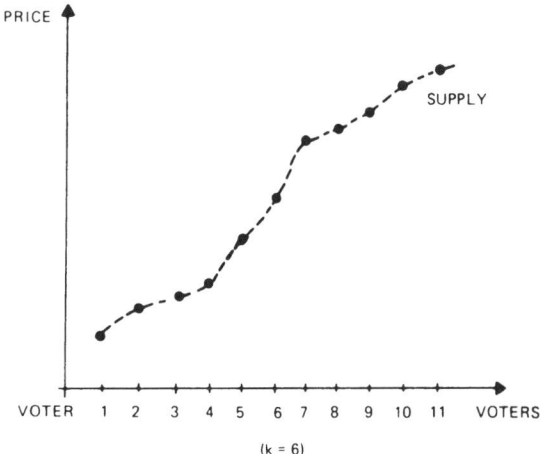

Fig. 1

If the lobbyist knew the supply curve he could engage in price discrimination and, moving up from the low end of the curve, pay just enough to each voter until he secured a majority. But in this context it may be difficult, if not impossible, for the lobbyist to gain much knowledge of the supply curve.[2]

Suppose instead that he acts as a price taker. Then the payoff function for the sellout game is the following modified form of (3): i gets his asking price p_i if he is bought, and p_i^0 otherwise.[3] In this case the voters at the low end of the curve can strategically raise their prices, and command a surplus. If a majority of k is required to win, $n/2 < k < n$, and M is sufficiently large $(M \geqslant k\, p_{k+1}^0)$ then each of the first k players can raise his price to $p = p_{k+1}^0$, the opportunity cost of the $(k+1)^{st}$ player; moreover these prices, $(p, \ldots, p, p_{k+2}^0, \ldots, p_n^0)$, constitute the unique marginal value for the sell-out game. The lobbyist's demand curve is a "spike" of height M at voter $k + 1$, and his profit of $M - kp$ represents ordinary economic surplus. (Figure 2).

This model of political bribery was first described in *Young* [1978a]. Various other approaches to competitive bribery may be found in *Young* [1978b, 1978c], *Shubik/ Young* [1978], and *Shubik/Weber* [1978].

Both of the above examples illustrate the proposition that a game without a core may be exploited for profit. Moreover it is precisely this exploitation that introduces stability into the system, since the removal of surplus allows a core to exist on what is

---

[2]) In addition, there may well be competition from other lobbyists who are proposing other bills for this same slot on the agenda.

[3]) In an earlier version of this model [*Young*, 1978a], the payoff function was defined only in terms of *direct* payments to the players: thus i's payoff was p_i if i is bought and *zero* otherwise. Also, the value of the bill, M, was treated as infinite. These differences lead in some cases to different equilibrium solutions than obtain in the present model. They also result in a distinction between "price" and "income" which is not necessary if opportunity costs are treated as income. In the earlier version the term 'canonical equilibrium' was used instead of 'core marginal value'.

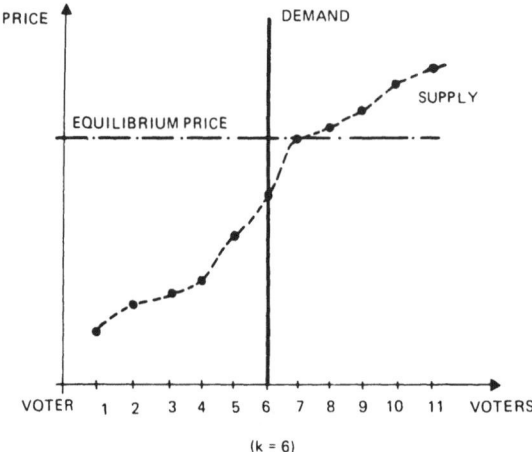

Fig. 2

"left over." The minimum "exploitable surplus" is computed as in (7) and can be interpreted as the ordinary economic surplus realized by organizers of production.

References

Shapley, L.S.: A Value for *n*-Person Games. Contributions to the Theory of Games. Ed by H.W. Kuhn and A.W. Tucker. Annals of Mathematics Studies **28**, Princeton 1953.

Shubik, M., and *R.J. Weber*: Competitive Valuation of Cooperative Games. Cowles Foundation Discussion Paper 482. New Haven 1978.

Shubik, M., and *H.P. Young*: The Nucleolus as a Noncooperative Game Solution. Cowles Foundation Discussion Paper 478. New Haven 1978.

Young, H.P.: Power, Prices and Incomes in Voting Systems. Math. Programming **14**, 1978a, 129–148.

– : The Allocation of Funds in Lobbying and Campaigning. Beh. Science **23**, 1978b, 21–31.

– : A Tactical Lobbying Game. Research Report 78–10. Laxenburg 1978c.

– : The Market Value of a Game, mimeographed. Laxenburg 1978d.

Applied Game Theory. 1979 ©Physica-Verlag, Wuerzburg/Germany

Structural Power and Satisfaction in Simple Games

By *Chr. H. Nevison*, Hamilton[1])

Abstract: This paper explores the relationship between power, the capacity to affect action, and satisfaction, a result of voting action, in voting situations which can be modelled as simple games. The properties of the Banzhaf, Coleman, and Shapley/Shubik indices of power have been thoroughly discussed elsewhere. The properties of the first two are compared in detail with the properties of two measures of satisfaction, the Zipke index defined here and a measure defined by Brams and Lake.

An alternative definition of power which is more appropriate for generalization is suggested, based on the relationships between the measures of power and satisfaction. Study of the "paradoxes" of power reveal that power tends to be more sensitive to manipulation than satisfaction. The limiting behavior of the indices of satisfaction for large games is also discussed.

1. Introduction

The concept for power has been the object of much study by political scientists and other social scientists. A fascinating aspect of the resulting body of literature is the wide variety of interpretations of this concept. In this paper we study power in a very specific context: a voting situation which can be modelled as a simple game. Even in this narrow context the notion of power is not without ambiguity. We confine our study to power, or that component of power, which derives strictly from the structure of the game.

The strictly limited context which we have chosen for study is, perhaps, the easiest context in which one can come to terms with the concept of power. Indeed, a number of authors have developed quantitative measures of power for simple games, among them *Shapley/Shubik* [1954], *Banzhaf* [1965, 1968], and *Coleman* [1971]. Each of these authors constructs a measure of structural power. We are primarily concerned with the measures of power devised by *Banzhaf* and *Coleman*, although we have occasion to refer to the Shapley/Shubik index of power.

We study power by looking at the closely related concept of satisfaction. Our first step is to define an index of satisfaction based on a model of the situation similar to the models which are the basis for the Banzhaf and Coleman indices of power. *Rae* [1969] and *Brams/Lake* [1977a] also discuss measures of satisfaction. We then examine

[1]) Prof. *Christopher H. Nevison*, Dept. of Mathematics, Colgate University, Hamilton, N.Y. 13346, USA

the relationship between structural power and satisfaction, as measured by these indices. This enables us to make a general definition of the power of individuals in a voting situation.

Having established the relationship between structural power and satisfaction, a relationship also discussed by *Brams/Lake* [1977a] and *Straffin* [1978], we explore some properties of power. These properties have been presented as "paradoxes" of power by various authors including *Affuso/Brams* [1976], *Brams* [1975, 1976], *Kilgour* [1974], and *Straffin* [1976]. By examining these "paradoxes" closely for both measures of power and measures of satisfaction, we determine which are characteristic of power and not satisfaction.

Finally, we discuss the limiting behavior of a measure of satisfaction for large games. Similar work has been done by *Shapley/Shapiro* [1960] with the *Shapley/Shubik* measure of power as a special case and by *Dubey* [1975] and *Dubey/Shapley* [1977] for the Banzhaf index. Combined with our work, these give a basis for the study of the relationships between structural power and satisfaction for large games.

2. Structural Power and Satisfaction

2.1 Power

We shall define certain measures of power and of satisfaction for simple games. A simple game may be defined as follows:

Let $I = \{1, 2, \ldots, n\}$ and let W be a family of subsets of I such that if $W \in W$ and $W \subset V$, then $V \in W$. The pair (I, W) then denotes a *simple game* with characteristic function, v:

$$v(S) = \begin{cases} 1 & \text{if } S \in W \\ 0 & \text{if } S \notin W \end{cases} \quad \text{for any } S \subseteq I.$$

The set I is interpreted as the set of *players* or *voters* and W is the set of winning coalitions. The condition on W insures that every superset of a winning coalition also wins.

If we add the condition that for any $W \in W$, $I - W \notin W$, then (I, W) is a *proper simple game* and v is superadditive.

For a simple game (I, W) we let $L = W^c$, the set of losing coalitions. A coalition, S, whose complement in I, S^c, is losing is called a *blocking coalition*. We call a blocking coalition which is also losing, a *strictly blocking coalition*. The indices of power which we shall define are based on the idea of a voter being critical to a winning coalition. For $W \in W$, i is *critical* to W if $i \in W$ and $W - \{i\} \in L$. We define a number of sets and constants which are used in the definitions:

(1) $W_i = \{W \in W: i \in W\}$, the winning coalitions including voter i;

   $w = |W|$, the total number of winning coalitions;

   $w_i = |W_i|$, the number of winning coalitions including voter i;

   $b_i = |\{S: i \in S \text{ and } S^c \in L\}|$, the number of blocking coalitions including voter i;

$c_i = |\{W: i \in W \in \mathcal{W} \text{ and } W - \{i\} \in L\}|$, the number of winning coalitions for which voter i is critical.

For any proper simple game, every winning coalition is also blocking, but this need not be the case for an improper simple game. A game is *decisive* if for any $W \in \mathcal{W}$, $W^c \in L$ and for any $S \in L$, $S^c \in \mathcal{W}$.

All of our examples will be simple games which can be represented as weighted voting games. A *weighted voting game* is denoted by $[c; w_1, w_2, \ldots, w_n]$, which defines the simple game $(I, \mathcal{W})$ in the following way:

$$W \in \mathcal{W} \text{ if and only if } \sum_{i \in W} w_i \geqslant c.$$

Thus each w_i is the *weight* accorded to player i's vote and c is the *quota* of votes needed for a winning coalition. Our notation here follows *Dubey* [1975]. If we change the requirement above to a strict inequality, we denote the game by $(c; w_1, w_2, \ldots, w_n)$.

Using the constants defined above we define four measures of an individual's power in a simple game. *Banzhaf* [1965, 1968] defines the power of voter i by

$$B(i) = c_i / \sum_{i \in I} c_i. \tag{2}$$

The Banzhaf index is a measure of the *share* of power by an individual, there being a total of one. This is useful for the study of the distribution of power.

Banzhaf also uses an absolute measure of power defined by

$$B'(i) = c_i / 2^{n-1}. \tag{3}$$

This measures the likelihood that a voter is critical to a coalition.

Coleman [1971] introduces two absolute measures of power. Coleman's measure of the power of an individual, i, to prevent action is defined by

$$CP(i) = c_i / w, \tag{4}$$

and his measure of an individual's power to initiate action is defined by

$$CI(i) = c_i / (2^n - w). \tag{5}$$

Coleman [1971] interprets the former as the proportion of times that an individual can block the action of a winning coalition by withdrawing from it. He interprets the latter as the proportion of times an individual can change a coalition from losing to winning by joining it. The Coleman indices are undefined, respectively, for the two uninteresting extreme cases where no coalition wins and where every coalition wins.

The indices of structural power are variations on a theme. They all count the coalitions for which a voter is critical and differ only in their normalizations. B and CI will attribute unit power to a voter only if he is a dictator. CP, on the other hand, gives a voter unit power if he holds a veto. All these indices give a dummy voter, a voter who is never critical, zero power.

2.2 Satisfaction

We shall define a measure of satisfaction for simple games based on work by *Nevison/Schoepke/Zicht* [1978]. The measure which we shall define applies to any simple game and is based on the following model.

We assume that in a voting situation the winning coalitions are those subsets of the set of voters, I, so defined for the simple game (I, W). Furthermore, we assume that the voters are voting for two alternatives so that on a vote, the set I is partitioned into two sets – the division. In order to assess the satisfaction which accrues to each voter due to the formal structure of the game, we assume that (i) each voter gets unit satisfaction if he is in a winning coalition and zero otherwise, and (ii) each of the 2^{n-1} divisions is equally likely. The latter can be deduced from the assumption that each voter is equally likely to vote for either alternative. The Zipke measure of satisfaction for voter i is then the probability of satisfaction for this voter under the above assumptions:

$$Z(i) = w_i / 2^{n-1}. \tag{6}$$

The values of Z can range from zero to one, although for a given game, the full range will not always be realized. The Zipke index is an absolute measure of satisfaction in the sense that CP and CI are absolute measures of power. It does not seem useful to measure shares of satisfaction.

Brams/Lake [1977a] define an alternative measure of satisfaction. It is based on a model essentially the same as the one used above for the Zipke index, except that property (i) is changed to the following: (i') each voter gets unit satisfaction if he votes with an alternative and it wins or if the votes against an alternative and it loses. This, along with the other assumptions, leads to the definition of the *Brams/Lake* index of satisfaction:

$$BL(i) = \frac{w_i + b_i}{2^n}. \tag{7}$$

Just as for the Zipke index, this measure can be interpreted as the probability of satisfaction for voter i. They differ in the definition of what constitutes satisfaction.

The *Brams/Lake* index of satisfaction would be most appropriate in situations where the question is one of yes or no, as in the passage of a law. An opponent of passage will derive satisfaction from its defeat. The Zipke index, on the other hand, would be appropriate when positive action must be taken, as in the election of one of two candidates, or the case of passage of a law where it is generally agreed that the status quo is unsatisfactory and essentially one of two actions must be taken.

A close relationship between the Zipke index and the *Brams/Lake* index can be derived from the concept of a dual game as defined by *Shapley/Shapiro* [1960] and adapted to our notation for simple games:

If (I, W) is a simple game, then the *dual game* associated with it is another simple game (I, W^*) where

$$W^* = \{S \subseteq I : S^c \nsubseteq W\}.$$

Thus a blocking coalition in the original game becomes a winning coalition in the dual game. If we denote the Zipke index for the original game by Z and for the dual game by Z^*, then it is evident from the definition of BL that

$$BL(i) = \frac{Z(i) + Z^*(i)}{2} \tag{8}$$

A simple game is *decisive* if it is self-dual, $(I, W) = (I, W^*)$, in which case $BL = Z$.

  Brams/Lake [1977b] suggest that a defect of the Zipke index is the paradox illustrated by the following example; Consider the weighted majority games $\Gamma_1 = [5; 5, 1, 1, 1, 1]$ and $\Gamma_2 = [9; 5, 1, 1, 1, 1]$. In Γ_1 the first player is a dictator while in Γ_2 each player holds a veto. For any measure of power, the dummies in the first game have power increase in the second game. For example, $B_1(2) = 0$ but $B_2(2) = .2$, where the subscripts indicate game 1 or 2. However, the satisfaction for these players as measured by the Zipke index decreases from 1 to game 2: $Z_1(2) = .5$ and $Z_2(2) = .063$. The decrease in power is due to the fact that the Zipke index measures only satisfaction with positive action and the probability of this decreases as the number of votes increases, offsetting the increase in power which occurs. Rather than being a defect, this "paradox" suggests that in situations where the model on which the Zipke index is based is appropriate, there is a clear divergence between power and satisfaction. In view of the other "paradoxes of power" which we shall discuss below, this does not seem surprising. We shall find it useful to use both the Zipke index and the Brams/Lake index in our comparisons of the properties of power and satisfaction.

2.3 Power derived from Satisfaction

  Nevison/Schoepke/Zicht [1978], using a slight variation of the Zipke index which we have defined here, show that, in our notation, the Coleman index of the power of an individual to block action can be derived from the Zipke index by

$$CP(i) = (2^n / w)(Z(i) - w / 2^n). \tag{9}$$

Equations (3) and (4) show that the absolute Banzhaf index can then the written as

$$B'(i) = 2(Z(i) - w / 2^n). \tag{10}$$

  Brams/Lake [1977a] prove an analogous result associating the absolute Banzhaf index of power to their index of satisfaction:

$$B'(i) = 2(BL(i) - 1/2). \tag{11}$$

  These results show that power as measured by any of the four indices presented here is *not* dependent on the concept of a voter critical to a coalition on which they were originally based. They can be based on a count of winning coalitions and are closely related to satisfaction. Equations (9), (10), and (11) enable us to offer an alternative basis for the definition of structural power in a simple game which avoids the complicated concept of a critical voter in a coalition. Observe that if a dummy, j, is added to a simple game, then the measure of satisfaction for the dummy in the modified game will be

$$Z(j) = w / 2^n \qquad (12)$$

or

$$BL(j) = 1/2. \qquad (13)$$

Equation (12) follows since each of the w winning coalitions in the original game will result in exactly one winning coalition with the dummy added in the modified game which has $n + 1$ players. Equation (13) follows by considering (12) for the original game and the dual of the original game, which has $2^n - w$ winning coalitions, and applying equation (8). In either case, we see that the derived indices of power can be regarded as the gain in satisfaction which accrues to a player over that which could be expected by a dummy, a powerless player, due to his strategic position in the game, appropriately normalized to a zero-one scale.

This interpretation of the relationship between the power indices and the measures of satisfaction lead to a general definition of structural power in a voting situation: The power of an individual is the gain in satisfaction he receives due to his strategic position over that satisfaction which would accrue to him by chance, relative to the greatest possible gain. *Straffin* [1978] also develops this approach to power. Further generalization suggests that a suitable definition for power derived from any source in a voting situation would be the gain in satisfaction over that which could be expected by a dummy. Since in voting situations we are able to measure satisfaction and structural power, we have a basis for further inquiry into power derived from other sources.

3. The Paradoxes of Power

In this section we discuss two "paradoxes" of power which have been presented in the literature. By examining these for power and satisfaction indices we elucidate whether they are characteristic of power.

This enables us to examine the roots of these paradoxes in somewhat more detail than has been done previously.

3.1 The Paradox of Quarreling

We give a more general formulation of the paradox of quarreling then *Straffin* [1976] or *Brams* [1975, 1976], but not as detailed as the framework used by *Kilgour* [1974] to study the Shapley value which has the Shapley/Shubik index as a special case.

If (I, W) is a simple game, then $(I, W_q, \mathcal{I})$ is an associated *simple game with quarreling* if it satisfies the following conditions:

(i) $\mathcal{I}$ is a family of subsets of I;

(ii) $W_q = \{W \in W: \text{for any } Q \in \mathcal{I}, Q \not\subseteq W\}$

$\mathcal{I}$ represents the set of quarreling groups and condition (ii) implies that no quarreling group will join the same winning coalition. W_q is the set of winning coalitions for the game with quarreling. (I, W_q) may not be a simple game in its own right, since supersets of a winning coalition may not be winning.

In order to appropriately extend the definitions of the power and satisfaction indices to games with quarreling, we define the following sets and constants:

$$W_i^q = W_i \cap W_q;$$

$$q = |W - W_q|, \qquad \text{the number of winning coalitions eliminated by the quarrel};$$

$$w_q = |W_q|;$$

$$w_i^q = |W_i^q|;$$

$$c_i^q = |\{W \subset W_i^q: W - \{i\} \notin W^q\}|,^0 \quad \text{the number of winning coalitions for which voter } i \text{ is critical in the game with quarreling.}$$

Using these constants, the definitions of the indices of power and satisfaction are extended to a game with quarreling in the natural way. The sub- or superscript "q" indicates a constant or an index defined for a game with quarreling. The Banzhaf index:

$$B_q(i) = c_i^q \Big/ \sum_{i \in I} c_i^q;$$

the Coleman indices:

$$CP_q(i) = c_i^q / w_q$$

and

$$CI_q(i) = c_i^q / (2^n - w);$$

the Zipke index:

$$Z_q(i) = w_i^q / (2^{n-1} - q).$$

The denominator in the Coleman measure of the power to initiate action CI, is unchanged since the number of potentially losing coalitions in the game is unchanged. We regard the coalitions in $W - W_q$ as being eliminated from the game altogether. The denominator in the Zipke index is modified to yield the number of possible divisions in the game with quarreling.

The *paradox of quarreling* is said to occur for a simple game (I, W) and an associated simple game with quarreling, (I, W_q, I) and a measure, M, suitably defined for the voters in both games, if the following conditions hold:

(i)   there is exactly one $Q \in I$;

(ii)   there is a voter $i \in Q$, such that $M(i) < M_q(i)$.

The paradoxical result is that player i has decreased his options and the number of winning coalitions he may join by his involvement in the quarrel, but the value attributed to him by the measure M has increased. Since there is no other quarrel which might offset the effect that the one including i has, one would expect that involvement in the quarrel would decrease the power or satisfaction which voter i might obtain.

Brams [1975, 1976], *Kilgour* [1974], and *Straffin* [1976] provide a number of examples of this paradox for the Banzhaf index, the Coleman index CP and the *Shapley/Shubik* index for power. Tables 1 and 2 give an example where the paradox occurs for the Banzhaf index with values for the Zipke index included.

	votes	B	B'	CP	CI	Z	BL
a	5	.308	.500	.571	.444	.688	.750
b	5	.308	.500	.571	.444	.688	.750
c	2	.231	.375	.429	.333	.625	.688
d	1	.077	.125	.143	.111	.500	.563
e	1	.077	.125	.143	.111	.500	.563

Tab. 1: [8; 5, 5, 2, 1, 1]

	B_q	CP_q	CI_q	Z_q
a	.300	.6	.333	.667
b	.300	.6	.333	.667
c	.200	.4	.222	.583
d	.100	.2	.111	.333
e	.100	.2	.111	.333

Tab. 2: [8; 5. 5. 2. 1. 1] with quarreling, $Q = \{d, e\}$.

In the case of the Zipke index, Corollary 3.3 to Theorem 3.2 demonstrates that the quarreling paradox cannot occur for this measure of satisfaction. Consequently, we may conclude that the paradox of quarreling is indeed a peculiarity of power which is not inherent in the satisfaction from which power is derived. However, the following result shows that we must be cautious in our interpretation of this result:

Proposition 3.1: The paradox of quarreling does not occur for the Coleman index of of the power of an individual to initiate action.

Proof: Suppose that (I, W) and (I, W_q, I) are a simple game and an associated simple game with quarreling with CI and CI_q the corresponding Coleman indices. Then

$$CI_q(i) = c_i^q / (2^n - w) \leqslant c_i / (2^n - w) = CI(i).$$

Q.E.D.

Thus the paradox of quarreling depends on the form of the power index which is used. Since the formulation of quarreling given here involves the elimination of winning coalitions only, the Coleman index of power to initiate action is an appropriate absolute measure of power, just as the Zipke index is the appropriate measure of satisfaction. However, the paradox of quarreling as formulated here does occur for the Banzhaf measure of shares of power, so that a cautious conclusion might be that the paradox of quarreling is characteristic of the relative power of voters in a simple game.

An explanation for the differing behavior of the absolute and relative power as measured by CI and B is provided by *Coleman's* [1971] measure of the capacity of a group to act. This measure does not apply to individual voters, but provides a single number which is an indication of the capacity of the group as a whole to take positive action. *Coleman* [1971] defines it as the ratio of winning coalitions to the total number of coalitions in the game, and we extend it to a game with quarreling in the natural way:

$$C = w / 2^n, C_q = w_q / (2^n - q).$$

In the case of quarreling, the numerator and denominator have both been reduced by q, so that, since $C \leqslant 1, C_q \leqslant C$, for any simple game and associated game with quarreling. Thus a quarrel decreases the ability of the group to act and the effect of this is dominant on the power of an individual to initiate action, even though his relative power may increase.

3.2 The Strong Paradox of Quarreling

We did not consider the *Brams/Lake* index for the paradox of quarreling because it is not clear how the blocking coalitions should be handled in this case. Indeed, the model of quarreling which we have considered has a disquieting lack of symmetry which one might feel is the source of the paradox. This lack of symmetry results from eliminating winning coalitions which include all the quarrelers but not losing, and possibly blocking coalitions which include them. Since one interpretation of a quarrel is that the participants will not agree on the vote, win, lose or draw, we formulate a stronger form of quarreling.

For a simple game (I, W) we call (I, W_s, I) an associated *game with strong quarreling* if the following conditions apply:

(i) I is a family of subsets of I;
(ii) $W_s = \{W \in W: \text{for any } Q \in I,$
        $Q \cap W \neq Q$ and $Q \cap W \neq \emptyset\}$.

Again I is the set of quarreling groups, but condition (ii) insures that a winning coalition cannot occur from a division where the quarreling group votes together. In defining our measures we also eliminate divisions of two blocking coalitions where the quarreling group votes togehter.

We can define the dual of an associated game with strong quarreling as follows:

If (I, W) and (I, W_s, I) are a simple game with an associated game with strong quarreling, then the *dual* of (I, W_s, I) is the game with strong quarreling $(I, W_s * I)$ associated with the dual of the original game (I, W^*).

The *strong paradox of quarreling* is said to occur under exactly the same circumstances as the paradox of quarreling with strong quarreling substituted in the definition. The suitable definitions for the measures for a game with quarreling are based on the constants as originally defined but modified to eliminate divisions where a quarreling group votes together. For the case of one quarreling group with k members, these are given as follows:

$$B_s(i) = c_i^s / \sum_{i \in I} c_i^s;$$

$$B_s'(i) = c_i^s / (2^n - 2^k);$$

$$CP_s(i) = c_i^s / w_s;$$

$$CI_s(i) = c_i^s / (2^n - 2^{n-k+1} - w_s);$$

$$Z_s(i) = w_i^s / (2^{n-1} - 2^{n-k});$$

$$BL_s(i) = (w_i^s + b_i^s) / (2^n - 2^{n-k+1}).$$

	B_s	$B_s' = CP_s = CI_s$	$Z_s = BL_s$
a	.250	.500	.750
b	.250	.500	.750
c	.250	.500	.750
d	.125	.250	.500
e	.125	.250	.500

Tab. 3: $[8; 5, 5, 2, 1, 1]$ with strong quarreling, $Q = \{d, e\}$.

Tables 1 and 3 give an example where the strong paradox of quarreling occurs for all the power indices, with the Zipke and Brams/Lake values given for comparison. We show that the strong paradox of quarreling never occurs for the Zipke of the Brams/Lake indices of satisfaction in the following.

Theorem 3.2: For a simple game (I, W) and an associated simple game with quarreling (I, W_s, I), which satisfy the conditions of the strong paradox of quarreling, for any $i \in Q \in I$:

$$Z(i) \geq Z_s(i).$$

Proof: Let k be the number of voters in the quarreling group. Then

$$w_i = w_i^s + w_Q,$$

where w_i^s is the number of winning coalitions in the game with strong quarreling which include i and w_Q is the number of winning coalitions in the original game which include the whole quarreling group. w_i^s can be partitioned into w_T:

$$w_i^s = \sum_{\substack{i \in T \subset Q \\ \neq}} w_T,$$

where w_T is the number of winning coalitions in the original game which include exactly that subset of Q, T, and no other members of Q. There will be exatcly $2^{k-1} - 1$ such subsets T, in the summation, since each must include i, leaving $k - 1$ other voters to choose from, and since Q itself is excluded. But $w_Q \geqslant w_T$, for any T in the summation, since if a particular coalition including T is winning in the original group, then the corresponding coalition with the other members of Q added also wins. Consequently

$$w_i \leqslant 2^{k-1} w_Q$$

so

$$w_i - \frac{w_i}{2^{k-1}} \geqslant w_i^s.$$

Dividing both sides of this inequality by $2^{n-1} - 2^{n-k}$ and simplifying, we obtain

$$Z(i) = \frac{w_i}{2^{n-1}} \geqslant \frac{w_i^s}{2^{n-1} - 2^{n-k}} = Z_s(i).$$

Q.E.D.

Corollary 3.3: The paradox of quarreling does not occur for the Zipke index of satisfaction.

Proof: Using the same framework as the theorem, the Zipke index for quarreling can be given by

$$Z_q(i) = \frac{w_i^q}{2^{n-1} - w_c}.$$

However, $2^{n-k} \geqslant w_c$, and $w_i^q = w_i^s$, since $i \in Q$, so that

$$Z_s(i) = \frac{w_i^q}{2^{n-1} - 2^{n-k}} \geqslant \frac{w_i^q}{2^{n-1} - w_c} = Z_q(i).$$

The corollary then follows from the theorem.

Q.E.D.

Corollary 3.4: The strong paradox of quarreling does not occur for the Brams/Lake index of satisfaction.

Proof: For the original game, we have observed that

$$BL\,(i) = \frac{Z\,(i) + Z^*\,(i)}{2}.$$ (14)

It is evident from the symmetry of the situation and the definition of the dual game with strong quarreling that

$$BL_s\,(i) = \frac{Z_s\,(i) + Z_s^*\,(i)}{2},$$

where Z_s^* is the Zipke index applied to the dual game with strong quarreling. By applying the theorem to both the original game and the dual game, the corollary follows.

Q.E.D.

We see that our stronger form of the paradox of quarreling occurs for all the indices of power and neither of the indices of satisfaction. Thus the behavior of the power indices is characteristic of power and cannot be attributed to its derivation from satisfaction.

3.3 The Paradox of Size

The second paradox of power is closely related to the paradoxes of quarreling which we have just considered. *Brams* [1975, p. 178] calls it the "paradox of (large) size." This paradox presupposes that voters gain an advantage by forming a coalition which acts as a unit in whatever voting takes place. The paradox occurs when a single coalition is formed and the voters in it have less power than they have acting independently.

We need to formulate more precisely what we mean by a coalition before we can adequately analyze this situation. If $(I,\, W)$ is a simple game, an associated *simple game with a coalition structure* is defined by $(I,\, W_c,\, C)$ where C is a family of disjoint subsets of I which represent groups which consistently vote together, and $W_c = \{W \in W: \text{for any } C \in C, C \cap W = C \text{ or } \emptyset\}$. Thus we only consider divisions where each group votes together. For the paradox of large size we consider only cases where there is a single group, $C \in C$.

Brams [1975] analyzes the game where there are $n - k + 1$ voters, k being the number in the group C, and counts the group C as a single voter. He provides an example reproduced in Tables 4 and 5. Note that in a weighted voting game, the structure of the derived game is determined simply by assigning the sum of the weights of its members as the weight of the coalition in the modified game. *Brams* suggests that in this situation one would expect power to be superadditive in the sense that the power of the coalition should be greater than or equal to the sum of the powers of its members. That this does not occur for any of the measures of power in this example illustrates the paradox of large size.

	votes	B	$B' = CP = CI$	$Z = BL$
a	2	.286	.500	.750
b	2	.286	.500	.750
c	1	.143	.250	.625
d	1	.143	.250	.625
e	1	.143	.250	.625

Tab. 4: [4; 2, 2, 1, 1, 1]

		B	$B' = CP = CI$	$Z = BL$
a	2	.333	.500	.750
b	2	.333	.500	.750
{c,d,e}	3	.333	.500	.750

Tab. 5: [4; 2, 2, 3] − [4; 2, 2. 1, 1, 1] with coalition structure

The same quantitative property occurs for the Zipke and Brams/Lake indices of satisfaction, identical in these decisve games. However, the appropriate interpretation is different. Since the measure of satisfaction of .75 for the coalition in the example indicates the probability that this group is satisfied, it also indicates that each member of a group is satisfied with the same probability. Consequently, the total satisfaction of a group in a game with a coalition structure should be the satisfaction measured for the group times the number of members of the group. Thus there is no paradox for satisfaction.

This method of measuring the satisfaction for a group is equivalent to analyzing the game (I, W_c, C) in the same way as we did games with strong quarreling. We calculate satisfaction for individuals as individuals after eliminating from consideration divisions where the groups of the coalition structure do not act together. Thus we eliminate exactly those divisions which were retained if the groups were construed as quarreling. A paradox of large size would occur if there were some voter i in the only group in the coalition structure C such that his satisfaction with the coalition structure, $Z_c(i)$ or $BL_c(i)$ were less than the corresponding satisfaction in the original game. That this cannot happen for either Z or BL is a consequence of the following:

Theorem 3.5: If (I, W) is a simple game with an associated game with a coalition structure (I, W_c, C) with only a single group $C \in C$, then the paradox of large size occurs for Z or BL if and only if the strong paradox of quarreling occurs for (I, W_s, C), where $I = C$.

Proof: We give the proof for the Zipke index. The proof for the Brams/Lake index is analogous. Let w_i, w_i^c, and w_i^s be the number of winning coalitions including voter i

for the original game, the game with coalition structure, and the game with strong quarreling, respectively. If k is the number of voters in the groups $C = Q$, then the Zipke indices for the three situations are respectively

$$Z(i) = w_i / 2^{n-1}, \; Z_c(i) = w_i^c / 2^{n-k}, \; Z_s(i) = w_i^s / (2^{n-1} - 2^{n-k}).$$

But the paradox of large size occurs if and only if for $i \in C$,

$$Z_c(i) < Z(i).$$

The following sequence of equivalent statements demonstrates that this occurs if and only if the paradox of strong quarreling also occurs:

$$w_i^c / 2^{n-k} < w_i / 2^{n-1} = (w_i^c + w_i^s) / (2^{n-1} - 2^{n-k}) + 2^{n-k},$$

$$w_i^c ((2^{n-1} - 2^{n-k}) + 2^{n-k}) < 2^{n-k} \, w_i^s,$$

$$(2^{n-1} - 2^{n-k}) \, w_i^c < 2^{n-k} \, w_i^s,$$

$$(2^{n-1} - 2^{n-k}) (w_i^c + w_i^s) < ((2^{n-1} - 2^{n-k}) + 2^{n-k}) \, w_i^s$$

$$Z(i) = (w_i^c + w_i^s) / ((2^{n-1} - 2^{n-k}) + 2^{n-k}) < w_i^s / (2^{n-1} - 2^{n-k}) = Z_s(i).$$

Q.E.D.

Corollary 3.6: The paradox of large size does not occur for the Zipke or the Brams/ Lake indices of satisfaction.

Proof: This is a consequence of Theorem 3.2, Corollary 3.4, and the previous theorem.

Q.E.D.

Thus the paradox of large size in *Brams'* [1975] sense is characteristic of power but does not apply to satisfaction.

However, *Brams* considers a coalition as a unit in his analysis, whereas our reformulation enables us to consider the members of the coalition as individuals in order to measure their satisfaction. We use this formulation and the generic definition of power described in section 2.3 to give a method for measuring the power of each individual in a game with a coalition structure.

We measure satisfaction for the game with coalition structure as above and we measure the power by using one of the equations (9), (10), or (11). In equations (9) and (10) the constants 2^n and w must be adjusted properly, so we shall analyze the single case where a version of the absolute Banzhaf index is derived from the Brams/ Lake index according to the equation:

$$B_c'(i) = 2 \, BL_c(i) - 1.$$

An extension of the Banzhaf index of shares of power can be derived for a game with a coalition structure by normalizing B'_c to have a total of unit power. Table 6 shows the resulting values of all the indices for the example. The other indices are simplified

	B_c	$B'_c = CP_c = CI_c$	$Z_c = BL_c$
a	.2	.500	.750
b	.2	.500	.750
c	.2	.500	.750
d	.2	.500	.750
e	.2	.500	.750

Tab. 6: $[4; 2, 2, 1, 1, 1]$ with coalition structure $C = \{c, d, e\}$

since this game is decisive. The paradox of large size for power no longer holds for either the absolute or the relative measure. Indeed, it is evident from the definition of B'_c and Corollary 3.6 that the paradox of large size will never occur for the absolute Banzhaf index of power.

If our formulation of power in a game with coalition structure is correct, then there seems to be no paradox (we have neither proof nor counterexample for the measure of relative power, B). This formulation uses an ego-oriented definition of power — one which concentrates on the capacity to control one's own satisfaction [*Riker*].

If *Brams'* formulation is correct, then this again is a paradox characteristic of power since the appropriate interpretation of satisfaction yields no paradox.

4. Power and Satisfaction for Large Games

The characteristics of large voting games are important for the study of voting situations involving large electorates. *Shapley/Shapiro* [1960] and *Dubey* [1975], and *Dubey/Shapley* [1977] have studied the limiting properties of the Shapley/Shubik and the Banzhaf indices of power. We present results on the limiting behavior of the Zipke and Brams/Lake indices of satisfaction for an important class of weighted voting games. We also discuss how these results are related to similar results on limiting properties of the absolute and relative Banzhaf indices. The presentation here follows the spirit and notation of *Dubey's* [1975] study. The proofs will appear elsewhere.

4.1 Limiting Behavior of the Zipke Index

We shall deal with weighted voting games as defined in section 2.1: $[c; w_1, \ldots, w_n]$, where the w_i are positive numbers. A coalition $W \subseteq I$ is a winning coalition if $\sum_{i \in W} w_i \geq c$. $(c; w_1, \ldots, w_n)$ denotes a variation where $W \subseteq I$ wins if $\sum_{i \in W} w_i > c$. We shall allow degenerate cases where $c \leq 0$ or $c \geq \sum_{i=1}^{n} w_n$. In the former case every coalition, including the empty coalition, is winning in the first type of game. The same holds for the second type of game if $c < 0$. If $c > \sum\limits_{i=1}^{n} w_i$, then no coalition wins in the first type of game, and the same holds for the second type if $c \geq \sum\limits_{i=1}^{n} w_i$. This extension of the notation for weighted voting games must be considered when counting blocking coalitions or coalitions for which a voter is critical in our measures. With this in mind, the statement of the theorems is simplified.

We shall study the limiting behavior of the Zipke index for a sequence of games (Γ^ν), $\nu = 1, 2, \ldots$, where the number of players increases to infinity while the games in the sequence retain a particular form. The total voting weight for every game is $w(\Gamma^\nu) = 1$ and the quota, c, is fixed for the sequence. The first k players are major players whose voting weight is the same in every game. There is also a set of minor players which grows as we advance in the sequence. We consider the symmetric case where in each game, Γ^ν, the voting weights of the minor players are all the same,

$$\alpha_j^\nu = \alpha / (n - k), \text{ where } \alpha = 1 - \sum\limits_{i=1}^{k} w_i. \text{ Thus the form of each } \Gamma^\nu \text{ is given by}$$

$$\Gamma^\nu = [c; w_1, \ldots, w_k, \alpha_{k+1}^\nu, \ldots, \alpha_n^\nu],$$

$$\sum\limits_{j=k+1}^{n} \alpha_j^\nu = \alpha, \text{ for every } \nu, \text{ and } \alpha_j^\nu = \alpha / (n - k) = \alpha^\nu,$$

and

$$\sum\limits_{i=1}^{k} w_i + \alpha = 1.$$

We use the notation $< x >$ to denote the least integer greater than or equal to x. We can then characterize the limiting behavior for the Zipke index by two theorems:

Theorem 4.1: If Γ^ν, $\nu = 1, 2, 3, \ldots$, is a sequence of symmetric weighted voting games as defined above, then for any major player i, the limiting value of the Zipke index is given by

$$Z_l(i) = \lim_{\nu \to \infty} Z^\nu(i) = (Z_1(i) + Z_2(i)) / 2$$

where $Z_1(i)$ and $Z_2(i)$ are the Zipke values for the games $G_1 = [c - \alpha / 2; w_1, \ldots, w_k]$ and $G_2 = (c - \alpha / 2; w_1, \ldots, w_k)$, respectively.

Theorem 4.2: For a sequence of games (Γ^ν) as defined for Theorem 4.1, the limiting value of the Zipke index for a minor player, j, will be

$$Z_l(j) = \frac{C_1 + C_2}{2},$$

where C_1 and C_2 are the Coleman measures of the capacity to act for the games G_1 and G_2.

The limiting values for the Zipke index will be the same if we use the sequence $(\Gamma^\nu) = (c; w_1, \ldots, w_k, \alpha^\nu_{k+1}, \ldots, \alpha^\nu_n)$ instead of the (Γ^ν) considered.

4.2 Limiting Values for BL, B', and B

The theorems in the preceding section lead to limiting values for the Brams/Lake index in the

Corollary 4.3: For a sequence of games (Γ^ν) as defined for Theorem 4.1, and for a major player i and a minor player j, the limiting values of the Brams/Lake index of satisfaction are given by

$$BL_l(i) = \frac{BL_1(i) + BL_2(i)}{2}$$

and

$$BL_l(j) = 1/2,$$

where BL_1 and BL_2 are the Brams/Lake index applied to the games G_1 and G_2 defined in Theorem 4.1.

We can now obtain limiting values for the absolute Banzhaf index with the same result as *Dubey* [1975] in

Corollary 4.4: For a sequence of games (Γ^ν) as defined in Theorem 4.1, the limiting values of the absolute Banzhaf index or power for a major player, i, and a minor player, j, are given by

$$B'_l(i) = \frac{B'_1(i) + B'_2(i)}{2}$$

and

$$B'_l(j) = 0,$$

where B'_1 and B'_2 are the absolute Banzhaf index applied to G_1 and G_2, as defined earlier.

Proof: This result follows immediately from Corollary 4.3 and equation (11) $B'(i) = 2(BL(i) - 1/2)$.

Q.E.D.

The limiting values for the Banzhaf index of shares of power do not follow so simply from the results on satisfaction, since they depend on rates of convergence. However, the approach which leads to the above results can form the basis for a proof of

Dubey's (1975) *Theorem 4.5*: For a sequence of games (Γ^ν) the limiting values of the Banzhaf index of shares of power for major players are given by

$$B_l(i) = B_1(i) \quad \text{if} \quad \alpha/2 < c < 1 - \alpha/2$$

and if there is no coalition of major players S such that $w(S) = c - \alpha/2$, otherwise $B_l(i) = 0$. $B_1(i)$ is the relative Banzhaf index applied to the game G_1. The relative power of minor players is always zero in the limit.

That proof is very similar to the one presented by *Dubey/Shapley* [1977]. Thus we see that by approaching the analysis of power through the measures of satisfaction, we are able to duplicate the results derived by *Dubey* [1975] and in addition we derive the limiting behavior for the Zipke and Brams/Lake indices of satisfaction.

5. Summary and Conclusions

We have defined the Zipke index of satisfaction for simple games and we have studied the properties of it and the Brams/Lake index of satisfaction. We have compared these to the various indices of structural power in simple games.

In Section 2 we found that the Zipke and the Brams/Lake indices are closely related, the latter being a combination of the former for a game and its dual. We also cited results showing that the indices of power could be derived from either of these in a way that suggests that power could be measured in terms of the gain in satisfaction for a voter due to the advantage he holds over a hypothetical dummy voter.

In Section 3 we analyzed the two "paradoxes" of power for both the indices of power and satisfaction. We saw that for the paradox of quarreling and the paradox of size, power seems to be more sensitive to manipulation than satisfaction as measured by our indices. We also saw that an alternate formulation of the situation leading to the paradox of size, based on the general definition of power given in Section 2, eliminated the paradox, at least for absolute indices of power.

In Section 4 we discussed the limiting behavior of the satisfaction indices for large games and used the results to duplicate *Dubey's* [1975] work for the Banzhaf and absolute Banzhaf indices of power. In addition to yielding important results in their own right, this analysis demonstrated the versatility of the technique of analyzing the power indices by means of their relationships to the satisfaction indices, and the importance of the relationship between the Brams/Lake and the Zipke indices for deriving theoretical results.

By detailing the properties of satisfaction as measured by the Zipke and Brams/Lake indices and by defining power in terms of satisfaction, we have established the measures of power on a firmer theoretical basis and have provided a baseline for further study.

References

Affuso, P.J., and *St. J. Brams:* Power and Size: A New Paradox. Theory and Decision 7, 1976, 29–56.

Banzhaf, J.F.III: Weighted Voting Doesn't Work: A Mathematical Analysis. Rutgers Law Review **19,** 1965, 317–343.

–: One Man, 3.312 Votes: A Mathematical Analysis of the Electoral College. Villanova Law Review **13,** 1968, 304–332.

Brams, St.J.: Game Theory and Politics. New York 1975.

–: Paradoxes in Politics. New York 1976.

Brams, St.J., and *M. Lake:* Power and Satisfaction in a Representative Democracy. Mimeographed paper presented at the Conference on Game Theory and Political Science, Hyannis, July 10–17, 1977a.

–: Private communication, 1977b.

Coleman, J.S.: Control of Collectivities and the Power of a Collectivity to Act. Social Choice. Ed. by B. Lieberman. New York 1971, 277–287.

Dahl, R.A.: The Concept of Power. Political Power: A Reader in Theory and Research. Ed. by R. Bell, D.E. Edwards, and Harrison Wagner. New York 1969, 79–93.

Dubey, P.: Some Results on Values of Finite and Infinite Games. Technical Report, Center for Applied Mathematics. Ithaca 1975.

Dubey, R., and *L. Shapley:* Mathematical Properties of the Banzhaf Power Index. Rand Paper P–6016. The Rand Corporation, 1977.

Feller, W.: An Introduction to Probability Theory and Its Applications. Vol. 1, third edition, New York 1968.

Kilgour, D.M.: A Shapley Value for Cooperative Games with Quarreling. Game Theory as a Theory of Conflict Resolution. Ed. by A. Rapoport. Boston 1974, 193–206.

Lucas, W.F.: Measuring Power in Weighted Voting Systems. Case Studies in Applied Mathematics, Chapter 3, Mathematical Association of America, 1976, 253–263.

Nevison, Chr. H., S. Schoepke, and *B. Zicht:* A Naive Approach to the Banzhaf Index of Power. Behavioral Science **23,** 1978, 130–131.

Rae, D.: Decision Rules and Individual Values in Constitutional Choice. American Political Science Review **63,** 1969, 40–56.

Riker, W.H.: Some Ambiguities in the Notion of Power. Political Power: A Reader in Theory and Research. Ed. by R. Bell, D.V. Edwards and R. Wagner. New York 1969, 110–119.

Shapley, L.S., and *N.Z. Shapiro:* Values of Large Games – I: A Limit Theorem. RM–2648, The Rand Corporation, 1960.

Shapley, L.S., and *M. Shubik:* A Method for Evaluating the Distribution of Power in a Committee System. American Political Science Review **48,** 1954, 787–792.

Straffin, Ph. D.: Power Indices in Politics. Modules in Applied Mathematics, Mathematical Association of America, 1976.

–: Probability Models for Power Indices. To appear in Game Theory and Political Science. Ed. by Ordeshook, 1978.

Wagner, R.H.: The Concept of Power and the Study of Politics. Political Power: A Reader in Theory and Research. Ed. by R. Bell, D.V. Edwards, and R.H. Wagner. New York 1969, 3–12.

Applied Game Theory. 1979 © Physica-Verlag, Wuerzburg/Germany

Voting Weights as Power Proxies: Some Theoretical and Empirical Results

By *A. Schotter*, New York[1])[2])

Abstract: It is widely known that the voting weights of members in a voting body are not good proxies for their influence or power within the body. The question that arises, however, is how bad a proxy are they?

In other words, if we were to increase (or decrease) a voter's voting weight within a voting body, would his resulting power within the organization always increase (or decrease), but possible not in proportion to the increase in his weight, or could his influence actually decrease (or increase)?

This note presents the results of two previous studies of this question done by *Fischer/Schotter* [1978] and *Dreyer/Schotter* [1978]. This results are clear. Voting weights can be extremely poor proxies for voting power. For instance, it is proven that no matter what voting weight distribution we are observing, it is always possible to find another distribution in which at least one voter has his percentage of the votes decreased and yet his percentage of the voting power increased when power is measured by either the Banzhaf or Shapley-Shubik power index. This type of result is proven to be inevitable for $n \geqslant 6$ in the Banzhaf case and $n \geqslant 7$ in the Shapley-Shubik case.

Finally, this result is illustrated by observing the power relationships at the International Monetary Fund after a recent redistribution in the votes of the organization. There it is shown that 38 countries had their percentage of the vote decreased and yet had their percentage of the organization's power increased while four major countries were allocated a larger percentage of the vote and yet had their power within the organization decreased.

It is widely known among game theorists that the voting weight a member has in a voting body is not a good proxy for his voting power or influence within the body. This fact is easily illustrated. For instance, consider the voting body $V = (50\%; 35\%; 45\%, 20\%)$ where 50 % is the voting rule used in the organization — if any members form a coalition with 50 % or more of the votes, they are winning and can make decisions that are binding on the entire body — and 35 %, 45 % and 20 % are the voting weights of voters 1, 2 and 3, respectively. In this body then, although the weights are not distributed equally, it is easy to see that all the members are indeed equally powerful under the 50 % majority rule, since any voter cannot win a vote by himself and any coalition of two or more voters is winning. Consequently, it is clear that voting weights

[1]) The author would like to thank Dietrich Fischer and Jacob Dreyer for their previous collaboration on the idea presented here. In addition, the support of the Office of Naval Research, Contract Number NO0014-78-C-0598 is gratefully acknowledged.

[2]) Prof. *Andrew Schotter*, New York University, Dept. of Economics, 538 Tisch Hall, Washington Square, New York, N.Y. 10003, USA.

are not perfect proxies for voting power. The question that arises, however, is how bad a proxy are they? In other word, if we were to increase (or decrease) a voter's voting weight within a voting body would his resulting influence within the organization always increase (or decrease), although possibly not in proportion to the increase in his weight, or could his influence within the organization actually decrease (increase)?

This note presents the results of two previous studies of this question done by *Fischer/Schotter* [1978] and *Dreyer/Schotter* [1978]. Our results are simple:

a) Voting weights are extremely bad proxies for voting power. In fact, they are so bad that it may be possible to increase a voter's voting weight within an organization and actually decrease his power when power is measured by both the Banzhaf and Shapley-Shubik power in indices. In addition, it is possible to decrease a voter's voting weight within a voting body and increase his power. This later result was termed the "Paradox of redistribution" by *Fischer/Schotter* [1978].

   While the result is interesting, it would only be of empirical significance if the likelihood of such an occurance were great. In other word, if the paradox were merely a theoretical possibility, it may not cause a problem from a policy point of view. To this end, *Fischer/Schotter* have proved the following.

b) Not only is the "Paradox of Redistribution" possible, but such a paradox is inevitable in the sense that for any voting body $V = (d, w)$ where d is a decision rule and w is a voting weight distribution, there always exists another voting body, $V' = (d, w')$, where d is held constant and w changed to w', in which at least one voter has his percentage of the vote decreased and his percentage of the power increased.

   This result is not quite satisfactory, however, because while it proves that there exist some other voting distribution w' which gives rise to our inverse relationship between voting power and voting weight, it gives us no indication as to how likely the paradox really is. To try to settle this question, *Fischer/Schotter* performed a Monte Carlo experiment. These results are summarized by the figure below and can be stated as follows:

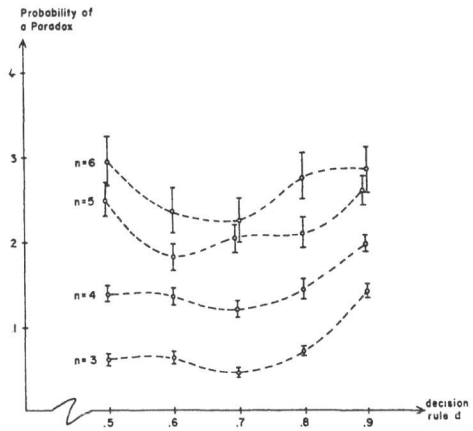

Occurrence of the "Paradox of Redistribution"

c) The paradox of redistribution has a surprisingly high likelihood of occurrence. In fact, for voting bodies with 6 voters using either a 50 or 90 percent voting rule, the probability of occurrence was 30 %.

All of these results would still be of no significance if the organizers of national and international organizations clearly understood the relationship between voting weight and voting power and acted accordingly. However, when one observes the real world, one realizes that in most major institutions the organizers of those institutions act as if voting weights were indeed perfect proxies for voting power. Consequently, it is not surprising to find institutions whose true power relationships are far different from the power relationships desired by the institution's planners.

To demonstrate this point, *Dreyer/Schotter* [1978] examined the power changes that occurred when some changes were made in the voting rules and voting weight distribution of the International Monetary Fund. They found that in the recent redistribution of the votes:

d) 38 countries had their percentage of the votes in the organization decreased and yet had their percentage of the power within the organization increased, while,

e) 4 countries had their percentage of the power within the organization increased and yet suffered decreases in their power within the organization.

Finally, the *Dreyer/Schotter* results indicate that smaller countries within the IMF have power that is out of proportion to their percentage of votes and that the recent voting changes at the Fund exacerbated this fact.

This paper tries to synthesize these theoretical and empirical results. We will proceed as follows:

Section 1 will briefly discuss two commonly used power indices, the Banzhaf and Shapley-Shubik index. Section 2 will present some simple theorems demonstrating the inevitability of the "paradox of redistribution." Section 3 will discuss the voting procedures of the International Monetary Fund and present some results indicating the extent to which voting weights are poor proxies for voting power. Finally, Section 4 will offer some conclusions and present some problems that might arise when we try to adjust the voting weights in a voting body to better represent the desired power relationships of the voting body.

1. Power Indices

Before we proceed to discuss our results, let us briefly review the power indices that we will be using. To do this, let N be the set of voters in a voting body indexed $i = 1, \ldots, n$ and let $w = (w_1, \ldots, w_n)$ be a vote distribution normalized such that $w_i > 0$ and

$$\sum_{i=1}^{n} w_i = 1.$$

The voting body is then fully described by an $n + 1$-tuple $v = (d; w_1, \ldots, w_n)$, where d is the decision rule of the body indicating the minimum fraction of votes that must be exceeded for the voting body to take collective action binding on all members, and $(w_1, \ldots, w_n)$ is a vote distribution. Let S be any subset of voters $S \subseteq N$. Then we can define the value of coalition S as

$$v(S) = 0 \text{ if } \sum_{i \in S} w_i \leqslant d,$$

$$v(S) = 1 \text{ if } \sum_{i \in S} w_i > d.$$

A voter is "critical" in a coalition S if his defection from that coalition changes the coalition from a winning to a losing coalition [i.e., $v(S) = 1$ and $v(S - \{i\}) = 0$].
    The Banzhaf index for member i is then defined as

$$p_B^i = \frac{\sum_S [v(S) - v(S - \{i\})]}{\sum_j \sum_S [v(S) - v(S - \{j\})]} \cdot$$

This index, then, describes the proportion of critical defections of member i. It follows that

$$\sum_{i=1}^{n} p_B^i = 1.$$

The Shapley-Shubik index for a member i is slightly more complex. It concerns itself with the proportion of permutations of the n members in which i's defection from a winning coalition is critical. More formally, it can be written as

$$p_{SS}^i = \sum_S \left[\frac{(s-1)! \, (n-s)!}{n!} \right] [v(S) - v(S - \{i\})]$$

where s is the number of members in the subset S and n is the total number of members in the voting body.

The Paradox of Redistribution for a Three-Member Voting Body

Consider the following voting body:

$$v = \left(\frac{70}{100}; \frac{55}{100}, \frac{35}{100}, \frac{10}{100} \right)$$

where $70/100$ is the decision rule and $(55/100, 35/100, 10/100)$ is the vote distribution. The Banzhaf and Shapley-Shubik power indices associated with this voting body are both $(1/2, 1/2, 0)$. Now let us redistribute the votes, keeping the decision rule the same, so that the following voting body is determined:

$$v' = \left(\frac{70}{100}; \frac{50}{100}, \frac{25}{100}, \frac{25}{100} \right)$$

Here the Banzhaf index is $(3/5, 1/5, 1/5)$, while the Shapley-Shubik index is $(2/3, 1/6, 1/6)$, showing that although member 1's voting weight decreased from $55/100$ to $50/100$, his power increased.

Now it might be interesting to ask whether in a three member voting body there exist vote distributions which are paradox proof in the sense that any redistribution of the votes which starts at one of these distributions must give a voter less voting power if it gives him less weight. The answer is yes, and as a matter of fact there are an infinite number of such vote distributions. For the purpose of illustration, however, consider the following voting body, $v = (70/100; 55/100, 23/100, 22/100)$. Here the power distribution is $(3/5, 1/5, 1/5)$ if the Banzhaf index is used, and $(2/3, 1/6, 1/6)$ if the Shapley-Shubik index is used. The reader can check for himself that it is not possible to give a voter more power by diminishing his voting weight.

2. The Inevitability of the Paradox of Redistribution

The question we investigate in this paper, then, is the circumstances under which the paradox illustrated above is inevitable. To do this, we must demonstrate how, given a vote distribution $w = (w_1, \ldots, w_n)$ and a voting rule d, we can construct a new vote distribution $w' = (w'_1, \ldots, w'_n)$ which gives at least one voter a smaller proportion of the vote yet gives him more power, when power is measured by either the Banzhaf or the Shapley-Shubik power index. The following propositions do just that. In order to conserve space, however, we merely state the propositions here and refer you to *Fischer/Schotter* [1978] for a constructive full proof.

Proposition 1: For voting bodies with $n \geqslant 6$, a paradox is always possible no matter what initial vote distribution exists, when power is defined by the Banzhaf index.

Proposition 2: For voting bodies with $n \geqslant 7$, a paradox is always possible no matter what initial vote distribution exists, when power is defined by the Shapley-Shubik index.

The import of these theorems is that for any voting body $v = (d; w_1, \ldots, w_n)$ — where $n \geqslant 6$ for the Banzhaf case and $n \geqslant 7$ for the Shapley-Shubik case — any initial vote distribution is subject to the paradox of redistribution in that there always exists another vote distribution such that in the redistribution of voting weights at least one player's weight is decreased while his voting power is increased. It is in this sense that the paradox is inevitable.

One shortcoming of our results is that they do not give us any indication of the minimum size of the redistribution necessary to create the paradox. Clearly our results would be more disturbing if for any given initial distribution of votes there was another distribution within a small neighborhood of the original which would create the paradox. It is interesting to note that in the case of the International Monetary Fund (to be discussed in the next section), the redistribution of voting weights was not a "drastic" one. Also, our simulation results given above demonstrate that the occurrence of the paradox is not a rare event.

Our results so far are rather general in that we do not constrain the decision rule used in our voting bodies in any way. However, the most common voting rule is the simple majority voting rule and when this rule is used, it is possible to show that both for the Banzhaf and the Shapley-Shubik power indices no paradox proof vote distributions exist for $n \geq 4$.

Proposition 3: If the voting rule used is the simple majority voting rule (i.e., if $d = 1/2$), then for $n \geq 4$ a paradox is always possible.

Instead of restricting the voting rule as we did in Proposition 3, we could have placed a restriction on the vote distribution. One restriction would be to constrain the weight of the smallest voter. If we make the restriction that the voter with the smallest voting weight (say voter i) has a weight $w_i > 2d - 1$ initially (which in the case of $d = 51\%$ merely requires that he have more than 2% of the vote), then it is possible to prove that no paradox proof vote distributions exist for $n \geq 4$ for both the Banzhaf and Shapley-Shubik power indices.

Proposition 4: If the decision rule is d and if the voter with the smallest voting weight in the voting body $v = (d; w_1, \ldots, w_n)$ has a weight greater than $2d - 1$, then with $n \geq 4$ (and either the Banzhaf or the Shapley-Shubik index), a paradox is always possible.

3. Power in the International Monetary Fund

As was stated in the introduction, all of the results derived so far would be of no significance if the types of phenomena we are discussing never occurred in the real world. However, at least in the International Monetary Fund, the planners of that organization seem to ignore the fact that voting weights are not perfect proxies for voting power. Consequently, when they desire to change the power relationships within the fund to reflect the changing importance of their members in international trade, they rely totally on changes in voting weights without giving any consideration to the effects these new weights actually have on the members' power.

Probably the best example of this myopia can be found in the recently ratified "Second Amendment to the Articles of Agreement of the International Monetary Fund" which along with the sixth review of the Fund's quota distribution was instituted to make the fund better reflect the changing importance of its members in the world trade. These amendments and quota revisions altered the voting rules and vote distribution among the members of the fund and the consequences of these revisions were analyzed by *Dreyer/Schotter* [1978]. Their conclusions were as follows:

1. Under the proposed changes 38 countries in the IMF had their percentage of the total vote decreased and yet their voting power within the organization increased when power is measured by the Banzhaf power index.
2. With the new vote distribution and voting rules, four major countries — Belgium, Holland, West Germany and Japan — had their percentage of the total vote increased and yet had their percentage of the total power decreased.
3. Under the previous distribution of votes and voting rules, smaller countries had a voting power that was out of proportion to their voting weights, and the newly introduced changes would generally aggravate this disproportion.
4. The power of the United States within the Fund increases substantially on issues where most countries vote through their Executive Directors (as groups), as opposed to issues where they vote through their Governors (individually).
5. Under both the previous and current voting system, significant diminishing returns to voting weights exist and the tendency is more pronounced under the new system. In most voting situations power is a concave function of voting weight with large linear segments.
6. Although these results are in many instances not quantitatively substantial, qualitatively they indicate noticeable discrepancy between what one would think the consequences of the voting changes would be and what they actually are.

Before we formally present these results, let us briefly discuss the voting rules and procedures at the International Monetary Fund.

Voting in the IMF

The organization of the IMF is very simple. All powers of the Fund are vested in its Board of Governors which is composed of the Fund's 131 member countries. Each country has 250 votes plus one additional vote for each part of its quota which is equivalent to one hundred thousand S.D.R.'s. The Board of Governors may, however, delegate certain decisions to be made to the Fund's body of Executive Directors which is composed of one representative from each of the five members of the Fund having the largest quotas plus 15 other representatives each of whom represents a certain subset or coalition of countries. There are then twenty Executive Directors, each one having the number of votes equal to the sum of the votes contained in the subset of countries it represents. Thus, voting by the Directors is, in fact, a two stage process. First each coalition meets and agrees (using a simple majority rule) on how its representative (Executive Director) will cast his vote in the body of Executive Directors. Then the Executive Directors themselves meet and, using a decision rule that is not a simple majority rule, cast their votes.

Essentially, decisions binding on all IMF members can be taken either by the required majority of votes cast by the voting body of Governors or the required majority of votes cast by the voting body of Executive Directors. The majorities required by these two bodies depend upon the type of issue to be decided. One type (which for want of a better description we shall denote as issues of procedure) requires a 70

percent majority in both voting bodies. This rule was left unchanged by the new amendments. Another class of issues (which we shall denote as issues of substance), however,
required in any of the two voting bodies an 80 percent majority under the old rules
and requires an 85 percent majority under the new ones. The required majority for
deciding issues of substance was raised from 80 to 85 percent at the insistence of the
United States to retain its veto power: the U.S. made this change a pre-condition for
its agreement to having its voting share lowered from above to below 20 percent of the
total vote.

One apparent curiosity that needs to be mentioned is that, in fact, no decisions
within the Fund are made by means of formal balloting. It is an institution operating
on the basis of consensus reached through informal consultations among members. In
this context, possession by a member, or a group of members, of a given share of the
total vote has to be viewed as an indication of its strength during the process of informal negotiations when a compromise among differing views on a certain issue is being
forged. It is the threat by a country (or a group of countries) of bringing an issue to a
formal vote in which its view would prevail that renders the actual vote unnecessary
and decision making by consensus roughly approximate to decision making by actually
counting ballots.

Previous Power Distribution

Before we investigate the consequences of the recent changes in the voting rules,
distribution of voting weights, or the shares of the members of the IMF, let us first
investigate the power relationship existing prior to April 1, 1978.

This data is illustrated in Diagrams 1 through 4 (upper solid curves) of the appendix.
From these diagrams, several interesting features appear. First, from Diagram 1 (upper
solid curve), we see that when individual countries voted on issues employing the 70
percent decision rule, the relationship between power and voting weights was practically log-linear except for the two largest countries, the United States and the United
Kingdom, for which the relationship flattened out considerably. Put differently, while
all members had a voting power proportional to their voting weights, the United States
and the United Kingdom had voting powers considerably below their shares of the
total vote. For issues involving the 80 percent rule (see upper solid curve in Diagram 2),
the linear relationship failed to hold for the ten countries with largest voting shares:
the United States, the United Kingdom, Germany (Federal Republic), France, Japan,
Canada, Italy, India, Netherlands, Australia. This result is interesting since, in the Fund,
votes are allocated on the basis of a country's contribution to the Fund. Consequently, say in the 70 percent case, while all countries may have claimed to be getting their
"money's worth" in terms of power, this could not be said for the United States and
the United Kingdom. For instance, while the United States contributed almost 16
percent more to the Fund than West Germany, it only received approximately 4 percent more of the power, as measured by the Banzhaf index, for that contribution[3].

---

[3]) *Olson/Zeckhauser* [1966] demonstrate, through a model that describes the services of international organizations such as NATO or the UN, as public goods, that in most organizations of this
type larger countries (notably the U.S.) wind up making a disproportionately large contribution
to the financing of the organization.

Similar results hold for issues which were voted upon by Executive Directors (see upper solid curves in Diagrams 3 and 4), again with the concavity more pronounced for issues involving the 80 percent decision rule than for issues involving the 70 percent rule. In this voting body, however, the United States had relatively more power than it did in the voting body consisting of Governors; e.g., on issues involving the 70 percent decision rule, the U.S. had 13.54 percent of voting power as opposed to 9.07 percent when the same rule was used by the Board of Governors. In other words the United States' power within the Fund increased when the other countries voted in blocks rather than separately. This seems somewhat counter-intuitive, of course, since one would expect that a "large" member would be hurt by the formation of syndicates, each of which would act in unison on particular votes. However, as several recent studies in game theory have shown [see *Aumann; Postlewaite/Rosenthal; Schotter*], syndication need not always be advantageous for the members who syndicate and here is an example of that phenomenon.

To present a picture of past power relationships in the IMF from a different angle, consider the "power Lorenz curves" (lower solid curves in Diagrams 5 to 8), constructed to represent the degree of inequality of the distribution of power in the IMF among Executive Directors and individual Governors for issues involving the 70 percent and 80 percent voting rules.

In all of these diagrams, the cumulative percentage of the voting weights is plotted against the cumulative percentage of the voting power. As we might expect from our previous discussion, since large countries had voting powers that were less than proportional to their voting weights, the resulting Lorenz curves should demonstrate a certain over-representation for smaller countries. This was more pronounced on issues requiring 80 percent of the total vote than on issues calling for a 70 percent majority. As we would expect, the inequality diminished when countries voted through their Executive Directors instead of individually which is easily explained by the increase in power of large countries on issues requiring Executive Directors.

Consequences of Quota Changes

1. Some Surprises

When we analyze the consequences of the recent quota changes, we find some surprises and some expected results. Among the surprises are the paradoxical results described before. Specifically, some of the power relationships resulting from the changes contradict the intentions of their initiators. These paradoxical results are summarized in Tables 1 and 2 where we see (in Table 1) that, as a result of the quota changes, 38 countries have their percentage of power increased within the Fund even though their voting share decreases, while four countries (Table 2) have their percentage of power decreased even though the new vote distribution awards them a larger share of the total vote[4]). This result is not surprising to students of voting power since, as we have seen, the power of a member depends not only on the number of votes

---

[4]) Since we are dealing with a very large voting body, it is not surprising that many of our results hold true at a third decimal place; many countries have virtually no power to start with. However, the qualitative result still holds: increases or decreases in voting percentages do not necessarily imply increases or decreases in corresponding voting powers.

or percentage of the total vote that he has, but also on the way the remaining votes are distributed amongst the other $n - 1$ voters.

2. Power Effects

If we were to give a general assessment of the recent changes, we could state that under the new distribution of voting weights larger countries have unequivocally less power than they had under the old distribution. This fact is demonstrated in Diagrams 1 through 4 below, where we superimpose the voting power/voting weights relationships under the new scheme over the relation we found under the previous scheme.

The overall impression from these diagrams is that the voting changes did not drastically alter the basic relationship between voting weight and voting power. There is a universal tendency for the power of the large countries to be less than proportional to their voting shares and this tendency is more pronounced for issues of substance than it is for issues of procedure.

When we superimpose our power Lorenz curves constructed for the new rules and distribution of voting shares over the Lorenz curves previously presented (see Diagrams 5 through 8), we see that there is practically no change in the degree of power inequality in the Fund except for issues of substance voted upon by Governors. In fact, for the body of Executive Directors, the Lorenz curves for the 70 percent voting rule under the new system coincides with the Lorenz curve generated under this rule for the old system (see Diagram 7). For issues of substance, however, power is further redistributed in favor of the smaller countries.

To summarize our results, the recent changes in the voting rules and distributions of voting shares in the International Monetary Fund have somewhat increased the power of its smaller members. In addition, in many instances, these changes have produced outcomes opposite to the intentions of the drafters of the Amendments.

There is one more question that deserves investigation. It is whether or not the power relationships in the Fund fairly reflect the relative economic importance of its members. This question will be discussed in the following section.

World Trade Shares and Power Within the IMF

One purpose of periodic adjustments in the IMF members' quotas is to bring in line a member's importance in the fund with its importance in world trade. To see if this purpose was achieved, *Dreyer/Schotter* [1978] investigated the relationship between the share of any particular country in total exports of all IMF members with its power within the organization. These relationships are presented as a set of two-power Lorenz curves (Diagrams 9 and 10) for 80 percent decision rules.

For the purpose of comparison, we have superimposed these Lorenz curves over the power Lorenz curves presented before. These diagrams demonstrate that for small countries, trade shares are even worse proxies for power within the Fund than are their voting weights, that is, the distribution of power is even more biased in favor of the smaller countries when the members' trade shares, instead of voting weights, are used as a yardstick. In addition, if we compare power Lorenz curves under previous rules and quotas with the ones that result from the recent change, we see that this inequality is magnified (Diagrams 11 and 12).

4. Conclusions and Implications

Our conclusions are simple. First, voting weights are extremely poor proxies for voting power as one might have expected. However, as was seen by our example of the International Monetary Fund, real world organizations seem not to want to realize this fact. If they did, however, the problem would not be solved, since it would then be mandatory for the planners of these institutions to prescribe the power distribution they desire and the analyst's problem would then be to choose a vector of voting weights whose associated power distribution was the prescribed one. This is not easy for two reasons:

1. First, it would assume that the members could decide on one power index as the one they were going to use to measure power and there is nothing that says that they will be able to agree on one unanimously. If they do not, then with what voting weights are they going to decide on one?
2. Second, even if the index could be agreed to, the mapping from voting weight distributions to voting power distributions is a many-to-one mapping. Consequently for any power distribution, there may exist many vote distributions yielding it. The choice among these distributions will also have to be decided and while each of these distributions yield identical power distributions, some may involve veto powers for members while others may not.

Consequently, it would appear that more work is needed in this area which would apply some results from the social choice literature and allow the members in voting bodies to rationally choose among alternative vote distributions.

References

Aumann, R.: Disadvantageous Monopolies. Journ. of Econ. Th. 6 (1), 1973, 1–12.

Brams, S., and *M. Lake*: Power and Satisfaction in a Representative Democracy. Game Theory and Political Science. Ed. by P. Ordeshook. New York 1978.

Dreyer, J., and *A. Schotter*: Power Relationships in the International Monetary Fund: The Consequences of Quota Changes. Discussion Paper 78–09, Center for Applied Economics, New York University, April 1978. (Forthcoming, Rev. of. Econ. and Stat.)

Fischer, D., and *A. Schotter*: The Paradox of Redistribution in the Allocation of Voting Weights. Public Choice. Summer 1978.

International Monetary Fund: A Report by the Executive Directors to the Board of Governors. Washington, D.C., March 1976.

Lucas, W.: Measuring Power in Weighted Voting Systems. Technical Report # 227, School of Operations Research and Industrial Engineering, Cornell University, September 1974.

Olson, M., and *R. Zeckhauser*: An Economic Theory of Alliances. Review of Econ. and Stat. 48 (3), 1966, 266–79.

Owen, G.: Multilinear Extensions and the Banzhaf Value. Nav. Res. Log. Quart. 22, 1975, 741–50.

Postlewaite, A., and *R. Rosenthal*: Disadvantageous Syndicates. Journ. of Econ. Th. 9, 1974, 324–26.

Raanan, J.: The Inevitability of the Paradox of New Members. Technical Report # 311, School of Operations Research and Industrial Engineering, Cornell University, September 1976.

Schotter, A.: Disadvantageous Syndicates in Public Goods Economies. Discussion Paper No. 75–43, Center for Applied Economics, New York University, October 1975. (Forthcoming Amer. Econ. Rev.)

Appendix

	The Voting Body of Governors			
	Previous Voting System d = .80		Current Voting System d = .85	
Country	% of Vote	% of Power	% of Vote	% of Power
Luxembourg	.14	.20	.13	.22
Papua New Guinea	.14	.20	.13	.25
Jordan	.15	.21	.13	.22
Honduras	.15	.22	.14	.23
Cyprus	.16	.22	.14	.23
Malagasy Republic	.16	.22	.14	.23
Ethiopia	.16	.23	.14	.23
Liberia	.17	.24	.14	.24
Yemen (P.D.R.)	.17	.24	.16	.25
Costa Rica	.18	.25	.16	.25
Cameroon	.19	.27	.17	.27
Guatemala	.19	.27	.18	.30
Panama	.19	.27	.17	.27
Bahamas	.14	.20	.13	.22
Dominican Republic	.21	.30	.19	.31
Kenya	.23	.33	.22	.37
Tunisia	.23	.33	.21	.34
Syria	.23	.33	.21	.34
Jamaica	.24	.35	.23	.39
Burma	.26	.38	.23	.39
Trinidad and Tobago	.27	.39	.25	.41
Uruguay	.29	.42	.26	.43
Sudan	.30	.43	.27	.44
Ghana	.35	.50	.31	.51
Sri Lanka	.38	.55	.34	.56
Iraq	.41	.60	.39	.64
Morocco	.43	.61	.41	.68
Zaire	.43	.61	.42	.69
Ireland	.45	.64	.43	.69
Peru	.46	.66	.45	.73
Bangladesh	.46	.67	.42	.66
Turkey	.54	.78	.53	.86
Egypt	.66	.94	.60	.97
Romania	.66	.95	.64	1.03
Pakistan	.80	1.14	.73	1.17
Norway	.82	1.17	.76	1.20
Denmark	.88	1.25	.79	1.26
Austria	.91	1.29	.84	1.32

Tab. 1: Occurence of the Paradox of Redistribution in the International Monetary Fund [from *Deyer/Schotter*]

A. Schotter

The Voting Body of Governors

Country	Previous Voting System D = .80		Current Voting System D = .85	
	% of Vote	% of Power	% of Vote	% of Power
Belgium	2.08	2.76	2.17	2.62
Netherlands	2.24	2.92	2.30	2.68
Japan	3.78	4.06	3.99	2.99
Germany (Fed. Rep.)	5.01	5.39	5.16	3.01

Tab. 2: Occurence of the Paradox of Redistribution in the International Monetary Fund [from *Dreyer/Schotter*]

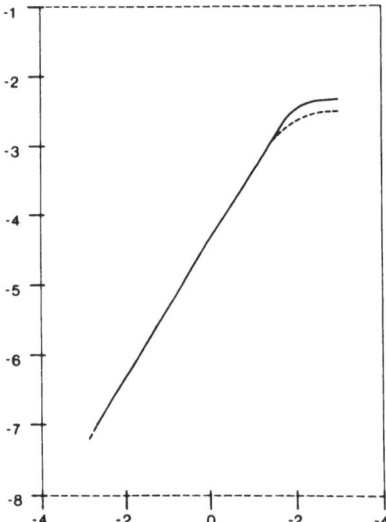

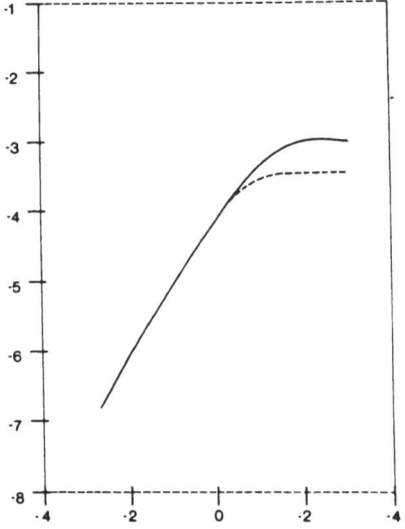

Diagram 1:
The relationship between voting weight and voting power under previous (——) and current (– –) quotas. Logs of weights (horizontal axis) vs. logs of Banzhaf power indices (vertical axis). The body of Governors voting on issues requiring a majority of 70 percent.

Diagram 2:
The relationship between voting weight and voting power under previous (——) and current (– –) quotas. Logs of weights (horizontal axis) vs. logs of Banzhaf power indices (vertical axis). The body of Governors voting on issues requiring a majority of 80 and 85 percent under previous and current rules, respectively.

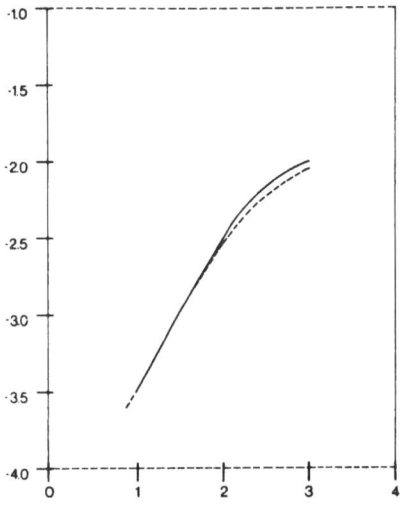

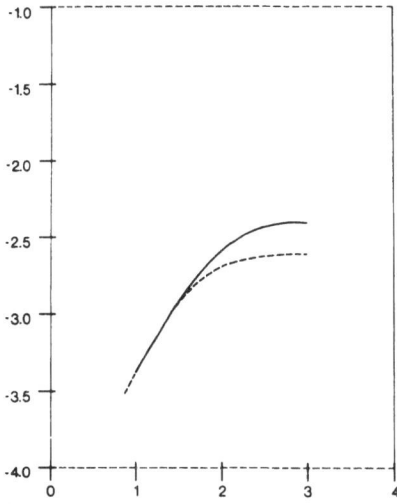

Diagram 3:
The relationship between voting weight and voting power under previous (——) and current (– –) quotas. Logs of weights (horizontal axis). vs. logs of Banzhaf power indices (vertical axis). The body of Executive Directors voting on issues requiring a majority of 70 percent.

Diagram 4:
The relationship between voting weight and voting power under previous (——) and current (– –) quotas. Logs of weights (horizontal axis) vs. logs of Banzhaf power indices (vertical axis). The body of Executive Directors voting on issues requiring a majority of 80 and 85 percent under previous and current rules.

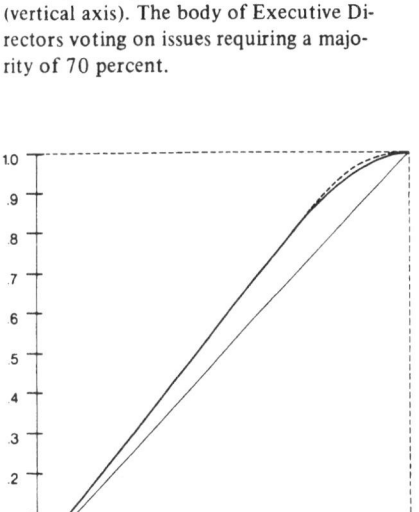

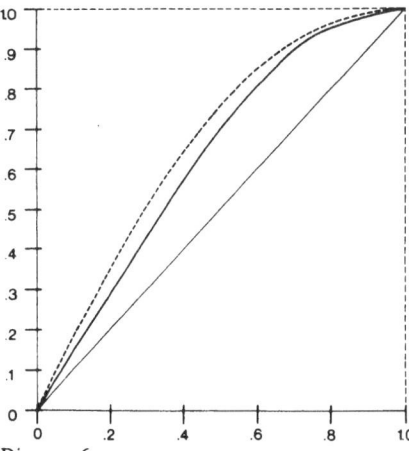

Diagram 5:
Power Lorenz curves. Cumulative voting weights (horizontal axis) vs. cumulative Banzhaf power indices (vertical axis) under previous (——) and current (– –) quotas. The body of Governors voting on issues requiring a majority of 70 percent.

Diagram 6:
Power Lorenz curves. Cumulative voting weights (horizontal axis) vs. cumulative Banzhaf power indices (vertical axis) under previous (——) and current (– –) quotas. The body of Governors voting on issues requiring a majority of 80 and 85 percent under previous and current rules, respectively.

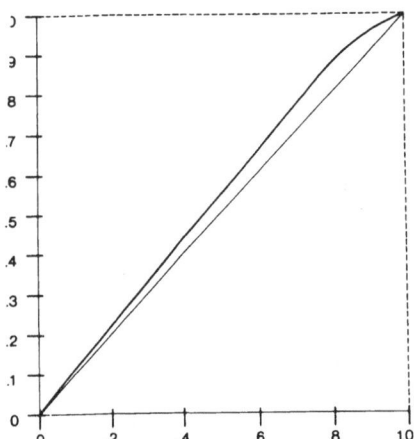

Diagram 7:
Power Lorenz curves. Cumulative voting
weights (horizontal axis) vs. cumulative
Banzhaf power indices (vertical axis) under
previous (——) and current (– –) quotas.
The body of Executive Directors voting on
issues requiring a majority of 80 and 85 percent under previous and current rules, respectively.

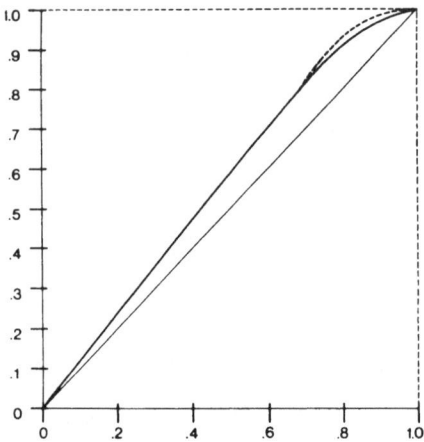

Diagram 8:
Power Lorenz curves. Cumulative voting
weights (horizontal axis) vs. cumulative
Banzhaf power indices (vertical axis) under
previous (——) and current (– –) quotas.
The body of Executive Directors voting on
issues requiring a majority of 80 and 85 percent under previous and current rules, respectively.

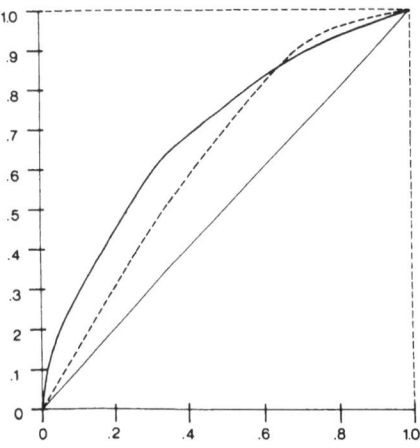

Diagram 9:
Power Lorenz curves. Cumulative export
shares (——) and cumulative voting weights
(– –) vs. cumulative Banzhaf power indices
(vertical axis) under previous quotas. The
body of Governors voting on issues requiring a majority of 80 percent.

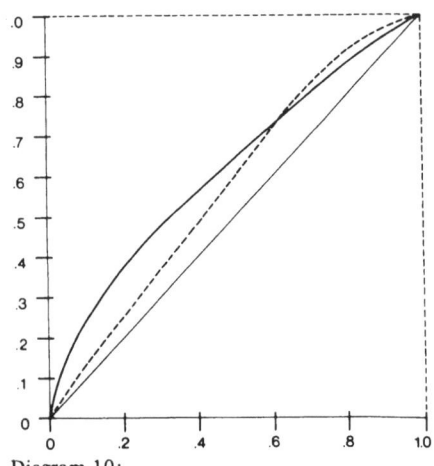

Diagram 10:
Power Lorenz curves: Cumulative export shares
(——) and cumulative voting weights (– –) vs.
cumulative Banzhaf power indices (vertical axis)
under previous quotas. The body of Executive
Directors voting on issues requiring a majority
of 80 percent.

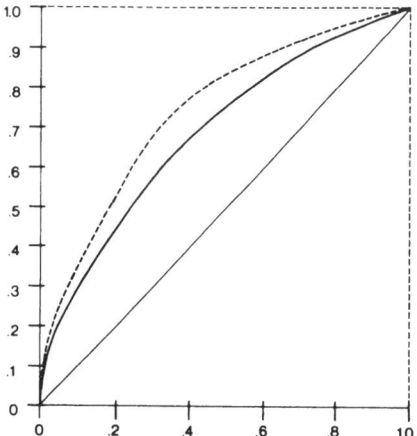 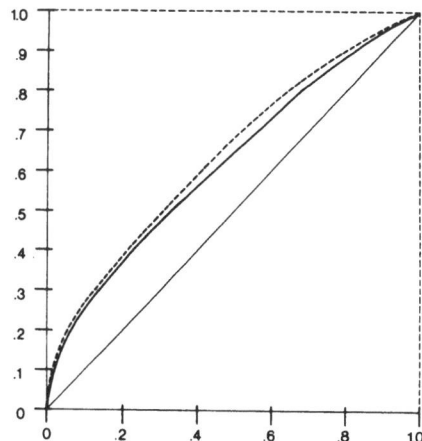

Diagram 11:
Power Lorenz curves. Cumulative export shares (horizontal axis) vs. cumulative Banzhaf power indices (vertical axis) under previous (——) and current (– –) quotas. The body of Governors voting on issues requiring a majority of 80 and 85 percent under previous and current rules, respectively.

Diagram 12:
Power Lorenz curves. Cumulative export shares (horizontal axis) vs. cumulative Banzhaf power indices (vertical axis) under previous (——) and current (– –) quotas. The body of Executive Directors voting on issues requiring a majority of 80 and 85 percent under previous and current rules, respectively.

Applied Game Theory. 1979 ©Physica-Verlag, Wuerzburg/Germany

Standards of Fairness in 4-Person Monopolistic Cooperative Games[1])

By *Am. Rapoport* and *J.P. Kahan*, Los Angeles[2])

Abstract: Six quartets of players participated in a sequence of 16 4-person nonconstant-sum characteristic function games in which one player (the monopolist) was necessarily a member of any coalition that could form. The 16 games were four repetitions each of four game types, arranged in a 2 X 2 design depending on (i) whether or not the grand coalition of all four players was allowed to form and (ii) whether or not the monopolist players were symmetric. The outcomes of these games were compared to predictions derived from the $M_1^{(i)}$ bargaining set [*Maschler*, 1963b], and from a modification of *Maschler's* [1963a] power model.

The results indicated that each quartet established a standard for the range of the monopolist's payoff early in the sequence, and that this standard took on the function of a norm in defining the legitimate bargaining ground for that quartet for the remainder of the sequence. Two quartets' standards were well within the prediction of the bargaining set; one quartet supported the power model. Analysis of the patterns of coalition structure and rudimentary analyses of the bargaining processes indicate that the central concept of standards of fairness underlying the power model is appropriate and timely.

Coalition formation situations presented as games in characteristic function form have attracted the attention of game theorists, resulting in a proliferation of proposed solution concepts. Although no single volume contains a summary of all of these solution concepts, major summaries may be found in, e.g. *Davis* [1970], *Luce/Raiffa* [1957], *Owen* [1968], and *Rapoport* [1970]. Social scientists have received these solution concepts with varying degrees of interest. On the one hand, some social scientists have been content to relegate these theories to the arcane realm of mathematics and to dismiss them as irrelevant to their own experimental research. In contrast to this group is a growing number of scientists who argue that it is the interpretability of an abstract mathematical model or the applicability of the interpreted theory rather than the formal structure that is important, and who have produced a sizeable corpus of coalition formation experiments that attempt to test the predicitive power of the mathematical solution concepts.

[1]) Research reported in this paper was supported by National Science Foundation Grant BNS76–84285. This paper was written while both authors were Fellows-in-Residence at the Netherlands Institute for Advanced Study in the Humanities and Social Sciences. We wish to thank S.G. Funk and A.D. Horowitz for their assistance in data collection and analysis. We wish particularly to thank Michael Maschler, with whom we have had a long and lively interaction on the power model, for his many helpful comments and suggestions for improving the paper.

[2]) Dr. *James P. Kahan*, Psychological Research and Service Center, University of Southern California, 734 West Adams Boulevard, Los Angeles, California 90007, USA.

Prof. *Amnon Rapoport*, Univ. of North Carolina at Chapel Hill.

The bargaining set [*Aumann/Maschler*] and its variants have enjoyed the focus of attention of social scientists, because they are founded on assumptions that are psychologically meaningful [*Kahan/Rapoport*, 1974], and because they are applicable to both superadditive and nonsuperadditive characteristic function games. An impressive array of studies can be summoned to examine coalition formation and bargaining from the bargaining set orientation [*Albers; Funk; Horowitz; Horowitz/Rapoport; Kahan/ Rapoport*, 1974, 1977, 1979; *Levinsohn/Rapoport; Maschler*, 1978; *Medlin; Michener/ Sakurai/Yuen/Kasen; Michener/Yuen/Ginsberg; Murnighan/Roth; Rapoport/Kahan*, 1976; *Rapoport/Kahan/Wallsten; Riker*, 1967, 1972; *Riker/Zavoina; Selten/Schuster*]. The consensus judgment of these studies is mixed; whereas most of them have supported the bargaining set predictions of payoff disbursements as a good first order approximation, others have shown systematic discrepancies as a function of the structure of relationships among the players and as a function of the communication conditions of the particular experiment.

Maschler[3]) [1978] introduced a 3-person game with a very simple structure that provides perhaps the strongest evidence against the uniform validity of the bargaining set. In this game, there is one monopolistic player who may join in coalition with either of two players; they, in turn, must either join with the monopolist or remain out of any coalition. *Maschler's* data caused him to question whether the value of a coalition is an adequate measure of its power, and led him to propose a new measure of power for coalitions [*Maschler*, 1963a] based not only on the values to the particular coalitions, but also on the ability of the members of a coalition to block other players from realizing gains. *Maschler* [1963a] termed this new solution concept the *power model.*

In the present paper, we shall compare the $M_1^{(i)}$ bargaining set with a revision of the power model that ameliorates some of the problems inherent in it. We shall then present an experiment that serves as a prototype of certain monopolistic conflict situations that are of interest to both economists and psychologists. This experiment will provide data for a comparative test of the two models that overcomes some problems inherent in previous comparisons [*Maschler*, 1978; *Michener* et al., 1977, 1978]. Editorial limitations on space force us to presume that the reader is generally familiar with game theoretical terminology and the notion of bargaining sets; the references cited in the first paragraph can serve to remedy such a lack of familiarity.

The following games in characteristic function form will serve as useful examples in the argument presenting the power model.

G1: $v(12) = 95, v(13) = 90, v(23) = 65, v(1) = v(2) = v(3) = v(123) = 0.$

G2: $v(12) = v(13) = 50, v(23) = v(1) = v(2) = v(3) = v(123) = 0.$

G3: $v(14) = v(24) = v(34) = 62, v(124) = v(134) = v(234) = 80,$
   $v(1234) = 95,$

   $v(1) = v(2) = v(3) = v(4) = v(12) = v(13) = v(23) = v(123) = 0.$

---

[3]) *Maschler's* paper, although published in 1978, is the first appearance in print of an experiment performed in the early 1960s [*Maschler*, 1965].

Maschler's Critical Experiment and the Bargaining Set

The bargaining set $M_1^{(i)}$ [*Maschler*, 1963b; *Peleg*] achieves a stability in the allocation of payoffs such that for any two members of a given coalition, every threat by one member against another can be met by a counterthreat by the latter against the former. For example game G1, the bargaining set solution is given by,

$$M_1^{(i)} = \begin{cases} (60, 35, \ 0; \ 12, 3) \\ (60, \ 0, 30; \ 13, 2) \\ (\ 0, 35, 30; \ 1, 23) \\ (\ 0, \ 0, \ 0; \ 123) \\ (\ 0, \ 0, \ 0; \ 1, 2, 3). \end{cases}$$

This solution is arguably and empirically [*Kahan/Rapoport*, 1974] reasonable. But consider the solution to G2, given by,

$$M_1^{(i)} = \begin{cases} (50, 0, 0; \ 12, 3) \\ (50, 0, 0; \ 13, 2) \\ (\ 0, 0, 0; \ 1, 23) \\ (\ 0, 0, 0; \ 123) \\ (\ 0, 0, 0; \ 1, 2, 3). \end{cases}$$

Here, players 2 and 3 are unlikely to be satisfied with zero points, and $M_1^{(i)}$ is neither arguably nor empirically [*Maschler*, 1978] reasonable.

Maschler [1978] employed game G2 in 13 instances involving different subsets of 38 Isreali high school students as subjects. Players bargained face-to-face, with minimal control over their interaction. Twice, no agreement was formed, and in the remaining 11 plays, player 1's mean payoff was 37.82, as opposed to the $M_1^{(i)}$ prescription of 50. *Maschler's* informal observations are quite enlightening and merit direct quotation:

> "At first, the games indeed ended in such a way that player 1 received almost 50, giving another player a very small amount. . . . However, as more games were played, some of the weak players realized that it is worthwhile to flip a coin under the condition that the loser would 'go out of the game,' thus 'forcing' a split 25:25 between the winner and player 1. Eventually, player 1 realized that he ought to offer player 2 or player 3 some amount around 12 1/2, in order that the 'coalition' 23 will not form, because 12 1/2 was the expectation for each of the weak players.. It appears from the accounts of the players that many were guided by some 'justice' feelings, that the 'right thing to do' in a 2-person game $v(1) = v(2) = 0, v(12) = 50$, is to split equally, and on the basis of this knowledge they acted as if the value of the coalition {23} was 25 and not 0 [*Maschler*, 1963a, p. 9]."

It is easy to verify that if $v(23) = 25, v(1) = 25$, and the other values remain as stated in G2, $M_1^{(i)}$ prescribes 37 1/2 points for player 1 in either of his 2-person coalitions.

The Power Model

Maschler [1963a] argued that the "power" of a coalition resides not only in the gains that accrue jointly to the members of the coalition but also in the losses it may inflict on other players by not cooperating with them. He attempted to capture some of the various notions of the power of a coalition by proposing a power function that would serve as a more adequate representation of the game than would the original characteristic function. A solution could be derived by translating the characteristic function into the power function, solving the derived game thus determined, and then inverting the solution back to the original characteristic function. The determination of the power function was embodied in *Maschler*'s term "standards of fairness," which, like the "standards of behavior" of *von Neumann/Morgenstern* [1947], refers to moral or conventional rules imposed by society or to social norms by which individual players determine or evaluate a particular disbursement of payoffs.

The formidable task of empirically determining standards of fairness and the conditions that affect them has been graciously relegated to the social sciences by both *Maschler* and *von Neumann/Morgenstern*. Previous research has employed two examples of such standards, a "cooperative standard of fairness" based on the principle of equal sharing among negotiating bodies [*Maschler*, 1978; *Michener* et al., 1978], and a "Shapley value standard of fairness" in which the Shapley values of the negotiating bodies the principle of sharing [*Michener* et al., 1977]. In the modification of *Maschler*'s [1963a] model presented below, we adopt the cooperative standard for the basis of our calculations, although the Shapley value standard yields identical results.

Generally, a standard of fairness for a game $(N; v)$ may be expressed as a vector valued function

$$\phi\left([P]\right) = \{\phi_1\left([P]\right), \phi_2\left([P]\right), \ldots, \phi_t\left([P]\right)\},$$

defined for each partition $P = (P_1, P_2, \ldots, P_t)$ of N into t negotiation groups ($t \leqslant n$). P_j denotes a negotiation group in the partition $[P]$, and $\phi_j\left([P]\right)$ designates the payoff to negotiation group P_j in the partition $[P]$ based on a given standard of fairness. The function $\phi\left([P]\right)$ is assumed to satisfy two conditions:

$$\phi_j\left([P]\right) \geqslant v\left(P_j\right), \quad j = 1, \ldots, t,$$

$$\phi_1\left([P]\right) + \phi_2\left([P]\right) + \ldots + \phi_t\left([P]\right) = v\left(N\right).$$

The first condition simply states that a negotiation group will refuse to obtain less than its value if the negotiation group is a coalition in G. The second condition claims that the maximum joint reward $v\left(N\right)$ is the amount to be shared by all the t negotiation groups that form. Finally, the pair $(N; \phi\left([P]\right))$, where N is the set of players as before and $\phi\left([P]\right)$ is a standard of fairness, is called the *game space*.

The implicit requirement that the grand coalition's value $v\left(N\right)$ be the largest in the game deserves comment. It is often the case in experimental games that the grand coalition technically cannot form, in that the experimenter does not permit an explicit agreement among all of the players on how to divide their joint winnings. In such cases, for purposes of calculation of the game space, $v\left(N\right)$ is taken as the maximum of the various other coalitions. We note here that even when the grand coalition is explicitly prohibited by the experimenter's rules of the game, researchers [e.g. *Maschler*, 1978; *Riker*, 1972] have recorded striking examples of how subjects transform the situation in which they find themselves into a game permitting the grand coalition.

Although *Maschler* [1963a] permitted all possible partitions of players to be considered in constructing the game space, we have simplified the procedure by specifying that, for any negotiation group S that forms, its complement will form to negotiate against it. This modification greatly reduces the number of partitions that may be considered, and makes the power model more reasonable from the point of view of cognitive constraints about information processing on the part of human players. The modification also alleviates a conceptual problem arising in the original formulation as to the nature of side payments both within and across coalitional boundaries. Table 1 presents the game space and the cooperative standard of fairness for game G3, which is employed in the present experiment. Instead of the 15 partitions that a full partition function would require, we need only consider eight possible breakdowns into coalition structures.

$v(14) = v(24) = v(34) = 62, \ v(124) = v(134) = v(234) = 80, \ v(1234) = 95$

$v(1) = v(2) = v(3) = v(4) = v(12) = v(13) = v(23) = v(123) = 0$

Game space	Cooperative standard of fairness			
$\{12, 34\}$	$\phi_{12} = 16.5,$		$\phi_{34} = 78.5$	
$\{13, 24\}$	$\phi_{13} = 16.5,$		$\phi_{24} = 78.5$	
$\{14, 23\}$	$\phi_{14} \approx 78.5,$		$\phi_{23} = 16.5$	
$\{1, 234\}$	$\phi_1 = 7.5,$		$\phi_{234} =$	87.5
$\{2, 134\}$	$\phi_2 = 7.5,$		$\phi_{134} =$	87.5
$\{3, 124\}$	$\phi_3 = 7.5,$		$\phi_{124} =$	87.5
$\{4, 123\}$	$\phi_4 = 47.5$		$\phi_{123} =$	47.5
$\{1234\}$		$\phi_{1234} = 95$		

$u(1) = u(2) = u(3) = 7.5, \ u(4) = 47.5, \ u(12) = u(13) = u(23) = 16.5,$

$u(14) = u(24) = u(34) = 78.5, \ u(123) = 47.5, \ u(124) = u(134) = u(234) = 87.5,$

$u(1234) = 95$

Tab. 1: Characteristic function, game space, and power function for game G3

Note that the cooperative standard of fairness assumes that the grand coalition will form, and that the two negotiation groups split the surplus over their individual coalition values equally.

The results of the cooperative standard of fairness may be read directly to produce a *power function* for game G3, which we denote by u, and show in the bottom of Table 1. Directly following *Maschler* [1963a], we would now find the bargaining set for the derived game expressed by $(N; u)$. But *Maschler*'s model implicitly assumes that the grand coalition will always form, while experimental studies of coalition formation have clearly demonstrated that subjects do not always maximize joint reward; systematic differences in the propensity to maximize joint payoff as a function of the characteristic function v have been shown in 3-person games both with [*Medlin*], and without [*Levinsohn/Rapoport*] grand coalitions. Even in *Maschler's* [1978] experiment, where the communication conditions strongly favored joint reward maximization, 19 % of the games that could terminate with a single $(n-1)$-person coalition did not maximize joint reward.

There are various tactics available to accommodate this difficulty. For example, the difference between the sum of the payoffs to the members of a coalition S prescribed by the solution, given by $u(S)$, and its actual value, $v(S)$, could be considered as a loss shared equally among the members of the coalition. However, *Maschler*[4]) has proposed a more elegant resolution to this difficulty, developing a modification of the power model that maintains its central idea that the game is played in terms of the power function while acknowledging that the players within any coalition structure are rewarded according to the original characteristic function of the game. Like $M_1^{(i)}$, this modification retains solutions for coalition structures and no longer requires group rationality. We present this model immediately below.

Assume that the outcome of a game is an individually rational payoff configuration (PC) that satisfies the two requirements of $M_1^{(i)}$,

$$x_i \geqslant v(i), \quad \text{for all } i \in N$$

$$\sum_{i \in B_j} x_i = v(B_j) \quad \text{for every } B_j \in \underset{\sim}{B}, \quad j = 1, 2, \ldots, m.$$

The *power bargaining set* M^P is then defined as the set of all PCs satisfying these requirements, such that for all $B_j \in \underset{\sim}{B}$, and all distinct players $k, l \in B_j$, for every objection of k against l in $(N; u)$, there is a counterobjection of l against k in $(N; u)$. Objections and counterobjections are defined as in $M_1^{(i)}$, in *Aumann/Maschler* [1964]. That is, although the original PC is in $(N; v)$, the objections and counterobjections are in $(N; u)$.

The power bargaining set M^P is based on the idea that players negotiate as if the value of each coalition S is $u(S)$. However, if a coalition structure B is formed, there is a *reality barrier* in the form of $v(B_j)$, which constrains the members of the coalition B_j from jointly gaining more than this value while in this coalition structure. Thus, M^P may be interpreted as a procedure for reaching stability in dividing the real values $v(B_j)$ for each $B_j \in \underset{\sim}{B}$ by basing objections and counterobjections, but not payoffs, on the players' perceived game $(N; u)$.

Assuming that $u(123) = \max \{v(12), v(13), v(23)\}$, the reduced power function for game G1 is given by $u(1) = 15$, $u(2) = 2\,1/2$, $u(3) = 0$, $u(12) = 95$, $u(13) = 92\,1/2$, $u(23) = 80$, and $u(123) = 95$.

---

[4]) *Maschler*, in a careful, lengthy review of an earlier draft of this paper, provided the M^P model as his resolution of the difficulty presented above. To our knowledge, he has not yet formally published this variation on the power bargaining set.

The power bargaining set for this game is:

$$M^P = \left\{ \begin{array}{c} 53\frac{3}{4}, \ 41\frac{1}{4}, \ 0 \ ; \ 12, 3 \\[4pt] 52\frac{1}{2}, \ 0 \ \ 37\frac{1}{2}; \ 13, 2 \\[4pt] 0 \ , \ 33\frac{3}{4}, \ 31\frac{1}{4}; \ 23, 1 \\[4pt] 0 \ , \ 0 \ , \ 0 \ , \ 1, 2, 3 \\[4pt] 0 \ , \ 0 \ , \ 0 \ ; \ 123. \end{array} \right\}$$

The corresponding operations for game G2 yield the reduced power function:
$u\,(1) = 25, u\,(2) = u\,(3) = 0, u\,(12) = u\,(13) = 50, u\,(23) = 25,$ and $u\,(123) = 50$.
This produces the power bargaining set

$$M^P = \left\{ \begin{array}{c} 37\frac{1}{2}, \ 12\frac{1}{2}, \ 0 \ ; \ 12,3 \\[4pt] 37\frac{1}{2}, \ 0 \ \ \ 12\frac{1}{2}; \ 13, 2 \\[4pt] 0 \ , \ \ 0 \ , \ 0 \ , \ 23, 1 \\[4pt] 0 \ , \ \ 0 \ , \ 0 \ ; \ 1, 2, 3 \\[4pt] 0 \ , \ \ 0 \ , \ 0 \ ; \ 123. \end{array} \right\}$$

It is interesting to note that M^P agrees better with *Maschler's* [1978] informal analysis of G2 which motivated the power model than his original proposal [*Maschler, 1963a*], and even is in better agreement with his data.

Method

Subjects

Subjects were 24 male volunteers, mostly undergraduate students at the University of North Carolina at Chapel Hill, who were recruited via posters placed around campus advertising cash to be earned by participation in a multi-session coalition formation and bargaining experiment. Subjects were sorted randomly (within the constraints of time of volunteering) into six quartets; close friends were not allowed to participate in the same quartet. Volunteers with previous experience in computer-controlled coalition formation experiments were not allowed to participate, although those who had been in other computer-controlled experiments for monetary gain were accepted.

Procedure

Subjects played 16 successive characteristic function games in a computer-controlled experimental procedure called *Coalitions*. As *Coalitions* has been repeatedly described in a variety of publications [*Kahan/Helwig; Kahan/Rapoport,* 1974; *Rapoport/Kahan,*

1974], we shall not repeat it here. Very briefly, the game consists of a threestage process: a negotiation stage, in which the potentials of various coalitions may be explored, an acceptance stage, in which a particular PC is seriously considered, and a ratification stage, in which the agreement on a division of coalitions' values whithin a given coalitional structure becomes binding. Messages are typed and the game protocols remain available throughout the game. Communication among the players takes place through the use of the following keywords: *offer; accept, reject, pass, solo,* and *ratify.* all communications are transmitted to all players except *offers,* which may instead be sent secretly to selected players involved in a proposed coalition, at the option of their authors. Ratification of a PC cannot occur until two rounds of communication, which amount to eight messages, have taken place in its acceptance stage.

At the beginning of the experiment, subjects in each quartet were gathered together for a three hour practice session of written and verbal instructions of how to play the *Coalitions* game. Following the instructions and a question and answer period, subjects participated in three or four characteristic function practice games to acquaint them with the teletypewriters, the nature of the messages and keywords, and various tactics of bargaining. For this experimenter-supervised training session, subjects were paid $4.50 each.

Several days later, quartets returned individually for three experimental sessions, each lasting approximately three hours, spaced from two to seven days apart. Within this period, each quartet played each of four different games four times, for a total of 16 games per quartet. Although players participated as a unified quartet for all 16 games, the actual identity of the other players within any game was not known, as role identification letters (A, B, C, D) were shifted between successive games within and between sessions.

Experimental games

Four different 4-person games were played in a 2×2 design. The characteristic functions of these four games are presented in Table 2. All four games share an important property, namely, that any coalition of two or more players must include player D. Player D assumes, therefore, the role of a monopolist; he may veto any PC offered

Game	v (DA)	v (DB)	v (DC)	v (DAB)	v (DAC)	v (DBC)	v (DABC)
I	62	62	62	80	80	80	0
II	62	62	62	80	80	80	95
III	54	62	70	69	77	85	0
IV	54	62	70	69	77	85	90

Tab. 2: Characteristic functions for four games*)

*) v (A) = v (B) = v (C) = v (D) = v (AB) = v (AC) = v (BC) = v (ABC) = 0   for each game.

by A, B, or C except the trivial case of all players in their 1-person coalitions. The two factors of the design are symmetry to the weak players' (A, B, C) payoffs and super-

additivity vs. non superadditivity. Thus, games I and II are symmetric and III and IV
are not; games II and IV are superadditive, whereas I and III are not.

The 16 games were played in the following order by each quartet: II, I, IV, III, I,
II, III, IV, III, IV, I, II, IV, III, II, I. Five to seven games were played within a session,
depending on the time spent per game. In all cases, the monopolist communicated in
each round after all three weak players (i.e., $A, B, C, D, A, \ldots$).

In addition to the $4.50 earned at the practice session, subjects were paid according
to the number of points they accumulated, at the rate of 5 cents per point. No other
payment was provided. In accordance with agreements made with the subjects at the
beginning of the experiment, no earnings were paid until the entire experiment terminated. The mean and standard deviation of payoff per subject for the experimental
games were $ 16.87 and $1.41, respectively.

Results

Presentation of the results of the experiment will first be oriented directly at the
different predictions of the $M_1^{(i)}$ and M^P models, and then shall turn to such features
as coalition frequencies and bargaining processes in an attempt to cast light on the
paths by which various outcomes might be obtained.

A comparison of models

Table 3 presents the $M_1^{(i)}$ prescriptions for the experimental games. The prescriptions for 2- and 3-person coalition outcomes are not affected by the presence of non-
zero 4-person coalition values, so games types I and II, and III and IV are collapsed for
economy of space. Note that the addition of the grand coalition creates a wide span in
the $M_1^{(i)}$ prescribed range for player D's payoffs if this coalition is the outcome; the
payoff range corresponds exactly with the core, and its low boundary (from player
D's point of view) coincides with the kernel [*Davis/Maschler*] solution.

Table 4 presents the M^P solutions for the reduced power function for all four game
types separately. A comparison of Tables 3 and 4 shows that the range of outcomes in
the solution M^P is small for each coalition structure, while the corresponding ranges
for $M_1^{(i)}$ are either point predictions or rather large spans. The reduced power model
generally produces, as expected, a more egalitarian prediction than the bargaining set
model.

In moving from predicted to actual outcomes, some difficulties present themselves.
First, for the $M_1^{(i)}$ solution, some players are assigned zero points while in coalition,
an outcome that appears realistically infeasible. The point of the bargaining set argument is that players can be forced to the limit of zero; in any play of the game, the
"costs" of entering into the coalition would certainly be taken into consideration.
Second, the predictions of the models are typically small ranges in the real numbers,
whereas the players actually bargained in integer values, and showed a pronounced
tendency to think in units of five points. Third, if a player withdrew from the game
via a *solo* move, the prescriptions of the models for the remaining players were changed
from those presented in Tables 3 and 4.

Game Types I and II

0,	0,	62;		AD, B, C
0,	0,	62;		BD, A, C
0,	0,	62;		CD, A, B
0,	0,	80;		ABD, C
0,	0,	80;		ACD, B
0,	0,	80;		BCD, A
$0 \leqslant x \leqslant 15$,	$0 \leqslant y \leqslant 15$,	$0 \leqslant z \leqslant 15$,	$50 \leqslant w \leqslant 95$;	ABCD[1])
0,	0,	0;		A, B, C, D

Game Types III and IV

0,	0,	0,	54;	AD, B, C
0,	0,	0,	62;	BD, A, C
0,	0,	$70 - w$,	$69 \leqslant w \leqslant 70$;	CD, A, B
0,	0,	0,	69;	ABD, C
0,	0,	$77 - w$,	$69 \leqslant w \leqslant 77$;	ACD, B
	$0 \leqslant y \leqslant 8$,	$0 \leqslant z \leqslant 16$,	$61 \leqslant w \leqslant 85$;	BCD, A[2])
$0 \leqslant x \leqslant 5$,	$0 \leqslant y \leqslant 13$,	$0 \leqslant z \leqslant 21$,	$51 \leqslant w \leqslant 90$;	ABCD[3])
0,	0,	0,	0;	A, B, C, D

Tab. 3: $M_1^{(i)}$ payoff configurations for all four game types

[1]) For game type II only; we require that $x + y + z + w = 95$.
[2]) Such that $y + z + w = 85$.
[3]) For game type IV only; we require that $x + y + z + w = 90$.

Game Type I

				Configuration
$7\frac{1}{3} < x < 11$,	0 ,		62−x;	AD, B, C
0 ,	$7\frac{1}{3} < x < 11$,		62−x;	BD, A, C
0 ,	0 ,	$7\frac{1}{3} < x < 11$,	62−x;	CD, A, B
$11.6 < x < 12\frac{1}{4}$,	x ,	0 ,	80−2x;	ABD, C
$11.6 < x < 12\frac{1}{4}$,	0 ,	x ,	80−2x;	ACD, B
0 ,	$11.6 < x < 12\frac{1}{4}$,	x ,	80−2x;	BCD, A
0 ,	0 ,	0 ,	0;	A, B, C, D

Game Type II

				Configuration
$4\frac{5}{6} < x < 11$,	0 ,		62−x;	AD, B, C
0 ,	$4\frac{5}{6} < x < 11$,		62−x;	BD, A, C
0 ,	0 ,	$4\frac{5}{6} < x < 11$,	62−x;	CD, A, B
$10.1 \leqslant x < 12\frac{1}{4}$,	x ,	0 ,	80−2x;	ABD, C
$10.1 < x < 12\frac{1}{4}$,	0 ,	x ,	80−2x;	ACD, B
0 ,	$10.1 < x < 12\frac{1}{4}$,	x ,	80−2x;	BCD, A
$11.5 < x < 12.8$,	x ,		95−3x;	ABCD
0 ,	0 ,	0 ,	0 ;	A, B, C, D

Game Type III

				Configuration
$0 < x < 5\frac{3}{4}$,	0 ,	0 ,	54−x;	AD, B, C
0 ,	$6\frac{1}{2} < x < 11\frac{3}{4}$,	0 ,	62−x;	BD, A, C
0 ,	0 ,	$13\frac{1}{6} < x < 17\frac{3}{4}$,	70−x;	CD, A, B
$4.3 < x < 6.5$,	x + 4 ,	0 ,	65−2x;	ABD, C
$5.9 < x < 7.5$,	0 ,	x+8 ,	69−2x;	ACD, B
0 ,	$11.5 < x < 12.5$,	x+4 ,	81−2x;	BCD, A
0 ,	0 ,	0 ,	0 ;	A, B, C, D

Game Type IV

				Configuration
$0 < \ddot{x} < 5\frac{3}{4}$,	0 ,	0 ,	54−x;	AD, B, C
0 ,	$5\frac{2}{3} < x < 11\frac{3}{4}$,	0 ,	62−x;	BD, A, C
0 ,	0 ,	$12\frac{1}{3} < x < 17\frac{3}{4}$,	70−x;	CD, A, B
$3.8 < x < 6.5$,	x+4 ,	0 ,	65−2x;	ABD, C
$5.4 < x < 7.5$,	0 ,	x+8 ,	69−2x;	ACD, B
0 ,	$12.5 < x < 15$,	x+4 ,	81−2x;	BCD, A
$6\frac{3}{7} < x < 7.8$,	x+4 ,	x+8 ,	78−3x;	ABCD
0 ,	0 ,	0 ,	0 ;	A, B, C, D

Tab. 4: M^P payoff configurations for the derived games $(N; u)$

Each of these difficulties were accommodated in considering the accuracy of the $M_1^{(i)}$ and M^P models. First, if a player's prescribed outcome for a given game was zero, any outcome to him of five or fewer points was regarded as supporting the model. In this instance, even if such a five point payoff reduced another player's outcome below prescription, the model was counted as supported. Second, the prescribed ranges for all players were extended both downwards and upwards to the nearest integer evenly divisible by five, and any outcome giving *all* of the players in the coalition outcomes within their extended ranges was counted as supporting the model. It is important to note with respect to this adjustment that we are not extending the ranges in the service of a theoretical adjustment such as *Riker's* [1967] point-shaving, but rather as an accommodation to a consistent and easily understood cognitive simplification in arithmetic that virtually all experimentally played games have shown. This is why all players must be within their prescribed ranges, instead of all but one of them. Third, if a player chose *solo* and there was subsequent bargaining among the remaining players (as occurred 16 times out of 96 games), the revised prescriptions of the models were calculated, and judgments of support or nonsupport were made on these additional calculations; space precludes their presentation here. With these three adjustments, then, the outcomes of all 96 games could be compared with the prescriptions of the $M_1^{(i)}$ and M^P models.

Table 5 presents the outcomes of the 96 plays of the four game types, classified by iteration and by quartet. The 16 games in which a player departed the game before a coalition was affirmed are indicated by footnotes. The most striking feature of Table 5 is the non-independence of the games; the within-quartet variance for the outcome to the monopolist player D is quite small, whereas the between-quartet variance for that value is quite large. Thus, the consistent finding of previous work with *Coalitions* [*Kahan/Rapoport*, 1974, 1977; *Rapoport/Kahan*, 1976; *Rapoport* et al., 1978] that players encounter each game as a separate task independent of its predecessors is clearly contradicted in the present case. Moreover, the actual game type appears to have little effect on the payoff to the monopolist. This is not to say that game type was completely ignored. In the two superadditive games II and IV, the 4-person outcome was frequently chosen. More importantly, in the two symmetric games I and II, out of 46 outcomes with payoffs to two or more nonmonopolist players, 37 were equal divisions. For the two nonsymmstric games III and IV, the corresponding figures are 11 equal divisions in 39 games. In all 28 cases of unequal payoffs, the player with the stronger position *vis à vis* the characteristic function received the higher payoff.

The high incidence of identical solutions for a game, particularly in quartets 3 and 4, indicates that each quartet separately arrived at a standard range for an allocation to the monopolist, and then played the game within that standard. Results presented later will attempt to investigate the nature of those standard. But first, a comparison of $M_1^{(i)}$ and M^P is in order. Table 6 presents a breakdown of the outcomes for each game and quartet separately according to whether either of both of the models were supported by the outcomes. Overall, 58 % of the outcomes supported $M_1^{(i)}$, while 38 % supported M^P. More than twice as many outcomes supported $M_1^{(i)}$ but not M^P than vice-versa, but because of the nonindependence of games, such results cannot be accepted on face value. A look at the results by game type in Table 6 reveals that for type I and II,

Game	Iteration	Quartet 1 A	B	C	D	Quartet 2 A	B	C	D	Quartet 3 A	B	C	D	Quartet 4 A	B	C	D	Quartet 5 A	B	C	D	Quartet 6 A	B	C	D
I	1		25		37	5		11	64	6		9	65	15	15	15	50[3]	12		12	56[2]		8	8	64
	2	18	18		44	8	8	8	64	10		10	60		17	15	48		4	4	72	1	1		78
	3	11	12		57	8		8	64	7		5	68	18	18		44[3]	5	5	5	70[3]		4	4	72[1]
	4	15		10	55	5		5	70	6	7		67	20		20	40[2]		5	5	70		1	1	78
II	1	22	22	22	29	9		9	62	10	10	10	65	17	17	17	44	16	16	16	47	10	10	10	60
	2			15	47		11	11	58	10	10	10	65	15	15	15	50	7	7	7	74	5	5	5	80
	3	15	15	15	50	11	11	13	60	10	10	10	65	15	15	15	50	6	6	6	77	6	6	6	77
	4	15	15	15	50	10	10	10	65	10	10	10	65	15	15	15	50	8	8	8	71		6	4	80
III	1			25	45		10	10	65		5	15	65		15	25	45[1]	8	8	8	69[1]		9	9	67[1]
	2			20	50		15	15	55		5	15	65			25	45[2]	4	4	8	73		4	4	77
	3			15	55		12	14	59		5	15	65			25	40[1]	4	4	8	73		5	5	75
	4		15	18	52		8	12	65		10	10	65	4			50	27			27[4]	4	4	7	74[1]
IV	1			30	40	5	15	15	55			11	59		25	25	35[1]	7	7	7	69		11	11	63[1]
	2	7		18	52	8	10	10	62	5	10	10	65	10	12	18	50	4	4	6	76	5	5	5	75
	3			20	50	8	10	12	60	5	10	10	65	10	12	18	50	4	5	7	74		5	6	74[1]
	4	10	15	15	50	5	7	13	65	5	10	10	65	10	14	18	48	3	5	8	74	2	5	5	78

Tab. 5: Outcomes by game type, iteration, and quartet

[1] Following a solo by player A
[2] Following a solo by player B
[3] Following a solo by player C
[4] Following a solo by player B and C

without grand coalitions, the power model suffered with respect to the bargaining set model, but for the superadditive games type II and IV, the two models fare about equally well.

	Neither model supported	Both models supported	$M_1^{(i)}$ alone supported	M^P alone supported	$M_1^{(i)}$ supported	M^P supported
All games	22	18	38	18	56	36
Game type I	10	1	10	3	11	4
Game type II	4	11	6	3	17	14
Game type III	4	2	14	4	16	6
Game type IV	4	4	8	8	12	12
Quartet 1	6	2	0	8	2	10
Quartet 2	5	3	4	4	7	7
Quartet 3	3	7	4	2	11	9
Quartet 4	5	5	3	3	8	8
Quartet 5	2	1	13	0	14	1
Quartet 6	1	0	14	1	14	1

Tab. 6: A comparison of the $M_1^{(i)}$ and M^P predictions, by game type and by quartet

Two of the six quartets (numbers 5 and 6) showed substantial support for $M_1^{(i)}$. In these two quartets, the monopolist was able to extract his full value (according to the bargaining set) on nearly every trial. Additionally, quartet 3 supported $M_1^{(i)}$, but the range of payoff was such that it also supported M^P; in seven of the 16 games, the outcome was in the supporting range of both models. Quartet 4 also supported each model equally often, although less frequently than quartet 3. Quartet 2 similarly supported both models equally, but an examination of Table 5 for this quartet shows that $M_1^{(i)}$ would have been more supported except that when the grand coalition formed, the weakest player received about 8 points rather than the model's limit of 5. Finally, quartet 1 provides the strongest support for M^P to the exclusion fo $M_1^{(i)}$; the outcomes to the monopolist were the lowest in this quartet.

These results mildly but certainly not conclusively favor $M_1^{(i)}$ over M^P. Most of the outcomes for the superadditive games type II and IV were 4-person coalitions, in which the range of the bargaining set is four to five times the range of the power model; just by chance more outcomes favorable to the bargaining set will occur in these instances. But the effect of this large range should not necessarily discount the favorability of the bargaining set. The characteristic functions for all four game types produced about the same range for the monopolist payoffs for the power model, and defined a similar interval for non-4-person coalitions in the bargaining set into which fell almost all the outcomes of quartets 3, 5, and 6. In a sense, then, the part of the monopolist's predicted range that was appropriate for nonsuperadditive games was the useful range for the superadditive games as well, and this part of the range was not larger than the range prescribed by the power model.

Finally, we note that other power models [*Maschler*, 1963a] were also tested, but none fared better than M^P.

Coalition frequencies

The principal question in examining coalition strcutures is how often a Pareto optimal coalition structure obtained. Pareto optimality required the formation of the grand coalition in game types II and IV, coalition BCD in game type III, and any of the possible 3-person coalitions in game type I. Over all 96 games, the Pareto optimal coalition structure occurred 78 times. The 18 counterinstances were distributed evenly over quartets except for quartet 1, who formed 8 of them. For the remaining five quartets, there were from one to three non-Pareto optimal coalitions. If we exclude the 16 games with *solo*, then 67 of the 80 games ended in a Pareto optimal coalition structure, with 8 of the 13 deviants in quartet 1. Thus, overall, Pareto optimality was achieved.

Differences among quartets

In the present experiment, as well as in previous experiments employing the *Coalitions* program, deliberate and elaborate attempts were made to achieve independence between successive games. The entire sequence of 16 games was broken into three separate sessions, the identity of the players was kept secret, and role assignment changed according to a balanced experimental design over games. Moreover, players were specifically instructed to play each game independently of the results of previous games. Whereas this experimental manipulation achieved its goal in the previous experiments, Table 5 clearly shows its failure in the present instance. Further light on this difference among quartets is shown in Figure 1, where the outcomes to the monopolist player D are displayed in order of the 16 games played, ignoring game type and iteration categorizations.

Inspection of Figure 1 shows three classes of quartets, one consisting of quartets 1 and 4, a second of quartets 2 and 3, and the third of quartets 5 and 6. After the third game played by the quartets, there is only one overlap in player D's payoff among these three classes, and that is the single instance of two players choosing *solo*, leaving players A and D to divide 54 points equally. Overall, there was an increase for all six quartets in the payoff to player D over the first four games, after which a norm was established which, within a range of only a few points, contained player D's payoff for the remaining 12 games.

Bargaining analyses

The observation of striking quartet differences, where each quartet achieved its own consistent mode of solving the game, indicates the need for analysis of the path by which the outcomes were achieved. To this end, the individual protocols of each game, as recorded by the computer, were examined in an *ad hoc* fashion, using constructed indicators suggested by the quartet differences in outcomes.

Although there were a considerable number of times that a quartet would come to an outcome identical to a previous one far a given game this did not mean that all of the negotiations were done in the first iteration through the four game types. Indeed, as is typically the case, later games did proceed more quickly, with means of 41.58,

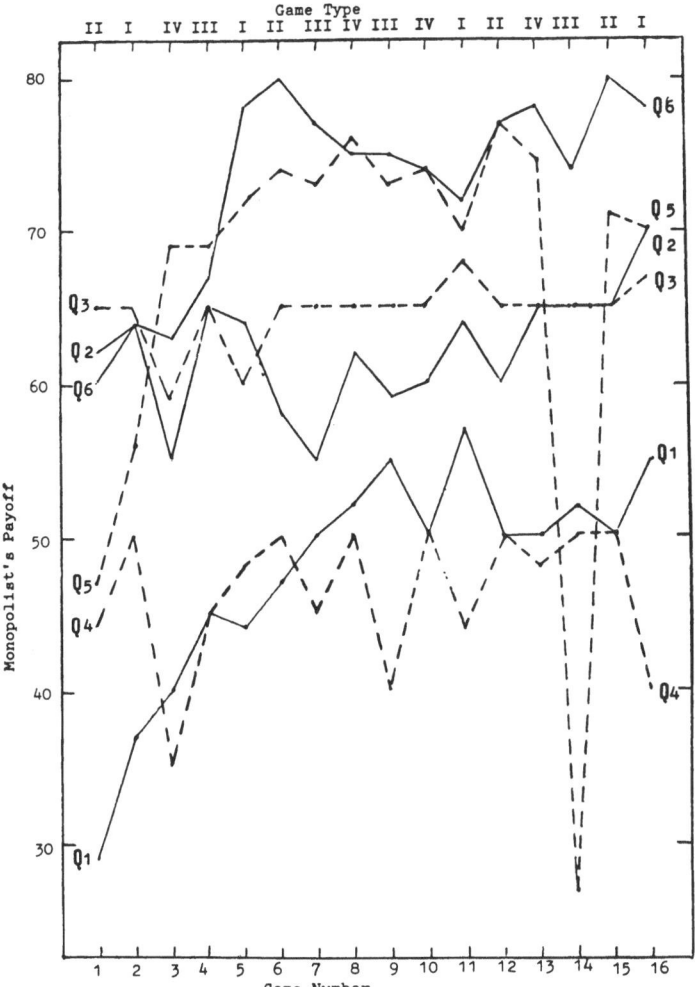

35.38, 30.38, and 27.88 messages per game for iterations one through four, respectively. However, even this last value is more than twice the minimum of 12 messages needed for the play of a game. Only in quartet 3 could it be legitimately said that the outcome was virtually determined at the beginning of a game. The range of outcomes that each quartet considered legitimate for the monopolist player D was typically not just one point; furthermore, particularly in the nonsuperadditive game type I and III, the outcomes to the nonmonopolist players could be heavily contested.

One very useful bargaining indicator was how Pareto optimality was achieved, when it did occur. For this analysis, each game was assigned to one of four mutually exclusive categories based on how or if Pareto optimality was arrived at. The four categories

were: *N* (Not Pareto): Pareto optimality not achieved; *I* (Immediate): Pareto optimal
coalition formed by ratification of an offer made on the first round of bargaining;
P (Pareto only): Only the Pareto optimal coalition that was formed was considered by
those players that did eventually form it; and *B* (Bargained): Other coalitions than the
one that eventually formed were offered and accepted during the course of the game.

Table 7 presents the incidence of these four categories collapsed over the 16 games,
for each quartet separately. From the incidence data, a bargaining index of the proportion of Pareto optimal coalitions in category *B* is shown, and is compared to the mean

Category	Quartet						over
	1	2	3	4	5	6	quartet
Not Pareto	8	2	1	3	1	1	18
Immediate	2	4	5	2	2	2	17
Pareto only	1	4	4	3	9	8	29
Bargained	5	6	6	8	4	3	32
B/(B+I+P)	.62	.43	.40	.62	.27	.23	.41
Mean payoff to D in Pareto optimal Coalitions	48.38	62.36	65.00	46.85	69.67	75.00	62.36

Tab. 7: Paths of Pareto optimal outcomes, by quartet

payoff for player *D* in Pareto optimal games (last row of Table 7). The coefficient of
correlation between this index and this mean is $r = -0.99$, indicating that the more
coalitions other than the Pareto optimal one were considered, the lower was the payoff
to player *D*. It is particularly interesting to note that the most active bargaining, in the
sense of different coalition structures considered, occured in the two quartets 1 and 4,
which least supported the bargaining set model; only in more egalitarian PCs is there
room for players to make offers and counteroffers against each other. For game type I,
about two-thirds of the Pareto optimal coalitions were bargained; this ratio was
reversed for the other three game types. This finding is not surprising in view of the
fact that in game type I, three coalitions were equivalently Pareto optimal, as compared to a single coalition in the remaining game types.

The second bargaining analysis attempts to examine the first games played in
order to trace the development of the norms demonstrated in Figure 1. Because there
were only six quartets, a formal analysis of bargaining style is impossible. This section
should be regarded as suggestive of both what accurred in the present experiment and
of what could grow to be an analytical method of studying bargaining processes. quartets will be considered individually with respect to their bargaining on the first four
games.

Quartet 1 provides the only clear example of a coalition of the three nonmonopolist
player against player *D*. The first game started out with a sequence of accepted 2-person
offers, each one giving player *D* a few points more than the previous one. However, first
player *A*, then *B* and *C*, came to insist on the PC (22, 22, 22, 29; *ABCD*), and player

D bowed to their pressure. Player *D* only tried once more for a high payoff on game 3, when again reference by player *C* to a coalition of the whole against him forced him down. Thus, in this quartet, there appears to have developed a low payoff for the monopolist because of the union of the other players against him. *Maschler's* [1978] argument is clearly manifest for this quartet.

Quartet 4, the other quartet with low payoffs for player *D*, arrived at this outcome by a different method. In the first round, player *D* was very passive, and eventually settled for 47 points. In the second round, a new player *D* was about to ratify a 3-person PC giving him 63 points when the player not in the coalition played *solo*. The other two players immediately re-negotiated the coalition to 50 points for player *D*. A nearly identical sequential process occurred on the next two rounds, and by then the low norm of 45–50 for player *D* had been accepted. In any future game, if a player *D* tried to get more points, one of the other players chose *solo*, causing a re-negotiation. Thus, the monopolist was held in line by the emphatic refusal of one of the other players to participate further in the escalation of the monopolist's payoffs. Whether this move was seen as a warning of other *solo* moves in the game or not can only be conjectured; what is known is that for this quartet, the *solo* did reduce player *D*'s payoff every time.

Quartet 2 and 3 were characterized by smooth bargaining among the parties resulting in a compromise in the range of 60 (for quartet 2) or 65 (for quartet 3). For quartet 3, the point solution of 65 for player *D* was established on the first round of the game, and held for 11 of the remaining rounds. Attempts at deviation from this norm were met by solid opposition from the remaining players; player *D* only got fewer points if he did not seek his 65 points (the same player twice), or more when the other three could not agree on how to split the remainder (two games). Quartet 2, unlike all the others, showed a suggestion of an overall increase in player *D*'s gains over all 16 rounds; the other players did not unite against him at any time, and his payoff limit was apparently only set by his own interpretation of an appropriate payoff.

Quartets 5 and 6 were characterized by player *D* setting a high value for himself, and then holding to it, insisting that the other players do more towards meeting his requests than he to meeting theirs. Although *solo* moves did occur in these games, they did not have the effect they had for quartet 4. Both these quartets were characterized by round of repeated identical offers with no concessions, until one and then more players gave in to most of the demand of player *D*. When the non-monopolist objected to the monopolist, he apparently did not feel their complaints to be justified.

Thus, through this decidedly informal scrutiny of the bargaining, there appears to be a breakdown of types of bargaining process that corresponds to the low, middle, and high payoffs to player *D* demonstrated in Figure 1. Quartets adopting low norms were characterized in the early bargaining by the nonmonopolist not conceding to the demands of the monopolist, albeit by various methods. Quartets adopting the middle norms were characterized by a traditional pattern of bargaining, where all parties gave reciprocal concessions. Finally, quartets adopting high norms were characterized by the monopolist insisting on, and eventually gaining, the amounts he wanted. Again, this analysis is based on the first four rounds played; typically later rounds had incorporated the norms, and bargaining proceeded more quickly to a solution in what the

particular quartet had defined as its appropriate range of payoffs; players would stead-
fastly hold against encroachments of this range, whether by the monopolist or a nonmonopolist.

Discussion

The major finding of the present study is that the variations in game type took second place to the search for a norm for the monopolist's payoff that was idiosyncratic
to the individual quartets who participated in the game. That is, early in the sequence
of games, the players established a standard range of payoff for the monopolist, and
this standard, plus the game type, served as the framework for their negotiations in the
remaining games. Characteristics of the performance of the players that give clues as
to the genesis of the monopolist's standard range come from the relationship of the
outcomes (expressed as PCs) to the $M_1^{(i)}$ and M^P prescriptions, the incidence of Pareto
optimal coalitions, and the characteristics of the bargaining leading to the outcomes
of the earlier games in the sequence. The discussion will elaborate on each of these
characteristics.

The bargaining set $M_1^{(i)}$ was supported by 56 of the 96 games played; more importantly, two of the quartets had 14 of 16 games in the bargaining set, while a third had
11 of 16 games in the prescribed range. Moreover, the support for the bargaining set
comes from the upper range of monopolist payoffs in its solution space. Although this
support corresponds to previous findings [Kahan/Rapoport, 1977, 1978; Rapoport/
Kahan, 1967; Rapoport et. al., 1978], the fact that only half of the groups supported
$M_1^{(i)}$ must be taken into consideration.These non-confirmatory quartets should not be
cause to reject the bargaining set, however. They were characterized by normative
ranges for the monopolist that gave the nonmonopolists some "bargaining room" in
which to negotiate among each other and with the monopolist. In other word, for these
quartets, there was room for the use of coalitions other than a single Pareto optimal
one to make claims and counter-claims in the sense of objection and counter-objection
used in the bargaining set. The result that the lower the payoff to the monopolist, the
more games ended in a Pareto optimal outcome bargained from many alternative outcomes (Table 7) supports this argument. In other word, for quartets with norms that
accord the monopolist a high outcome, there are no coalitions available for use as objections, while for quartets assigning the monopolist a lower outcome norm, objecting
coalitions may be found.

This line of reasoning supports the sense, if not the substance, of Maschler's [1963a]
reasoning leading to M^P. The central idea of the establishment of "standards of fairness" is correct, but in the absence of compelling extraneous conditions, a consistent
standard over groups does not materialize. Each quartet rapidly generated its own
"standard of fairness," which for some was the objective standard leading to $M_1^{(i)}$, but
for others was different.

Still, there is the difference between Maschler's [1978] and the present findings to
reconcile. An obvious conjecture as to a source of this difference is the difference between Maschler's communication conditions and those of the present experiment. Here,

yet a third experiment with monopolists in which communication was specifically manipulated can be of assistance. *Murnighan/Roth* [1977] used a 3-person monopolist game in which $v(12) = v(13) = 100$, $v(S) = 0$ otherwise, in which the payoffs were either public or secret, offers were either public or secret, and messages were either forbidden, secret or public. As they used no monetary or other extrinsic incentive to maximize points, a direct comparison to either *Maschler* or the present experiment is impossible. *Murnighan/Roth* did find that there was a relationship between openness and flexibility of communication and payoff that in general was in the direction of more equitable payoffs with more open communication, but this relationship was not monotonic for their six communication conditions. The most striking result they bring to bear on this issue is that, in their terminal three trials, games with messages allowed (whether secret or public) ended with the monopolist receiving an average of 56 points, while games without messages yielded the monopolist an average of 76 points. Here, the more restrictive communications support the power model, but, again, the incentives are not clear. The larger point, that the presence or absence of open communications does make a difference, however, is clearly established, and calls for further study of this factor in bargaining.

With only six quartets to examine, a true analysis of how different norms for the monopolist's payoffs arise cannot be ascertained; there is an indication that, however they are obtained, they rapidly take on the status of group norms, and define the reasonable bargaining range within which each game would be separately negotiated. It is both true and trivial to say that individual personality differences determine a large part of the establishment of such norms; our scrutiny of the bargaining protocols indicates that, for example, the quality of stubbornness might well be advantageous to a monopolist. When the nonmonopolists were united and steadfast, they could win their case against all but the most stubborn of monopolists.

That the standard of fairness for the present quartets was largely centered on the payoff of the monopolist can be attributed to two causes, one structural, and one artifactual. In the construction of the four game types, one basic game was used, and so the actual characteristic functions of the games were so similar that a common point value to the monopolist serves as a very convenient orientation that will facilitate bargaining in all four game types. In retrospect, modifying the characteristic functions of the individual game types slightly would have made interpretation of the norms easier. Even given this artifact, though, the necessity of the monopolist makes it logical that his payoff should be central to the negotiations, and the other allocations secondary. In apex games, where the weak players may form a coalition in the absence of the strong player, the structure dictates a central role to their payoffs in that weak coalition, against which the strong player must focus his attention. Experimental results from apex games [*Horowitz/Rapoport; Kahan/Rapoport*, 1979; *Rapoport* et al., 1978] show that this structural logic is followed by bargainers.

The conclusions that may be drawn from the present results are of course limited. Future experiments that employ more groups and investigate a wider range of payoffs will cast more light on the types of solutions and paths of bargaining that are mandated by negotiation situations where one party holds the lion's share of the power. But our results are strongly supportive of *Maschler's* [1963a] conclusion that the power of a coalition lies not only in its characteristic function, but also in what the members of the coalition can do by *not* entering into coalitions. We have evidence that the definition of that true power will vary as a function of features other than the characteristic function itself; future experiments must therefore scrutinize bargaining processes as well as outcomes.

References

Albers, W.: Bloc forming tendencies as characteristics of the bargaining behavior in different versions of apex games. Beiträge zur experimentellen Wirtschaftsforschung. Ed. by H. Sauermann. Vol. VIII, Tübingen 1978.

Aumann, R.J., and *M. Maschler*: The bargaining set for cooperative games. Advances in game theory. Ed. by M. Dresher, L.S. Shapley and A.W. Tucker. Princeton 1964.

Davis, M.: Game theory: A nontechnical introduction. New York 1970.

Davis, M., and *M. Maschler*: The kernel of a cooperative game. Naval Research Logistics Quarterly 12, 1965, 223–259.

Funk, S.G.: Value Power and positional power in *n*-person games. L.L. Thurstone Psychometric Laboratory Report No. 152, Chapel Hill 1978.

Horowitz, A.D.: A test of the core, bargaining set, kernel, and Shapley models in *n*-person quota games with one weak player. Theory and Decision 8, 1977, 49–65.

Horowitz, A.D., and *Am. Rapoport*: Test of the kernel and two bargaining set models in four- and five-person games. Game theory as a theory of conflict resolution. Ed. by An. Rapoport. Dordrecht 1974.

Kahan, J.P., and *R.A. Helwig*: Coalitions: A system of programs for computer-controlled bargaining games. General Systems 16, 1971, 31–41.

Kahan, J.P., and *Am. Rapoport*: Test of the bargaining set and kernel models in the three-person games. Game theory as a theory of conflict resolution. Ed. by An. Rapoport. Dordrecht 1974.

–: When you don't need to join: The effects of guaranteed payoffs on bargaining in three-person cooperative games. Theory and Decision 8, 1977, 97–126.

–: The influence of structural relationships on bargaining in 4-person apex games. European Journal of Social Psychology, 1979.

Levinsohn, J.R., and *Am. Rapoport*: Coalition formation in multistage three-person cooperative games. Beiträge zur experimentellen Wirtschaftsforschung. Ed. by H. Sauermann. Vol. VIII. Tübingen 1978.

Luce, R.D., and *H. Raiffa*: Games and decisions: Introduction and critical survey. New York 1957.

Maschler, M.: The power of a coalition. Management Science 10, 1963a, 8–29.

–: *n*-Person games with only 1, $n - 1$, and *n*-person permissible coalitions. Journal of Mathematical Analysis and Applications, 6, 1963b, 230–256.

–: Playing an *n*-person game: An experiment. Princeton: Economic Research Program, Research Memorandum No. 73, Princeton 1965.

–: Playing an *n*-person game: An experiment. Beiträge zu experimentellen Wirtschaftsforschung. Ed. by H. Sauermann. Vol. VIII. Tübingen 1978.

Medlin, S.M.: Effects on grand coalition payoffs on coalition formation in 3-person games. Behavioral Science 21, 1976, 48–61.

Michener, H.A., M.M. Sakurai, K. Yuen, and *T.J. Kasen*: A competitive test of the $M_1^{(i)}$ and $M_1^{(im)}$ bargaining set solutions in three-person conflicts. Journal of Conflict Resolution 23, 1979, 102–119.

Michener, H.A., K. Yuen, and *I.J. Ginsberg*: A competitive test of the $M_1^{(im)}$ bargaining set, kernel, and equal share models. Behavioral Science 22, 1977, 341–355.

Murnighan, J.K., and *A.E. Roth*: The effects of communication and information availability in an experimental study of a three-person game. Management Science 23, 1977, 1336–1348.

Owen, G.: Game theory. Philadelphia 1968.

Peleg, B.: Existence theorem for the bargaining set $M_1^{(i)}$. Essays in mathematical economics in honor of Oskar Morgenstern. Princeton 1967.

Rapoport, Am., and *J.P. Kahan*: Computer controlled research on bargaining and coalition formation. Behavior Research Methods and Instrumentation 6, 1974, 87–93.

–: When three isn't always two against one: Coalitions in experimental three-person games. Journal of Experimental Social Psychology 12, 1976, 253–273.

Rapoport, Am., J.P. Kahan, and *T.S. Wallsten*: Sources of power in 4-person apex games. Beiträge zur experimentellen Wirtschaftsforschung. Ed. by H. Sauermann. Vol. **VIII**. Tübingen 1978.

Rapoport, An.: *n*-person game theory: Concepts and applications. Ann Arbor, Michigan 1970.

Riker, W.H.: Bargaining in a three-person game. American Political Science Review **61**, 1967, 642–656.

–: Three-person coalitions in three-person games: Experimental verification of the theory of games. Mathematical applications in political science VI. Ed. by J.F. Herndon and J.L. Bernd. Charlottesville 1972.

Riker, W.H., and *W.J. Zavoina*: Rational behavior in politics: Evidence from a three-person game. American Political Science Review **64**, 1970, 48–60.

Schelling, T.C.: The strategy of conflict. Cambridge 1960.

Selten, R., and *K.G. Schuster*: Psychological variables and coalition forming behavior. Risk and uncertainty. Ed. by K. Borsh and D. Mossin. London 1968.

von Neumann, J., and *O. Morgenstern*: Theory of games and economic behavior, 2nd. ed. Princeton N.J. 1947.

Models and Applications in Political Science

Applied Game Theory. 1979 © Physica-Verlag, Wuerzburg/Germany

A Model of the U.S. Presidential Primary Campaign[1])

By *J.H. Aldrich*, East Lansing[2])

Abstract: A game theoretic model of the decision by candidates for the U.S. Presidential nominations concerning which set of presidential primaries to contest is developed. Using the two candidate contest for the 1976 Republican nomination as a prototype, interest centers on the effects of variations in the institutional context in affecting candidate strategy. Variables studied include the dates of the primaries, their size, the rules translating votes into delegate allocations, and variations in each primary's electorate. Dynamic elements, e.g., "momentum," are included, a number of propositions produced, and these propositions are tested successfully against the choices made by Reagan and Ford in the 1976 Republican campaign.

1. Introduction

The campaign for a major party nomination for President takes place in a unique institutional context. Rarely, in recent years, have nominations been decided at national nominating conventions. Rather, they have been determined in the fifty state primaries and caucuses designed to select delegates to attend the national conventions. Recent reforms, notably the proliferation of primaries and the growth of media attention to these pre-convention campaigns, have made the Presidential primaries the major battleground for the nomination. In 1968, there were only 17 primaries, selecting less than half of Republican and Democratic delegates. In 1976, there were 30 primaries, choosing over 70 percent of the delegates for each party. Moreover, by 1976, virtually all delegates selected in primaries were chosen as committed to one candidate or another.

Even given that attention here will be restricted to presidential primaries, the institutional context remains rich. Laws and procedures governing primaries are established through individual state legislatures and state party organizations. As a consequence, each state's primary was unique in 1976, and the two parties' primaries within a state often differed. Combined with the fact that each primary electorate is disjoint and

[1]) I would like to thank Robert Axelrod, Steven Brams, John Ferejohn, Morris Fiorina, Kenneth Shepsle, and Dennis Simon for their comments on earlier versions of this paper. This research was supported with funds provided by the National Science Foundation, Grant No. SOC 76-24218. Derivations of propositions stated in the text are available from the author on request. I retain responsibility for the content of this paper.

[2]) *John H. Aldrich*, Michigan State University, Department of Political Science, East Lansing, Michigan 48824, USA.

that the 30 primaries were spread over 112 days in 1976, this complex institutional context ought to be expected to play a major role in influencing candidate strategy and consequent behavior.

The purpose of this paper is to extend a game theoretic model of candidate choice in this institutional context, a model originally developed in *Aldrich* [1978]. The empirical referent is the 1976 Republican campaign, pitting then President Ford against the former California Governor Reagan. Both candidates were active contestants before the first primary, and each was a powerful figure in the party in his own right. Therefore, the model is developed in the context of a two candidate contest, and propositions derived from it will be tested against the behavior of these two candidates.

The strategy set facing each candidate is whether or not to contest actively in each given primary. Clearly, candidates can and do choose among many alternative behaviors. Yet, the fundamental choice is whether or not to contest a given primary. Failing to contest a primary in 1976 guaranteed that an active opponent would sweep that state's delegation.[3]) It also should be noted that candidates can file a petition to have their name placed on the primary ballot and yet withdraw from active competition subsequently.[4]) At the same time, it is possible for a candidate who failed to have his name placed on the ballot to conduct a last minute write-in campaign, as Governor Brown did in Oregon in the 1976 Democratic campaign and as Henry Cabot Lodge did in winning the 1964 Republican primary in New Hampshire. The key is to communicate one's competitive status to the media and hence to the public. As a consequence, the actual choices made by Ford and Reagan must be inferred by media portrayal. Theoretically, the implication is that candidates could — and did — modify their behavior after the immediately preceding primary. Thus, it is assumed that candidates employ behavioral strategies over the sequence of primaries, modifying their behavior if they so choose between the dates of one primary and the next.

In sum, the ensuing model will be that of a two candidate game in which the players choose to compete or not in each primary *seriatim*. The institutional context, broadly conceived, will provide the substance for deriving propositions that will be tested against the behavior manifested in the 1976 Republican primary campaign.

2. The Institutional Structure

Some simplifying assumptions have already been made, specifically that of two candidates competing for delegate support through presidential primaries. In addition, it is assumed that no uncommitted delegates are selected nor are there any selected committed to minor candidates (an assumption essentially true in 1976). As a consequence, delegate selection is necessarily "zero-sum," and delegates are not independent players.[5])

---

[3]) Ford won 91.6 % and Reagan won 95.5 % of the delegates in states where that candidate was the sole active competitor.

[4]) Two of several examples in 1976 are Ford in California and Reagan in Wisconsin. In each case, the candidate also expended vast resources before withdrawing.

[5]) Models viewing delegates as independent players include *Straffin* [1977] and those reviewed in *Brams* [1978].

In the empirical case, Ford and Reagan faced 27 primaries scattered among 14 different dates over the 16 week primary season (see Table 1 for details). There were 1,483 delegates so selected with a majority point of 742. The timing of particular primaries is a key variable.

State	Date	Size[3])	D_j	Type[3])	Competitiveness	Cell in[3]) Table 2	Activity Ford	Activity Reagan
New Hampshire	2/24	Large	21	non-PR	Competitive	4	Y	Y
Massachusetts	3/2	Large	43	PR	Pro-Ford	2	N	N
Florida	3/9	Large	66	non-PR	Competitive	4	Y	Y
Illinois	3/16	Large	96	non-PR	Competitive	4	Y	Y
North Carolina	3/23	Large	54	PR	Pro-Reagan	3	Y	Y
New York }	4/6	Large	117	non-PR	Pro-Ford	5	Y	N [2])
Wisconsin }		Large	45	non-PR	Competitive	4	Y	N
Pennsylvania	4/27	Large	84	non-PR	Pro-Ford	5	N	N
Texas	5/1	Large	96	non-PR	Pro-Reagan	6	Y	Y
Alabama		Small	37	non-PR	Pro-Reagan	12	N	Y
Georgia	5/4	Large	48	non-PR	Pro-Reagan	6	Y	Y
Indiana		Large	54	non-PR	Competitive	4	Y	Y
Nebraska }	5/11	Small	25	non-PR	Competitive	10	Y	Y
West Virginia }		Small	28	non-PR	Competitive	10	N	N
Maryland }	5/18	Small	43	non-PR	Competitive	10	N	N
Michigan }		Large	84	PR	Competitive	1	Y	Y
Arkansas		Small	27	PR	Pro-Reagan	9	N	N
Idaho		Small	17	PR	Pro-Reagan	9	N	Y
Kentucky	5/25	Small	37	PR	Competitive	7	Y	Y
Nevada		Small	18	PR	Pro-Reagan	9	N	Y
Oregon		Small	30	PR	Competitive	7	Y	Y
Tennessee		Small	43	PR	Competitive	7	Y	Y
Rhode Island }	6/1	Small	19	non-PR	Pro-Ford	11	N	N
South Dakota }		Small	20	PR	Competitive	7	N	N
California		Large	167	non-PR	Pro-Reagan	6	N	Y
New Jersey	6/8	Large	67	non-PR	Pro-Ford	5	Y	N
Ohio		Large	97	non-PR	Competitive	4	Y	Y

Tab. 1: Values of Republican Primaries in 1976 on Key Variables[1])

[1]) The determination of the particular values is described in the text.
[2]) The brackets enclose primaries that fall on the same date.
[3]) Refers to categorization described in section 8.

A second important variable is due to the fact that the primaries differed considerably in terms of the number of delegates each was to select. Idaho Republicans selected but 17 delegates, while Californians chose 167. While variation in size was large, no one state was necessary to form a winning coalition.

State primaries differed along a third crucial dimension: the manner in which the electorates' votes were translated into delegate allocations to the candidates. We will examine three classes of transformation rules. At one extreme was California all of whose delegates were allocated to the state-wide plurality winner. This method was called "winner-take-all" (denoted "WTA"). Proportional "representation" (or PR) forms the other extreme, where a perfect matching of vote to delegate percentages will be assumed [but see *Rae; Tufte*]. Two other procedures, the "Democratic Loophole" (or DL), whereby sub-state units could allocate delegates via a winner-take-all proce-dure or "plurality winner" (or PL), whereby delegate hopefuls who win a plurality of votes in a sub-state unit won a trip to the convention, were intended to provide some-thing "in between" these two extremes. Empirically, they did [see *Aldrich*, forthcom-ing, for details], and they will be so designated here.

Since each primary electorate is disjoint, each presents the two candidates with a different probability distribution over outcomes. Reagan, as a wellknown conservative, could be assumed to be favored by conservative electorates, Ford by more liberal Re-publican electorates. Viewed as a probability distribution, the inherent riskyness of election outcomes can be conceived of as presenting a random move (the election) drawn from a known probability distribution, a priori. Thus, both candidates are as-sumed to know and agree upon each state's a priori probability over vote outcomes.

3. The Game Theoretic Model

The following notation will be used. The two candidates will be denoted by C_i and C_k (k being suppressed for stylistic convenience where relevant). The set S, $S = S_1, \ldots, S_j, \ldots, S_n$, will denote the n different dates on which primaries occur in order. If there is only one primary per day, there will be n primaries. With two or more primaries on one day, they will be denoted with numerical superscripts, e.g., S_j^1, S_j^2, etc. Matched with each primary is the size of delegation to be selected, D_j, with $D = \sum_j D_j$ being the total number of delegates selected through primaries. T_j denotes the rule transforming votes into delegates, varying over the three rules, WTA, PR, and DL/PL. P_{ij} denotes the probability distribution over vote results for C_i in S_j. If $E(P_{ij})$ represents C_i's expected vote percentage a priori in S_j, it is assumed that $E(P_{ij}) = 1 - E(P_{kj})$ for all j.

Combining P_{ij} with T_j implies that both candidates can agree on an estimate of the expected delegate returns for each S_j. For example, if $T_j = PR$ for S_j, then $E(D_{ij})$, the expected number of delegates for C_i is:

$$E(D_{ij}) = E(P_{ij}) D_j \quad \text{or} \quad E(D_{ij}) / D_j = E(P_{ij}) \tag{3.1}$$

If $T_j = WTA$, the relevant calculation for C_i may be written as:

$$E(D_{ij}) / D_j = \text{Prob}(P_{ij} > .5) \tag{3.2}$$

In general, the expected number or percentage of delegates is a function of the trans-formation operator, T_j, and P_{ij}:

$$E(D_{ij}) = E[T_j(P_{ij}) D_j] = T_j(D_j) E(P_{ij}) \quad \text{or} \quad E(D_{ij}) / D_j = T_j E(P_{ij}) \tag{3.3}$$

If primaries were independent events (i.e., P_{ij} does not depend on the results of earlier primaries) candidates could calculate the expected number of delegates each would receive if both candidates competed in all primaries, viz:

$$E(D_i) = \sum_j E(D_{ij}) = \sum_j T_j E(P_{ij}) D_j \tag{3.4}$$

At each primary both candidates choose from a dichotomous set of alternative behaviors, say $A_i = (a_{1i}, a_{2i})$. Let a_{1i} denote that C_i chose to be an active competitor, while a_{2i} indicates he was not. The choice space for each primary is then mapped into the outcome space, say O_j, for primary j. Thus, for S_j, there are four possible pure strategy pairs: both candidates compete, neither candidate competes, C_i competes but not C_k, and vice versa.

The value of the four possible outcomes can be only partially specified at this point. The following general assumptions are made. It is assumed that if a candidate competes in a state, there is some fixed cost of mounting the campaign that varies from primary to primary. Call these fixed costs b_j, and assume them equal for both candidates. These costs are discussed in detail in the earlier paper [Aldrich, 1978]. Next, it is assumed that if only one candidate competes in a state, that candidate wins D_j, all delegates to be selected from that state, at a cost b_j. The final simplifying assumption is that if both candidates select the same strategy pair (both compete or both do not), the primary is held subject to the random move assumed to be the same for O_1 and O_4. O_1 and O_4 differ, therefore, over costs only.

4. The Game for Primaries That Occur on Different Dates

This section examines the behavioral strategies of candidates when primaries are assumed to be independent events and when each primary is held on a different date. The next two sections relax the first assumption, while the subsequent section extends the analysis to include the case of two primaries on the same date with all primaries interdependent.

A single stage of the game under the conditions of this section is shown in normal form in Figure 1. There are three moves; both candidates choose whether or not to compete actively in the primary, and, if matching strategies are chosen, a random move (the election) is drawn from P_{ij}. The full game, then, can be represented as a repeated 2 × 2 matrix game with varying parameters.

If candidates adopt behavioral strategies, it is possible to evaluate the strategy choices for a typical stage. Actually, with independent primaries, the assumptions made about information sets, etc., there is little to be gained from employing the implied extensive form. In particular, the following solution concepts to each stage will be general to the full game, because the game has "perfect recall" (i.e., there are not signalling information sets). Therefore, by Kuhn's Theorem [1953], any behavioral strategy can be associated with at least one mixed strategy, in terms of payoffs, and the behavioral solution is a solution in the usual sense. Unfortunately, all of this breaks down if the results of the random move modifies the outcome of succeeding stages in

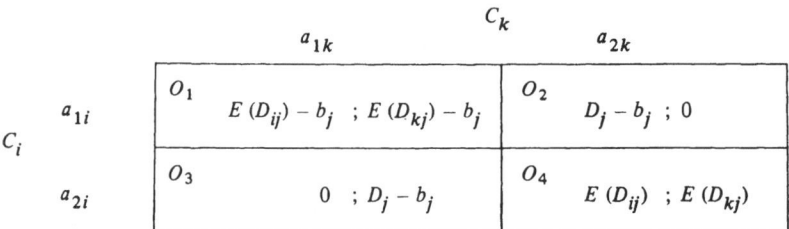

Fig. 1: A Representative Stage for Primaries on Unique Dates and as Independent Events

the game. Also eliminated is the usual approach of moving backward through the game tree. Rather, a forward approach is necessary. Therefore, it will be useful to start off with the analysis of independent primaries, adding in complications as they arise.

In Figure 1, it can be seen that competition will dominate non-competition when:

$$E(D_{ij}) - b_j > 0 \text{ and } D_j - b_j > E(D_{ij}) \tag{4.1}$$

Given the conditions of this section, equation (4.1) yields the following substantive propositions:

Proposition (4.1): If a primary is expected to be competitive ($E(D_{ij})$ "near" $D_j/2$), then competition is likely to dominate non-competition. Conversely, if the primary is non-competitive ($E(D_{ij})$ "near" 0 or "near" D_j), competition is unlikely to be dominant.

Proposition (4.2):
a) If b_j is low, competition will tend to be the dominant strategy, and if $b_j = 0$, competition will be dominant for both candidates.
b) If costs are high relative to D_j (e.g., in small states), non-competition will tend to dominate competition for both candidates.

There are two cases where a mixed behavioral strategy is optimal for a given stage:

Case I $E(D_{ij}) - b_j > 0$ but $E(D_{ij}) > D_j - b_j$

$$\tag{4.2}$$

Case II $E(D_{ij}) - b_j < 0$ but $E(D_{ij}) < D_j - b_j$

Case I corresponds to instances where C_i expects to do well a priori, while Case II represents the opposite, each with non-trivial costs. If C_i faces a Case I primary, C_k faces a Case II primary. Examples of this situation in 1976 might be Massachusetts and Maryland (Case I for Ford) and Idaho, Nevada and California (Case I for Reagan).

Proposition (4.3): If equation (4.2) holds, then the candidate with high $E(D_{ij})$ will be likely to compete in that state, while his opponent will be relatively unlikely to do so. That is, the mixed strategy will be skewed towards active competition for C_i.

The general solution to the primary problem with independent events and one primary at a time consists of the "summation" of these separate stage solutions. The candidates can calculate the expected result of the nomination as, simply, the sum of the delegates expected from each stage if both candidates play optimal (mixed or pure) strategies throughout.

The "problem of entry," i.e., should a candidate run for the nomination, can be addressed. The probability of attaining the nomination a priori can be compared to costs involved.[6]) The "value" to the play of the game, thus constructed, would indicate that the game is worth playing if the value is non-negative. Candidates who are considered "long shots" nonetheless run. However, this "value" might explain why "when a realist like Senator Walter F. Mondale of Minnesota, after more than a year of actively 'exploring' his chances, recognized in November 1974 that the price was more than he was willing to pay, he dropped out." [*Witcover*, pp. xi–xii].

This section analyzed the comparative statics of the primaries. The primary game, however, is not a sequence of static games, but one dynamic game.

5. Payoffs with Interdependent Events

The goal of the candidates is to win the nomination. Viewing each stage in isolation models candidates as maximizers of delegates in each primary, but ignores the progress the candidates have made toward their ultimate goals and what opportunities lie ahead.

The value of the marginal delegate will not be identical over the primary campaign. If C_i is far ahead of C_k in terms of committed delegates, he may be more willing to forfeit some delegates than if the two delegate totals are virtually identical. Similarly, if there are many delegates yet to be selected, the value of an additional delegate will be less than if there are only a few yet to be selected, *ceteris paribus*. To measure this, let r_i and r_k denote the number of delegates supporting C_i and C_k, respectively. Then $\Delta r_{ik} = r_i - r_k$ indicates how far ahead (behind) his opponent C_i is. There is a pool of $D - r_i - r_k$ delegates yet to be selected. The "competitive standings" of the two candidates, R_i, may be written:

$$R_i = \frac{r_i - r_k}{D - r_i - r_k} = \frac{\Delta r_{ik}}{D - r_i - r_k} \tag{5.1}$$

Clearly, R_i equals $-R_k$. Further, if r_i equals $D_j/2$, then R_i equals one. It is assumed that the utility function of C_i is a strict monotone increasing function of R_i.

Several implications follow from this assumption. First, it is not necessarily the case that R_i is a monotone increasing function of r_i. Second, this formulation necessarily builds a type of dynamic component into the model. To see this, denote the expected value of R_{ik} given strategy pair O_j as $E(R_i/O_j)$. Since the denominator is constant across outcomes of S_j, denote it as m_j. Then the expected value of R_i is:

$$E(R_i/O_j) = (1/m_j)[\Delta r_{ik} + E(D_{ij}) - E(D_{kj})] \quad \text{for } O_1, O_4$$

$$= (1/m_j)[\Delta r_{ik} + D_j] \quad \text{for } O_2 \tag{5.2}$$

---

[6]) The total expected costs would be the sum of b_j in primaries where competition is dominant, plus the expected costs from mixed strategy cases.

$$= (1/m_j) [\Delta r_{ik} - D_j] \qquad\qquad \text{for } O_3$$

Note that this term is the same if both candidates choose the same strategy, and that the value of non-competition if the opponent chooses to compete is negative. In particular, the candidate loses a competitive advantage as his opponent picks up all delegates. At the same time the denominator decreases, representing a smaller pool of available delegates remaining. The implications are that there is a general increase in the incentive to compete, that Case II conditions are less likely to be met, and that Case I conditions for a mixed strategy are more likely. For example, one Case II condition in eq. (4.2) was $E(D_{ij}) - b_j < 0$. If θb_j denotes the transformation of costs to units comparable to R_j, this condition with interdependent events is the much less likely:

$$E(D_{ij}) - E(D_{kj}) - \theta b_j < -D_j \qquad\qquad (5.3)$$

Thus, the failure to compete when one's opponent does is more costly (and the reverse more beneficial). Note that the value placed on the delegate return component increases over time, *ceteris paribus*.

The new form of candidates' utility functions imposes one type of dynamic (a "long term" component). A "short term" dynamic effect is generated from the outcome of specific primaries. The news media have been criticized for their emphasis on the "horse race" aspect of nomination campaigns; who's ahead or behind, who is likely to win the nomination, etc. Critics have been bothered even more by the media's interpretations of specific primaries; who "won" or "lost," did candidates live up to vote "expectations," etc. These two practices are assumed to generate what the media call "momentum" and modify voter behavior.

"Momentum" here refers to changes in P_{ij} due to prior events, and has a separate "short-term" and "long-term" component to it. The "long-term" component is the R_{ik} term and need not be discussed further.

The "short-term" factor is developed as follows. Media personnel often indicate that a candidate would be considered "doing well" in a primary if he obtains a certain percentage of the preference vote cast by the electorate. Exceeding that level is seen as a "victory" and widely proclaimed as such. For example, McCarthy's 42 percent of the vote in the 1968 New Hampshire primary was viewed as a "victory" and Johnson's 50 percent a "defeat." The basic idea is that there is some vote percentage that serves as a baseline of measurement. Exceeding that level generates favorable publicity as a "winner," while failing to attain that level makes the candidate appear a "loser." One important point is that this baseline is not necessarily 50 percent (e.g., the McCarthy example), nor is it necessarily the most probable vote distribution (i.e., $E(P_{ij})$). If the actual vote exceeds the baseline the probability function over suceeding primaries is assumed to increase and to decrease if the vote differential is negative. Call this vote differential DV_{ij} for C_i in S_j and, of course, the expected differential prior to the actual play of the random move is $E(DV_{ij})$. I assume that this term is non zero only for those states in which both candidates are active competitors. If they are not, media attention is sparse [see *Aldrich*, forthcoming]. For simplicity, I assume that this factor operates on succeeding primaries only until the next competed primary (and hence next DV term, which supplants prior terms). Thus, a "victory" in this sense after a string of "defeats" will "turn momentum around," *ceteris paribus*, just as Reagan's Texas victory, following a series of defeats, turned his campaign around [see *Aldrich*, forthcoming]. Thus, for primary j, the relevant primary is the last one competed by both candidates, denoted $j - m$. Finally, in a two candidate contest, it seems reasonable to assume that $DV_{ij} = -DV_{kj}$.

The "revised" probability distribution over outcomes in S_j for C_i, P_{ij}, is assumed to be a function of P_{ij}, the a priori vote probability, DV_{ij-m}, and R_{ij-1}:

$$\hat{P}_{ij} = f(P_{ij}, DV_{ij-m}, R_{ij-1}) \tag{5.4}$$

Even assuming that candidates employ behavioral strategies, a larger $\hat{P}_{ij}$ than P_{ij} must be considered a positive payoff from earlier stages in the game, since it represents an increase in the likelihood of attaining the nomination, *ceteris paribus*. In particular, if $E(DV_{ij})$ is expected to be positive, C_i would have further incentives to compete actively, while if it is negative, C_i would have an incentive to not compete. Note that DV_{ij} and R_{ik} are not necessarily monotonically related, and that in the first stages of the primary campaign, R_{ik} will tend to be small. Therefore, a given DV_{ij} term will be relatively more consequential earlier than later when R_{ik} generally will be larger. At this point, I simply assume that there is some function, f, consistent with eq. (5.4), and I will examine the consequences of its arguments separately on candidate strategy (especially $E(DV_{ij})$ prior to S_j).

6. Non-Independent Events: Propositions for Primaries on Different Dates

The game matrix is now changed to that of Figure 2. The differences from

	C_k	
	a_{1k}	a_{2k}
a_{1i}	$E(R_i/O_1), -b_j$, $E(DV_{ij})$;	$E(R_i/O_2), -b_j$, 0 ;
	$E(R_k/O_1), -b_j$, $E(DV_{kj})$	$E(R_k/O_2)$, 0 , 0 ,
a_{2i}	$E(R_i/O_3)$, 0 , 0 ;	$E(R_i/O_4)$, 0 , 0 ;
	$E(R_k/O_3), -b_j$, 0	$E(R_k/O_4)$, 0 , 0

Fig. 2: A Normal Form Stage for Primaries on Unique Dates with Interdependent Events

the original matrix are in the addition of $E(DV_{ij})$ (as an argument in the revision of P_{iq} for $q > j$) to O_1 and in the substitution of $E(R_{ik})$ for the delegate allocations in all outcomes.

For C_i, $E(R_i/O_1)$ will be larger than $E(R_i/O_3)$ if:

$$E\,(\hat{D_{ij}}) - E\,(\hat{D_{kj}}) \geqslant -D_j \quad \text{but, since } E\,(\hat{D_{kj}}) = D_j - E\,(\hat{D_{ij}})$$

$$2E\,(\hat{D_{ij}}) - D_j \geqslant -D_j \quad \text{or} \quad E\,(\hat{D_{ij}}) \geqslant 0$$

(6.1)

The weak inequality must hold and O_1 will be preferred to O_3 in terms of delegates. Counteracting this is the cost term and $E\,(DV_{ij})$ if it is expected to be negative. The comparable equations of O_2 and O_4 are:

$$E\,(R_i/O_2) > E\,(R_i/O_4) \quad \text{if}$$

$$D_j > E\,(\hat{D_{ij}}) - E\,(\hat{D_{kj}}) \quad \text{or} \quad D_j > E\,(\hat{D_{ij}})$$

(6.2)

Again, the weak form of the inequality must hold. Here, though, the only other term is b_j which works in the opposite direction.

The conclusion, then, is that in comparison to the independent events case, interdependence makes active competition dominant in larger sets of circumstances. The only exception is the possibility of a negative $E\,(DV_{ij})$ term — a condition that must hold for one of the candidates in contemplating the possibility of O_1. Therefore Case I inequalities are more likely to describe one of the candidates than are those of Case II. Indeed, especially early in the play of the game when R_i is small, the possibility of a "bad showing" may be the dominant factor. Presumably, large negative values of $E\,(DV_{ij})$ are likely only when P_{ij} is particularly small, and/or the number of delegates to be selected in S_j is small. This point conforms to the inordinate attention paid to the small — but first — New Hampshire primary and the media focus on the votes won there, but not the delegates.

A number of substantive propositions about the effects of varying institutional parameters are presented in *Aldrich* [1978]. To summarize these propositions, larger primaries (large D_j) are more likely to be actively competed than small ones, smaller primaries are more likely to be contested early in the campaign, and the effects of WTA (and less so of DL/PL) transformation rules exaggerate the effects of other variables (size, P_{ij}, etc.). PR primaries with large thresholds are more similar to DL/PL than PR type T_j's. Finally, the dispersion of P_{ij} or "riskyness" of the primary (presumably larger at the outset of the campaign when information is rarer) offsets some of the exaggeration of non-PR T_j's.

7. Two Primaries on the Same Date

The 27 Republican primaries in 1976 were held on 14 different dates. Thus, Ford and Reagan had to make a more complicated decision when there were two or more primaries on the same day. In this section, I will examine the simplest case of exactly two primaries on the same day. The superscripts "1" and "2" denote the two primaries on day j. Further, $b^{1,2}$ for example denotes the costs of competing actively in S^1 and S^2 simultaneously. Finally the hats will be deleted for convenience.

It is possible that conducting two campaigns at one time induce some "economies of scale," so that $b^{1,2} \leqslant b^1 + b^2$. For example, media advertisements might be cheaper in overlapping media markets if the two states are contiguous (as many were in 1976). However, conducting two campaigns might affect the candidate's expected delegate return in one or both states due to the necessary "spreading" of resources (e.g., organization and staff) over two states. If C_i competes in both states, $E(R_i)$ in each state will be denoted by a prime, and it is assumed that:

$$E(R_i/O^1)' \leqslant E(R_i/O^1) \tag{7.1}$$

The four by four game matrix is shown in Figure 3.

		C_k Competes in:			
		a²) S^1 and S^2	b S^1	c S^2	d Neither
C_i Competes in:					
S^1 and S^2	a	$E(R_i/O^{1,2})', -b^{1,2},$ $E(DV_i^{1,2})'$	$E(R_i/O^{1,2})', -b^{1,2},$ $E(DV_i^1)'$	$E(R_i/O^{1,2})', -b^{1,2},$ $E(DV_i^2)'$	$E(R_i/O^{1,2}), -b^{1,2},$ 0
S^1	b	$E(R_i/O^1), -b^1,$ $E(DV_i^1)$	$E(R_i/O^1), -b^1,$ $E(DV_i^1)$	$E(R_i/O^1), -b^1,$ 0	$E(R_i/O^1), -b^1,$ 0
S^2	c	$E(R_i/O^2), -b^2,$ $E(DV_i^2)$	$E(R_i/O^2), -b^2,$ 0	$E(R_i/O^2), -b^2,$ $E(DV_i^2)$	$E(R_i/O^2), -b^2,$ 0
Neither	d	$E(R_i/O), 0,$ 0	$E(R_i/O), 0,$ 0	$E(R_i/O), 0,$ 0	$E(R_i/O), 0$ 0

Fig. 3: The Normal Form of a Stage for Two Primaries on the Same Day[1])

[1]) Entries are for C_i only, entries for C_k parallel those listed here. The specific values for the various $E(R_i)$ terms are detailed in the text. The conditional notation $O^{1,2}$, etc. denotes only where C_i competes actively and associated outcomes.

[2]) Note that C_k has "spread" his resources in this column. Thus, for example, if C_i chooses row b, the $E(DV_i^1)$ term in column a is larger (i.e., more favorable to C_i) than in column b.

Using a, b, c, and d as in Figure 3, I will compare C_i's choices within columns and address two questions: when is active competition in two states preferable to one and competition in one preferable to none, and when is competition in S^1 only preferable to competition in S^2 only?

Note that costs generate an incentive to fail to compete in two states as compared to either one and to either one state to neither. General statements about the expected

DV_i terms depend on the specific parameter values involved. However, it may be assumed that any primed DV_i term is strictly less than its unprimed counterpart.

The $E(R_i/O^j)$ terms may be compared. Suppose first that C_k does not compete in either state. By simple algebraic manipulations, it is easy to see that a is greater than b or c and b or c is greater than d:

$$a \geqslant b \text{ since } D_i^2 \geqslant E(D_i^2), \quad a \geqslant c \text{ since } D_i \geqslant E(D_i^1)$$

$$a \geqslant d \text{ since } D^1 + D^2 \geqslant E(D_i^1) + E(D_i^2) \tag{7.2}$$

$$b \geqslant d \text{ since } D^1 \geqslant E(D_i^1), \quad c \geqslant d \text{ since } D^2 \geqslant E(D_i^2)$$

Thus, the $E(R_i)$ term yields exactly the opposite order of costs, implying incentives for C_i to compete in two states over one and one over none, if C_k is inactive.

If C_k competes in only one state, say S^2, the following inequalities hold concerning $E(R_i)$:

$$a \geqslant b \text{ since } E(D_i^2)' \geqslant 0, \quad a \geqslant d \text{ since } D^1 + E(D_i^2)' \geqslant E(D_i^1)$$

$$a \geqslant c \quad \text{if} \quad D^1 - E(D_i^1) \geqslant E(D_i^2) - E(D_i^2)' \tag{7.3}$$

$$b \geqslant d \text{ since } D^1 \geqslant E(D_i^1), \quad c \geqslant d \text{ since } E(D_i^2) \geqslant 0$$

Thus, the same conclusions hold when C_k competes in one state as when C_k competes in none. However, notice that a generally but *not* always will be greater than c. The inequality will be reversed if spreading of resources is costly in S^2, and if C_i expects to win almost all delegates in S^1. Admittedly these are rare but not logically impossible circumstances.

If C_k competes in both states, the same general pattern emerges:

$$a \geqslant b \text{ if } E(D_i^2)' \geqslant E(D_i^1) - E(D_i^1)', \quad a \geqslant c \text{ if } E(D_i^1)' \geqslant E(D_i^2) - E(D_i^2)'$$

$$a \geqslant d \text{ since } E(D_i^1)' + E(D_i^2)' \geqslant 0 \tag{7.4}$$

$$b \geqslant d \text{ since } E(D_i^1) \geqslant 0, \quad c \geqslant d \text{ since } E(D_i^2) \geqslant 0$$

All of these inequalities will hold generally, except with relatively large costs due to resource spreading.

The following substantive propositions (with appropriate *ceteris paribus* contingencies) follow:

Proposition 7.1: In general, there are incentives to compete in two primaries as compared to competing in only one and one as compared to none.

Proposition 7.2: If "economies of scale" are less than the costs of resource spreading competition in only one of the two states will tend to be preferable.

Proposition 7.3: Proposition 7.2 is more likely to hold when the two state primaries are relatively large and/or when the two states are quite distinct (geographically, etc.).

To exemplify proposition 7.3, Alabama and Georgia held their primaries on the same day. These two states are quite similar and geographically contiguous. Therefore, the economies of scale are relatively high and resource spreading relatively low in contrast to, say, the New York and Wisconsin primaries of April 6.

The more interesting question concerns the choice of which one state to compete. As has been shown, there is some reason to suspect that competing in one state will be preferable to competing in two states under fairly common conditions. Obviously, whether b^1 or b^2 is larger is an empirical question. So, too, is the direction and magnitude DV_i terms. However, $E\,(R_i)$ can be studied closely. Competing in S^1 will be preferable to competing in S^2 with respect to $E\,(R_i)$:

If C_k competes in S^1, S^2 $\quad b \geqslant c$ if $E\,(D_i^1) \geqslant E\,(D_i^2)$ $\hspace{2cm}$ (7.5.1)

If C_k competes in S^1 $\quad\quad b \geqslant c$ if $E\,(D_i^1) \geqslant D^2 - E\,(D_i^2)$ $\hspace{1.5cm}$ (7.5.2)

If C_k competes in S^2 $\quad\quad b \geqslant c$ if $D^1 - E\,(D_i^1) \geqslant E\,(D_i^2)$ $\hspace{1.5cm}$ (7.5.3)

If C_k competes nowhere $\quad b \geqslant c$ if $D^1 - E\,(D_i^2) \geqslant D^2 - E\,(D_i^2)$ $\hspace{1cm}$ (7.5.4)

Note that in equation (7.5.1), $E\,(D_i^1)$ and $E\,(D_i^2)$ are larger than their counterparts in other equations since C_k had "spread" his resources.

Equation (7.5) can be used to generate a number of propositions about when C_i is more likely to compete in S^1 than S^2. Recall that $E\,(D_{kj}) = D_j - E\,(D_{ij})$. Upon substitution, the conditions under which active competition in S^1 dominates S^2 are:

$$E\,(D_i^1) \geqslant E\,(D_i^2); E\,(D_i^1) \geqslant E\,(D_k^2); E\,(D_k^1) \geqslant E\,(D_k^2); E\,(D_k^1) \geqslant E\,(D_i^2) \quad (7.6)$$

These conditions state that competition in S^1 will be dominant for C_i only when both candidates expect to obtain more delegates in S^1 than either candidate expects to receive in S^2, *ceteris paribus*. If equation (7.6) is satisfied for C_i, it will be for C_k.

Proposition 7.4: In general, mixed strategy solutions are more likely to apply to the choice of competing in one primary rather than the other, *ceteris paribus*.

However, the conditions of these equations can be satisfied, leading to the following propositions:

Proposition 7.5: Ceteris paribus, active competition in the state selecting the smaller delegation can *not* be the dominant strategy for either candidate, while competition in the larger state can be.

Proposition 7.6: A dominant strategy is more likely if both states are "competitive" (i.e., $P_i^1 \cong P_i^2 \cong .5$). If so, proposition 7.5 applies.

Proposition 7.7: A dominant strategy is more likely if both states use PR than if one or both use WTA (and DL/PL falls "in between").

Proposition 7.8: If one or both states use WTA (DL/PL), competition in one (namely the larger) state primary will be dominant for both candidates only if there is substantial "risk" (i.e., if $E(D_i) > .5, P(P_i > .5)$ is less in a "risky" than more "certain" primary).

Proposition 7.9: If one or both of the PR states has a very large threshold (λ), then a dominant strategy is unlikely.

The above propositions rest on the condition that the opponent might choose any of the four options, and the results have both candidates matching strategies (i.e., competing in the larger state). Suppose, however, that, as with C_i, the payoffs for C_k imply that he, too, is choosing in which one state to campaign. Again, the large state is favored. However, suppose $E(DV_i^1)$ and $E(DV_k^2)$ are positive, then:

Proposition 7.10: Under the above conditions, C_i is likely to compete in S^1 where $E(DV_i^1)$ is positive and C_k in S^2 where $E(DV_k^2)$ is positive.

The irony of proposition 7.10 is that the candidates choose to compete in states where they expect to obtain the short-term "momentum" gain, and thus compete in different states, receiving no short-term gain.

Suppose now that C_k has a dominant strategy of competing in both states. Then, if C_i does not have a dominant strategy of competing in both states:

Proposition 7.11: Under the above conditions, C_i is likely to have a dominant strategy of competing in the state that yields the greater expected delegates.

The general implication of these propositions is that large states are more likely to be actively competed than small ones, that PR (and *not* WTA nor DL/PL) exaggerates this tendency, that the degree of activity will increase over the campaign, and that the degree of competitiveness of the state is perhaps more crucial than in earlier sections.

8. An Empirical Test

Candidates and media recognize overtly many of the assumptions of and propositions from this model [see *Aldrich*, 1978 and forthcoming]. Here the revealed behavior of Ford and Reagan in the 1976 Republican campaign will be examined.

a) *Operational Measures.* At this point, I will focus on only a few key variables. Others are more difficult to measure and are the source of continuing research. The basic data are in Table 1, discussed above.[7]) First, state primary delegations have been

---

[7]) Three of the 30 initial primaries are excluded. Montana and Vermont conducted so-called "beauty contests" in which voters expressed their presidential preference in primaries wholly unrelated to delegate selection. As such, the candidate should have been expected to ignore these contests, and Ford and Reagan did. Reagan failed to file for a position on the District of Columbia's primary ballot which consequently was not held and Ford delegates declared the winners. The D.C. primary was one in which Reagan would be particularly likely to have a dominant strategy of non-competition there (it was small, pro-Ford, and aligned with three more attractive primaries for Reagan). Thus, while excluded from further analysis, the actions of the candidates were consistent with the model in these three states.

dichotomized into "small" (less than the median or 45) and "large" ($D_j \geqslant 45$). Propositions indicated that some "small" states are strategically similar to large, particularly those in the earliest stages. In practice, this means that the two first primaries (New Hampshire and Massachusetts) are considered "large." The third distinction is states that use PR vs. those that use any T_j. Here, Rhode Island ($\lambda = 1/3$) is the exception. Based on proposition 7.9, this state is considered as "other than PR."

More problematic are measures of a priori probabilities over outcomes. Region of the state can be used as a first approximation. This measure is the traditional one [cf., Lengle/Shafer]. It seems particularly appropriate here. It would seem plausible that eastern states would be most likely to support Ford, southern and western states would be good ones for Reagan, and the rest would be competitive, a priori. These expectations are borne out. Of the convention votes cast for the two candidates from non-primary states, Ford won 93.2 percent of the vote from eastern states ($N = 90$), 51.2 percent from the midwest ($N = 215$, all border states used primaries), 23.4 percent from the south ($N = 128$) and 11.9 percent from the west ($N = 175$). Therefore, region is used as a first discrimination on P_{ij}, trichotomizing into "competitive" (midwest and border states), pro-Ford (east), and pro-Reagan (south and west).[8]

The final variable is the behavior of the candidates. An assistant coded what you see on the last column of Table 1 from contemporary media accounts, especially CQ Weekly Reports' pre- and post-primary analyses. Details may be found in Aldrich [forthcoming].

b) *Predictions.* The three categorizations generate twelve possible combinations of characteristics as indicated in Table 2, with individual assignments listed in Table 1.

| | Large[1]) | | Small | | N |
	PR	non-PR[2])	PR	non-PR	
Competitive[3])	[1] 1	[4] 6	[7] 4	[10] 3	14
Pro-Ford	[2] 1	[5] 3	[8] 0	[11] 1	5
Pro-Reagan	[3] 1	[6] 3	[9] 3	[12] 1	8
N	3	12	7	5	27[4])

[1]) A primary is large if it selects (1) at least 45 delegates or (2) if it is small but early.
[2]) Non-PR primaries include all WTA, DL and PL primaries and the large threshold PR primaries.
[3]) Competitive primaries include all Midwestern and border state primaries, plus New Hampshire and Florida. Pro-Ford primaries include all other eastern primaries, while Pro-Reagan primaries include all remaining southern and western primary states.
[4]) The two "beauty contest" and the D.C. primary are excluded.

Tab. 2: Distribution of 1976 Republican Primaries over Categories of Independent Variables

[8]) New Hampshire and Florida are considered "competitive" since early and widely reported public opinion polls of each state indicated that the primary would be close.

Different predictions are made about primaries on unique dates than about several primaries on the same day. For single primaries on the same date, the propositions in Section 6 are relevant. There are observations only for cells 2–7, inclusive. The predictions, therefore, are that primaries in cell 4 are more likely to be entered by both candidates than cell 7 and both cells are more likely to be competed by both than the remaining instances. Ford is more likely to contest a cell 5 than cell 2 primary, and Reagan a cell 6 than cell 3. Conversely Ford is predicted to be somewhat more likely to enter a cell 3 than 6 (and Reagan a cell 2 than 5) primary. Since there are no small, non-competitive primaries, there are not strong examples of primaries in which neither candidate is predicted to compete.

In the case of multiple primaries on the same date, the role of the T_j distinction is reversed. Here the prediction is that both are most likely to compete in primaries in cells 1, 4, 7, and 10, roughly in that order. Ford only is more likely to compete in a cell 2 than 5 primary and Reagan in a cell 3 than 6. Both are least likely to enter cell 8, 9, 11, and 12 primaries. In this case we can make the prediction that two primaries are fairly likely to be contested, and also predict which one (or more) primary on a given date is more likely to be contested by one or both candidates.

c) *Results: Unique Date Primaries.* Table 3 compares predicted with actual candidate behavior for unique date primaries. Table 3a examines where both were active. If a primary falls in cell 4 or 7, it is predicted that both would actively contest it. Both

		Actual	
		Both Competed	At least One Did Not
Predicted	Both[1]) Competed	3	0
	At Least[2]) One Did Not	2	2

[1]) Cell 4 in Table 2
[2]) Cells 2, 3, 5, 6 in Table 2

Tab. 3a: Both Compete versus At Least One Failure to Compete

Cell in Table 2		Ford Competed	Did Not	Cell in Table 2		Reagan Competed	Did Not
Most Likely To Compete	4	3	0	Most Likely To Compete	4	3	0
	5	1	0		6	1	0
	2	0	1		3	1	0
	3	1	0		2	0	1
Least Likely To Compete	6	1	0	Least Likely To Compete	5	0	1

Tab. 3b: Predictions About Individual Candidates

Tab. 3: A Test of the Propositions for Single Primaries on a Date

candidates did compete in all three such cases and were less likely to compete jointly elsewhere. As expected, each competed in more primaries than not.

Table 3b contains the predictions on an individual candidate basis. Reagan fits the predictions perfectly. Ford does not. He might have been expected to compete in Massachusetts and Pennsylvania. He was expected to be especially unlikely to challenge Reagan in Texas and only somewhat less so in North Carolina. In both cases, Ford lost and suffered an unusually severe set back. Before North Carolina, Republican leaders were calling for Reagan's withdrawal, and the result surprised both candidates. *NY Times* reporter Naughton said [*Witcover*, p. 414], "they (Reagan and his staff) were dumbfounded. They didn't know how to handle it." Ford strategist Teeter has said [*Moore/Fraser*, pp. 39, 44]: "(Ford's initial strategy was that) he was entering all the primaries . . . North Carolina changed it immensely . . . At one point we thought we had a substantial lead in North Carolina, and decided not to incur the cost of doing another poll closer to the election, or of sending the president in during the last week or so of the campaign. There is no guarantee that either move would have changed the result, of course, but I think the decisions were based on political inexperience." Texas was an even more severe blow to Ford, completely changing the campaign [see *Aldrich*, forthcoming]. In short, while these two might be errors in prediction, it may have behooved Fort to fail to challenge Reagan in these primaries.

d) *Multiple Primaries on the Same Date.* Assuming that propositions in section 7 generalize to the case of more than two primaries on the same date, there are three types of predictions that can be made. The first two are the date-aggregated analyses as above. Third, each date with two or more primaries can be examined individually. Doing so, several apparent anomalies can be attributed to the specific features of the primaries on the same date.

The results for the aggregation are found in Table 4. The first predictions are fairly strong (see Table 4a). Reagan competed in three states (Idaho, Alabama, and Nevada) where competition was predicted to be unnecessary, while Ford did so in Georgia. Neither competed in the very small South Dakota primary (cell 7), possib-

		Actual		
Prediction	Cell in Table 2	Both Competed	One Competed	Neither Competed
Both Compete	1	1	0	0
	4	2	1	0
	7	3	0	1
	10	1	0	2
Only One	3, 5, 6	1	3	0
	9	0	2	1
	11	0	0	1
Neither Compete	12	0	1	0

Tab. 4a: Date-Aggregated Predictions of Where Both, One or No Candidates are Likely to Compete

Cell in Table 2	Ford Competed	Did Not	Cell in Table 2	Reagan Competed	Did Not
Most Likely 1	1	0	Most Likely 1	1	0
4	3	0	4	2	1
7	3	1	7	3	1
10	1	2	10	1	2
5	2	0	6	2	0
11	0	1	9, 12	3	1
Least Likely 3, 6 9, 12	1	5	Least Likely 5, 11	0	3

Tab. 4b: Date-Aggregated Predictions by Candidate

Tab. 4: A Test of Propositions for Multiple Primaries on a Date

ly due to the three very large primaries held the next week. Basically both candidates competed where both were so predicted. Three "exceptions" (Wisconsin,
Maryland, and West Virginia) are due to aggregating by date.

Table 4b is even clearer. Reagan did not compete in any of the unlikely primaries,
and Maryland, South Dakota, West Virginia, and Wisconsin are the only exceptions
to the "more likely to compete" category. Ford was predicted to be likely to compete in 11 states and failed to do so only in Maryland, South Dakota, and West
Virginia. His competition in Georgia provides the only exception in the other direction. Since five of the eight exceptions are due to date-aggregation, the results are
strongly supportive of the model.

Some of the exceptions can be accounted for by examining the set of coincidential
primaries individually. The first such pair is Wisconsin and New York. Both candidates were expected to be active in Wisconsin, and both concentrated personal and
campaign resources there. After the preceding primary, Reagan reassessed his position and decided to withdraw from Wisconsin, appeal for support on national television (an unprecedented move), and concentrate on Texas. In general, the rewards
of this extraordinary behavior were substantial, but the Wisconsin campaign staff
lamented his last minute withdrawal. Even Ford aide Kay felt that "Wisconsin was
(Reagan's) major mistake of the whole year" and said "We were damned lucky and
it put us back on the tracks after North Carolina" [Witcover, p. 416].
Maryland, as a small, non-PR primary is predicted to be less likely to be contested
than the large, PR primary in Michigan on the same day. Proposition 7.5 states that
in choosing which one of two primaries to contest, both candidates should tend to
choose to compete in the same state. Thus, the behavior predicted is found exactly
in this case.

This proposition is also supported in the Nebraska-West Virginia case. However, the
two primaries are identical on all variables considered here. Why, then, did they
contest Nebraska and not West Virginia? The most plausible answer may be costs.
West Virginia had the unusual feature of listing the names of citizens desiring to become delegates on the ballot, but *not* indicating what presidential aspirant they sup-

ported. The organizational costs required to communicate this information were unusually high.

In sum, the model's predictions are generally well supported in the one campaign analyzed here. Moreover, the exceptions fall into several categories; those that appear to be attributable to weaknesses in the measurements available (e.g., West Virginia) those that may be considered as "errors" in actual candidate behavior (e.g., North Carolina and Texas) or to unusual decisions by the candidates (e.g., Wisconsin), and finally those that are erroneous predictions (or perhaps the result of a sort of mixed strategy calculation).

9. Conclusions

The model of the presidential primary campaign developed in this paper indicates the richness and texture of this unusual mechanism for electoral choice. While the primaries present a bewildering array of institutional features, it is clear that some general and testable propositions are derivable from the major structures encountered. The propositions derived from the model account for the behavior of the actors in this complex game exceedingly well. They are by no means exhaustive of anything but the data. Nonetheless, there are two clear questions that further research must address. First, is it possible to produce (by assumption or deduction) a utility function that reasonably evaluates the costs and benefits of the components here considered separately? If so, more general solutions to this game become possible. More importantly, is it possible to generalize the model to the n-candidate case? If so, the domain of applicability would be generalized substantially.

References

Aldrich, J.: Candidate Behavior in the Presidential Primaries. Paper delivered at Public Choice Society Meetings, 1978.
– : A Dynamic Model of Pre-Convention Campaigns. Paper prepared for delivery at Midwest Political Science Association Meetings, 1979.
– : Before the Convention. A Theory of Campaigning For the 1976 Presidential Nominations. Forthcoming.
Brams, S.: The Presidential Election Game. New Haven 1978.
Congressional Quarterly Weekly Reports 34. Weekly throughout the campaign. 1976.
Kuhn, H.: Extensive Games and the Problem of Information. Contributions to the Theory of Games, II. Ed. by H. Kuhn and A. Tucker. Annals of Mathematical Studies **28**, Princeton 1953.
Lengle, J., and *B. Shafer*: Primary Rules, Political Power, and Social Change. APSR 70, 1976, 25–40.
Moore, J., and *J. Fraser* (Eds.): Campaign For President Campaign Managers Look At '76. Cambridge 1977.
Rae, D.: The Political Consequences of Electoral Laws. Rev. ed. New Haven 1971.
Straffin, P.: The Bandwagon Curve. AJPS 21, 1977, 695–710.
Tufte, E.: The Relationship Between Seats and Votes in Two-Party Systems. APSR 67, 1973, 540–554.
Witcover, J.: Marathon. The Pursuit of the Presidency 1972–1976. New York 1977.

Applied Game Theory. 1979 © Physica-Verlag, Wuerzburg/Germany

A New Campaign Resource Allocation Model

By *M. Lake*, Brooklyn[1])

Abstract: A model of campaign resource allocation in a two candidate race is developed under the assumption that the candidates wish to maximize their probability of winning the election. If the candidates have equal budgets it is locally optimal for them to allocate in proportion to the Banzhaf power index of each voter in the electorate. When this result is applied to the United States Electoral College, it is seen that the disparity in the power between voters in small and large states induces presidential candidates to spend a disproportionately large amount of their funds in the large states.

1. Introduction

Brams/Davis [1973] develop a model of resource allocation in presidential campaigns which we examine briefly. Then, we will modify their model so that the goal of the candidates is to maximize their probability of winning the election rather than to maximize their expected electoral vote. We shall extend the model so that it applies to all elections which are representable as simple *n*-person voting games. We will show that in the case where the election can be represented by a decisive voting game (that is, one in which two complementary coalitions of voters can be neither both winning nor both losing) there exist local equilibrium strategy solutions for the candidates, provided that they have equal budgets. We will then apply our results to the case of the U.S. Electoral College.

In the Brams/Davis model, there are two competing candidates, Republican and Democrat. The electorate is divided into two classes, committed and uncommitted. The committed voters have already made up their minds as to which candidate they will vote for. The uncommitted voters randomize their choice between the two candidates, where

$$p_i = \frac{r_i}{r_i + d_i}$$

is assumed to be the probability that an uncommitted voter in state *i* votes for the Republican candidate, and r_i and d_i are the respective expenditures in state *i* by the Republican and Democrat. That is, the probability that an uncommitted voter in state *i* votes Republican is equal to the proportion of the total resources allocated to state *i* spent by the Republican.

---

[1]) *Mark Lake*, 400 Argyle Road, Brooklyn, N.Y. 11218, USA.

The candidates are assumed constrained by fixed budgets, so that $\sum_i r_i = R$, and $\sum_i d_i = D$, where R and D are the budgets of the two candidates. It is further assumed that the committed voters in each state are split 50–50 between the two candidates. Thus, winning a majority of uncommitted voters in a state is equivalent to winning that state since, under the Electoral College, the candidate receiving a majority of votes in a state wins all of the state's electoral votes.

Brams/Davis assume that the goal of each candidate is to maximize his expected electoral vote. Specifically, if n_i is the number of uncommitted voters in state i (n_i is assumed to be even for all i), the Republican maximizes

$$E = \sum_{i=1}^{50} v_i q_i,$$

where v_i is the number of electoral votes of state i, and

$$q_i = \sum_{k=n_i/2+1}^{n_i} \binom{n_i}{k} p_i^k (1-p_i)^{n_i-k}$$

is the probability that the Republican wins state i. The Democrat seeks to minimize E.

Assuming that the candidates have equal budgets, $R = D$, *Brams/Davis* show that the strategies

$$r_i = d_i = \frac{v_i \sqrt{n_i}}{\sum_i v_i \sqrt{n_i}} R \tag{1}$$

are at a local equilibrium. That is, given that one candidate adopts the strategy (1), it is nonoptimal for the other candidate to choose a strategy deviating by very small amounts form (1).

If the number of uncommitted voters is assumed proportional to the number of electoral votes in each state, (1) becomes

$$r_i = d_i = \frac{v_i^{3/2}}{\sum_i v_i^{3/2}} R.$$

Brams/Davis term this the "3/2's rule."

One of the main difficulties with the Brams/Davis model is their assumption that the candidates wish to maximize their expected electoral votes. Actually, one would expect that the candidates wish to maximize their probability of winning the election, or, equivalently, receive a majority of electoral votes, which is quite different from the Brams-Davis goal assumption.

As an example, consider a system consisting of only two states, one with three electoral votes and one with two electoral votes. A simple majority of three electoral votes is needed by a candidate to win the election. Clearly, the two-vote state has no effect on the outcome of the election. Thus, if the candidates wish to maximize their probability of winning the election, they will spend all of their resources in the three-vote state and ignore the two-vote state. In fact, given the goal of maximizing one's probability of winning, this strategy can easily be verified as the unconditionally best strategy for each candidate no matter what the other candidate does; it is independet of any assumptions as to the relative size of the candidates' budgets or the distribution of committed and uncommitted voters.

The case is quite different if the candidates wish to maximize their expected electoral vote. For suppose the candidates have equal budgets. If the Republican spends everything in the three-vote state, it is optimal for the Democrat to spend a very small amount in the two-vote state and the remainder in the three-vote state. With this strategy, the Democrat will win the two-vote state with certainty and still have an almost 50–50 chance of winning the three-vote state. Thus, the Democrat has an expected electoral vote of (approximately) 3 1/2 as compared to 1 1/2 for the Republican. So the strategy of spending everything in the larger state is nonoptimal when the candidates seek to maximize their expected electoral vote.

2. The Model

We will assume that the candidates wish to maximize their probability of winning the election[2]). Rather than restrict our attention to U.S. presidential elections, we shall consider a more general setting. Again, we have two candidates, Republican and Democrat, with fixed budgets R and D, respectively.

Definition: An n-person voting game is a pair $\Gamma = (N, W)$ where N contains the n voters of Γ and W is a set consisting of the winning coalitions of Γ.

A voter $i \in N$ is said to be critical for a coalition $S \in W$ if $i \in S$ and $S - \{i\} \notin W$.

The election is an n-person voting game Γ in which the candidates make expenditures directly to the individual voters, all of whom are assumed to be uncommitted. As before, the probability that voter i votes Republican is

$$p_i = \frac{r_i}{r_i + d_i},$$

where r_i and d_i are now the expenditures made to voter i by the Republican and Democrat, respectively.

Given that the voting decisions of all voters are independent, the Republican seeks to maximize

$$P = \sum_{S \in W} \pi(S),$$

---

[2]) A quite different model based on this goal assumption is discussed by *Young* [1978]. However, our results are not directly comparable to Young's as he only analyses the 51 player state game and not the composed game in which the players are the individual voters.

where

$$\pi(S) = \prod_{i \in S} p_i \prod_{i \notin S} (1 - p_i)$$

is the probability that S is the set of voters who vote Republican[3]). We will assume that $\Gamma = (N, W)$ is a simple voting game, that is, if $S_1 \in W$ and $S_1 \subset S_2$, then $S_2 \in W$.

This condition means that a winning coalition remains winning when new voters are added to it.

For any $i \in N$, we can partition W into three classes:

W_i = the set of all coalition in W for which voter i is critical,
$\bar{W}_i$ = the set of all coalitions in W of which voter i is a member but is not critical, and
W_i' = the set of all coalitions in W of which voter i is not a member.

Note that W_i, $\bar{W}_i$, and W_i' are pairwise disjoint sets. There is a very close relationship between the coalitions of $\bar{W}_i$ and W_i'.

Lemma: $\displaystyle\sum_{S \in \bar{W}_i} \pi(S) = \sum_{S \in W_i'} \pi(S \cup \{i\})$.

That is, the probability that the Republican wins the election without voter i's support is exactly equal to the probability that he wins the election with voter i's support and voter i is not critical.

Proof: It is easily seen that if $S \in \bar{W}_i$, then $S - \{i\} \in W_i'$, and that if $S \in W_i'$, then $S \cup \{i\} \in \bar{W}_i$. Thus we see that

$$\bar{W}_i = \{S \cup \{i\} \mid S \in W_i'\},$$

and the lemma follows immediately.

Now suppose $S \in W_i'$, and that $x_1, x_2, \ldots, x_k$ are the elements of S and i, $x_{k+1}, \ldots, x_{n-1}$ are the members of $N - S$.

Then,

$$\pi(S) = (1 - p_i) p_{x_1} p_{x_2} \cdots p_{x_k} (1 - p_{x_{k-1}}) \cdots (1 - p_{x_{n+1}}), \text{ and}$$

$$\pi(S \cup \{i\}) = p_i p_{x_1} \cdots p_{x_k} (1 - p_{x_{k+1}}) \cdots (1 - p_{x_{n-1}}).$$

---

[3]) The function P is very similar to what *Owen* [1975a, 1975b] calls the "multilinear extension" of a voting game Γ. *Owen* defines this multilinear extension as

$$\sum_{S \in W} \prod_{i \in S} x_i \prod_{i \notin S} (1 - x_i)$$

where the x_i are independent variables.

Thus,

$$\pi (S) + \pi (S \cup \{i\}) = p_{x_1} \cdots p_{x_k} (1 - p_{x_{k+1}}) \cdots (1 - p_{x_{n-1}}),$$

which does not depend on p_i.

Now M_i' the probability that the Republican wins the election with voter i being noncritical is given by

$$M_i' = \sum_{S \in \bar{W}_i} \pi (S) + \sum_{S \in W_i'} \pi (S) = \sum_{S \in W_i'} \pi (S) + \sum_{S \in W_i'} \pi (S \cup \{i\})$$

$$= \sum_{S \in W_i'} \pi (S) + \pi (S \cup \{i\})$$

and is independent of p_i.

Consider $\sum_{S \in W_i} \pi (S)$. Since $i \in S$ for all $S \in W_i$, p_i divides $\sum_{S \in W_i} \pi (S)$. Let

$$M_i = (\sum_{S \in W_i} \pi (S)) / p_i.$$

M_i is the probability that the Republican wins the election and i casts a critical vote. Then $P = \sum_{S \in W} \pi (S) = \sum_{S \in W_i} \pi (S) + \sum_{S \in \bar{W}_i} \pi (S) + \sum_{S \in W_i'} \pi (S) = p_i M_i + M_i'$, where M_i and M_i' are independet of p_i.

To maximize P we use the method of Lagrange multipliers and from the function

$$P_\lambda = P - \lambda (\sum_{i=1}^{n} r_i - R).$$

Then

$$\frac{\partial P_\lambda}{\partial r_i} = \frac{\partial P}{\partial r_i} - \lambda = \frac{d_i}{(r_i + d_i)^2} M_i - \lambda.$$

Letting $\dfrac{\partial P_\lambda}{\partial r_i} = 0$ for all i, we have

$$\frac{d_i}{(r_i + d_i)^2} M_i = \lambda, \quad i = 1, 2, \ldots, n. \tag{2}$$

With the assumption we have made so far, we cannot progress any further. Even if the d_i are known, we cannot, in general, find closed form solutions for the r_i because not enough is known about M_i. To solve for the r_i, we assume that the candidates have equal budgets, $R = D$. It is clear that the game played by the candidates is zero-sum if the voting game Γ is decisive.

Definition: A voting game $\Gamma = (N, W)$ is said to be *decisive* if the complement of every winning coalition is losing and the complement of every losing coalition is winning.

Thus, a decisive game is one in which any vote is guaranteed to produce a clear-cut outcome, i.e., no ties are possible.

We shall henceforth assume that Γ is decisive. Coupled with the assumption that $R = D$, the game played by the candidates becomes *symmetric*. That is, if A and B are any two possible allocations, then the payoff to the Republican if he chooses strategy A and the Democrat chooses strategy B is the same as the payoff to the Democrat if he chooses strategy A and the Republican chooses strategy B. Thus, under these new assumptions, if a solution in pure strategies exists, the optimal strategies for the two candidates will be identical. So to find these strategies, we may set $r_i = d_i$, for all i. Then $p_i = 1/2$ for each i and equations (2) reduce to

$$\left(\frac{1}{4\,r_i}\right) \left(\sum_{S \in W_i} 2^{1-n}\right) = \lambda,$$

$$\sum_{S \in W_i} (1) = 2^{n+1}\, r_i\, \lambda,$$

or

$$c_i = 2^{n+1}\, r_i\, \lambda, \quad i = 1, 2, \ldots, n, \tag{3}$$

where $c_i = |W_i|$ is the number of critical defections for voter i.

Summing (3) over i, we have

$$\sum_{i=1}^{n} c_i = 2^{n+1}\, R\, \lambda.$$

Dividing (3) by this, we obtain

$$r_i^* = \frac{c_i}{\displaystyle\sum_{i=1}^{n} c_i}\, R = \frac{c_i}{\displaystyle\sum_{i=1}^{n} c_i}\, D = d_i^*,$$

where the asterisks denote optimal values.

Now $\beta_i = \dfrac{c_i}{\displaystyle\sum_{i=1}^{n} c_i}$ is just the Banzhaf power of voter i, and we have

$$r_i^* = \beta_i R = \beta_i D = d_i^*. \tag{4}$$

The Banzhaf power of a voter is just proportional to the probability that he is critical on a vote, given that all voters have a probability of $1/2$ of voting for either candidate [see *Banzhaf*, 1965; *Brams*, 1975; *Lucas; Straffin*, 1976a].

We next have that the strategies given by (4) are in local equilibrium when the voting game Γ is decisive. Setting $d_i = \beta_i D$ for all i, we must show that P, the probability that the Republican wins, considered as a function of the n variables r_i, $i = 1, 2, \ldots, n$, is

indeed at a local maximum when $r_i = \beta_i R$, for all i, given the constraint $\sum_{i=1}^{n} r = R$.

We will use the Hessian method. Let $x_0 = (\beta_1 R, \ldots, \beta_n R)$, and let A be the set of all $x = (x_1, x_2, \ldots, x_n)$ such that $\sum_{i=1}^{n} x_i = R$. Any $x \in A$ can be written as $x_0 + h$ for some $h = (h_1, h_2, \ldots, h_n)$ such that $\sum_{i=1}^{n} h_i = 0$. Similarly, for any such h, $(x_0 + h) \in A$. Let B be the set of all h such that $\sum_{i=1}^{n} h_i = 0$.

The Hessian of P is the $n \times n$ matrix $H = \dfrac{\partial^2 P}{\partial r_i \, \partial r_j}$. It is easily seen that all the second order partial derivatives of P are continuous in a neighborhood of x_0, so that

$$\frac{\partial^2 P}{\partial r_i \, \partial r_j} = \frac{\partial^2 P}{\partial r_j \, \partial r_i} \quad \text{for all } i \text{ and } j. \text{ Thus}$$

$$H_{x_0} = \left(\frac{\partial^2 P}{\partial r_i \, \partial r_j} \right) \bigg|_{x_0}$$

is a symmetric matrix and determines a quadratic form.

To show that P restricted to the set A is at a local maximum at x_0, we must show that H_{x_0} is negative definite of B, that is $H_{x_0}(h) < 0$ for all $h \in B$, $h \neq 0$.

To calculate $\dfrac{\partial^2 P}{\partial r_i \, \partial r_j}$ when $i \neq j$, we must examine how voter i and voter j interact in the game Γ. Given $i, j \in N$ with $i \neq j$, we can partition W_i into four classes:

W_{ij}^1 = the set of all coalitions in W_i for which voter j is critical;

W_{ij}^2 = the set of all coalitions in W_i of which voter j is a member but is not critical;

W_{ij}^3 = the set of all coalitions S in W_i of which voter j is not a member and where, if voter i is replaced by voter j in S, S remains a winning coalition;

W_{ij}^4 = the set of all coalitions S in W_i of which voter j is not a member and where, if voter i is replaced by voter j in S, S becomes a losing coalition.

Note that W_{ij}^1, W_{ij}^2, W_{ij}^3 and W_{ij}^4 are pairwise disjoint sets. Analogous to Lemma 1, we have.

Lemma 2: $\displaystyle\sum_{S \in W_{ij}^2} \pi\,(S) = \sum_{S \in W_{ij}^4} \pi\,(S \cup \{j\})$.

Proof: If $S \in W_{ij}^2$, then clearly $S - \{j\} \in W_{ij}^4$. On the other hand, if $S \in W_{ij}^4$, then $(S - \{i\}) \cup \{j\} = (S \cup \{j\}) - \{i\}$ is losing. But S is winning, so $S \cup \{j\}$ is winning, and

$S \cup \{j\} \in W_{ij}^2$. Thus we see that

$$W_{ij}^2 = \{S \cup \{j\} \mid S \in W_{ij}^4\},$$

and the lemma follows.

It is easy to see that $W_{ij}^4 \subset W_j'$. So if $S \in W_{ij}^4$, we have $\pi(S) + \pi(S \cup \{j\})$ is independent of p_j. Applying Lemma 2, we have $\sum_{S \in W_{ij}^2} \pi(S) + \sum_{S \in W_{ij}^4} \pi(S) = \sum_{S \in W_{ij}^4} \pi(S \cup \{j\}) +$

$+ \sum_{S \in W_{ij}^4} \pi(S) = \sum_{S \in W_{ij}^4} \pi(S) + \pi(S \cup \{j\})$ is independent of p_j.

Lemma 3: If $\Gamma = (N, W)$ is a decisive game, $\mid W_{ij}^1 \mid = \mid W_{ij}^3 \mid$ for all $i, j, i \neq j$.

Proof: We will establish a one-to-one correspondence between the elements of W_{ij}^1 and W_{ij}^3. Let $S \in W_{ij}^1$. Since Γ is decisive, $N - S$ is a losing coalition. Since i is critical for S, $(N - S) \cup \{j\}$ is winning. Thus $(N - S) \cup \{i\} \in W_{ij}^3$.

Now let $S \in W_{ij}^3$. Then $S - \{i\}$ is a losing coalition. Since (N, W) is decisive, $(N - S) \cup \{i\}$ is winning and i is critical because $N - S$ must be losing. But j is also critical for $(N - S) \cup \{i\}$ because $(S - \{i\}) \cup \{j\}$ is winning (since $S \in W_{ij}^3$). This means that

$$N - [(S - \{i\}) \cup \{j\}] = [(N - S) \cup \{i\}] - \{j\}$$

is losing. So we have $(N - S) \cup \{i\} \in W_{ij}^1$.

Thus we have a one-to-one correspondence between W_{ij}^1 and W_{ij}^3, and $\mid W_{ij}^1 \mid = \mid W_{ij}^3 \mid$.

Theorem 1: The strategies given by (4) are in local equilibrium when Γ is decisive.

Proof: We will first show that $\dfrac{\partial^2 P}{\partial r_i \, \partial r_j}\bigg|_{x_0} = 0$ for $i \neq j$. Consider

$$M_i = \frac{1}{p_i} \sum_{S \in W_i} \pi(S)$$

$$= \frac{1}{p_i} [\sum_{S \in W_{ij}^1} \pi(S) + \sum_{S \in W_{ij}^2} \pi(S) + \sum_{S \in W_{ij}^3} \pi(S) + \sum_{S \in W_{ij}^4} \pi(S)].$$

It is clear that $p_i p_j$ devides $\sum_{S \in W_{ij}^1} \pi(S)$, and that $p_i(1 - p_j)$ divides $\sum_{S \in W_{ij}^3} \pi(S)$. Let

$M_{ij} = \dfrac{1}{p_i p_j} \sum_{S \in W_{ij}^1} \pi(S)$ and $M_{ij}' = \dfrac{1}{p_i(1 - p_j)} \sum_{S \in W_{ij}^3} \pi(S)$. We have

$$M_i = p_j M_{ij} + (1 - p_j) M_{ij}' + \frac{1}{p_i} [\sum_{S \in W_{ij}^2} \pi(S) + \sum_{S \in W_{ij}^4} \pi(S)],$$

where M_{ij}, M'_{ij}, and $\dfrac{1}{p_i} [\sum_{S \in W^2_{ij}} \pi(S) + \sum_{S \in W^4_{ij}} \pi(S)]$ are all independent of p_i and p_j.
Now,

$$\frac{\partial^2 P}{\partial r_j \, \partial r_i} = \frac{\partial}{\partial r_j} \left(\frac{\partial P}{\partial r_i} \right) = \frac{\partial}{\partial r_j} \left(\frac{d_i}{(r_i + d_i)^2} M_i \right) = \frac{d_i}{(r_i + d_i)^2} \cdot \frac{d_j}{(r_j + d_j)^2} [M_{ij} - M'_{ij}].$$

Evaluating the above expression at x_0 we get,

$$\frac{\partial^2 P}{\partial r_i \, \partial r_j} \bigg|_{x_0} = \frac{1}{2^{n+2} d_i \, d_j} [\,|W^1_{ij}| - |W^3_{ij}|\,] = 0.$$

Thus H_{x_0} is a diagonal matrix.

If a voter happens to be a dummy in Γ, then it is clear that both candidates should spend nothing on him. So a dummy voter has no effect on the outcome of the election and no effect on the spending of the candidates. We may then assume without loss of generality that there are no dummies in Γ.

Since $d_i = \beta_i R$, $d_i = 0$ if and only if $\beta_i = 0$. But then voter i would have to be a dummy for d_i to be zero. Since we are assuming no dummies in Γ, $d_i > 0$ for all i. Thus,

$$\frac{\partial^2 P}{\partial r_i^2} = \frac{-\partial d_i}{(r_i + d_i)^3} < 0 \text{ for all } i.$$

Now we have shown that H_{x_0} is a diagonal matrix with negative numbers down the diagonal. This shows that H_{x_0} is negative definite over R^n and is negative definite over B.

Q.E.D.

This local equilibrium result can be extended to the class of all simple games, but we shall not prove this here because it is difficult to interpret nondecisive voting games in the context of elections. One problem which arises immediately is how to deal with the possibility of ties.

It is not known under that conditions the local equilibrium result given by (4) is a global equilibrium. We shall return to this question later.

3. Modifying the Model

Definition: Two voters $i, j \in N$ are said to be *perfectly symmetric* if the set of winning coalitions W remains unchanged when i and j are permuted in N.

The usual definition of symmetry between two voters in N is that there exists some permutation of N, carrying i into j and leaving W fixed. For example, two voters in a representative democracy are symmetric if they are in equally sized districts. However, two voters in a representative democracy are perfectly symmetric if and only if they are in the same district.

The definition of perfect symmetry can be restated in a more usable form. Two voters i and j are perfectly symmetric if and only if $S \in W$ and $j \notin S$ implies $(S - \{i\}) \cup \{j\} \in W$, and $S \in W$ and $i \notin S$ implies $(S - \{j\}) \cup \{i\} \in W$.

Lemma 4: If voter i and voter j are perfectly symmetric,

$$W_{ij}^2 = W_{ij}^4 = W_{ji}^2 = W_{ji}^4 = \emptyset.$$

Proof: This lemma follows immediately from the restated definition of perfect symmetry.

Lemma 5: $M_{ij} = M_{ji}$, and $M'_{ij} = M'_{ji}$.

Proof: It is obvious that $W_{ij}^1 = W_{ji}^1$ since each is by definition the set of winning coalitions for which both i and j are critical. We then have

$$M_{ij} = \frac{1}{p_i p_j} \sum_{S \in W_{ij}^1} \pi(S) = \frac{1}{p_j p_i} \sum_{S \in W_{ji}^1} \pi(S) = M_{ji}.$$

Now it is easy to see that if $S \in W_{ij}^3$, $(S - \{i\}) \cup \{j\} \in W_{ji}^3$, and if $S \in W_{ji}^3$, then $(S - \{j\}) \cup \{i\} \in W_{ij}^3$. Thus,

$$M'_{ij} = \frac{1}{p_i(1 - p_i)} \sum_{S \in W_{ij}^3} \pi(S) = \frac{1}{p_j(1 - p_i)} \sum_{S \in W_{ji}^3} \pi(S) = M'_{ji}.$$

We now come to the main theorem of this section.

Theorem 2: If $i, j \in N$ are perfectly symmetric, then it is globally optimal (in the minimax sense) for each candidate to spend equal amounts on i and j. (Note that we are not assuming that Γ is decisive or that $R = D$.)

Proof: Suppose that the Democrat spends equally on i and j, allocating an amount b to both. Let the Republican spend $r_i = a + \epsilon$ on i and $r_j = a - \epsilon$ on j. Let all other allocations be fixed. Then,

$$\frac{dP}{dr_i} = \frac{d_i}{(r_i + d_i)^2} M_i - \frac{d_j}{(r_j + d_j)^2} M_j.$$

From Lemma 5 it follows that

$$M_i = p_j M_{ij} + (1 - p_j) M'_{ij},$$

and

$$M_j = p_i M_{ji} + (1 - p_i) M'_{ji}.$$

It follows that

$$\frac{dP}{dr_i} = \frac{d_i}{(r_i + d_i)^2} \, [p_j M_{ij} + (1 - p_j) M'_{ij}] - \frac{d_j}{(r_j + d_j)^2} \, [p_i M_{ji} + (1 - p_i) M'_{ji}]$$

$$= \frac{d_i}{(r_i + d_i)^2} \, [p_j M_{ij} + (1 - p_j) M'_{ij}] - \frac{d_j}{(r_j + d_j)^2} \, [p_i M_{ij} + (1 - p_i) M'_{ij}]$$

$$= \left[\frac{d_i}{r_i (r_i + d_i)} - \frac{d_j}{r_j (r_j + d_j)} \right] p_i p_j M_{ij} +$$

$$\left[\frac{1}{r_i + d_i} - \frac{1}{r_j + d_j} \right] \; (1 - p_i)(1 - p_j) M'_{ij}$$

$$= b \left[\frac{1}{(a + \epsilon)(a + b + \epsilon)} - \frac{1}{(a - \epsilon)(a + b - \epsilon)} \right] p_i p_j M_{ij}$$

$$+ \left[\frac{1}{a + b + \epsilon} - \frac{1}{a + b - \epsilon} \right] (1 - p_i)(1 - p_j) M'_{ij} .$$

Now it is easy to see that $\dfrac{dP}{dr_i} = 0$ when $\epsilon = 0$, $\dfrac{dP}{dr_i} < 0$ when $\epsilon > 0$ and $\dfrac{dP}{dr_i} > 0$

when $\epsilon < 0$. Thus, P is at a global maximum when $\epsilon = 0$ and $r_i = r_j$.

$$\text{Q.E.D.}$$

It follows immediately from Theorem 2 that if the election is determined by any form of majority rule, then the optimal strategy for each candidate is to spend equally on all voters.

The strategies given by (4) have an intuitive appeal; they say that the candidates should spend in proportion to the Banzhaf power of each voter. However, there is a difficulty in the model. This lies in the assumption that the candidates make their allocations directly to the individual voters. Theorem 2 allows us to eliminate this restrictive assumtion in the class of compound games where it is probably most suspect. Namely, suppose we have an election in which voters are divided into districts within each of which they are equally weighted, e.g., the U.S. Electoral College. It is clearly optimal under our model for the candidates to spend equally on all voters within the same district.

Now assume that the candidates play optimally. Then suppose that district i has n_i voters. Let the Republican allocate r_i to district i, and the Democrat allocate d_i to district i. Since the candidates spend equally on all voters in district i, the probability that any voter in district i votes Republican is

$$p_i = \frac{r_i / n_i}{\dfrac{r_i}{n_i} + \dfrac{d_i}{n_i}} = \frac{r_i}{r_i + d_i} . \tag{5}$$

Without in any way altering the mathematical model, we may simply assume that the candidates make lump sum expenditures to the districts, as in the Brams/Davis model, and define the probability that a voter in district i votes Republican by equation (5). Then Theorem 1 tells us that it is locally optimal to allocate to each district in proportion to the *sum* of the Banzhaf powers of the voters in the district.

Theorem 2 also allows us to answer, in part, the question of when the strategies given by (4) are globally optimal. Again suppose that the election is a compound game (with districts), where the outcome in each district is determined by simple majority rule. We will assume that no district is a dictator, that is, if W_D is the set of winning coalitions of districts and i is any district, $\{i\} \notin W_D$, though one or more (but not all) districts may have veto power. The probability that the Republican wins district i is then q_i defined in Section I.

As the number of voters in district i grows very large, q_i is approximated very closely by

$$\overline{q_i} = \begin{cases} 1, & \text{if } r_i > d_i \\ 1/2, & \text{if } r_i = d_i \\ 0, & \text{if } r_i < d_i. \end{cases}$$

[See also *Colantoni/Levesque/Ordeshook*, 1975a]. In fact q_i approaches $\overline{q_i}$ very rapidly as the number of voters increases.

Now suppose that the candidates have equal budgets, and that the Republican allocates according to (4). Then the Democrat can allocate nothing to the district with the least power, match the Republican's expenditures in the remaining districts and then use the funds that would have gone to the weakest district to outspend the Republican in all the remaining districts. These remaining districts must constitute a winning coalition because the weakest district cannot have veto power by our previous assumption. If the number of voters in each district is large enough, the Democrat will be almost certain of winning all but the weakest district and thus almost certain to win the election[4]). A similar argument is given by *Brams/Davis* [1973] to show that the 3/2's rule is in general not a global equilibrium solution.

This shows that whenever the number of voters in a district is large enough, the strategies given by (4) are not globally optimal. Because of the rapidity with which q_i approaches $\overline{q_i}$, we can say that the strategies given by (4) will in general not be globally optimal if Γ is a compound game. Thus, we must look to weighted voting games if we are to have any hope of showing that (4) gives a global equilibrium result. The only exception occurs when one of the districts is a dictator, as in the example of Section 1. In this degenerate case, the strategies given by (4) are in equilibrium.

4. The Electoral College

We can now apply our results to the case of the U.S. Electoral College. When there are only two candidates the Electoral College is essentially a decisive voting game. In the event of an even split of electoral votes, which is a current possibility due to the

---

[4]) The above argument need only be slightly modified to handle the case where all districts have veto power.

present allocation of electoral votes,, the election goes to the House of Representatives. Thus, the House may be considered as a 1-electoral vote state. Since it is so small
a body relative to the 50 states, its effect on the allocation of campaign resources by
the candidates may be considered as negligible. Also, the possibility of ties within a
state due to an even number of voters is remote enough to be ignored. Thus, we will
assume that the Electoral College is decisive.

From the results at the end of the last section we can be sure that no pure strategy
solutions exist for the candidates under the Electoral College when the candidates have
equal budgets. However, we can look to our local equilibrium result as a guide to how
candidates should behave in the absense of such solutions. Then our results reinforce
the argument that the Electoral College is biased in favor of large states. This is so because the Banzhaf power of voters in large states is greater than that of voters in small
states [see *Banzhaf*, 1978; *Owen*, 1975a].

Thus a voter in the largest state (California) has more than three times the Banzhaf
power of a voter in the District of Columbia. If the candidates allocated their resources
according to the sum of the Banzhaf powers of voters in a state, then California would
get over 15 % of the candidates' combined resources though it makes up less then
10 % of the population of the U.S. In fact, candidates would tend to spend over 90 as
much in California as in North Dakota even though California has only about 30 times
the population. Under our model, as in the Brams-Davis model, small states are seriously disadvantaged by the Electoral College bias.

It should be noted that for the Electoral College the Banzhaf and 3/2's rules give virtually the same results. The reason for this is that the 3/2's rule says the candidates should
spend in proportion to $v_i \sqrt{n_i}$, i.e. electoral votes times the square root of the population. It has been shown [*Owen*, 1975b; *Brams/Lake*, 1978] that the Banzhaf index
decomposes when applied to compound games. That is, Banzhaf power of a voter in
a compound game is the product of his power within his district and the power of his
district in the game played by the districts. In the U.S. Electoral College a voter's
power within a state is proportional to $1/\sqrt{n_i}$, while the power of a state in the state
game is close to proportional with its electoral votes. Thus, the power of an individual
in state i is given very closely by $v_i/\sqrt{n_i}$, and the sum over all voters in state i is
$n_i (v_i/\sqrt{n_i}) = v_i \sqrt{n_i}$ which coincides with the 3/2's rule.

Another scheme which has been proposed to describe how candidates should allocate their resources in presidential campaigns is the proportional theory [see *Colantoni*,
et. al., 1975a; *Young*]. Under the proportional theory the candidates allocate in proportion to the electoral votes of each state. A detailed comparison of the 3/2's (and
thus Banzhaf) and proportional allocation rules against actual campaign data for
recent elections is given by *Brams/Davis* [1974]. As a sample comparison Table 1 gives
the predictions of the different theories and the actual trip data for the 1976 campaign.
The states are grouped in the categories of small (less than 10 electoral votes), medium
(10 to 19 electoral votes), and large states (20 or more electoral votes). Also included
in the table are the allocations given by the formula devised for Jimmy Carter by
Hamilton Jordan [see *Schram*]. Also for a more complete analysis of spending in the
1976 campaign see *Brams* [1978].

	Small States	Medium States	Large States
Proportional rule	30.9	29.9	39.2
Banzhaf rule	15.7	25.9	58.4
3/2's rule	16.1	26.4	57.5
Jordan formula	41.9	28.8	29.3
Democratic slate	16.1	30.4	53.6
Republican slate	27.1	28.9	44.0

Tab. 1: Percent Allocations

Looking at the table we see that Democratic slate allocations come very close to the Banzhaf and 3/2's rules. The Republican slate allocations are close to the proportional rule though they are somewhat biased in favor of the large states. It is interesting to note that Jimmy Carter's original plan as given by the Jordan formula was biased in favor of the small states. However, in the end Carter's strategy was strongly biased in favor of the large states.

From Theorem 2 it follows that under direct popular election of a president it is globally optimal for the candidates to spend directly in proportion to the population of each state regardless of their respective budgets. Under direct popular election, candidates are induced to treat all voters equally. It has been argued that the Electoral College should be abolished because of the power disparities it creates. In point of fact the actual power of any single voter is negligible, so that while some voters may have as much as three times the Banzhaf power as other voters, this difference in itself is not really meaningful. But under our model these disparities among the powers of voters induce a substantial bias in the behavior of candidates toward the larger states. In addition to campaign funds and trips, we can also include planks in party platforms as part of campaign resources. We can then argue that the policies adopted by the candidates will tend to overrepresent the interests of large states.

Direct popular election of a president would avoid all these difficulties as well as another one. Since no pure strategy solutions exist under the Electoral College, it pays for the candidates to be secretive about their campaign strategies or to announce one plan and actually follow another. With direct popular election, pure strategy solutions always exist and are given by Theorem 2. Thus, candidates would find it less necessary to be secretive about their campaign moves.

5. Conclusion

In Section 2 we introduced an important modification to the Brams-Davis model that took into account that the goal of candidates is to maximize their probability of winning the election. This led to a local equilibrium result which said that the candidates should spend in proportion to the Banzhaf index of each voter. This result was modified in Section 3 to say that given perfect symmetry of voters within districts, the candidates should allocate to each district in proportion to the sum of the Banzhaf powers of the voters in each district. Finally, in Section 4 we applied our results to the Electoral College and argued that it should be abolished and replaced by direct popular election of a president.

While the empirical results for the Banzhaf allocation rule very closely matched the 3/2's rule in the special case of the Electoral College, our model is applicable in the general setting of all simple games; the Brams-Davis model is only applicable to elections which are structured like the Electoral College.

References

*Banzhaf, J.F.*III: Weighted Voting Doesn't Work: A Mathematical Analysis. Rutgers Law Review 19, 1965, 317–345.

–: One Man, 3.312 Votes: A Mathematical Analysis. Villanova Law Review 13, 1968, 304–332.

Brams, S.J.: Game Theory and Politics. New York 1975.

–: Resource Allocations in the 1976 Campaign. American Political Science Review 72, 1978.

Brams, S.J., and M.D. Davis: Resource-Allocation Models in Presidential Campaigning: Implications for Democratic Representation. Annals of the New York Academy of Sciences 210, 1973, 105–123.

–: The 3/2's Rule in Presidential Campaigning. American Political Science Review 68, 1974, 113–134.

–: Comment on 'Campaign Resource Allocations under the Electoral College. American Political Science Review 69, 1975, 155–156,

Brams, S.J., and M. Lake: Power and Satisfaction in a Representative Democracy. Proceedings on the Conference of Game Theory and Political Science, Ed. by P.C. Ordeshook. New York 1978, 529–562.

Colantoni, C.S., T.J. Levesque, and P.C. Ordeshook: Campaign Resource Allocations under the Electoral College. American Political Science Review 69, 1975a, 141–154.

–: Rejoinder to 'Comment' by Steven J. Brams and Morton D. Davis. American Political Science Review 69, 1975b, 947–953.

Lucas, W.F.: Measuring Power in Weighted Voting Systems. Case Studies in Applied Mathematics. 1976, 42–106.

Owen, G.: Evaluation of a Presidential Election Game. American Political Science Review 69, 1975a, 947–953.

–: Multilinear Extensions and the Banzhaf Value. Naval Research Logistics Quarterly 22, 1975b, 741–750.

Schram, M.: Running for President: A Journal of the Carter Campaign. New York 1978.

*Straffin, P.D.*Jr.: Power Indices in Politics. MAA Module in Applied Mathematics, Cornell University, 1976a.

–: Probability Models for Measuring Voting Power. Technical Report No. 320, School of Operations Research and Industrial Engineering, Cornell University, 1976b.

Young, H.P.: The Allocation of Funds in Lobbying and Campaigning. Behavioral Science, 1978.

Applied Game Theory. 1979 © Physica-Verlag, Wuerzburg/Germany

Cabinet Coalition Formation:
A Game-Theoretic Analysis

By *M. Winer*, Pittsburgh[1])[2])

Abstract: The use of N-person game theory solution concepts to predict cabinet coalitions in European parliaments has been fairly limited. This is because most concepts predict only the alternatives in the solution space and not the coalitions which achieve these alternatives. In order to make coalition predictions, additional ad-hoc assumptions are necessary. The Competitive Solution attaches a coalition structure to each alternative in the solution space. It is thus extremely applicable to this problem.

The cabinet formation game is formulated spatially with parties viewed as the active players. The parties, cohesive units, negotiate over possible policy positions for the potential governing coalitions. It is this bargaining that results in particular coalitions being able to form and others being unable to compete. The Competitive Solution for 11 cases from five countries is computed, and the results are compared to those of other prediction techniques. Of the solution concepts tested, the Competitive Solution and *Axelrod's* connected coalition theories are the most successful.

Introduction

Many democratic countries have multiparty political systems in which no single party is able to maintain sole leadership of the government. In parliamentary democracies, once members of parliament are elected, an attempt is made to form some coalition of seatholders large enough to control the government. This is usually accomplished by parties, rather than individual members, coalescing into a majority coalition. They then form a cabinet and run the government until the coalition breaks apart or a new election is held. Political scientists have long been interested in this process of coalition formation, one primary objective being the prediction of which coalitions might form. This paper is a continuation of that attempt.

One useful way to model this process is as an N-person 2-spatial game with the committee's alternatives corresponding to an N dimensional Euclidean space and each party's preferences described by a metric over this space (e.g., Euclidean distance). If one interprets the dimensions as issues (e.g., ideology or percent of the budget to be spent

[1]) *Mark Winer*, School of Urban and Public Affairs, Carnegie-Mellon University, Pittsburgh, Pennsylvania 15213, USA.

[2]) The author wishes to thank Richard McKelvey and Peter Ordeshook for their helpful comments and assistance. Any errors are the responsibility of the author.

on education), points in the space are then potential policy proposals. The parties are then bargaining over the policies of potential governments.

Unfortunately, much of cooperative game theory concerns the special case of side-payments with transferable utility. To satisfy the transferable utility assumption there must be a commodity that, when traded, preserves the total utility for the coalition. Legislative or parliamentary processes are more accurately modeled as non-sidepayment games. In these games conflict can be assumed to be over issues and ideology, and while coalition or party leaders might attempt to form winning coalitions by manipulating policy or allocating ministerships, we cannot assume that the set of outcomes correspond to alternative allocations of an infinitely divisible, unidimensional commodity such as money [c.f., *DeSwaan; McKelvey/Ordeshook/Winer*, 1978].

If we adopt this more general formulation of parliamentary activity, we find that game theory is itself deficient as an analytical tool. In the transferable utility case, numerous solution concepts such as the *V*-solution, the kernel, the nucleolis, and various bargaining sets exist to treat ("solve") games without cores (i.e., games without undominated imputations). One is faced then with the problem of choosing from among these competing definitions. For games that do not assume transferable utility these concepts are either undefined or, like the core, typically do not exist. Hence, for such games, rather than offering a plethora of competing definitions of a solution, game theory often provides no hypothesis about outcomes. Further, much of the work in game theory focuses on predicting the payoffs to the players and is silent on the question of what coalitions might form. This observation, in fact, leads Michael Taylor to conclude that formal solution theory is largely irrelevant to the study of government coalition formation. For example, one well-known attempt at coalition prediction in a game theoretic framework is *Riker*'s size principle. Unfortunately, since all existent theoretical justifications of that principle assume transferable utility it can not be applied to a simple spatial game modelling legislative or parliamentary processes. In addition, concepts which could be applied to spatial games, such as the *V*-solution, are principally mathematical abstractions without behavioral rationale. Hence, it is difficult to assess their applicability when rules constrain bargaining or negotiation procedures in a committee. The competitive solution makes both policy and coalition predictions and is, thus, uniquely applicable to the problem of predicting cabinet formations.

The paper begins with some essential notation and definitions of relevant game-theoretic concepts. Section 2 reviews the competitive solution, while Section 3 reviews several other hypotheses, drawn from earlier studies of parliamentary coalition formation, to which the competitive solution will be compared. Section 4 details our methodology and Section 5 presents the general results.

Section 1

We turn first to some basic notation:

$N = \{1, 2, \ldots, n\}$: the set of players.

$C \subseteq N$: a coalition.

A = set of all possible outcomes, here it is assumed $A \subseteq R^m$.

V_i = the weight of player i.

$\theta \in A$: a particular outcome.

$|C| = \Sigma V_i$: the combined weight of the players in the coalition;

$|N|$ then, is the total weight of all players.

$U = (U_1(\theta), \ldots, U_n(\theta)) \in R^n$: an ordered payoff vector, where U_i denotes
   the utility to player i.

$H \subseteq R^n$: the set of all possible payoff vectors.

In the context of cabinet coalition formation, the class of games that concern us is *Strong Simple Games* — games in which a coalition can guarantee any point in H or it can guarantee nothing more than the players can get by themselves. A coalition is said to be *winning* if it can guarantee any point in H and *losing* otherwise.[3]) In particular, if we let W denote the set of winning coalitions, then $W = \{C \subseteq N \mid |C| > |N|/2\}$ and the set of minimal winning coalitions is $W^* = \{C \in W \mid C - \{i\} \notin W \text{ for all } i \in C\}$. In this analysis we assume that the political parties are the players and the weights represent the number of seats each party holds. The actual legislators are not treated as independent actors but merely as members of a unified and centralized party. A winning coalition is one in which the total number of seats held by the member parties exceeds a majority of all seats.

Individual players (parties) have preferences over the alternatives, as determined by the utility they associate with particular outcomes. For example: for any $\theta_1, \theta_2 \in A$, $\theta_1 \underset{i}{\geqslant} \theta_2$ iff $U_i(\theta_1) > U_i(\theta_2)$. A coalition's preference is defined by the preferences of its members:

$$\theta_1 \underset{C}{\geqslant} \theta_2 \text{ iff } \theta_1 \underset{i}{\geqslant} \theta_2 \text{ for all } i \in C$$

$$\theta_1 \underset{C}{\geqslant} \theta_2 \text{ iff } \theta_1 \underset{i}{\geqslant} \theta_2 \text{ for all } i \in C.$$

The set of alternatives, A, can be given a number of substantive interpretations. Here we view the alternatives as the set of potential policies for the government.

If we view each policy issue as a dimension in a multidimensional space, then a point in this space represents a policy position on each issue. The players have preferences among the policies and are attempting to have the new government adopt policy position as favorable to them as possible. When a coalition with a majority of the parliamentary seats agrees to a set of policy positions, they can form a government.

A party's preferences over policies is represented by a utility function, $U_i : A \to R$, corresponding to the Euclidean metric. That is, there is a point $X_i \subset R^m$ s.t. $U_i(\theta) = f(\| \theta - X_i \|)$ for all $\theta \in A$, where f is a monotone decreasing function. The

---

[3]) We assume that there is no advantage to forming a blocking coalition, a coalition in which $|C| = |N|/2$. It will be treated like any other losing coalition.

point X_i is called player i's ideal point, the point that maximizes his utility, while the utility he associates with any other alternative decreases as one moves further away from X_i. For any given alternative, then, there is an associated payoff vector so that we can view the players as bargaining over points in the space rather than over payoff vectors. The utility a player receives from the formation of a particular coalition is solely a function of the policy agreement of the coalition members, not the player's membership in the coalition.

Unfortunately, it is not an easy matter to observe the policy outcomes of a governing coalition. What we can observe is the actual coalition that forms. While most solution concepts do not predict which coalition forms, the competitive solution offers not only a set of possible policy positions but also includes a coalitional structure in each proposal.

Section 2

While the competitive solution is an entirely general solution concept, it is defined here only for simple spatial games.

The competitive solution is based on the idea that a potential coalition must bid for its members in a competitive environment. Several coalitions are simultaneously attempting to form and each of these coalitions must attempt to bid for its critical members. Thus, if one player or set of players is pivotal between two coalitions and each has a chance to form, the pivotal players should be indifferent between the offers of both, lest their preferences insure that one of them does not form. This environment can result in some coalitions being unable to compete. Thus we must define not only the policy offers made by competing coalitions but also the coalitions that make "successful" offers.

To formalize this notion we define a *proposal*, p, by $C \subseteq N$ as an ordered pair (θ, C) such that $\theta \in A$ and $|C| > |N| / 2$. Let K be some set of proposals and m_K the set of coalitions making the proposals (i.e., $m_K = \{C \subseteq N \mid$ for some $(\theta', C') \in K, C' = C\})$.

For any $C_1, C_2 \in M$ with proposals $p_1 = (\theta_1, C_1)$, and $p_2 = (\theta_2, C_2)$.

> p_1 is *viable* against p_2 if it is *not* the case that
> $\theta_1 \underset{i}{\leqslant} \theta_2$ for all $i \in C_1 \cap C_2$.

If p_C is viable against all $p_{C'} \in K$ it is said to be *viable in K.*

> K is *balanced*, if for all $p_1, p_2 \in K$
> i) $C_1 \neq C_2$;
> ii) every (θ, C) is viable in K.
> A proposal (θ, C) *upsets* K if
> i) (θ, C) is viable in K;
> ii) for some $(\theta', C') \in K, \theta \underset{i}{\gg} \theta'$ for all $i \in C \cap C'$.
> K is *competitively balanced* if it is balanced and if for no $C \in m_K$
> is there a proposal (θ, C) that upsets K.
> K is a *competitive solution* if there is no proposal (θ, C) that upsets K.

The definition can be made clearer with the example in Fig. 1, which displays the calculated spatial positions of the six parties in the 1961 Norwegian parliament and their relative weights.

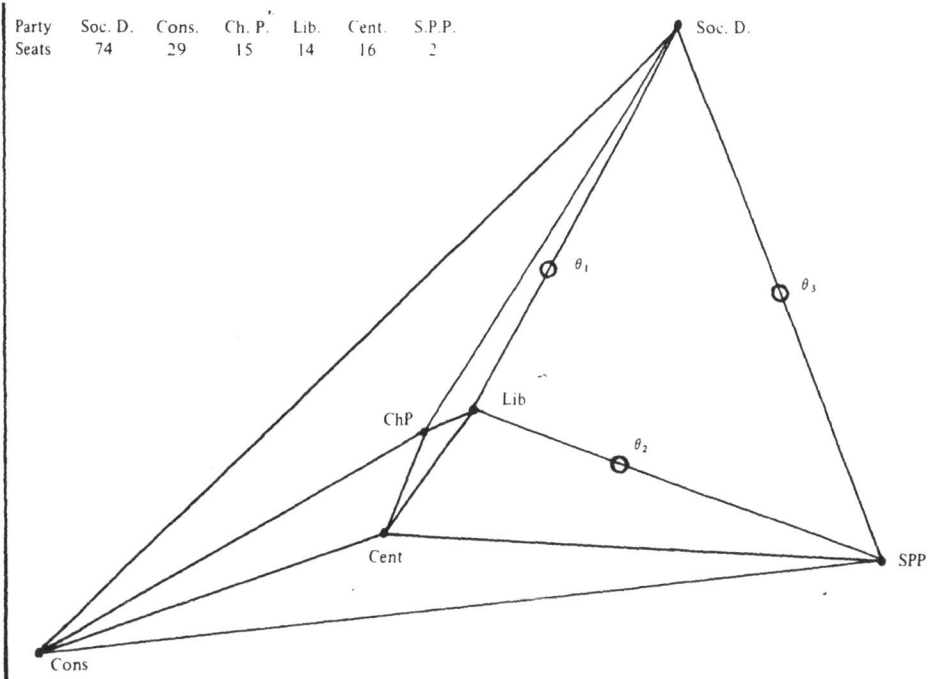

Party	Soc. D.	Cons.	Ch. P.	Lib.	Cent.	S.P.P.
Seats	74	29	15	14	16	2

Fig. 1

The set of minimum winning coalitions are either a coalition of the Social Democrats with any one of the other 5 parties or a coalition of all the other parties. Assuming that the parties evaluate any point in this 2-dimensional space in terms of its distance from their ideal point, the proposals

$$p_1 = (\theta_1; \{\text{Soc.D., Lib.}\})$$
$$p_2 = (\theta_3; \{\text{Soc.D., S.S.P.}\})$$
$$p_3 = (\theta_2; \{\text{Lib., S.S.P., Cons., Ch.P., Cent.}\})$$

are a competitive solution, K, for this game.

To see this, note first that p_1 is *viable* against p_2 since Soc.D. (the pivotal player) is indifferent between the two. Similarly, the Liberal party is indifferent between p_1 and p_3 so p_1 is *viable in* K. The same argument can be made for the other coalitions, so K is *balanced*.

To confirm that $\{p_1, p_2, p_3\}$ is *competitively balanced*, consider the possibility of another proposal by the coalition of the five small parties. Since the new proposal, θ^* must be viable in K, the Liberal party must be indifferent or prefer θ^* to θ_1 and the

S.P.P. must be indifferent or prefer θ^* to θ_3. The only point for which this is possible is θ_2. Finally, K is a *competitive solution* because no new coalition can upset the existing structure. If, for example, a (Soc.D., Ch.P.) coalition were to attempt to enter, it must have a proposal θ^* that is viable in K. To be viable Soc.D. must not prefer θ_1 to θ^* and Ch.P. must not prefer θ_2 to θ^*. No point where both these conditions hold exists.

Section 3

Previous attempts to model cabinet coalition formation can be placed in one of three categories. 1. Those that are game theoretic but, in which utility is assumed to be transferable; 2. those that use each party's desire for possession of cabinet ministerships to draw conclusions on the preferred size and number of parties in the coalition; and 3. those that use a one-dimensional policy space to search for some equilibrium or stable set of coalition structures. This section briefly surveys the work using each of these approaches.

1. A simple game with *sidepayments and transferable utility* is one in which a winning coalition gets a fixed payoff that can be divided in any way among the coalition members, while losing coalitions receive payoffs of zero. Assuming transferable utility presumes that there is some infinitely divisible commodity that players can transfer among each other that preserves the "total utility" of the coalition. Three major theories make use this condition.

   a) *Minimum winning coalitions: Riker's* size principle is the best known theory of this type. It can be summarized as: "in N-person, zero-sum games, where sidepayments are permitted, where players are rational, and they have perfect information, only minimal winning coalitions occur" [*Riker*, p. 32].

   b) *Minimal size:* Later in the same book, *Riker* argues that only those coalitions of smallest total weight above a majority will form. This hypothesis is more expressly made by *Gamson* [1961], and it is based on the additional assumption that "any participant will expect others to demand from a coalition a share of the payoff proportional to the amount of resources which they contribute" [*Gamson*, p. 376]. The minimal size theory predicts that the winning coalition with the smallest total weight will form.

   c) *Bargaining proposition: Leiserson* [1968, p. 775] hypothesizes that: "As the number of actors increase there is a tendency for each actor to prefer to form a minimal winning coalition with as few members as possible." The players, then, try to minimize the number of other players in the winning coalition.

2. The research that tests the view that ministerships are a linear representation of the rewards for coalition membership is best summarized in *Browne* [1971, 1973a, 1973b] and *Schofield* [1976]. *Browne* demonstrates the nearly one-to-one relationship between % ministerships and % seats for parties in the governing coalition. However, *Schofield* finds that, viewing the payoffs to a coalition as the distribution of portfolios, the differences between the actual distributions and those predicted by the Kernel are quite substantial. *Schofield* hypothesizes, then, that the explanation for such discrepancies is the importance of the policy space.

3. Several hypotheses use the policy position of parties to predict cabinet coalitions. Most of these predictions are based on the ordering of the parties in a one-dimensional space.

   a) *Connected coalitions: Axelrod*, in *Conflict of Interest*, predicts the formation of "minimal connected winning coalitions." Connected coalitions consist of coalitions of parties that are adjacent on the policy scale and, hence, minimize the "conflict of interest" of the members. Minimal connected coalitions are connected coalitions for which the deletion of either extreme member makes the coalition losing. This coalition can be greater than minimal winning.

   This concept can be extended to the multi-dimensional case if the players have a particular type of utility function. If for each player his utility for a given proposal by a coalition is a function of the Euclidean distance of the proposal from his ideal point then the "connected coalition theory" can be extended to a "minimum variance theory." For all winning coalitions the variance along each dimension of the members' ideal points is computed.[4] The dimensional variances are summed for each coalition producing a set of total variances. The winning coalition with the minimum variance is the one predicted to form.[5]

   b) *Policy distance: DeSwaan* [1973] determines the expected policy of a coalition from the policy position of its members and their weights. The preferences of the actors over the set of coalitions is then determined by hypothesizing that:
   1. a party always prefers a coalition he is a member of to one he is not, and
   2. among the remaining proposals he prefers the ones closer to his ideal point to ones further away and does not make comparisons between those to his right in the issue space.

   The predicted coalitions are then chosen to be those in which "there is no other winning coalition that is preferred by all actors who are members of both." [*DeSwaan*, glossary]

   The solution concepts of *DeSwaan* and *Axelrod*, as well as those of *Riker, Gamson* and *Leiserson* will be compared to the Competitive Solution in the empirical examples.

Section 4

To calculate the competitive solution we must locate the parties in a multidimensional issue space. In many countries more than one spatial map is used, but the basic data all comes from ICPR national surveys of the country's electorate. The spatial maps of the parties are obtained in one of two ways: 1. A quasi-unfolding technique [c.f., *Poole*], in which voters' evaluations of the political parties are used to find the

---

[4] The utilities must be a function of Euclidean distance for the combination of unidimensional variances to be reasonable.

[5] It is, however, unreasonable to always limit the predicted set to this single coalition. Differences may be so slight as to be unperceivable. As a rough rule of thumb, then, we assume that all coalitions with variances within 1/2 standard deviation of the minimum variance coalition are consistent with *Axelrod*'s original hypothesis.

best position for the ideal points of the parties; or 2. the results from research where spatial maps are obtained using different scaling techniques on similar data. A party's utility is assumed to decrease as a function of the Euclidean distance from its ideal point.[6]) In the countries analyzed, two or three spatial dimensions were recovered. A goodness of fit statistic is reported for each scaling based on the correlation between two sets of distances. 1. The voter's evaluation of the parties and 2. the distances in the recovered space from the voters' ideal points to the parties.

The competitive solution is found with an iterative program, based on the relationship between a Nash equilibrium and the competitive solution [see *McKelvey/Ordeshook*, 1977]. In this algorithm, however, the maximized function is non-differentiable; moreover the algorithm is rather complicated and time consuming. It is also possible to converge to a local optimum which is not a competitive solution. If this occurs, or if no Nash equilibrium exists, a competitive solution still might exist − it simply must be found some other way. In cases where numerous parties exist (i.e., Netherlands, Denmark, and Israel), the number of iterations necessary to search for a competitive solution is prohibitive. Therefore, a number of assumptions are made to ease computational difficulties.

Assumption 1: In large games, a competitive solution including only minimal winning coalitions exists.

When between 25 and 100 minimal winning coalitions exist, it is computationally impossible to consider any other coalition as having a potential to form. The assumption of limiting the search to minimal winning coalitions can be best justified by pointing out that in all simple spatial games examined to date the competitive solution consists of minimal winning coalitions. In fact, the following is conjectured to be the case:

Conjecture: If there exists a competitive solution to a simple spatial game, then there exists a competitive solution that predicts only minimal winning coalitions.

Assumption 2: Only large parties are involved in coalitions when there are many players.

When over 100 minimal winning coalitions exist, it is impossible to compute the competitive solution even by limiting the allowable coalitions to minimal winning coalitions. *DeSwaan* encountered the same unmanageability in his analysis and handled the problem by eliminating all parties with less than $2-1/2$ % of the legislative seats. I adopt this technique and consider only the larger parties as potential coalition members. The justification for this assumption is that in no case since World War II has one of these parties participated in the cabinet coalition. Once this assumption is made, the number of minimal winning coalitions generally is reduced to about 35, which, using assumption 1 is manageable.

Under Assumption 2, the size of a winning coalition is still defined as $|C| > |N|/2$ where $|N|$ is the total number of seats *after* the application of this assumption. The set of possible winning coalitions is therefore considered to be chosen in reference to a new N, N', where N' is the total number of seats held by parties who have more than $2-1/2$ % of the total. However, the minimal winning coalitions in this revised game are

---

[6]) The nature of the scaling routine makes this assumption more reasonable [see *Poole*].

often losing in the actual legislative game. To correct this, the competitive solution is said to predict all coalitions that are minimal winning in the actual parliament and contain one of those predicted in the new game.

Assumption 3: For the one-dimensional ideological scale, we use *DeSwaan*'s orderings of the parties.

Using the survey data and quasi-unfolding technique, a one dimensional issue space does not seem to fully describe the partys' positions. *DeSwaan*, however, develops a one dimensional space based on an historical study of the policies of the various political parties. Based on a variety of other types of evidence others [e.g., *Morgan*] have argued that one dimension is reasonable.

Section 5

In the analysis of particular elections, it is our hope that the competitive solution will more frequently predict the government coalition than the other hypotheses do. Eleven elections from five countries are examined.

Norway

To examine the coalitions that formed after the 1961, 1965, and 1969 elections of the Norwegian Sorting, three configurations of ideal points for the six political parties are considered. A 2-dimensional and a 3-dimensional configuration are obtained using the quasi-unfolding program. Additionally, a two-dimensional configuration, the result of a non-metric scaling routine reported by *Groenning* [1970], is included. The data for all of these mappings comes from a 1963 survey in which the voters are asked to order their preferences of the political parties. Using the quasi-unfolding program in two dimensions yields an average correlation between the original distances individuals perceive between parties and those distances between the estimated coordinates of only .581. This is still used because of arguments in the literature that there were only two major dimensions present at this time. The three-dimensional configuration produces a much higher correlation of .767.

In 1961 there were six parties in the Sorting and six possible minimum winning coalitions. In 1965 the same parties held seats but now only five minimum winning coalitions existed, none including the Socialist Peoples Party. In 1969 there were only five political parties represented and the same minimum winning coalitions as in 1965. The election results and the ideological rankings of the parties are shown in the tables below. The coalitions which actually formed are indicated by the starred entries.

	Labor	Cons.	Cent.	Christ.Peo.	Lib.	Soc.Peo.
ideological order	2	6	5	3	4	1
# of seats	74	29	16	15	14	2
coalition	*					*
# of seats	68	31	18	13	18	2
coalition		*	*	*	*	
# of seats	74	29	20	14	13	
coalition		*	*	*	*	

The predictions of the various models are summarized in the table below:

Predictions:	M.Win.	M.Size	B.P.	Con.	P.D.	M. Var.			Comp. Sol.		
						N.M.	2d.	3d.	N.M.	2d.	3d.
1961	$1/6^7)$	1/2	1/5	1/4	1/1	1/7	0/2	1/5	$X^8)$	1/3	1/4
1965	1/5	1/1	0/4	1/4	1/3	1/1	1/1	0/2	1/3	1/3	1/3
1969	1/5	1/1	0/4	1/4	1/3	1/1	1/1	0/2	1/3	1/3	1/3

Despite quite a shift in seats between 1965 and 1969, each theory predicts the same possible coalitions in the two cases. The same coalition also continued in power. The coalitions predicted by the competitive solution in the 1961 case for the 2-dimensional configuration were illustrated in Section 2. All of the theories, except the bargaining proposition predict correctly.

Germany

For the 1969 and 1972 elections, a two-dimensional configuration of ideal points for the four major parties is computed using the quasi unfolding procedure on data from a 1972 survey of the German electorate. Respondents of the survey are asked, "What do you think of the . . . . . party?" They respond by selecting a point on a 1–11 scale ranging from "not much" to "a great deal." The question is asked in 3 phases of the survey and the result of each phase is scaled separately. The resulting configurations yield a correlation of .488, .661, and .679 respectively to the original distances. Only the second and third two-dimensional configuration are used to compute the competitive solution. These àre included in the Appendix. The four parties are assumed to be independent actors, even though the CDU and CSU have always acted on a national level as if they were one party. However, they possess separate party organizations and have often talked as if they were close to splitting. There is no formal agreements stopping the CDU from acting independently and the FDP, which was originally part of this coalition, did eventually split away. The fact that the CDU and CSU do not separate is probably due to long-term considerations that are not considered here.

In addition to the two-dimensional configurations, an additional configuration is produced to find positions for the parties in a unidimensional space. The results, using data from any of the three phases, gives the ordering CSU, CDU, FDP, SPD.

In the two governments, the SPD and FDP ended up forming an alliance. The actual seat allocations were:

---

[7]) The first number in each column reports whether or not the theory was successful. The second number reports the number of coalitions in the predicted set. The abbreviations for the theories should be clear: M. Size = minimum size; B.P. = bargaining proposition (fewest parties); Con. = connected coalitions; P.D. = policy distance minimization; and M. Var. = minimum variance.

[8]) In cases where a large X appears, I was unable to find a competitive solution using the Nash equilibrium equivalence. The solution may still exist, but for now it is unknown.

	CSU	CDU	FDP	SPD
1969				
# of seats	49	201	31	237
coalition			*	*
1972				
# of seats	48	177	41	230
coalition			*	*

In both of these elections the same eight coalitions are possible, with four of these minimal winning. Using either the phase 2 or phase 3 location of ideal points the competitive solution has proposals by coalitions {SPD, CSU}, {SPD, FDP}, {CSU, CDU, FDP}. This result is compared to the other theories in the following tables.

	M.Win.	M.Size	B.P.	Con.	P.D.	M.Var.	Comp.Sol.
1969	1/4	1/1	1/3	1/2	1/4	1/1	1/3
1972	1/4	0/1	1/3	1/2	1/4	1/1	1/3

Denmark

The data from a 1973 electoral survey was unavailable to us, but *Damgaard* [1974] has published a two-dimensional configuration of ideal points for ten political parties. He uses a non-metric multidimensional scaling technique on the ranking of the political parties by the respondents. The elections of 1971 and 1973 are analyzed using this map. In 1971 only five parties obtained seats in the legislature:

	Social Dem.	Cons.	Lib.Dem.	Rad.Dem.	Soc.Rep.
ideological order	2	4.5	4.5	3	1
# seats	73	31	30	27	17
coalition	*				*

The predictions are:

M.Win.	M.Size	B.P.	Con.	P.D.	M.Var.	Comp.Sol.
1/5	1/1	1/4	1/2	1/2	1/2	1/3

The actual coalitions predicted by the competitive solution are {S.D., Rad., Dem.}, {S.D., S.R.}, {Cons., Lib.Dem., Rad.Dem., S.R.}.

In 1973 all 10 parties held legislative seats. The electoral returns for this election are:

party	S.D.	Cons.	Lib.D.	Rd.D.	Soc.Peo.	Prog.	Soc.D.	Ch.P.	Com.	Tax
seats	46	16	22	20	11	28	14	7	6	5
coalition		*	*	*		*	*			

Since no one-dimensional configuration of the parties is available, it is impossible to rank the parties on an ideological scale. Therefore, there is no prediction available for the connected coalition theory or the policy distance theory. The other results are:

M.Win.	M.Size	B.P.	M.Var.	Comp.Sol.
1/67	0/11	0/4	1/4	1/10

Netherlands

The survey data for the Netherlands is of the form most appropriate for the quasi-unfolding program. A random sample of voters were asked how sympathetic they felt towards the various political parties. They picked a point on a 0–100 scale corresponding to their feelings. There were three waves to the survey, conducted in 1970, 1971, and 1972, and each was scaled separately. In all cases three dimensions were recovered. The average correlation between initial distances and the distances of recovered points are .651, .625, and .653 respectively. At this point the 1971 wave is applied to the 1971 election and the 1972 wave to the 1972 election. There are between twelve and fourteen political parties in each of these legislatures and hundreds of minimal winning coalitions. Therefore, the small parties are eliminated and winning conditions are viewed as a majority of the remaining players. A search for a competitive solution is made considering only minimal winning coalition. Table 1 shows all of the parties which won seats in the 1971 election with the parties in parenthesis those that are dropped from the analysis.

	Comm.	(Rad.)	Labor	D'66	Cath.	Anti-Rev.	CHU
ideological order	1	2	3	4	5.5	5.5	7
# of seats	6	2	39	11	35	13	10
coalition					*	*	*

	(Calvin)	DS'70	F&D	(Reform Pol.)	(Ind.)
ideological order	8	9	10		
# of seats	3	8	16	2	3
coalition		*	*		

The three religious parties — Catholic Peoples, Anti-Revolutionary and Christian Historical Union — had been part of the governing coalition since 1952. The Labor and Democrat '66 parties had collaborated in the election and campaigned as an opposition group. It would, therefore, be unlikely to see one or both of them join with the Catholic parties in a coalition. The Catholic parties turned to their right and ended up forming a coalition with the two viable parties — the Democrat Socialists '70 and Peoples Party for Freedom and Democracy. The DS'70 party defected from the coalition after only one year, and new elections were held.

There are eight parties with a total of 138 seats large enough to enter into our analysis. In calculating a competitive solution a winning coalition is considered to be one with over half of these partys' total seats. There are twenty-six such minimal winning coalitions. Two of these, {Catholic, ARP, CHU, F and D} and {Catholic, ARP, F and D, Demo, Soc.'70} are subsets of the coalition that actually formed. The competitive solution contains proposals by seven coalitions including these two. The seven coalitions are subsets of thirteen of the 26 minimal winning coalitions of the game requiring 76 seats to win. Therefore, the competitive solution is said to predict correctly with a choice of thirteen out of the possible 26 minimal winning coalitions and 105 winning coalitions. The results are summarized, per usual, below:

Win.	M.Win.	M.Size	B.P.	Con.	P.D.	M.Var.	Comp.Sol.
1/105	1/26	0/4	0/6	1/5	1/21	1/1	1/13

The 1972 election returns were:

	Comm.	Rad.	Labor	D'66	Cath.	ARP	CHU	DS'70	LIB
ideological order	1	2	3	4	5.5	5.5	7	8	9
# of seats	7	7	43	6	27	14	7	6	22
coalition		*	*	*	*	*			

The coalition took almost a year to form after the election was held. The progressive parties — Radical, Labor, D'66 — had committed themselves before the election to a coalition. The government was so difficult to form because the religious parties only alternative to these three parties was to return to a coalition with DS'70 — whose defection caused the election. Eventually, the progressive and religious parties formed together. In this case, the pre-election coalition of the progressive parties and the desire to have some method of comparison led us to accept a prediction of a {Labor, Catholic, Anti-Revolutionary Party} coalition as correct.[9]) The two smaller parties are assumed to go along with any coalition which Labor forms.[10]) Given these assumptions, the results are:

Win.	M.Win.	M.Size	B.P.	Con.	P.D.	M.Var.	Comp.Sol.
1/230	1/40	0/14	1/8	1/4	0/28	1/1	1/18

Israel

The spatial maps of the Israeli parties comes from the responses of a sample of voters in a survey conducted in 1973. They stated their first, second, third, third from last, next to last, and last voting choices from a list of ten parties. From this data only an incomplete ordinal ranking of respondents' voting preferences can be obtained. The

[9]) It is tempting to view the extra player as being a surplus player who joined the coalition but required no concessions. Unfortunately, the ministership division does not support this since all members received at least one post.

[10]) An alternative approach could be to limit the set of possible coalition structures to take into account the pre-election coalitions. At this time, unfortunately, no results using this technique have been found.

desired data for the quasi-unfolding technique would be interval level data, over all political parties, and over fondness not voting preference. Nevertheless, the results from this technique are used to compute the competitive solution and minimum variance coalitions. Distances are created for each individual by ranking the parties from one to ten. The most preferred party is scaled at one and the least preferred at ten. Unmentioned parties are placed at five on the scale. The average correlation between these distances and those recovered by the scaling routine is .546. The meaning of this statistic is unclear since the input distances are inaccurate.

Further difficulties were encountered in the use of the 2−1/2 % cut-off rule. The Arab party (Marakh) and Citizen's Rights party both won three seats in the 120-seat Knesset in 1973 and are not included on the list of possible choices. Other parties (Communist, Workers, Orthodox) were listed although they got fewer seats. While these parties could have been eliminated with the 2−1/2 % rule, the number of minimum winning coalitions is sufficiently small that this approximation is not required; and all parties for which information was obtained are included.

An additional difficulty is the emergence of a greater than minimal winning coalition in both elections. In both the 1971 and 1973 coalitions, the Independent Liberals are an unnecessary coalition member. *Mahler/Trilling* [1975] explain this by Israel's uniquely unstable political situation. The external threat of war makes it strongly desirable that civil disruptions be minimized. Since religious groups have a capacity to cause difficulties in the administration of the government far beyond their political numbers, it is important that they be a part of any governing coalition. Analytically, then, we use the same technique for judging a prediction correct here as was used with the Netherlands' 1972 election. The prediction of an Alignment (Labor-Mapan) and National Religious Party coalition is interpreted as a correct choice.

Finally, in 1973 there are 340 possible winning coalitions and it is computationally too costly to find all predictions of the policy distance minimization hypothesis. Therefore, minimum winning coalitions which had positions undominated by other minimum winning coalitions are selected to fit in this category.

The membership positions in 1971 were:

	Maki	Labor	State	Relig.	Agudat	Ind.Lib.	Gahal
ideological order	1	2	3	4	5.5	5.5	7
# of seats	4	56	4	12	6	4	26
coalition		*		*		*	

In the search for a competitive solution in the revised game, no winning coalition excluding Labor exists. Therefore, a core exists at Labors' ideal point and any coalition that includes Labor is predicted. The results of the analysis are:

Win.	M.Win.	M.Size	B.P.	Con.	P.D.	M.Var.	Comp.Sol.
1/60	1/6	0/1	1/3	0/2	0/5	1/2	1/6

In 1972 the Labor party lost some of its strength and many more winning coalitions are possible. The election returns are:

	Comm.	Maki	Lab.	NRP	Ag.	Rel.Lib.	I.Lib.	Gahal
ideological order	1.5	1.5	3	4	6.0	6.0	6.0	8.0
# of seats	4	1	51	10	3	2	4	39
coalition			*	*		*	*	

All of the theories but policy distance correctly predicted the outcome. The competitive solution was calculated for the revised game which has 12 minimum winning coalitions. Four had proposals in the competitive solution which translated into four coalitions in the legislative game.

Win.	M.Win.	M.Size	B.P.	Con.	P.D.	M.Var.	Comp.Sol.
1/340	1/8	1/4	1/2	1/1	0/5	1/3	1/4

Summary

Two statistical tests, based on the degree to which the number of correct predictions differ from chance, are developed to compare the accuracy of the different theories. By difference from chance we mean that given a certain number of predicted coalitions for a particular theory, the probability that the observed number of successful predictions is greater than one would expect by a random selection of that number of winning coalitions. The probability of predicting correctly by chance is $(np_{i,k}/nc_{i,k})$, where $np_{i,k}$ is the number of coalitions predicted by theory i in election k and $nc_{i,k}$ is the number of winning coalitions in the same election. These probabilities are presented in the table below.

Probability theory predicts correctly by chance

	M.Win.	M.Size	B.P.	Con.	P.D.	C.S.**	M.Var.**
Norway '61	.1875	.0625	.1563	.1250	.0313	.084	.1563
Norway '65	.3125	.0625	.25*	.25	.1875	.1875	.125*
Norway '69	.3125	.0625	.25*	.25	.1875	.1875	.125*
Germany '71	.50	.125	.375	.25	.5	.375	.125
Germany '72	.50	.125*	.375	.25	.5	.375	.125
Denmark '71	.3125	.0625	.25	.1250	.1250	.187	.0625
Denmark '73	.133	.022*	.008*	X	X	.020	.008
Netherlands '71	.248	.038*	.057*	.048	.20	.124	.010*
Netherlands '72	.174	.061*	.035	.017	.122*	.078	.004
Israel '71	.10	.017*	.05	.033*	.083*	.10	.033
Israel '73	.023	.012	.006	.003	.015*	.012	.009

* = The model did not include the actual result in the set of possible outcomes.

X = Missing values.

** = Represents results based upon the 3-dimensional configurations of ideal points (in Germany only 2-dimensional).

The matrix is then partitioned as shown so that a comparison can be made separately for situations with small and large numbers of parties. Two tests of the models' accuracy are developed. The first test was a standard Chi-square test for the difference between expected and observed results. That is $\chi^2_{j,n-1} = \sum_{i=1}^{n} ((0_i - e_1)^2/e_i)$ where e_i is the number in the matrix (expected probability of success) and 0_i is the observed value (1 or 0).[11] The larger the score the better the concept has done in predicting correctly. A second test is reported based on the probability that the observed number of successful predictions, or more, could occur if one were to choose randomly. This statistic is binomially distributed and the smaller the number the better the accuracy of prediction for a given concept. The major difference between the two tests is that the Chi-square test takes into consideration where the incorrect prediction was made while the binomial test does not. For example, the binomial test score for the minimum size theory and a small number of parties would be identical regardless of where the one incorrect prediction occurred. In the Chi-square test, however, if the incorrect prediction had been for Denmark in 1971 the score would decrease from 62.5 to 54.6. The results of these tests are shown in Table 2.

Small No. Parties	Binomial*	Chi-Square	Large No. Parties	Binomial	Chi-Square
M.Win.	1.43	9.05	M.Win.	1.32×10^{-5}	61.6
M.Size	.016	62.5	M.Size	.13	81.5
B.P.	43.1	9.39	B.P.	1.94×10^{-4}	209.4
Con.	.062	21.40	Con.	3.56×10^{-5}	407.1
P.D.	.034	42.18	P.D.	.366	3.4
C.S.	.087	21.40	C.S.	1.16×10^{-5}	156.6
M.V.	4.70	32.86	M.V.	5.83×10^{-8}	492.4

* The actual score is 10^{-3} times those shown.

Tab. 2

For countries with a small number of parties, then, the minimum size, policy distance, and competitive solution theories all do reasonably well as coalition predictors. The minimum variance theory does well by one of the tests but not the other. This is because one of the cases predicted correctly has a very small probability of being correctly selected by chance. However, two of the errors in prediction by the minimum variance theory occur in the cases of the Norwegian 1965 and 1969 cabinets. In these cases we choose the three dimensional spatial configurations as the best representation for the parties. If either of the two dimensional configurations had been selected, the theory would have predicted correctly. Thus, in all fairness, we must include this theory in the set of successful predictors.

---

[11]) This test is not wholly appropriate for the situation in the table. If one views the matrix as a contingency table then $(0_i - e_i)$ is supposed to be a normally distributed random variable. Here the distribution is clearly, not normal. In general, it takes more than five observations per cell before the normal begins to approximate a binomial distribution. However, this measure is used elsewhere and the numerical results do have an interpretation in relation to each other.

The relative weakness of the competitive solution among the successful theories is due to the concept's inability to make unique predictions. In any game with three or more active players there will always be at least three coalitions predicted. This becomes less of a disadvantage when the number of parties is large and, as might be expected, the statistical results support this. In the cases where the number of political parties is large all of the theories except minimum size and policy distance do well. Thus, overall, the connected coalition theory, our interpretation of *Axelrod*'s minimum variance theory, and the competitive solution seem to be the best predictors.

The real disadvantage of the competitive solution among these three "successful" theories is the increased number of assumptions made to avoid computational difficulties. Nevertheless, the competitive solution is the only technique which can successfully make predictions while invoking only game theoretic assumptions. If one views coalition formation as a spatial game, with parties negotiating government policy positions, the competitive solution puts to rest *Taylor*'s comment on the irrelevance of game-theory to the study of government coalition formation. It is the only tested theory to predict correctly in every case.

One additional problem with all of the theories tested is that minority coalitions are assumed not to form. However, in the eleven European countries where governments are formed by legislative coalitions, twenty percent of the coalitions since World War II have been minority coalitions. That is, according to our formulation, the coalition that forms is a losing one. At this moment there is no way of predicting when this will occur. However, if there were a way to determine the cases where the competitive solution does not exist, these cases can be taken to predict a minority government. This is a capability that none of the other theories examined have.

Appendix

Ideal Points Recovered for the Political Parties

Norway

Social Democrats	− 2.69	3.25	− 3.90
Conservatives	4.20	3.48	2.94
Centrists	− .91	− .89	1.94
Christian Peoples	− 6.01	− 1.92	1.94
Liberal	− .73	.74	.42
Socialist Peoples	.92	− 1.53	− 1.91

Netherlands 1971

Labor	51.61	− 8.80	− 12.42
Catholic Peoples	− 17.04	− 12.24	− 15.30
People for F&Dem.	− 25.62	− 55.36	5.29
Anti-Revolution	− 38.10	16.44	− 24.13
Democrat '66	38.31	− 21.17	− 20.34
Christ. Hist. Un.	− 39.52	7.90	− 12.65
Dem. Socs. '70	− 9.31	− 64.42	− 13.99
Communists	50.85	17.50	17.74

Netherlands 1972

Labor	62.84	− 9.15	− 6.05
Catholic Peoples	− 24.12	− 6.20	− 4.24
People for F&Dem.	− 45.52	5.00	38.26
Anti-Revolution	− 38.90	− 11.83	− 7.06
Christ. Hist. Un.	− 38.72	− 5.71	− 5.97
Communist	57.05	33.73	4.96
Radical	36.82	− 25.09	− 2.76
Democrat	45.14	− 19.68	.97
Dem. Soc. '70	− 6.77	− 19.41	26.83

Israel

Free Center	− .50	.26	− 2.09
Gahal	1.44	2.22	1.04
Independent Lib.	1.12	2.74	− .23
Alignment (Mapam)	− .75	4.20	2.92
Religious Lab.	− 1.69	− 3.14	− .03
Nat. Rel. Party	− 2.32	− 3.70	1.92
Agudath	− 1.48	− 3.48	− .20
Israel Communists	9.03	− .55	− .94
Arab Communists	.29	− 1.25	.05
New Force	− 5.13	3.15	− 2.42

Germany 1971

SPD	4.85	− 1.40
CDU	− 3.88	− .30
CSU	− 5.03	.12
FDP	4.06	1.56

Germany 1972

SPD	4.48	− 1.43
CDU	− 3.85	− .32
CSU	− 4.12	.13
FDP	3.49	1.63

References

Axelrod, R.: Conflict of Interest. Chicago 1970.

Barnes, S.: Left, Right, and the Italian Voter. Comp. Pol. Studies 4, 1971, 157–176.

Blondel, J.: Party Systems and Patterns of Government in Western Democracies. Can. Jour. of Pol. Sci. 1, 1968, 180–204.

Browne, E.: Testing Theories of Coalition Formation in an European Context. Comp. Pol. Studies 4, 1971, 391–413.

– : Coalition Theories: A Logical and Empirical Critique. Sage Professional Papers in Political Science 4, 1973.

Browne, E., and *M. Franklin*: Aspects of Coalition Payoffs. Amer. Pol. Sci. Rev., 1973, 453–469.

Damgaard, E.: The Parliamentary Basis of Danish Governments: Patterns of Coalition Formation. Scan. Pol. Studies 4, 1969, 30–58.

Damgaard, E., and *J. Rusk*: Cleavages, Structures, and Representational Linkages. Presented at EPCR Workshop, March 1974.

Gamson, W.: Theory of Coalition Formation. Amer. Soc. Rev. 26, 1961, 373–382.

Groennings, S.: Patterns, Strategies and Payoffs in Norwegian Coalition Formation. The Study of Coalition Behavior. Ed. by Groennings/Kelly/Leiserson. New York 1970, 60–79.

Inglehart, R., and *H. Klingemann*: Party Identification, Ideological Preference and the Left-Right Dimension Among Western Mass Publics. Party Iden. and Beyond. Ed. by Budge/Crewe/Farlie. London 1976, 243–276.

Leiserson, M.: Fractions and Coalitions in One-Party Japan. Amer. Pol. Sci. Rev. **57**, 1968, 770–787.

– : Game Theory and the Study of Coalition Behavior. The Study of Coal. Behav. Ed. by Groennings/Leiserson/Kelly. New York 1970.

Luce, D., and *H. Raiffa*: Games and Decisions. New York 1957.

Mahler, G., and *R. Trilling*: Coalition Behavior and Cabinet Formation: The Case of Israel. Comp. Pol. Studies **8**, 1975, 200–234.

McKelvey, R., and *P. Ordeshook*: Competitive Coalition Theory. Game Theory and Pol. Sci. Ed. by P. Ordeshook. New York 1978, 1–38.

McKelvey, R., P. Ordeshook, and *M. Winer*: The Competitive Solution for N-Person Games Without Sidepayments. Amer. Pol. Sci. Rev. **72**, July 1978, 599–615.

Morgan, M.J.: The Modelling of Government Coalition Formation. Ph.D. Dissertation, University of Michigan, 1976.

Nachimas, D.: Coalition Politics in Israel. Comp. Pol. Studies **7**, 1974, 316–333.

Poole, K.: A Method for Testing the Equilibrium Theories of the Spatial Model of Party Competition. Ph.D. Dissertation, University of Rochester, 1978.

Riker, W.: The Theory of Political Coalitions. New Haven 1962.

Sani, G.: Mass Constraints on Political Realignments. Brit. Jour. of Pol. Sci. **6**, 1976, 1–31.

Schofield, N.: The Kernel and Payoffs to European Government Coalitions. Pub. Choice **26**, 1976, 29–51.

Solstad, A.: The Norwegian Coalition System. Scan. Pol. Studies **6**, 1969, 160–167.

DeSwaan, A.: Coalition Theories and Cabinet Formations. San Francisco 1973.

Taylor, M.: Party Systems and Government Stability. Amer. Pol. Sci. Rev. **65**, 1971, 28–38.

– : On the Theory of Government Coalition Formation. Brit. Jour. of Pol. Sci. **2**, 1972, 361–373.

Taylor, M., and *M. Laver*: Government Coalitions in Western Europe. Eur. Jour. of Pol. Res. **1**, 1973, 205–248.

Applied Game Theory. 1979 © Physica-Verlag, Wuerzburg/Germany

An Experimental Test of Several Theories
of Committee Decision Making Under Majority Rule[1])

By *R.D. McKelvey* and *P.C. Ordeshook*, Pittsburgh[2])

Abstract: This paper reports a series of thirty-four experiments that are designed to test the adequacy of several alternative theories of group choice under "majority rule" when no Condorcet point (core) exists. The games we use in these experiments seek, in particular, to establish as much separation as possible between the predictions of several alternative solution concepts and, thereby, to provide a critical test of these competing theories. Briefly, our results strongly support the competitive solution notions for *n*-person game theory [c.f., *McKelvey/Ordeshook/Winer*], but they provide little if any support for the *V*-set, the $M_1^{(i)}$ Bargaining Set, *Asscher*'s [1976] Ordinal Bargaining Set, M^0, and several solutions based on vulnerability. The first section of this essay defines and discusses the various solution theories. Section 2 describes our experimental design, and Sections 3 and 4 review the results of these experiments.

1. Alternative Solution Theories for Group Decision Making

Consider a group of individuals, $N = \{1, 2, \ldots, n\}$, confronted with a finite and mutually exclusive set of alternatives, $X = \{x_1, x_2, \ldots, x_m\}$. Let individual i's preferences be represented by the binary relation $R_i \subset X \times X$, where strict preference and indifference are denoted, in the usual manner by P_i and I_i. We assume R_i is a weak order. Let $u_i: X \to R$ be an utility representation of R_i for the i-th individual, i.e., for any $x, y \in X$, $u_i(x) \geq u_i(y) \Leftrightarrow x R_i y$.

A subset $C \subseteq N$ is called a coalition, with $|C|$ denoting the number of members of C. For any $x, y \in X$, and C, individual preferences define coalitional preferences:

$$x P_C y \Leftrightarrow x P_i y \text{ for all } i \in C.$$

Assuming simple majority rule, now, the set of *winning* coalitions is
$\underline{W} = \{C \subseteq N \mid |C| > n/2\}$, and the set of *minimal winning coalitions* is

---

[1]) This research is supported by an NSF Grant #SOC77–15267 to Carnegie-Mellon University. We also wish to thank Rod Gretlein for technical assistance in computation of the various solutions reported herin and Ed Packel and M. Maschler for comments on an earlier draft.

[2]) *Richard D. McKelvey* and *Peter C. Ordeshook*, Carnegie Mellon University, Schenley Park, Pittsburgh, Pennsylvania 15213, USA.

$\underline{W}^* = \{C \in \underline{W} \mid C - \{i\} \notin \underline{W} \text{ for all } i \in C\}$. We can then define a social ordering, based on majority rule, as follows:

$$xPy \Leftrightarrow xP_Cy \text{ for some } C \in \underline{W}$$

$$xRy \Leftrightarrow \sim xPy.$$

The preceding definitions also define an n-person characteristic function game without sidepayments or transferable utility. In this instance, the characteristic function $v : 2^N \rightarrow X$ is given by

$$v(C) = \begin{cases} X & \text{if } C \in \underline{W} \\ \Phi & \text{if } C \notin \underline{W}. \end{cases}$$

The usual dominance relation in this n-person game is exactly equivalent to the social relation. Thus x *dominates* y, written $x \gg y$, iff xPy.

Some solution notions (such as the Core and Von Neumann-Morgenstern Solution) isolate subsets of X as "solutions." Others (such as the bargaining set and the competitive solution) isolate coalition outcome pairs, or what we call "proposals" in the context of simple games. Formally, a *proposal* is a pair $p = (x, C)$ such that $C \subseteq N$ and $x \in v(C)$. In simple majority games a proposal is any pair $p = (x, C)$ satisfying $C \in \underline{W}$ and $x \in X$. We let P denote the *set of proposals*.

The most familiar solution, called either a Condorcet point or a *Core*, depending on whether it is applied in the social choice or game theoretic framework is, $C = \{x \in X \mid yPx \text{ for no } y \in X\}$.

Much experimental work already exists to show that, if the core is nonempty, it predicts fairly well the outcomes of committees [c.f., *Berl* et al.; *Fiorina/Plott; Isaacs/Plott*]. In this essay, however, we are concerned with solutions to games in which the core is empty, and several competing solution notions have been advanced to explain the outcomes of such games. Some of these notions are related specifically to a game theoretic conceptualization of committee decisionmaking, as described by a characteristic function, while others, more in a social choice tradition, do not explicitly view the process in characteristic function terms.

We define a number of these, starting with the least restrictive solutions, which merely eliminate the most obviously implausible alternatives, and proceed to the more restrictive solutions, which eliminate large subsets of alternatives.

The Pareto Optimal Set $\underline{P}$: The Pareto Optimal set, $\underline{P}$, is the set of alternatives that are not Pareto dominated by any other alternative. Formally

$$\underline{P} = \{x \in X \mid yP_N x \text{ for no } y \in X\}.$$

(Note that $C \subseteq \underline{P}$.)

The Top Cycle Set: Let P^* be the transitive closure of the majority relation. Thus, for any $x, y \in X$, we say that xP^*y iff there is a finite set of alternatives $\{\theta_i\}_{i=0}^K \subseteq X$ such that

i) $\theta_0 = y$

ii) $\theta_K = x$

iii) $\theta_i P \theta_{i-1}$ for all $1 \leqslant i \leqslant K$.

The relation P^* is transitive, and the top cycle set is the set of alternatives that are unbeaten under the relation P^*. Thus

$$\underline{G} = \{x \in X \mid \text{ for no } y \in X \text{ is } yP^*x \text{ and } \sim xP^*y\}.$$

The set G is also called the GOCHA set by *Schwartz* [1972]. Note moreover that the top cycle set is a generalization of the core. Thus, the top cycle set always contains the core, and if P is complete (i.e., xPy or yPx whenever $x \neq y$), then $\underline{G}$ collapses to the core if $C \neq \Phi$, i.e., $C \subseteq \underline{G}$ and; P complete and $C \neq \Phi \Rightarrow \underline{G} = C$. Unfortunately, if the core is empty, G can be a large if not exhaustive set.

The Minimax Set: For any $x, y \in X$, let $n(x, y)$ be the number of individuals preferring x to y: $n(x, y) = |\{i \mid xP_iy\}|$. We define, for any $y \in X$, $n(y) = \sup_{x \in X} n(x, y)$ and let $n^* = \inf_{y \in X} n(y)$. Then, the minimax set $\underline{M} \subseteq X$ is

$$\underline{M} = \{y \in X \mid n(y) = n^*\}.$$

This set is defined originally by *Simpson* [1969], and is studied extensively by *Kramer* [1975]. Briefly, $\underline{M}$ is the set of alternatives that are least vulnerable in the sense that the number of individuals who simultaneously benefit from a change is minimized. Note that the minimax set is a subset of the core if the core is non-empty; but, if $C = \Phi$, $\underline{M}$ is a subset of both the Pareto Optimal Set and the Top Cycle Set. Thus, if $\underline{M} \subseteq \underline{P} \cap \underline{G}$, and $C \neq \Phi \Rightarrow C \supseteq \underline{M}$.

The Least Alternative, or Least Objectionable Set: If X is finite, let $t(x) = |\{y \in X \mid yPx\}|$ be the number of alternatives that defeat x. Then, let

$$t^* = \min_{x \in X} t(x)$$

and

$$\underline{A} = \{x \in X \mid t(x) = t^*\}.$$

Hence, $\underline{A}$ is the set of alternatives that are "least objectionable," in the sense that they can be defeated by the fewest number of alternatives. Again, if the core is nonempty, C is identical to $\underline{A}$, i.e., $C \neq \Phi \Rightarrow C = \underline{A}$. For further discussion of A see *Ferejohn/ Fiorina/Packel* [1978].

The Least Vulnerable Set: For any $x \in X$, let

$$l(x) = |\{C \in W^* \mid yP_C x \text{ for some } y \in X\}|$$

and

$$l^* = \min_{x \in X} l(x).$$

Then the least vulnerable set, $\underline{L}$, is,

$$\underline{L} = \{x \in X \mid l(x) = l^*\}.$$

The least vulnerable set, then, consists of the members of X that minimize the number of coalitions that can find a better alternative. Note again that $\underline{L}$ reduces to C if $C \neq \Phi$. The Least Vulnerable Set is developed more fully in *Ferejohn/Fiorina/Weisburg* [1978].

Probabilistic Versions of Vulnerability: *Ferejohn/Fiorina/Packel* [1978] propose a Markov model of committee choice that uses the notions of vulnerability embodied in the definition of $\underline{L}$ to define the transition probabilities. Specifically, for any $x, y \in X$, let $c(x, y) = |\{C \in \underline{W}^* \mid yP_Cx\}|$, and $c(x) = \sum_{y \in X} c(x, y)$, and define

$$p(x, y) = \begin{cases} c(x, y) / c(x) & \text{if } c(x) > 0 \\ \delta_{ij} & \text{otherwise} \end{cases}$$

(where $\delta_{ij} = 1$ if $i = j$, and 0 if $i \neq j$). Letting $p(x, y)$ be the transition probability between x and y, one can generate a limiting distribution (given a particular starting state) over X in the usual manner.

Turning now to game-theoretic concepts, various solution notions seek to treat characteristic function form games, but several of these are developed specifically for sidepayment games with transferable utility and cannot be extended readily to games without transferable utility [c.f., *McKelvey/Ordeshook/Winer*] for a discussion of these difficulties). Consequently, we limit ourselves here to four solution notions that can be defined without transferable utility: the Von Neumann-Morgenstern Solution, (V-set), the Bargaining Sets $M_1^{(i)}$ and M^0, and the Competitive Solution.[3])

The Von Neumann-Morgenstern Solution: A Von Neumann-Morgenstern Solution, or V-set, is any set $V \subseteq X$ satisfying: (Internal Stability) For any $x, y \in V$, it is not the case that xPy; and (External Stability) For all $y \notin V$, $\exists x \in V$ such that xPy.

The $M_1^{(i)}$ *Bargaining Set*: Although *Aumann/Maschler's* [1964] initial development of the bargaining set is for games with sidepayments in a transferable utility, *Peleg* [1963, 1969], *Billera* [1970], and others extend their definitions to general games without transferable utility. We can simplify these definitions here, since we are concerned only with simple games, although our definitions are equivalent to *Peleg's* [1963] if we substitute "individually rational payoff configurations" for "proposal."

Letting $p_1 = (x_1, C_1)$ be a proposal, and $i, j \in C_1$, then a proposal $p_2 = (x_2, C_2)$ is an *objection* of i against j if: (1) $j \notin C_2$; (2) $x_2 R_k x_1$ for all $k \in C_2$; (3) $x_2 P_i x_1$.

Letting $p_1 = (x_1, C_1)$ be a proposal, with $i, j \in C_1$, and $p_2 = (x_2, C_2)$ be an objection of i against j, then $p_3 = (x_3, C_3)$ is a *counter objection* by j if: (1) $i \notin C_3$; (2) $x_3 R_j x_1$; (3) $x_3 R_k x_2$ for all $k \in C_3$. The Bargaining Set, $M_1^{(i)}$ is then the set of all proposals (x, C) such that every objection of $i \in C$ against $j \in C$ can be met by a counter objection.

---

[3]) While all of the non-game theoretic solution notions are guaranteed to exist if X is finite, existence is not assured for any of the game theoretic solutions.

The Ordinal Bargaining Set: Owing to the potential problems of extending the classical bargaining set to games without transferable utility, *Asscher* [1976] generalizes the definition of $M_1^{(i)}$ thus: Given a proposal $p = (x, C)$, and $j, k \in C$, we say that j is stronger than k iff j has a justified objection against k. Then a proposal p is in the ordinal bargaining set, M^0, iff for all $j, k \in C$, whenever j is stronger than k, there exists a sequence $\{i_l\}_{l=0}^L \subseteq C$ with $i_0 = j$, $i_L = k$, such that i_l is stronger than i_{l-1} for all $1 \leqslant l \leqslant L$.

. Asscher proves that for any winning coalition, there exists at least one proposal in M^0. His proof, however, assumes the comprehensiveness of $v(C)$ – that is, it assumes that for any $x \in v(C)$, if $w \in R^n$ satisfies $w_i \leqslant u_i(x)$ for all $i \in C$, then there is a $y \in v(C)$ with $u_i(y) = w_i$ for all $i \in C$. But, this assumption makes little sense in our experimental context (as well as most real-world applications) since it supposes that players can "throw away" a commodity, qua utility, they do not possess. Consequently, in testing M^0, we apply the definition only to those alternatives that the committee can actually choose. This explains why in our experimental games, we do not get existence of a point in M^0 for every winning coalition.

The Competitive Solution: The Competitive Solution is developed by *McKelvey/Ordeshook/Winer* [1978] specifically for games without sidepayments or a transferable utility. The general idea is that coalitions bid for members by offering proposals that are as attractive as possible against any conceivable counter offers by other coalitions. Thus, each coalition bids up to the pivotal players (members who are also members of opposing coalitions), and the competitive solution can be thought of as an equilibrium (if it exists) to this bidding war among the coalitions.

Let $p_1 = (x_1, C_1)$, $p_2 = (x_2, C_2)$ be two proposals. Then p_1 is *viable* against p_2 if it is not the case that $p_2 \; P_{C_1 \cap C_2} \; p_1$. If K is a set of proposals, and $p \in K$, we say that p is *viable* in K if it is viable against any $q \in K$. Moreover, K is *balanced* if,

1. for all $p_1 = (x_1, C_1)$, $p_2 = (x_2, C_2) \in K$, $C_1 \neq C_2$

2. for all $p \in K$, p is viable in K.

A proposal p *upsets* K if

1. p is viable in K

2. for some $q \in K$, q is not viable against p.

A set of proposals, K is a *competitive solution* if K is balanced, and no proposal upsets K [c.f., *McKelvey/Ordeshook/Winer*, and *McKelvey/Ordeshook*, 1978b for a more complete treatment of K].

2. Experimental Design

To test the several solution notions outlined in the previous section, we designed a series of five-person majority rule games, which we then ran experimentally. First, we describe the particular games, and then we detail our experimental procedures.

Games Run: Two different games are considered, and the preference orders describing these games are presented in Tables 1 and 2. In the actual experiments the alternatives are permuted to different lettering schemes from those given here to disguise the similarities between games. Here, we simply describe the features that are of interest in each of the basic games.[4])

Games 1 and 2 are designed simply to obtain existence and to secure as much separation as possible between the four game theoretic solution notions: the V-set, $M_1^{(i)}$, M^0, and K. Thus, in these games, all four of these solutions exist, are unique and further, both the V-set, $M_1^{(i)}$, and M^0 are disjoint from the Competitive Solution, K. The only differences between Games 1 and 2 is the inclusion of two additional alternatives, P and Q in Game 2, and a change in status of G for player 3. Game 1 can be viewed as a game in which $M_1^{(i)}$, M^0, and the V-set include "fair" outcomes, while in Game 2 they do not. The initial rationale for designing two separate games is that, while we want to give the V-set and bargaining set a reasonable probability of success, we are also concerned that subjects might, in Game 1, choose points in the V-set or either bargaining set because of equity considerations.

Experimental Procedures: Our experiments use a subject pool of approximately 100 undergraduate and graduate students at Carnegie-Mellon University. Subjects are seated at a round circular table. At the beginning of the experiment, each subject is given a sheet that lists the monetary value to him of each alternative, as well as the preference *order* of the alternatives for all five subjects. The subjects are instructed that they must choose one and only one alternative from this list, and that each subject will be paid for his participation in the experiment in accordance with the value, to him, of the alternative chosen. Subjects are allowed to communicate freely among themselves. However, to prevent sidepayments, subjects are not permitted to discuss the value to them of any alternative or to otherwise mention money. Nor are they permitted, during the experiment, to discuss agreements for dividing their winnings after termination of the game. As a further guarantee against any possible collusive arrangements for transferring utility, their final payoff from the game is determined by adding (or subtracting if negative) a random, but predetermined constant to their value from the alternative selected by the committee (see Appendix A for more details). This constant is then added (or subtracted if negative) to the value of the alternative chosen. If this sum is less than one, the subject is paid one dollar; if it is greater than one, the subject is paid that amount in dollars.

One of three possible payoff schedules is assigned to each player, with the assignment being made randomly with replacement. Thus, there are $3^5 = 243$ possible configurations of payoff schedules across all 5 players, and one of these is drawn randomly with replacement. Since the theories we test involve only ordinal assumptions on in-

---

[4]) Several games with cores were run also to test our experimental procedures and to contrast these procedures with the results of earlier studies. These games are reported in *McKelvey/Ordeshook* [1978c].

Preference Orders of Players

Player 1	Player 2	Player 3	Player 4	Player 5
N	J	B	L	B
J	O	H	E	A
F	M	A, F	D	E, G, H
I	E, F	I	A, G, O	K
K	I	K, M. D, J	M	D
G, O, H	K, D	G	I	M
D, M	G, B	E, C	K, B	O, C
B	H, C	L, N	F, C	L, N, J
A, C	L, N	O	N, J	F, I
L	A		H	
E				

Alternatives Predicted By

1. Pareto Optimals $\quad\quad\quad\quad \underline{P} = \{A, B, D, E, F, G, H, I, J, K, L, M, N, O\}$
2. Top Cycle Set $\quad\quad\quad\quad \underline{G} = \{A, B, D, E, F, G, H, I, J, K, M, O\}$
3. Minimax Set $\quad\quad\quad\quad\quad \underline{M} = \{B, D, E, F, G, I, J, M, O\}$
4. Least Vulnerable Set $\quad\quad \underline{L} = \{B, J\}$
5. Least Alternative Set $\quad\quad \underline{A} = \{J\}$
6. Von Neumann-Morgenstern Set $\quad V = \{D, J, K, M\}$
7. Bargaining Set $\quad\quad\quad\quad M_1^{(i)} = \{D, G, M\}$
8. Ordinal Bargaining Set $\quad\quad M^0 = \{D, G, K, M\}$
9. Competitive Solution $\quad\quad K = \{A, E, F, H, O\}$

Proposals Predicted By

1. $M_1^{(i)} = \{(D, \{1, 3, 4\}), (G, \{1, 4, 5\}), (G, \{1, 2, 5\}), (G, \{1, 2, 4, 5\}), (M, \{2, 3, 5\})\}$
2. $M^0 = \{(D, \{1, 2, 3, 5\}), (D, \{1, 2, 5\}), (D, \{1, 3, 4\}), (D, \{2, 3, 5\}), (G, \{1, 2, 4, 5\}),$
   $\quad\quad (G, \{1, 2, 4\}), (G, \{1, 2, 5\}), (G, \{1, 4, 5\}), (G, \{2, 4, 5\}), (K, \{2, 3, 5\}), (M, \{1, 3, 4, 5\}),$
   $\quad\quad (M, \{1, 3, 4\}), (M, \{2, 3, 5\})\}$
3. $K = \{A, \{3, 4, 5\}), (E, \{2, 4, 5\}), (F, \{1, 2, 3\}), (H, \{1, 3, 5\}), (O, \{1, 2, 4\})\}$

Tab. 1: Game 1

dividual preferences, the actual payoff intervals between alternatives should be irrelevant. In the absence of any reason for doing otherwise, but to prevent subjects from inferring the payoff schedules of others from their own, on any one schedule the intervals are chosen according to a uniform distribution across the following six values: .50, 1.00, 1.50, 2.00, 2.50, 3.00.[5]) In general, final payoffs averaged $ 25 per experiment, or $ 5 per subject per experiment.

---

[5]) The magnitudes and ranges of the specific payoff schedules correspond roughly to those reported in *McKelvey/Ordeshook* [1978b].

Preference Orders of Players

Player 1	Player 2	Player 3	Player 4	Player 5
N	J	B	L	B
J	O	H	E	A
F	M	A, F	D	E, G, H
I	E, F	I	A, G, O	K
K	I	L, N	M	Q
G, O, H	K, D	P	I	D
P, Q	G, B	Q	P	M
D, M	H, C, P	K, M, D, J	K, B	O, C
B	L, N, Q	E, C	F, C	P
A, C	A	G	N, J, Q	L, N, J
L		O	H	F, I
E				

Alternatives Predicted By

1. Pareto Optimal Set $\quad\underline{P} = \{A, B, D, E, F, G, H, I, J, K, L, M, N, O, P, Q\}$
2. Top Cycle Set $\quad\underline{G} = \{A, B, C, D, E, F, G, H, I, J, K, L, M, O, P, Q\}$
3. Minimax Set $\quad\underline{M} = \{B, D, E, F, G, I, J, M, O\}$
4. Least Vulnerable Set $\quad\underline{L} = \{B, J\}$
5. Least Alternative Set $\quad\underline{A} = \{B, J\}$
6. Von Neumann-Morgenstern Set $\quad\underline{V} = \{D, J, K, M\}$
7. Bargaining Set $\quad M_1^{(i)} = \{D, G, M\}$
8. Ordinal Bargaining Set $\quad M^0 = \{D, G, K, M\}$
9. Competitive Solution $\quad K = \{A, E, F, H, O\}$

Proposals Predicted By

1. $M_1^{(i)} = \{(D, \{1, 3, 4\}), (G, \{1, 4, 5\}), (M, \{2, 3, 5\})\}$
2. $M^0 = \{(D, \{1, 2, 3, 5\}), (D, \{1, 2, 5\}), (D, \{1, 3, 4\}), (D, \{2, 3, 5\{), (G, \{1, 2, 5\}),$
   $(G, \{1, 4, 5\}), (K, \{2, 3, 5\}), (M, \{1, 3, 4, 5\}), (M, \{1, 3, 4\}), (M, \{2, 3, 5\})\}$
3. $K = \{(A, \{3, 4, 5\}), (E, \{2, 4, 5\}), (F, \{1, 2, 3\}), (H, \{1, 3, 5\}), (O, \{2, 3, 5\})\}$

Tab. 2: Game 2

The committee deliberations proceed in any manner the subjects choose (with the restrictions mentioned above). However, a fifteen minute discussion period is required before any final decision can be made. At any point after this period, if three or more players come to an agreement on some alternative, they can secure this outcome and terminate the experiment. This is done by each subject signing and handing in an "agreement" card, listing on it the alternative he or she has agreed to. If a majority (3 or more) of these cards agreed, the experiment terminates.

Game 1: Alternatives in Set

Solutions		A	B	C	D	E	F	G	H	I	J	K	L	M	N	O	Success	Failure	Expected Success	Actual Success	Likelihood
1. Pareto Optimal Set	P	X	X	X		X	X	X	X	X	X	X	X	X	X	X	14	0	.933	1.000	.37874
2. Top Cycle Set	G	X	X	X		X	X	X	X	X	X	X	X	X	X	X	14	0	.933	1.000	.37874
3. Minimax Set	M			X	X	X	X	X	X	X	X						9	3	.600	.642	.48586
4. Least Alternative Set	A		X														1	13	.066	.071	.61554
5. Least Vulnerable Set	L									X							1	13	.133	.071	.86439
6. Von Neumann Morgenstern Set	V				X				X	X			X		X		2	12	.267	.142	.91998
7. Bargaining Set	$\underline{M}_1^{(i)}$				X		X							X			1	13	.200	.071	.95601
8. Ordinal Bargaining Set	M^o				X		X				X			X			1	13	.267	.071	.9870
9. Competitive Solution	K	X					X	X		X						X	12	2	.333	.857	.00008
Actual Outcomes		3			4	3		2		1		1		1							

Game 2: Alternatives in Set

Solutions		A	B	C	D	E	F	G	H	I	J	K	L	M	N	O	P	Q	Success	Failure	Expected Success	Actual Success	Likelihood
1. Pareto Optimal Set	P	X	X	X	X	X	X	X	X	X	X	X	X	X	X	X	X	X	14	0	.941	1.000	.42683
2. Top Cycle Set	G	X	X	X	X	X	X	X	X	X	X	X	X	X	X	X	X	X	14	0	.941	1.000	.42683
3. Minimax Set	M			X	X	X	X	X	X	X	X	X							11	3	.529	.786	.04603
4. Least Alternative Set	A			X															1	13	.118	.071	.82759
5. Least Vulnerable Set	L			X															1	13	.118	.071	.82759
6. Von Neumann Morgenstern Set	V				X				X	X			X						0	14	.235	.000	1.000
7. Bargaining Set	$\underline{M}_1^{(i)}$				X		X						X						0	14	.176	.000	1.000
8. Ordinal Bargaining Set	M^o				X		X				X		X						0	14	.235	.000	1.000
9. Competitive Solution	K	X				X	X	X									X		13	1	.294	.928	.00000
Actual Outcomes		3	1			4	4		4								2						

Tab. 3: Non Core Experiments

3. Experimental Outcomes

The outcomes of our experiments are summarized in Table 3. This table lists the predicted outcomes for the various solution theories in each of the two games, as well as the actual outcomes of the experiments, the number of successes and failures, the expected success rate (assuming that the actual experimental outcomes are distributed uniformly across all alternatives), the actual success rate, and the likelihood of the model doing as well as it did. This final figure, labeled "likelihood" in the tables, re-presents the probability of observing at least as many successes as are observed if the actual distribution is the null model of equally likely alternatives.[6])

As Table 3 shows, all alternatives fall in the Pareto optimal set and in the top cycle set. These predictions, however, are not very restrictive (all other predictions are a sub-set of these). The only other solutions that do better than expected under the null

	GAME 1			GAME 2	
		FFP Model			FFP Model
	Actual	Expected		Actual	Expected
Alternative	Proportion	Proportion	Alternative	Proportion	Proportion
A	.214 (3)	.105 (1.5)	A	.214 (3)	.084 (1.2)
B	.000 (0)	.176 (2.5)	B	.071 (1)	.152 (2.1)
C	.000 (0)	.000 (0)	C	.000 (0)	.001 (.01)
D	.000 (0)	.091 (1.3)	D	.000 (0)	.081 (1.1)
E	.286 (4)	.069 (1.0)	E	.286 (4)	.069 (1.0)
F	.214 (3)	.057 (.8)	F	.286 (4)	.057 (.8)
G	.000 (0)	.023 (.3)	G	.000 (0)	.020 (.3)
H	.143 (2)	.063 (.9)	H	.000 (0)	.057 (.8)
I	.000 (0)	.101 (1.4)	I	.000 (0)	.101 (1.4)
J	.071 (1)	.086 (1.2)	J	.000 (0)	.079 (1.1)
K	.000 (0)	.037 (.5)	K	.000 (0)	.042 (.6)
L	.000 (0)	.000 (0)	L	.000 (0)	.014 (.2)
M	.071 (1)	.083 (1.3)	M	.000 (0)	.086 (1.2)
N	.000 (0)	.000 (0)	N	.000 (0)	.000 (0.0)
O	.000 (0)	.098 (1.3)	O	.143 (2)	.095 (1.3)
			P	.000 (0)	.028 (.4)
			Q	.000 (0)	.033 (.5)

$x^2 = 25.55$ $x^2 = 33.79$
with 13 d.f.   with 15 d.f.
reject at level $p = .05$ reject at level $p = .01$

Tab. 4: Test of the Markov Model

---

[6]) Letting n be the number of trials (i.e., $n = 14$), k_i be the number of observed successes under model i, p_i be the expected success rate of model i, and $q_i = 1 - p_i$, the likelihood figure is

$$L = \sum_{i=k_i}^{n} \binom{n}{k_i} p_i^{k_i} q_i^{(n-k_i)}.$$

model are the minimax set and the competitive solution. Further, if all solutions are ranked in terms of the likelihood figures, the competitive solution completely overwhelms the others: It is the only solution with which we can reject the null hypothesis of equally likely outcomes at (actually below) the .01 level.

The Markov Model: To assess the adequacy of the markov model of *Ferejohn/Fiorina/Packel* [1978], its limiting probabilities are presented in Table 4. Note that competitive solution points (especially *A, E* and *F*) tend to be under-predicted, but that points outside of *K* (in particular, points in the least vulnerable set) tend to be over-predicted. The model is tested under the unrestricted multinomial distribution, and the values of χ^2, assuming that the Markov model is in fact correct, are 25.55 and 33.79 for Games 1 and 2 respectively. Thus, we can reject this model at the .05 level in both cases.

Coalition Predictions: The competitive solution, M^0, and $M_1^{(i)}$ make coalition predictions as well as predictions about what alternative in X a coalition should choose (since these solution notions isolate sets of proposals rather than sets of alternatives). Tables 1 and 2 present these coalition predictions, and Table 5 describes the proposals we observe for Game 1 and 2. Note first, that in *every instance*, when the alternative chosen corresponds to an outcome in the competitive solution, the coalition is also the appropriate coalition as predicted by K. Further, for the one case in which an alternative corresponds to M^0 and $M_1^{(i)}$, the proposal is supported by the *wrong* coalition. In this experiment (experiment # 1), player 4 supported the proposal even though he had a valid objection against player 2. Player 4 in fact, was the primary spokesman for this proposal, and this outcome would undoubtedly not have prevailed if he had not supported it.

GAME 1						GAME 2					
Experiment #	Outcome	Coalition	in K ?	in $M_1^{(i)}$?	in M^0 ?	Experiment #	Outcome	Coalition	in K ?	in $M_1^{(i)}$?	in M^0 ?
1	M	{2, 3, 4, 5}	N	N	N	15	F	{1, 2, 3}	Y	N	N
2	E	{2, 4, 5}	Y	N	N	16	B	{1, 2, 3, 5}	N	N	N
3	F	{1, 2, 3}	Y	N	N	17	F	{1, 2, 3}	Y	N	N
4	E	{2, 4, 5}	Y	N	N	18	A	{3, 4, 5}	Y	N	N
5	E	{2, 4, 5}	Y	N	N	19	O	{1, 2, 4}	Y	N	N
6	J	{1, 2, 3}	N	N	N	20	E	{2, 4, 5}	Y	N	N
7	H	{1, 3, 5}	Y	N	N	21	F	{1, 2, 3}	Y	N	N
8	A	{3, 4, 5}	Y	N	N	22	E	{2, 4, 5}	Y	N	N
9	A	{3, 4, 5}	Y	N	N	23	E	{2, 4, 5}	Y	N	N
10	F	{1, 2, 3}	Y	N	N	24	O	{1, 2, 4}	Y	N	N
11	A	{3, 4, 5}	Y	N	N	25	A	{3, 4, 5}	Y	N	N
12	F	{1, 2, 3}	Y	N	N	26	E	{2, 4, 5}	Y	N	N
13	E	{2, 4, 5}	Y	N	N	27	A	{3, 4, 5}	Y	N	N
14	H	{1, 3, 5}	Y	N	N	28	F	{1, 2, 3}	Y	N	N

Tab. 5: Coalition Outcomes of Non Core Experiments

Although the competitive solution receives considerable support in these experiments, further refinements in its definition seem warranted. The competitive solution does not make any predictions about the frequency of various of the solution points being observed, and, in the absence of such a prediction, the most reasonable assumption is that all proposals in K are equiprobable. If we pool the results of Games 1 and 2, however, there appears to be a significant difference between the actual and observed frequencies of the various solution points (see Table 6). In particular, note that of the coalitions that include player 1, the proposals in which 1 is an individual pivot, $(H, \{1, 3, 5\})$ and $(O, \{1, 2, 4\})$, occur less frequently than the proposal $(F, \{1, 2, 3\})$. We might attribute this to the fact that player 1 is in a good bargaining position: There are several proposals that are better for 1 than 0 and H, that beat other competitive solution points, and that come up fairly frequently in the bargaining. Hence, player 1 seems unwilling to lower his expectations to 0 and H, and, without 1's support for these proposals, other competitive solution points become more attractive and more probable. While the present definition of K does not take such strategic considerations into account, it is hoped that a future version might be able to do so.

Proposal	Frequency
$(A, \{3, 4, 5\})$	6
$(E, \{2, 4, 5\})$	8
$(F, \{1, 2, 3\})$	7
$(H, \{1, 3, 5\})$	2
$(O, \{1, 2, 4\})$	2

Tab. 6: Coalition Frequencies of Competitive Solution Points in Non Core Experiments

4. Information Effects

Given the preceding experimental support for K, it is useful to test K's sensitivity to various conditions. In particular, we report in this section a series of experiments in which subjects must play games 1 and 2 as before except that they are given information about only their own preferences. They are free to discuss these preferences (subject to the usual prohibition against talking about money or sidepayments) in any way and, hence, are also free to lie. Table 7 summarizes our experimental results.

As this table indicates, K again receives considerable support, although, as expected, the results are somewhat noisier — K's success rate drops from .893 to .722. None of the other game theoretic solution notions fare well, however, nor do the least alternative and least vulnerable sets. Like K, the support for the minimax set remains strong, but again we must caution that $\underline{M}$ here does not make particularly restrictive predictions and is, to a considerable extent, the beneficiary of K's successes.

5. Conclusion

While the experiments we report here strongly support the competitive solution as a useful solution notion (when it exists), we should ask why the competitive solution does so well, while the alternative solution notions do so poorly.

		Success	Failure	Expected Success	Actual Success	Likeli-hood
1. Pareto Optimal Set	$\underline{P}$	18	0	.937	1.000	.30996
2. Top Cycle Set	$\underline{G}$	18	0	.937	1.000	.30996
3. Minimax Set	$\underline{M}$	13	5	.5645	.722	.00398
4. Least Vulnerable Set	$\underline{L}$	0	18	.092	.000	1.00000
5. Least Alternative Set	$\underline{A}$	2	16	.126	.111	.54194
6. V-N Morgenstern Set	$\underline{V}$	1	17	.251	.056	.98251
7. Bargaining Set	$M_1^{(i)}$	0	18	.188	.000	1.00000
8. Ordinal Bargaining Set	M^0	1	17	.251	.056	.98251
9. Competitive Solution	K	13	5	.314	.722	.00000

Tab. 7: Summary of Games 1 and 2 with Minimal Information

First, it is reasonable to argue that the solutions derived from notions of vulnerability are put here to unfair tests: These solution notions are not developed in the context of a characteristic function form game, and should not be expected to do well in this experimental context. The minimax set and the least alternative set, for example, are developed in the context of a model that assumes a status quo and that changes of this status quo are made sequentially. It is likely, then, that these hypotheses would fare better in an experimental context that incorporates these features.

The game theoretic solutions, on the other hand, are all tested in an experimental setting that corresponds to a game in characteristic function form, and, hence, a priori there is no reason to suppose that one should fare better than another. The failure of the V-set seems attributable to the fact that it is simply an ad hoc mathematical invention without any behavioral rationale. Thus, even for sidepayment games, we are tempted to ascribe its past success to the fact that it coincides typically with other solution notions that do possess some behavioral justification.

The failure of the $M_1^{(i)}$ bargaining set, on the other hand, stems, we suspect, from the fact that it is a solution that is developed originally for sidepayment games with transferable utility and then imported to the non sidepayment context. In the context of sidepayment games, the definition of the bargaining set is reasonable: No player in a coalition should be expected to get more than he is "worth," because other players in the coalition can extract (transfer) some of that value from him and obtain it for themselves. For games *without* sidepayments and *without* transferable utility, however, there are inescapable positive externalities of coalition formation that necessarily bestow on some members payoffs that exceed what they can defend. And, since there is no transferable utility, there is no way for other members of the coalition to expropriate this benefit. Further, a player may do better in a coalition in which he possesses a justified objection against another player than he does in a coalition in which he has no justified objections. In short, the usual $M_1^{(i)}$ bargaining set seems an inappropriate solution for games without sidepayments or transferable utility, and the results of our experiments support this a priori judgement.

The failure of the ordinal bargaining set is, perhaps, somewhat more surprising since this is a solution notion designed specifically for games without sidepayments. And, while the general existence of M^0 presupposes the comprehensiveness of $v(C)$, the ideas underlying its definition seem reasonable. We can conclude only that these ideas do not in fact model the bargaining process and the strategic features of our experimental games – if not all simple majority rule games.

Although the competitive solution fares the best of the alternative solutions, some cautions are in order. First, as we note in the text, strategic considerations appear to affect the frequency with which various solution points are adopted. This suggests that K does not give us the whole story. Second, our experiments concern only two versions of the same game, in which all three of the game theoretic solution notions exist. In the context of choice over a finite alternative set, none of the game theoretic solutions are guaranteed to exist, and we have no evidence what might happen when a competitive solution fails to exist. Finally, we cannot reject the hypothesis that K succeeds here for the same reason V and $M_1^{(i)}$ receive support in earlier experiments entailing transferable utility – namely, that K's predictions correspond fortuitously to some more general solution notion.

APPENDIX: Instructions Read to Subjects

In this experiment, your task, as a committee, is to choose one and only one alternative from a list of alternatives labeled A, B, C, etc.

To assist you in understanding the experiment, each of you has been given a sample worksheet, and at the beginning of the experiment you will each be given a similar worksheet, although the entries there may differ from this sample. Your worksheet provides two types of information. First, it gives the payoff to you, in dollars, of each alternative. Thus, in this sample, which is a worksheet for a fictitious player 2, alternative H is worth \$ 9 to player 2, while A is worth \$ 2 to him. Second, your worksheet also summarizes the preference orders of the other players as determined by the payoffs assigned to the alternatives for them. Looking at the sample again, the payoff player 1 associates with C in this fictitious game exceeds that of I, I exceeds that of D and G, but D and G are assigned the same payoff, and so on.

How much you will receive ultimately from participation in this experiment depends, however, not only on the payoff you associate with the alternative the committee selects, but also on your "α-value." Specifically, each of you has been assigned a positive or a negative number, which we call alpha and which appears on the bottom of your worksheets in the space labeled α. In the actual experiment, this will be covered, and will only be revealed to you at the termination of the experiment. When the experiment ends, α will be added, or subtracted if it is negative, to the value to you of the alternative chosen by the committee. If this total is less than one, you will receive \$ 1. If it is greater than 1, you will receive that amount in dollars.

Looking at the sample worksheet, suppose alternative A is selected and that this fictitious player's α value equals plus \$ 4. Thus his final payoff is \$ 2 – the value of alternative A to him – plus \$ 4.00 or \$ 6.00. But, if alternative I is chosen, his final payoff would be $-\$2.00 + \$4.00$, or $\$2.00$. Notice, then, that if α is sufficiently large, each of you can win more than your minimum of $\$1$ even if the value of the alternative chosen by the committee is less than 1 or even negative. On the other hand, α may be negative so that a decision worth more than $\$1$ is not a guarantee that you'll win more than $\$1.00$.

In the actual worksheet that you receive at the beginning of the experiment, your payoffs will, of course, differ from those in this table. Further the payoffs that you associate with a given alternative may differ from those of another player. Thus, one alternative may be much more beneficial for one player than it is for another. There is no symmetry to the payoffs, so you cannot infer the payoffs of other players from your own. In particular, the payoffs on the sample worksheet refer *only* to the payoffs of player 2, and *not* to those of other players. The fact that there is a one dollar difference in this example between 2's best and second best alternative does not mean that the same is true for any one else. Finally, the value of α assigned to each of you may differ from player to player so that even if two of you associate the same value with the committee's decision, you need not ultimately receive the same payoff from participating in this experiment.

Note now that although your worksheet tells you the preference order of each player, it does *not* allow you to determine how much money each player associates with each alternative. In fact, one essential rule of this experiment is that *at no time are you to mention the value to you of any alternative, or to make any other mention of money*. Further, at no time are you permitted to discuss schemes for dividing your winnings at the termination of play. At the termination of the experiment, in fact, each of you will be paid in the absence of other players, so there is no need to *ever* reveal your total winnings, should you decide not to do so.

Any Questions?

Let me now discuss the way in which the experiment will proceed. The experiment is divided into two phases, the *discussion* phase and the *decision* phase. At the beginning of the experiment, I will hand out your actual worksheets and you will begin with a 15-minute discussion period, during which you may discuss what your final decision might be, but you may not implement any binding decisions. At the end of the discussion phase, you will begin the decision phase. During this phase, for which there is no time limit, you are permitted not only to continue your discussion, but also to terminate the experiment and implement a final decision. Specifically, when three or more of you can come to some agreement on an alternative, you may formalize this agreement with these 3 X 5 agreement cards (display cards). Each of you should sign one of your cards, indicating on it your player number and the alternative chosen. I will then collect these cards and if three or more *agree completely*, the experiment will terminate and you will each be paid as I described earlier in accordance with the alternative chosen.

Note that a majority is needed only to reach a final decision, and it is not necessary for you to be in this majority to receive a positive payoff. The payoff each of you receives will correspond to the payoff you associate with final alternative chosen, plus your α value, regardless of whether you were a party to this agreement.

References

Asscher, N.: An Ordinal Bargaining Set for Games Without Sidepayments. Mathematics of Operations Research 1 (4), 1976, 381–389.

Aumann, R.J., and *M. Maschler*: The Bargaining Set for Cooperative Games. Annals of Mathematical Studies 52, 1964, 443–476.

Berl, J., R.D. McKelvey, P.C. Ordeshook, and *M. Winer*: An Experimental Test of the Core in A Simple N-Person Cooperative Nonsidepayment Game. Journal of Conflict Resolution 20, 1976, 453–479.

Billera, L.J.: Existence of General Bargaining Sets for Games Without Side Payments. Bulletin of the American Mathematical Society 76, 1970, 375–379.

Ferejohn, J.A., M.P. Fiorina, and *E.W. Packel*: A Non Equilibrium Approach to Legislative Decision Theory. Social Science Working Paper 202, California Institute of Technology, 1978.

Ferejohn, J.A., M.P. Fiorina, and *H.A. Weisburg*: Toward a Theory of Legislative Decision. Game Theory and Political Science. Ed. by P.C. Ordeshook. New York 1978.

Fiorina, M., and *Ch. Plott*: Committee Decisions Under Majority Rule: An Experimental Study. American Political Science Review 72 (2), 1978.

Isaacs, R.M., and *Ch. Plott*: Cooperative Game Models of the Influence of the Closed Rule in Three Person Majority Rule Committees: Theory and Experiments. Game Theory and Political Science. Ed. by P.C. Ordeshook. New York 1978.

Kramer, G.H.: A Dynamical Model of Political Equilibrium. Cowles Foundation Discussion Paper No. 396, 1975.

Laing, J.D., and *S. Olmstead*: Policy Making by Committees: An Experimental and Game Theoretic Study. Game Theory and Political Science. Ed. by P.C. Ordeshook. New York 1978.

McKelvey, R.D., and *P.C. Ordeshook*: Competitive Coalition Theory. Game Theory and Political Science. Ed. by P.C. Ordeshook. New York, forthcoming, 1978a.

– : Vote Trading: An Experimental Study. Public Choice, forthcoming, 1978b.

– : Some Experimental Evidence for Complex Choice in Cooperative Games with Cores Using Simple Majority Rule, mineo, 1978c.

McKelvey, R.D., P.C. Ordeshook, and *M. Winer*: The Competitive Solution for N-Person Games Without Sidepayments. American Political Science Review, June 1978.

Peleg, B.: Bargaining Sets of Cooperative Games Without Side Payments. Israel Journal of Mathematics 1, 1963, 197–200.

– : Existence Theorem for the Bargaining Set $M_1^{(i)}$. Essays in Mathematical Economics in Honor of Oskar Morgenstern. Ed. by M. Shubik. Princeton 1967.

– : The Extended Bargaining Set for Cooperative Games Without Side Payments. Research Program in Game Theory and Mathematical Economics, Department of Mathematics, The Hebrew University of Jerusalem, Israel, Research Memorandum No. 44, March 1969.

Schwartz, Th.: Rationality and the Myth of the Maximum. Nous 6, 1972.

Simpson, P.: On Defining Areas of Voter Choice: Professor Tullock on Stable Voting. Quarterly Journal of Economics 83, 1969.

Applied Game Theory. 1979 ©Physica-Verlag, Wuerzburg/Germany

Iannucci and Its Aftermath:
The Application of the Banzhaf Index to Weighted Voting in the State of New York[1])

By *B. Grofman*, Irvine[2]) and *H.Scarrow*, Stony Brook[3])

Abstract: We review the use by New York State Courts of John Banzhaf III's game-theoretic inspired index of power as a measure of fair representation. We look at the extent to which game-theoretic arguments have been a) properly understood by the courts; b) integrated into constitutional and legal analysis; and c) properly applied. We pay particular attention to weighted voting in Nassau County, for which we provide a more detailed historical analysis.

1. Introduction

Recent decisions of the U.S. Supreme Court have stressed the requirement that apportionment and electoral systems at all levels of government approach the ideal of "one person, one vote." (See e.g., **Baker v.Carr, Reynolds v.Sims, Wesberry v.Sanders**.) In the 1960's unequally populated single-member legislative districts have been more or less eliminated due to court and legislative action. As a necessary consequence of adoption of a strict population standard, unit voting schemes which provided for one representative from each political subunit regardless of subunit population (such as those which had been used in 1960 at the county level in Michigan, Illinois, Wisconsin, New York and New Jersey and at the state level in eight state legislatures) have also been eliminated.

However, while single membership districting is the most common form of representation in the U.S., multimember districting and mixed single and multiple member

---

[1]) This paper is excerpted from "Game Theory and the U.S. Courts," delivered at the International Conference on Applied Game Theory, Institute for Advance Studies, Vienna, June 12–15, 1978. A copy of the longer paper is available from the authors upon request. We would like to thank the staff of the Word Processing Center of the School of Social Sciences, University of California, Irvine, for its secretarial assistance and Norman Jacobson of the Public Policy Research Organization of the University of California, Irvine, for the computer program used to generate Banzhaf values. This research was supported by NSF Grant SOC 77-24474, Political Science Program.

[2]) Prof. *Bernhard Grofman*, School of Social Sciences, University of California, Irvine, Calif. 92717, USA.

[3]) Prof. *Howard Scarrow*, Dept. of Political Science, State University of New York, Stony Brook, N.Y., USA.

apportionments are to be found in various levels of government in the U.S.; and in one state (New York) weighted voting is the most common of the various systems in use for county government. In the late 1960's and 70's such systems have come under increasing challenge as violating 14th Amendment "equal protection" standards. In the past decade there have been well over a dozen lawsuits in New York alone challenging the constitutionality of local weighted voting and multimember district apportionment schemes on one man, one vote grounds.

In measuring deviations from the ideal of "one person, one vote — one vote, one value" in the case of weighted voting systems and systems which involve multimember districts, U.S. Courts have been urged by various plaintiffs to judge the fairness of voter/group/unit/legislator representation and weightings in terms of game-theoretic indices of power such as the Banzhaf index [*Banzhaf*, 1965, 1966]. The U.S. Supreme Court has rejected *Banzhaf*'s reasoning as to the appropriate measure of **voter** power in the case of mixed single and multimember districts (**Whitcomb v. Chavis**); while New York State courts, on the other hand, have explicitly endorsed the use of the Banzhaf index as the appropriate measure of *legislator* power and as *the* criterion of fair representation in weighted voting schemes.

We propose to examine the nature and extent of the use by New York State courts of Banzhaf's game-theoretic inspired index of power as a measure of "fair" representation, making use of a number of criteria by which fair representation might be judged. We shall look at the extent to which game-theoretic arguments have been a) properly understood by the courts; b) integrated into constitutional and legal analysis; and c) appropriately applied.

2. The Banzhaf Index

In two articles which appeared in American law journals in the mid-1960's, *John Banzhaf* III, a lawyer and mathematician, proposed to evaluate representation systems in terms of the extent to which they allocated "power" fairly. *Banzhaf*'s analysis makes use of game-theoretic notions in which power is equated with the ability to affect outcomes.

2.1 Equal Voter Power

Consider a group of citizens choosing between two opposing candidates. To calculate the power of the individual voter, we generate the set of all possible voting coalitions among the district's electorate. If there are N voters in the district, then there will be 2^N possible coalitions. Then we ask, for each of these possible coalitions, whether a change in an individual voter's choice from Candidate A to Candidate B (or from Candidate B to Candidate A) would alter the electoral outcome. If so, that voter's ballot is said to be *decisive*. A voter's power is defined as the number of times, in all possible coalitions, that his vote could be decisive, and can best be expressed as a percentage — i.e., the number of his decisive votes divided by the total number of all the decisive votes of all the voters (including himself). The higher the percentage of voter coalitions in which *his* vote is *decisive,* the higher a voter's power score.[4]) The Banzhaf index has considerable intuitive appeal; power is based on ability to affect outcome.

For single member district systems, each district having equal populations, all voters have identical power; the ability of the voter in one district to affect his district's electoral outcome is identical with the ability of another voter in a neighboring district to affect the outcome there. But what about the case of multiple member districts, with some districts of one size and others of another size? Here, since the voters who elect k representatives have k times as much importance as voters who can elect only one representative, we might expect that to equalize voter power we should assign the districts with k representatives k times as many voters as well, since with all votes of equal weight, intuitively, we would expect a voter's ability to decisively affect outcomes should be inversely proportional to district size. *Banzhaf* [1966] pointed out that this argument is mathematically incorrect.

In a two-party candidate contest where all voters have equal weight, in order for a voter to be decisive in a district of size N, the rest of the voters (who are $N - 1$ in number) must split half for one candidate/party and half against. A straightforward combinatoric analysis reveals [*Banzhaf*, 1966; **Whitcomb v.Chavis** 403 U.S. at 145 n. 23; *Walther*, p. 11; *Lucas*, p. 52] that, *if all combinations of vote outcomes are equally likely* (i.e., each voter is equally likely to vote for either candidate/party), each member's decisive votes b) are given by:

$$b = \frac{2\,(N-1)\,!}{((N-1)/2)\,((N-1)/2)} \,! \tag{1}$$

We can examine the link between b and N by using Stirling's approximation [see *Feller; Banzhaf*, 1966; *Walther*, p. 12–13; *Lucas*, p. 53]

$$N\,! \approx e^{-N}\,N^N\,\sqrt{2\,\pi\,N} \tag{2}$$

to rewrite (1) as

$$b \approx \frac{2^N}{\sqrt{2\,\pi\,(N-1)}} \tag{3}$$

Thus, each member's Banzhaf Index, which we shall denote B_i, is simply

$$B_i \approx \frac{1}{\sqrt{2\,\pi\,(N-1)}} \tag{4}$$

This analysis can be applied to electoral systems involving both single and multimember districts. We see from expression (4) that B_i is approximately proportional to the

---

[4]) There are considerably more powerful mathematical tools to calculate the *Banzhaf* index than merely enumerating all 2^k possible coalition outcomes and identifying decisive voters in each. We shall not discuss such techniques here. [See *Walther; Brams/Affuso*.]

square root of N, district population. (This appears to have been first pointed out by *Penrose* [1946]; cf. also *Fielding* [1973]). Thus, if we wish to assign all voters equal power to affect outcomes, we should assign each district a number of representatives proportional to the *square root* of district population, rather than directly proportional to district population. Doing so, however, violates the norm of allocating an equal number of citizens an equal number of representatives.

If we assign one representative for every 100 population in the square root of district size, then if there are 20,000 population spread equally over 2 smds, these voters (10,000 per district) would be entitled to have 2 representatives, 1 per district, since the square root of 10,000 is 100. Similarly, if there are 40,000 citizens spread equally over 4 smds (10,000 each) they would be entitled to 4 representatives. However, a single mmd of size 40,000 would be allocated only 2 representatives, since the square root of 40,000 is only 200. Thus, in this example, 20,000 voters would be entitled to as many representatives as 40,000 voters. If we follow the square root rule, the allocation of representatives depends on how voters are divided among the districts.

In a case decided in 1970, **Whitcomb v. Chavis** 403 U.S. 143, the Supreme Court dealt directly with *Banzhaf's* concept of voter power. The case involved Indiana's scheme of single and multiple member districts for its state legislature. The plaintiffs, citing *Banzhaf's* work, argued that voters in the multiple member districts were over-represented, claiming that citizens in the larger district had a power disproportionate to their population.

We can make this argument explicit as follows: if votes in a multi-member district (with population mN) elected m representatives, each voter in such a district would have a power of $\dfrac{m}{\sqrt{m}} / \sqrt{N}$, while those in smds with population of N would have a power of $\dfrac{1}{\sqrt{N}}$. Since $\dfrac{m}{\sqrt{m}} > 1$, for $m > 1$, this would be denying to all citizens an "equally effective voice in the election of members of his legislature." (377 U.S. at 565.)

The court in **Whitcomb** rejected the *Banzhaf* argument, both in the majority opinion and in Justice Harlan's dissenting opinion. Only Harlan's opinion, however, dealt forthrightly with the intellectual merits of the *Banzhaf* argument. Harlan lampooned the absurdity of *Banzhaf's* simplifying assumptions (clearly articulated by *Banzhaf* himself) e.g., the assumption that there exist no ingrained voting habits and that therefore each voter is equally likely to vote for either candidate before him. He pointed out with glee how minor variations in *Banzhaf's* assumptions can lead to major variations in results. Harlan's opinion of the *Banzhaf* index is best summed up in one sly footnote which quotes Mark Twain: "There is something fascinating about science. One gets such wholesale returns of conjecture out of such a trifling investment of reality." (403 U.S. at 169 n.5.)[5]

---

[5]) *Banzhaf* [1965] was previously cited briefly in **Kilgarlin v. Hill** 386 U.S. 120, 125; and at greater length in **WMCA Inc. v. Lomenzo**, 246 F. Supp 953 at 959.

Banzhaf's own views on the reasonableness in political terms of his index are worth mentioning.
... Thus, in constructing a mathematical model, which must of necessity ignore many of the real problems of the system, one may hypothesize the representative to be no more than a vehicle for

2.2 Legislator Weight Proportional to Decisiveness

So long as each legislator has a single YEA or NAY vote on issues coming before the legislature, the question of legislative power does not have to be explicitly addressed. Thus, in the leading apportionment cases which have come before the U.S. Supreme Court, all of which have involved single or multiple-member districts with each elected representative eligible to cast a single vote, it seems to be simply *assumed* that the justification for examining the number of persons contained within each district is the fact that elected representatives by their vote wield decision-making power in the affairs of the polity; and that equality of apportionment thus indirectly results in equality of policy-making power among citizens.

But what about weighted voting schemes (also fractional voting schemes) where, say, a legislator from a district with 20,000 population casts two votes, while a legislator from a district with 10,000 population casts only one vote? Again, it was *John Banzhaf* III who pointed out the fallacy of such "common sense" apportionment schemes. Consider, for example, a three-member committee, with members A and B with two votes, and member C with only one vote. Despite the fact that *vote* shares (weights) are not equal, from the standpoint of *Banzhaf's* concept of decisive votes all committee members have equal *power* (1/3, 1/3, 1/3) when a majority (3 of 5 votes) is needed. This is shown in Table 1, which shows the 8 possible coalitions. When a two-thirds vote is necessary for passage, the power scores change. Now member C has no power at all (in the language of game theory, he is a *dummy*), while the other two members each hold 50 percent of the power. *Banzhaf's* argument is simply that when weighted voting schemes are designed, weights should be assigned in such a way that a legislator's *power* (as contrasted with the number of votes) should be made proportional to the number of citizens in his district.

Since the U.S. Supreme Court in **Whitcomb** rejected *Banzhaf's* "square root" argument regarding assigning representatives to multiple-member districts, one might have predicted that courts would also have rejected his closely related line of reasoning for weighted voting schemes as well. Such was not the case. Ruling on the constitutionality of a weighted voting scheme for Washington County and one for Saratoga County, the New York State Court of Appeals in **Iannucci v. Board of Supervisors** (1967) 282

---

reflecting as best he can the votes of his constituents on certain issues. In such a model of the representative system, each representative would in effect poll his district on each issue and cast his vote according to the majority vote. For the limited purpose of establishing the outer boundaries of a fair representative system, it seems reasonable to assume this type of representative as an over-simplified model [*Banzhaf*, 1968, p. 817].

What little is known about how legislator's votes are influenced tends to cast doubt on any theory which would have a constituent's ability to affect his representative's vote depend solely on the population of the district. Such a theory would ignore party alliances, ethnic blocs, regional differences and interests, lobbying, influence peddling, and other realities of political life. Yet, so far, the Supreme Court has looked no further than population figures in deciding reapportionment cases. Moreover, the justification offered for multimember district systems also depends upon such a theory. If influence and representation cannot with some reasonable degree of accuracy be approximated by such a theory, then the justification fails and multimember district systems should be abandoned. On the other hand, if any such numerical theory can give even a reasonable approximation to political reality, it is submitted that the analysis contained herein is at least mathematically consistent and therefore more likely to be correct than the inverse ratio theory offered as justification for such systems [*Banzhaf*, 1968, p. 817].

A B C	Votes	Needed to pass Majority = 3	Decisive	Needed to pass 2/3rd = 4	Decisive
1 Y Y N	5	P	–	P	A, B
2 Y Y N	4	P	A, B	P	A, B
3 Y N Y	3	P	A, C	F	B
4 Y N N	2	F	B, C	F	B
5 N Y Y	3	P	B, C	F	A
6 N Y N	2	F	A, C	F	A
7 N N Y	1	F	A, B	F	–
8 N N N	0	F	–	F	–

Tab. 1

N.Y.S. 2d 502 proposed that the districts be assigned weight such that the *Banzhaf* power index for each district's representative would be approximately equal to that district's population share.

In the context of multimember districts whose representatives were assumed to vote as a bloc, this criterion (applied in **Iannucci** only to the weighted voting case) would assign a number of representatives to each district such that the *Banzhaf* power of that district's bloc would be approximately equal to that district's population share. New York is a state where there are counties with both mmds and smds. However, as far as we are aware, New York courts have never seen the connection between the line of reasoning in **Iannucci** (which has been accepted in subsequent weighted voting cases) and the constitutionality of mixed mmd-smd systems. Thus, a weighted voting system with weights 3, 1, 1 would be unconstitutional under **Iannucci**, yet a mixed mmd and smd system with one mmd with three representatives and two single-member districts would be held to be perfectly all right.[6])

2.3 Equal Voter Decisiveness on Legislative Outcomes

Banzhaf [1966] has proposed that the most appropriate criterion of fair representation in a weighted voting system is neither equal voter power nor equal legislator pow-

---

[6]) In **Whitcomb** the court rejected the argument that representatives from multimember districts are *necessarily* more likely to vote as a bloc than representatives elected from the same area, elected from contiguous single-member districts (403 U.S. at 147–148), although it accepted the fact that "bloc voting tended to occur" in Marion county, and "defendant's own witnesses thought it was advantageous for Marion County's delegation to stick together." (403 U.S. at 147). Nonetheless the Court asserted that "nothing before us shows or suggests that any legislative skirmish ... would have come out differently had Marion County been subdistricted and its delegation elected from single-member districts" (403 U.S. at 148).

Moreover, the Court was not impressed with the notion that bloc voting led to influence for representatives from multimember districts more than proportionate to their numbers. *Brams/ Affuso* [1976] have shown that, in power index terms, coalitions are not always more powerful than their members taken singly and the Court may have had some dim intuition of this when it claimed that "the theory that plural representation ... unduly enhances a district's power and the influence of its voters remains to be demonstrated in practice and the day-to-day operation of the legislature" (403 U.S. at 147).

er but rather that the requirement that all voters have equal ability to affect the outcome of a vote in the legislature. *Walther* [1976] shows that to achieve this result requires that the ratio below should be the same for voters in each district.

$$\frac{\text{decisive outcomes for legislator(s) from district } j}{\text{square root of population of district } j} . \tag{5}$$

The probability that voter i will be decisive in his district election of legislative representatives multiplied by the probability that the legislators (or legislative bloc) from that district will be decisive in the legislature is the product we wish to be the same for all voters if we wish to equalize *voter* ability to affect legislative outcomes. The second term is proportional to the combined number of decisive outcomes for legislators from that district and the first term is approximately inversely proportional to the square root of district population. Because of the approximate correspondence between population weight and *Banzhaf* weights in most weighted voter systems this criterion will, in general, be incompatible with the criterion of equalizing legislator power, but will be compatible with the criterion of equalizing voter power.

As of 1960, most New York counties used a unit voting system for their County Boards of Supervisors in which each town/city ward was given one representative. This scheme was struck down in **Graham v. Board of Supervisors of Erie County** (1967) 267 New York Supplement 2d 383. As of 1960, only Nassau county used weighted voting. In Nassau, weights were assigned directly proportional to population but with certain other peculiar features (see Section 4, below). In the 1960's, in response to the voiding of unit voting systems, a number of New York counties sought to preserve township-based representation while still complying with Court directives on "one man, one vote" by shifting to weighted voting schemes similar to that in use in Nassau County. Two cases involving such counties (Saratoga County and Washington County) were combined and decided by the New York Court of Appeals in an important decision, **Iannucci v. Board of Supervisors of the County of Washington** 282 N.Y.S. 2d 502. In that case, as we noted above, the court held that weighted voting was permissible only if the weights led to *Banzhaf* values proportional to population. We shall quote the Court's opinion at some length:

> Although the small towns in a county would be separately represented on the board, each might actually be less able to affect the passage of legislation than if the county were divided into districts of equal population with equal representation on the board and several of the smaller towns were joined together in a single district. [See *Banzhaf*, 1965, p. 317] ... *The significant standard for measuring a legislator's voting power, as Mr. Banzhaf points out, is not the number or fraction of votes which he may cast but rather his ability ... by his vote, to affect the passage or defeat of a measure ... (Ibid, p. 318).* And he goes on to demonstrate that a weighted voting plan, while apparently distributing this voting power in proportion to population, may actually operate to deprive the smaller towns of what little voting power they possess, to such an extent that some of them might be completely disenfranchised and rendered incapable of affecting any legislation. (**Iannucci** 282 N.Y.S. 2d at 507, emphasis ours.)
>
> The principle of one man-one vote is violated, however, when the power of a representative to affect the passage of legislation by his vote, rather than by influencing his colleagues, does not roughly correspond to the proportion of the population in his constituency. Thus, for example, a particular weighted voting scheme would be invalid if 60 % of the population were represented by a single legislator who was entitled to cast 60 % of the votes. Although his vote would apparently

be weighted only in proportion to the population he represented, he would actually possess 100 % of the voting power whenever a simple majority was all that was necessary to enact legislation. Similarly a plan would be invalid if it was *mathematically impossible* for a particular legislator representing say 5 % of the population to ever cast a decisive vote. Ideally, in any weighted voting plan, it should be mathematically possible for every member of the legislative body to cast the decisive vote on legislation in the same ratio which the population of his constituency bears to the total population. Only then would a member representing 5 % of the population have, at least in theory, the same voting power (5 %) under a weighted voting plan as he would have in a legislative body which did not use weighted voting – e.g., as a member of a 20-member body with each member entitled to cast a single vote. This is what is meant by the one man-one vote principle as applied to weighted voting plans for municipal governments. A legislator's voting power, *measured by the mathematical possibility of his casting a decisive vote,* must approximate the power he would have in a legislature which did not employ weighted voting. (**Iannucci** 282 N.Y.S. 2d at 5U8; emphasis ours.)

The Court then went on to confess itself unable to determine whether the plans before it met the criterion proposed, and asserted that the Boards are not entitled to rely on the presumption that their legislative acts are constitutional. Rather,

... with respect to weighted voting ... a considered judgment is impossible without computer analyses and, accordingly, if the boards choose to reapportion themselves by the use of weighted voting, *there is no alternative but to require them to come forward with such analyses and demonstrate the validity of their reapportionment plans.* (**Iannucci**, 282 N.Y.S. 2d at 510; emphasis ours.)

With these words the Court ushered in the age of computerized weighted voting in New York county government and helped a New York mathematician and consultant, Lee Papayanopoulos, to supplement his income in the next decade by providing New York counties with weighted voting schemes acceptable under the **Iannucci** guidelines. In the next few years, New York counties were more likely to adopt weighted voting and mixed multimember and single-district systems than they were to shift to single-member districting. (See Table 2) In Most of these systems, a handful of towns controlled a majority of votes [see Tables 3 and 4 in *Grofman/Scarrow*].

Single Member Districts (20)

Albany	Franklin	Onondaga
Broome	Herkimer	Orange
Cayuga	Genessee	Otsego
Chemung	Lewis	St. Lawrence
Clinton	Monroe	Suffolk
Dutchess	Niagara	Westchester (P)[1])
Erie	Oneida	

Multi-Member of Uniform District Size (1)

Allegany

Multimember of Unequal District Size (12)

Cattaraugus	Rockland	Tioga
Chatauqua	Schenectady	Tom King
Greene	Schuyler	Ulser
Rensselaer	Steuben	Yates

Simple Weighted Voting (3)

Orleans[2])   Oswego   Putnam

Computerized Weighted Voting (21)

Chenango (P)	Jefferson	Schoharie
Columbia	Livingston	Seneca (P)
Cortland[3]) (P)	Madison	Sullivan (P)
Delaware	Montgomery	Warren
Essex	Nassau (P)	Washington
Fulton	Ontario	Wayne
Hamilton	Saratoga	Wyoming

[1]) (P) refers to counties for which we know Papayanopoulous to have prepared a weighted voting scheme.

[2]) Orleans' system of weighted voting is presently under court challenge.

[3]) Cortland uses weighted voting with single-member districts of approximately equal size. See Slater v. Board of Supervisors of Cortland County 330 NYS 2d 947.

Tab. 2: County Governing Bodies in New York State as of June 1977

In **Iannucci** New York's highest court has relied on the *Banzhaf* index as the measure of fair representation for legislators in weighted voting systems. In **Whitcomb** three years later, the U.S. Supreme Court considered and then rejected the relevance of voter power calculations based on a very similar line of reasoning (see discussion in Section 2 above). We might expect New York courts to have subsequently repudiated the **Iannucci** doctrine. They didn't. Instead, nearly two dozen New York counties shifted from unit voting to weighted voting; and with only a dwindling handful of exceptions (now only 3 in number), the apportionments rested on computerized schemes intended to satisfy the **Iannucci** doctrine. (Indeed in some cases, these apportionments were devised by New York district courts themselves in response to challenges to existing apportionments.) How can we account for this seeming divergence between New York and U.S. Supreme Court rulings?

First, in all the New York county apportionment cases involving weighted voting decided after **Iannucci**, it is **Iannucci** which is looked to for guidance; given the nature of our federal system, Supreme Court cases are treated as gloss. In **Iannucci** only the legislator power argument of *Banzhaf* [1965] is discussed; the voter power criteria of *Banzhaf* [1966] are not mentioned, despite the fact that *Banzhaf* himself entered an amicus curiae brief in the **Iannucci** case in which he sets forth the basic arguments of both *Banzhaf* [1965] and *Banzhaf* [1966].

Second, we should note that the conflict is more apparent than real. The argument before the Supreme Court in **Whitcomb** involved the square-root law as a means to maximizing *voter* equality. On the other hand, the argument before the New York Court of Appeals in **Iannucci** involved the use of the *Banzhaf* index as a measure of *legislator* strength. As pointed out previously, for weighted voting schemes (or mmd schemes with bloc voting) the two lines of argument lead to conflicting apportionment criteria. Thus, it would seem possible to accept one line of argument without accepting the other. The divergence between these two lines of reasoning was pointed out in an insightful article by *Johnson* [1969], and had even earlier been recognized by some New York courts.[7])

Third, there is an important difference between the line of reasoning used by the New York Court of Appeals in **Iannucci** and that used by the U.S. Supreme Court in **Whitcomb** which explains, in part, the divergent conclusions reached by the courts as to the usefulness of the *Banzhaf* index as a major component of a measure of "fair representation." In **Whitcomb**, as we previously pointed out, the Supreme Court rejected the use of the *Banzhaf* index on the grounds that it did not take into account "any political or other factors . . . " (403 U.S. at 406). However, in **Iannucci**, the New York Court of Appeals asserted that the sole criterion is the mathematical voting power which each legislator possesses in theory – i.e., the indices of representation – and not the actual voting power he possesses in fact – i.e., the indicia of influence." What was a sin for the U.S. Supreme Court was a virtue for the New York Court of Appeals. The one condemned *Banzhaf's* approach because of its lack of realism, the other applauded it because its abstractness and timelessness did not require constant revisions of apportionment decisions in the light of new election returns or changing political alignments.

3. Weighted Voting Case Study: Nassau County

Conflicting court rulings re the consitutionality of various Nassau County Legislature appointment schemes reveal how difficult determining the nature of "fair" representation in a weighted voting scheme can be, even if the **Iannucci** guidelines are followed.

---

[7]) The Supreme Court of the State of New York in Westchester County (**The Town of Greenburgh and the Town of Yorktown vs. the Board of Supervisors of Westchester County**, August 23, 1967), anticipating the Supreme Court in **Whitcomb**, held that the reasoning leading to the conclusions in *Banzhaf* [1966] "ignores all the realities of representative government and the conclusion reached can be no more reliable than the premise from which it starts."

In a slightly earlier Westchester case (town of **Greenburgh v. Board of Supervisors of Westchester County** 277 N.Y.S. 2d 855), the Westchester County Supreme Court reviewed and rejected *Banzhaf's* square root argument.

It is obvious and conceded that under any plan which employs electoral districts of substantially unequal populations, effective population-based legislative representation cannot be obtained by distributing legislative seats, or votes in proportion to population square roots. Whatever this may accomplish in giving a legislator voting power commensurate with that of the citizen, in his district, it will not give him a legislative influence, proportional to the population of the district which he represents. For instance, in the situation . . . with Districts A, B, and C having population of 400 each and District D a population of 2500, whether votes or legislative seats are to be distributed 2, 2, 2 and 5, the results will be the same. In either case the three small districts with a total population of 1200 will be able to outvote District D with a population of 2500 and control, by minority rule, the deliberations of a legislative body consisting of the four representatives (277 N.Y.S. 2d at 897–898).

In a post-**Iannucci** case which contrasts the arguments of *Banzhaf* [1965] and *Banzhaf* [1966], the Supreme Court of Seneca County in **Glessing and Glessing v. Board of Supervisors of Seneca County**, October 26, 1967, rejects the test of equalizing effective votes of citizens in electing their legislators, and asserts that "the test laid down by the Court of Appeals in **Iannucci** is the legislator's *decisive* voting power and not the effective voting power of a citizen in electing the legislator" (emphasis in original).

In Nassau County, from 1900 (the year in which the county was formed, being carved out of Queens County) until 1917, the County Board of Supervisors was composed of the three town supervisors – one from the town of Hempstead, one from the town of North Hempstead, and one from the town of Oyster Bay. Each cast a single vote. Beginning in 1917, however, a system of *weighted voting* was introduced awarding votes to a supervisor in proportion to the population of his town. The following year a second elected position of supervisor-at-large was created for the town of Hempstead, specifically for the purpose of doubling the town's representation, since the town by this time had over half the population of the county. The net result of these two changes was a voting scheme whereby the town of Hempstead's two supervisors were able to cast a total of eight votes, whereas the supervisors from the other two towns were able to cast only two votes each. Even though Nassau's two cities were incorporated about this time, with each city being awarded one vote, the town of Hempstead was still able to dominate the proceedings.

Thus, from 1922 until 1936 the distribution of votes was as follows:

Town of Hempstead #1	4
Town of Hempstead #2	4
Town of North Hempstead	2
Town of Oyster Bay	2
City of Glen Cove	1
City of Long Beach	1
	14

A new charter was adopted in 1936. This charter continued the system of weighted voting, awarding one vote for every 10,000 population, but contained a major constraint on weighted voting apportionment: *no town could cast a majority of the weighted votes.*

Applying the one vote per 10,000 voters formula to the 1936 population the distribution of legislative votes was:

Hempstead #1	9
Hempstead #2	9
North Hempstead	6
Oyster Bay	3
Glen Cove	1
Long Beach	1
	29

In this case, a majority is 15 $(\doteq (29 + 1)/2)$ votes and Hempstead's combined voting power was 18 votes. Clearly the one-vote-per-ten-thousand-citizens provision of the charter is incompatible with the charter provision which prohibits any town from wielding majority voting power. What was to be done? In December 1937, the Nassau Country attorney proposed an ingenious method to "reconcile" the conflict between the two clearly incompatible charter provisions. He proposed that the votes of any town having over a majority be reduced to just under a majority. Thus, Hempstead's total vote would be reduced from 18 to 14 (7 + 7). This would produce the following distribution:

Hempstead #1	7
Hempstead #2	7
North Hempstead	6
Oyster Bay	3
Glen Cove	1
Long Beach	1
	25

However, the majority would be left at 15 (not 13)! Similarly, when a two-thirds vote was required, the county attorney proposed that a two-thirds vote (a constitutional two-thirds) be calculated at 20 votes (not 17 votes), and that Hempstead would still only be allowed to cast 14 votes. This "interpretation" of the 1936 charter was accepted by the Board and went into effect in January 1938 although corrected census figures which excluded aliens changed the actual vote allocations slightly. The county attorney's recommended procedure was used to determine votes in the three subsequent reapportionments: 1942, 1962, and 1972.[8]) As of 1968, Hempstead, which had 57 % of Nassau's population, was given only the effective equivalent of 49,6 % of the votes on the Board of Supervisors in order to comply with the charter prohibition against any town having more than 50 % of the votes. (Of course, given the majority required to pass legislation, this scheme gave Hempstead's representatives the power to block legislation — since no bill could be passed without the support of at least one Hempstead representative.) In 1968 the Nassau County Supreme Court held that this apportionment scheme was unconstitutional in that it deprived "citizens of Hempstead of their right to substantial equality of representation." The New York Court of Appeals upheld this decision **Franklin v. Mandeville** (1970) 308 N.Y.S. 2d 375, but allowed the existing scheme to remain in effect until the 1970 census, at which time the Board was directed to draw up a reapportionment scheme consistent with the principle of one man, one vote.

In ruling the scheme unconstitutional, neither the Supreme Court nor the Court of Appeals addressed the question of whether the scheme satisfied the test that it allocated to each legislator "voting power, measured by the mathematical possibility of his casting a decisive vote, approximately equal to the power which he would have in a

---

[8]) These vote reduction and special majority requirement procedures are not well known; most authors who have discussed weighted voting in the Nassau County Board of Supervisors were unaware of them. For example, *Thomas* [1960], in a book on Nassau County government mistakenly asserts that the 1942 apportionment violates the Charter provision that no town be given voting majority in that "Hempstead has 18 out of the 30 votes ... clearly more than fifty percent of the total." The most cited article on weighted voting in the legal literature [*Banzhaf*, 1965] also makes this mistake, claiming that both the 1942 and the 1962 apportionments resulted in three of the six Nassau County Board members having zero voting power and the remaining three having equal power (as measured by the *Banzhaf* Index). Other authors [*Brams*] repeat *Banzhaf*'s mistake, some laboring under the misapprehension that these special procedures did not go into effect until 1971 *Lucas* [1974, p. 421] or 1972 *Andelman* [1972].

legislative body which did not employ weighted voting." The finding in **Franklin v. Mandeville** was based on more straightforward grounds. It simply held that a scheme which forever denied majority representation to residents of a town which had 57 % of the county population was, on the face of it, invalid.

In 1972, the Nassau County Board of Supervisors in accord with the Court's earlier directive, proposed a reapportionment scheme based on 1970 census data. The Board had employed a computer analyst, the aforementioned Lee Papayanopoulos, who reviewed over 2,000 different combinations of votes and voting. The final plan proposed involved weighted voting and, indeed, was substantively identical to that previously rejected by the Court. This plan was declared unconstitutional by the Nassau County Supreme Court (72 Misc. 2d 104, 338 N.Y.S. 2d 561). However, the New York Court of Appeals reversed this ruling. (**Franklin v. Krause** 1973 344 N.Y.S. 2d 885.)

The Appeals Court rejected the view offered by the Nassau County Supreme Court that weighted voting was, per se, unacceptable as a matter of law, and also rejected the claim that the new plan had the same flaw as the apportionment scheme previously rejected as unconstitutional. In **Franklin v. Krause** the Court cleverly finesses the question of what (if anything) is *different* about the 1972 plan to improve it over the 1962 scheme (rejected as unconstitutional in **Franklin v. Mandeville** 308 N.Y.S. 2d 375), other than the fact that it was drawn up by a computer analyst (**Franklin v. Krause** at 887). Instead, the court addresses the 1972 plan *de novo* on its merits. The principal test used by the Court was the difference between population share and power share (as measured by the *Banzhaf* index, for majority votes only) for each of Nassau's three towns and two cities, *under the assumption that the two Hempstead supervisors voted independently of one another.* The maximum deviation was +3.8, and the total deviation was only 13.9. The deviation range was 7.3 (−3.5 to +3.8).

The court held (**Franklin v. Krause** at 888) that these deviations were within the **Iannucci** guidelines.

There are a number of problems with the Appeals Court's decision in **Franklin v. Krause**.

a) By the criterion used in that decision, the plan rejected as unconstitutional in **Franklin v. Mandeville** is constitutional. The sum of the deviations in the 1962 plan was only 14.6 (as compared to 13.9 for the 1972 plan), and the maximum deviation difference was again only +3.8 while the range was 7.5 (−3.7 to +3.8). Furthermore, the two plans are *identical* in the power they assign representatives under majority voting. See Table 9, Column 4 in *Grofman/Scarrow* [1978].[9])

b) By the criterion used in **Franklin v. Mandeville** the plan accepted in the **Franklin v. Krause** decision should have been rejected. Under the 1962 plan, Hempstead's two representatives have 57.1 % of the population and 49.6 % of the vote share. Under the 1972 plan, Hempstead's two representatives have 56.2 % of the population and 50.0 % of the vote share. Under both plans, a township with over 50 % of the population is denied the possibility of ever obtaining a vote share of over 50 %. In **Frank-**

[9]) The data on power indices for the 1962 plan were made available to the court in the plaintiff's and cross-respondent's briefs. The court chose to disregard them and to decide **Franklin v. Mandeville** on other grounds. The court was reminded of these data in the cross-respondent's brief in **Franklin v. Krause**. Again, they disregarded them.

lin **v. Mandeville** the Court worried whether Hempstead was unconstitutionally *under*represented and concluded that it was. In **Franklin v. Krause**, with the same power distribution and virtually unchanged population fractions, the court worried whether Hempstead was unconstitutionally *over*represented, and concluded that it was not. Both the question and the means of answering it shifted.

In fairness to the Court of Appeals we should note that it defended its seeming inconsistency by pointing out (344 N.Y.S. 2d at 891) that, in the light of very recent Supreme Court cases extending the range of permissible deviations from strict population standards, the validity of standards in *local* apportionment decisions applied to its former decision had been very significantly altered.

c) The Court cannot make up its mind as to whether to treat Hempstead's two representatives as independent or as a voting bloc. For the purpose of calculating the power index, they are treated as voting independently of one another. Yet, for purposes of warding off the danger of a constituency unit having 100 % voting power, they are treated as a voting bloc, as illustrated in the following passage:

> It was also noted in **Iannucci** that a weighted voting plan would be invalid if over 50 % of the population were represented by a legislator entitled to cast over 50 % of the votes for them; in reality, he would possess 100 % voting power, at least as to measures requiring a majority vote for passage. The instant plan would violate that injunction, of course, were it not for its provision that for passage of a measure requiring a majority, 71, and not 66, votes are required; and for measures requiring a two-thirds vote, 92, and not 87, votes are required. *Thus, while the Town of Hempstead Supervisors together possess 70 votes, more than a majority of the total 130, they cannot have 55 % voting power which would ordinarily be 100 % voting power in a 'pure majority' situation.* This admittedly artificial voting requirement, in reality, gives the town of Hempstead a greater disenfranchisement than would otherwise be the case in certain voting combinations. (**Franklin v. Krause**, 888, emphasis ours.)[10])

If the court were to treat Hempstead's two representatives as a voting bloc, as they do in the above passage, then the *Banzhaf* power indices would be different — based on a five-member rather than a six-member board. Under the assumption that Hempstead's representatives vote together, in a majority vote under both the 1962 and 1972 plans, Hempstead has 89 % of the voting power, and the remaining four towns divide up in the remaining power equally, 3 % per town. With these assumptions, the sum of the differences between population and power are huge. In 1972 the sum is 68.2, and the maximum deviation is +32.7. In 1962 it was 66.9 with a maximum deviation of +31.3. [See *Grofman/Scarrow*, Table 9 for details.]

Of course it can be argued that to treat the Hempstead supervisors as the voting bloc they in fact *are* is to violate the stipulation offered in **Iannucci** as to the irrelevance of actual voting patterns.

---

[10]) Immediately after this passage comes a sentence of remarkable ambiguity: *"This* is precisely the point which caused our rejection of the former plan, which, although based on different scales and values, contained the same sort of bar preventing the town of Hempstead supervisors from having 100 % voting power." (**Franklin v. Krause**, 888, emphasis ours.) What does the "this" refer to? The court *rejected* the 1962 apportionment scheme. Furthermore, in 1962 the Court was worrying about whether Hempstead was *under*represented, not about whether they were *over*represented.

Nor will practical experience in the use of such plans furnish relevant data since the sole criterion is the mathematical voting power which each legislator possesses in theory – i.e., the indicia of representation – and not the actual voting power he possesses in fact – i.e., the indicia of influence. (20 N.Y. 2d at 252.)

We agree that observed patterns of voting coalitions, which effect legislators' power indices (since they lead to situations in which not all coalitions are equally likely) may be disregarded. If we were to look at observed coalition patterns, any legislature dominated by a single party with a very high index of party cohesion might be taken to be in violation of the **Iannucci** doctrine that no members be shut out from the possibility of ever being decisive, since minority members' votes would be irrelevant if the majority party always voted as a bloc. We believe, however, that the Hempstead case does not fall under the **Iannucci** rubric in that both supervisors represent the *same* constituency and are elected at the same election. Thus, to lump their votes together as *the* Hempstead vote seems the more reasonable procedure, *regardless* of how they actually vote. In fact, virtually no case is known in which they have ever voted differently. Furthermore, since the "not more than 50 %" clause of the 1936 Nassau County Charter was obviously intended to protect the smaller towns from domination, and since Hempstead had always had two representatives, even prior to this charter, the drafters of the charter clearly operated under the assumption that Hempstead's supervisor and deputy supervisor would indeed vote together as a single unit. The arithmetic gyrations called for in the 1938 county attorney's opinion are also based on that same premise. We do not see why the court did not take judicial notice of these facts, although we must admit that none of the briefs offered in either **Franklin v. Mandeville** or **Franklin v. Krause** call this issue to the court's attention. If the court had proceeded on the assumption that Hempstead's voting power should be based on the combined vote strength of its two representatives, it is inconceivable that the 1972 apportionment plan could have been found constitutional under the **Iannucci** guidelines – the discrepancies between population share and power share are simply too vast.

The question of how to treat multiple representatives of a single constituency is a troublesome one, and in none of the New York cases has there been any consideration of empirical evidence on the bloc voting tendencies of such representatives. [See further discussion of this point in *Grofman/Scarrow*, 59–60.]

d) Even if we treat Hempstead's representatives as separate from one another, the Court used an inappropriate measure to compare township population and township power. The New York Court of Appeals looked at the difference between population share and power share (measured in percents). We believe the appropriate measure should have been that used by the New York Supreme Court in 342 N.Y.S. 2d 189, to wit:

$$\frac{\text{population share} - \text{power share}}{\text{population share}} \tag{6}$$

For example, the U.S. Supreme Court in **Mahan v. Howell** 1973 410 U.S. 315 has looked at deviations from ideal representation using the formula

$$\frac{\text{mean district population share} - \text{population share}}{\text{mean district population share}} \tag{7}$$

and examined the range of differences. The calculation is based on the deviation from the ideal (= mean district population share − population share) measured in terms of (i.e., divided by) the ideal (= mean district population share). Ideally, each district would have identical population; that is, each district would have a population equal to the mean district population. The analogue of expression (7) for an assembly using weighted voting is given in expression (6). Ideally, each town would have a power share equal to its population share. The calculation used by the Court in **Franklin v. Krause** looks only at the numerator of this expression. Thus, for example, Glen Cove is found to deviate only 3.3 % (= .056 − .023) from its ideal representation (**Franklin v. Krause**, 889); yet, in actuality Glen Cove deviates 143 % = (.056 − 023/.023) from its ideal representation; that is, Glen Cove has more than twice the representation (measured in power share) that it is entitled to (measured in population share). This, we believe, indicates that the **Iannucci** guidelines, *when these are properly construed*, are violated − even when the Hempstead supervisors are treated separately [cf. *Johnson; Imrie*]. In Monroe County a small township with one vote was overrepresented in power terms (as measured by expression (6)) by more than 200 %, and this was held by the local Supreme Court to violate the principle of one man, one vote (342 N.Y.S. 2d 189). Had the Court of Appeals reasoned similarly in **Franklin**, the Nassau scheme would have been voted unconstitutional, since it involved a discrepancy of over 200 % for Long Beach. [For details see *Grofman/Scarrow*, Table 10.][11])

References

Andelman, D.: Nassau Supervisors Unveil Reapportionment Plan. New York Times, August 1, 1972.

Banzhaf, J. III.: Weighted Voting Doesn't Work: A Mathematical Analysis. Rutgers Law Review **19**, 1965, 317−343.

− : Multi-Member Electoral Districts − Do They Violate the 'One-Man, One-Vote' Principle? Yale Law Journal **75**, 1966, 1309−1338.

Brams, S.: Game Theory and Politics. New York 1975.

Brams, S., and *J. Affuso*: Power and Size: A New Paradox. Theory and Decision 7, 1976, 29−56.

Feller, W.: An Introduction to Probability Theory. Revised edition. New York 1970.

Fielding, G., and *H. Liebeck*: Voting Structures and the Square Root Law. British Journal of Political Science **5**, 1973, 249−256.

Grofman, B., and *H. Scarrow*: Game Theory and the U.S. Courts: One Man, One Vote, One Value. School of Social Sciences Research Report, University of California, Irvine, December, 1978.

Imrie, R.: The Impact of the Weighted Vote of Representation in Municipal Governing Bodies in New York State. Annals of the New York Academy of Sciences 219, 1973, 192−199.

Johnson, R.: An Analysis of Weighted Voting as Used in Reapportionment of County Governments in New York State. Albany Law Review 34, 1969, 317−343.

Lucas, W.: Measuring Power in Weighted Voting Systems. Technical Report No. 227, Department of Operations Research, College of Engineering, Cornell University, Ithaca, September, 1974.

Penrose, L.: The Elementary Statistics of Majority Voting. Journal of the Royal Statistical Society. 109, Part I, 1946, 53−57.

Thomas, S.: Nassau County: Its Governments and Their Expenditure and Revenue Patterns. New York 1960.

Walther, E.: An Analysis of Weighted Voting Systems Using the Banzhaf Value. Technical Report No. 809, School of Operations Research and Industrial Engineering, Ithaca, New York, 1976.

[11]) For discussion of power discrepancies when 2/3 vote is required see *Grofman/Scarrow* [1978, Table 11 and note 37].

Models and Applications in Economics

Applied Game Theory. 1979 © Physica-Verlag, Wuerzburg/Germany

Values of Large Market Games[1])

By *S. Hart*, Stanford[2])

Abstract: Three aspects of the application of the game theoretic concept of "value" to non-atomic economies – such as markets or production – are studied: first, the relation between value and equilibria; second, the problems of existence and non-existence of value; and third, a new way of defining value for these games, in order to guarantee its existence, which leads to interesting economic interpretations.

1. Introduction

The relations between game theoretic and economic concepts have been studied for a long time, trying to get a better insight into the laws governing the behavior of economic agents.

Much interest has been devoted to "large" economies[3]), where the individual is "negligible". Such situations are called "perfectly competitive", and the appropriate economic concept is that of "competitive equilibrium".

The first game theoretic solution studied in this context is the *core*, the main result being:

Core Equivalence Theorem: In a perfectly competitive economy, the core and the set of competitive allocations coincide [cf. *Debreu/Scarf; Aumann* [1964]; *Vind; Hildenbrand*, and others].

The next most used concept is the (*Shapley*) *value* – in particular, since it captures traditional economic ideas of "marginal contribution" (or, "worth"). The corresponding result is the following:

Value Theorem: In a perfectly competitive economy, every value allocation is competitive, and the two sets of allocations coincide if the economy is "sufficiently differentiable".

There are two main ways to model perfect competition. One is a limit approach, where sequences of finite economies, increasing in size, are considered (e.g., replicas). The other is using a non-atomic continuum as the space of agents.

[1]) This work was supported by National Science Foundation Grant SOC75-21820 at the Institute for Mathematical Studies in the Social Sciences, Stanford University. Presented at the International Conference on Applied Game Theory, Vienna, June 1978.

[2]) *Sergiu Hart*, Stanford University, Institute for Mathematical Studies in the Social Sciences. Stanford, California 94305, USA.

[3]) By "economy" we mean a market, or a production economy – as in Section 2.

Also, the two kinds of economic models are studied: Walrasian exchange (markets *without* transferable utility), and "monetary" markets (*with* transferable utility). As it will be pointed out in Section 2, the latter also represents production economies. The following table summarizes the research done on the Value Theorem:

	Limit of Finite Economies	Non-atomic Economies	
Monetary (with transferable utility)	*Shapley* [1964]	*Aumann/Shapley* [1974]	Differentiable
	Champsaur [1975]	*Hart* [1977b]	Non-differentiable
Walrasian (without transferable utility)	*Mas-Colell* [1977]	*Aumann* [1975]	Differentiable
	Champsaur [1975]	*Hart* [1977b]	Non-differentiable

Tab. 1

In this paper we deal with non-atomic economies. After presenting the basic models and defining the generalized asymptotic values in Section 2, we divide the results in three parts. The first one (Section 3) is devoted to the Value Theorem; the second one (Section 4), to the existence (and non-existence) of the asymptotic value; and in the last one (Section 5), we try to overcome these problems by defining a new measure-based value.

2. Preliminaries

This section includes the basic models of non-atomic economies, the definitions of (generalized) asymptotic values, and some preliminary results.

We start by describing a *non-atomic economy (market)* — as in *Aumann/Shapley* [1974], *Aumann* [1975], and *Hart* [1977b].

The *trader space* is a measurable space (I, C), which we assume to be standard[4]). A non-negative, σ-additive and non-atomic measure μ on C is given, called the *population measure*. To simplify our notations, we will sometimes write[5]) $\int\limits_{S} f d\mu$ and $\int\limits_{S} f$ to mean $\int\limits_{S} f(t) d\mu(t)$.

The *commodity space* is Ω, the non-negative orthant of the l-dimensional Euclidean space $\mathbf{R}^l$, where l is the number of commodities. For x in $\mathbf{R}^l$, x^j will denote its j-th coordinate.

The *initial allocation* a is an integrable function from I to Ω. We assume that every commodity is actually present in the market, i.e.,

---

[4]) I.e., isomorphic to the unit interval with the Borel σ-field. This assumption is not too restrictive, since any uncountable Borel subset of any Euclidean space, and indeed of any complete separable metric space, with the corresponding Borel σ-field, is standard — cf. Proposition (1.1) in *Aumann/Shapley* [1974].

[5]) Letters with 'wiggle' underneath will denote function defined on I.

$$\int_{\tilde{I}} a^j > 0, \quad \text{for all } 1 \leqslant j \leqslant l \tag{2.1}$$

(commodities with no initial supply can be obviously ingnored).

An *allocation* is an integrable function $\underset{\sim}{x}$ from I to Ω, such that $\int_I \underset{\sim}{x} = \int_I \underset{\sim}{a}$.

Here we distinguish between the two kinds of economies: monetary and Walrasian.

In the *transferable utility case* (monetary markets), to each t in I there corresponds a real-valued function u_t defined on Ω, called the *utility function of t*. All these functions are normalized by $u_t(0) = 0$, and they further satisfy:

(2.2) $x \geqslant y$ implies[6]) $u_t(x) \geqslant u_t(y)$ (*weak monotonicity*),

(2.3) u_t is a continuous function (*continuity*),

(2.4) $u_t(x)$, as a function of the pair (t, x), is measurable in the product field $C \times B^l$, where B^l denotes the Borel σ-field on Ω (*measurability*), and

(2.5) $u_t(x) = o(\|x\|)$ as $\|x\| \to \infty$, integrably in t (i.e., for every $\epsilon > 0$ there is an integrable function η defined on I, such that $|u_t(x)| \leqslant \epsilon \|x\|$ whenever $\|x\| \geqslant \eta(t)$).

Given the above economy, we will define the corresponding *market game* v by

$$v(S) = \max \{ \int_S u_t(\underset{\sim}{x}(t)) \, d\mu(t) \mid \int_S \underset{\sim}{x} = \int_S \underset{\sim}{a} \text{ and } \underset{\sim}{x}(t) \in \Omega \text{ for all } t \in S \}, \tag{2.6}$$

for all $S \in C$, the maximum being attained by the main theorem in *Aumann/Perles* [1965]. Then v is a non-atomic game with side payments, and further it belongs to the space H'_+, as defined in *Hart* [1977a, Section 2]; namely, it is the limit in the supremum norm of polynomials of non-atomic measures, it is superadditive, monotone, and homogeneous of degree one [see Proposition (3.4) in *Hart*, 1977b].

In terms of our exchange market, the interpretation of v(S) is as follows: there is an additional commodity, called "money", such that each trader's utility increases by one unit for each one unit of money. The maximum utility the coalition S can get, by reallocating its own initial resources between its members, is then exactly v(S).

The second interpretation of this model is that of a production economy. There are l "inputs" (i.e., raw goods), and one "output" (i.e., a finished good). Each participant t can produce out of a vector x (in Ω) of inputs, an amount $u_t(x)$ of the output good[7]). The initial allocation of raw goods being $\underset{\sim}{a}$, v(S) is then the maximal quantity of the finished good that S can produce, again by using its own resources alone.

A *transferable utility competitive equilibrium* (t.u.c.e.) is a pair $(\underset{\sim}{x}, p)$, where $\underset{\sim}{x}$ is an allocation and p is a *price vector* in Ω, such that

$$u_t(x) - p \cdot (x - \underset{\sim}{a}(t)) \leqslant u_t(\underset{\sim}{x}(t)) - p \cdot (\underset{\sim}{x}(t) - \underset{\sim}{a}(t)) \tag{2.7}$$

---

[6]) For x and y in Ω, $x \geqslant y$ means $x^j \geqslant y^j$ for all $1 \leqslant j \leqslant l$.

[7]) A more precise interpretation will be that a producer dt gets $u_t(x) \, \mu(dt)$ out of $x \, \mu(dt)$ – cf. Section 30 in *Aumann/Shapley* [1974].

for all x in Ω and (almost) all t in I. The corresponding *competitive payoff distribution* is the measure[8]) v_p defined by

$$v_p (S) = \int_S [u_t (\underset{\sim}{x} (t)) - p \cdot (\underset{\sim}{x} (t) - \underset{\sim}{a} (t))] \, d\mu (t), \tag{2.8}$$

for all $S \in C$.

The t.u.c.e. is actually a usual (Walrasian) competitive equilibrium, the price of "money" being normalized to 1 – see Section 32 in *Aumann/Shapley* [1974] for a more detailed discussion.

From now on, P will always denote *the set of all equilibrium prices*, i.e., the set of all p in Ω such that $(\underset{\sim}{x}, p)$ is a t.u.c.e. for some allocation $\underset{\sim}{x}$.

Proposition 2.9:
(i) P is a non-empty, convex and compact subset of Ω.
(ii) $\underset{\sim}{x}$ is a t.u.c.e. allocation if and only if $v (I)$ is attained[9]) at $\underset{\sim}{x}$. Moreover, $(\underset{\sim}{x}, p)$ is then a t.u.c.e. for *all* p in P, and the corresponding competitive payoff distribution does *not* depend on $\underset{\sim}{x}$ (i.e., is the same for all such $\underset{\sim}{x}$[10]).
(iii) The core of v and the set of competitive payoff distributions coincide.

Proof: Propositions 32.2 and 32.5 in *Aumann/Shapley* [1974], Propositions (2.10) and (2.20) in *Hart* [1977b], and Theorem 23.4 in *Rockafellar* [1970].

The second model is that of a *Walrasian non-atomic market*. Unlike the previous case, no (cardinal) utility functions are given. Instead, for each t in I there is an (ordinal) *preference relation* $\gg_t$ on Ω, satisfying:

(2.10) $x \geqslant y$ and $x \neq y$ imply $x \gg_t y$ (*desirability*),

(2.11) for each x in Ω, the sets $\{y \mid y \gg_t x\}$ and $\{y \mid x \gg_t y\}$ are open relative to Ω (*continuity*), and

(2.12) for any two measurable functions $\underset{\sim}{x}$ and y from I to Ω, the set $\{t \mid \underset{\sim}{x} (t) \gg_t y (t)\}$ belongs to C (*measurability*).

A *competitive equilibrium* is a pair $(\underset{\sim}{x}, p)$, where $\underset{\sim}{x}$ is an allocation and $p \neq 0$ is a *price* vector in Ω, such that $\underset{\sim}{x} (t)$ is maximal with respect to $\gg_t$ in the budget set of trader t

$$B_p (t) = \{x \in \Omega \mid p \cdot x \leqslant p \cdot \underset{\sim}{a} (t)\},$$

for (almost) all t in I.

We come now to the definition of value – first in the transferable utility case – using the asymptotic approach, due to *Kannai* [1966].

---

[8]) Because of Proposition 2.9, we can use the notation v_p (instead of $v_{(\underset{\sim}{x}, p)}$).
[9]) I.e., $v (I) = \int_I u_t (\underset{\sim}{x} (t))$ and $\int_I \underset{\sim}{x} = \int_I \underset{\sim}{a}$.
[10]) But it *does* depend on p.

Let v be a non-atomic game on the measurable space (I, C) (i.e., v is a real function on C with $v(\emptyset) = 0$).

A *partition* Π of (I, C) is a finite family of subsets of I, which are measurable (i.e., belong to C) and disjoint, and whose union is I. A sequence $P = \{\Pi_m\}_{m=1}^{\infty}$ of partitions is called *admissible* if it satisfies

(2.13) it is *decreasing*, i.e. for all m, each member of Π_m is a union of members of Π_{m+1}; and

(2.14) it is *separating*, i.e. for each s, t in I, $s \neq t$, there is m such that s and t are in different members of Π_m.

For a given partition Π, let v_Π denote the finite game derived from v, whose players are the members of Π, namely

$$v_\Pi(\Lambda) \equiv v\left(\bigcup_{B \in \Lambda} B\right)$$

for all $\Lambda \subset \Pi$.

Let $T \in C$ and let $P = \{\Pi_m\}$ be an admissible sequence of partitions whose first term is $\Pi_1 = \{T, I \setminus T\}$. For each m, let $T_m = T_{\Pi_m}$ be the coalition corresponding to T in v_{Π_m}, namely $T_m = \{B \in \Pi_m \mid B \subset T\}$. Let ϕv_{Π_m} denote the Shapley value of the finite game v_{Π_m}. If the numbers $(\phi v_{\Pi_m})(T_m)$ approach a limit as $m \to \infty$, and this limit is independent of the sequence P, then it will be denoted by $(\phi v)(T)$. It $(\phi v)(T)$ exists for all $T \in C$, then the function ϕv will be called the *asymptotic value* of v.

In view of the non-existence of the asymptotic value in some cases of interest (see Section 4), we will also consider *generalized asymptotic values*, where the limit of $(\phi v_{\Pi_m})(T_m)$ should exist and be the same *for all P in a given class of admissible sequences*. Examples of such values are the measure based values, to be defined in Section 5. The reason for this definition is that the Value Theorem holds for *any* generalized asymptotic value (and not only for the usual one), as shown in Section 3. A last immediate remark is that the asymptotic value exists if and only if all generalized asymptotic values are identical.

In what regards the non-transferable utility case, we will use the following Nash-Harsanyi-Shapley procedure [cf. *Harsanyi; Shapley* 1969; *Aumann*, 1975].

Let $U = \{u_t\}_{t \in I}$ be a family of utility functions, representing the given preferences $\{\gg_t\}_{t \in I}$, namely,

$$u_t(x) > u_t(y) \text{ if and only if } x \gg_t y, \tag{2.15}$$

for all t in I, x and y in Ω. If U also satisfies (2.5), a transferable utility market $v \equiv v_U$ can be defined by (2.6) (note that (2.2), (2.3) and (2.4) are implied by (2.10), (2.11) and (2.12), respectively).

An allocation $\underset{\sim}{x}$ is called a *(generalized) asymptotic value allocation* if there exists a family U of utilities, satisfying (2.5) and (2.15), such that v_U has a (generalized) asymptotic value ϕv_U, and

$$(\phi v_U)\,(S) = \int_S u_t\,(\underset{\sim}{x}\,(t))\,d\mu\,(t),$$

for all S in C.

For discussions of this approach, the reader is referred to the above noted papers of *Shapley* [1969] and *Aumann* [1975].

3. The Value Theorem

The results in this section are the same as those in *Hart* [1977b], extended here to the generalized asymptotic values (using essentially the same proofs).

We start with the monetary markets.

Theorem 3.1: Let v be a market game arising from a non-atomic economy with transferable utility, satisfying (2.1) — (2.5). Let ϕv be a generalized asymptotic value. Then ϕv is a competitive payoff distribution.

Proof: By Proposition (3.4) in *Hart* [1977b], v belongs to H'_+. Applying now Proposition (7.1) and (5.4) in *Hart* [1977a], we get

$$(\phi v)\,(T) = \lim_{m \to \infty} (\phi v_{\Pi_m})\,(T_m) \geqslant \partial v^*\,(\chi_I; \chi_T) \geqslant v\,(T),$$

from which it follows that ϕv belongs to the core of v, hence by Proposition 2.9 (iii) it is a competitive payoff distribution.

The second part of the Value Theorem assumes differentiability. The following theorem is actually stronger than the asymptotic part of the results of *Aumann/Shapley* [1974] (see Table 1).

Theorem 3.2: Let v be a market game arising from a non-atomic economy with transferable utility, satisfying (2.2) — (2.5) and

(3.3) for every t in I and $1 \leqslant j \leqslant l$, $\partial u_t\,(x) \,/\, \partial x^j$ exists at every x in Ω with $x^j > 0$.

Then the set of competitive payoff distributions and the set of (generalized) asymptotic values of v concide, and consist of one element only.

Proof: Theorem D in *Hart* [1977b].

We come now to the non-transferable utility case.

Theorem 3.4: In a non-atomic Walrasian market satisfying (2.1) and (2.10) — (2.12), every generalized asymptotic value allocation is competitive.

Proof: The same proof as that of Theorem E in *Hart* [1977b], using Theorem 3.1 above instead of Theorem A there.

4. Existence of Asymptotic Value

Since the value for the Walrasian markets, by its definition, depends on the existence of value in the transferable utility case, we will deal in the next two sections with the latter only.

We start with some "positive" results.

Theorem 4.1: Let v be a market game arising from a non-atomic economy with transferable utility, satisfying $(2.1) - (2.5)$. If there is a unique competitive payoff distribution, then v has an asymptotic value.

Proof: Theorem C in *Hart* [1977a], Proposition (3.4) in *Hart* [1977b], and Proposition 2.9 (iii).

The next theorem is a "generic existence theorem".

Theorem 4.2: Given utility functions $U = \{u_t\}_{t \in I}$ satisfying $(2.2) - (2.5)$, let $A \equiv A_U$ be the set of all vectors a in Ω such that there is a transferable utility non-atomic market $(\underset{\sim}{a}, U)$ with $\int_I \underset{\sim}{a} = a$, for which the asymptotic value does not exist. Then A is a set of Lebesgue measure zero in Ω.

Proof: See the proof of Theorem C in *Hart* [1977b].

Remark 4.3: Actually, a stronger result is proved as Theorem C in *Hart* [1977b], namely, that the set of competitive payoff distributions coincides with the asymptotic value "almost everywhere" (which is defined in the same way as in the above Theorem 4.2).

However, in general, the asymptotic value need not exist. A necessary condition is given in the next theorem. For a subset X of a linear space, x_0 is a *center of symmetry* of X if, for every x in X, its symmetric image with respect to x_0, $2x_0 - x$, belongs also to X.

To avoid inessential complications, we will assume that the excess demand in equilibrium has full dimension, namely, that given a t.u.c.e. allocation x, the linear (affine) subspace $L(\underset{\sim}{x} - \underset{\sim}{a})$ of $\mathbf{R}^l$ spanned by all vectors $\int_S (\underset{\sim}{x} - \underset{\sim}{a}) \, d\mu$, for all $S \in C$, has full dimension (i.e., dimension l). In case this is not satisfied, P (the set of equilibrium prices) should be replaced by its projection on $L(\underset{\sim}{x} - \underset{\sim}{a})$.

Theorem 4.4: Let v be a market game arising from a non-atomic economy with transferable utility satisfying $(2.1) - (2.5)$. If v has an asymptotic value, then the set of competitive payoff distributions and the set P of equilibrium prices each have a center of symmetry.

Proof: Theorem B in *Hart* [1977b] (see also Added in Proof (2) there).

As an example where this condition is not satisfied (hence, there is no asymptotic value) – one can consider the "three-handed glove market" [cf. *Aumann/Shapley*, p. 203]. It should be also noted that the above condition is necessary but *not suffi-cient* [cf. example 8.1 in *Hart*, 1977a].

5. Measure-Based Values

In order to get a value for all market games, we will define in this section a specific generalized asymptotic value [see *Hart*, 1978].

First, let us consider the reasons for the non-existence of the asymptotic value. The definition requires that for *all* admissible sequences, the limit of the Shapley values of the corresponding finite games should exist, and be *independent* of the particular sequence chosen. In all the cases studied in this context, admissible sequences with dif-ferent limits have been constructed. A deeper look reveals that all those partitions consisted of one (small) set which was "much bigger" than all the others. For example, consider the partition of $[0, 1]$ into one interval of length $1/n$ and n $(n - 1)$ intervals of length $1/n^2$, and let $n \to \infty$ (to ensure that the partitions get "finer", one can take $n = 2^m$ and $m \to \infty$).

In case the only "data" is a game v, there is nothing that can make the above se-quence of partitions "inadmissible". However, when one is considering an economy, and v is the derived "market game", additional data is given – namely, an underlying "population measure" μ. E.g., assume $[0, 1]$ to be the set of traders, and μ the Lebesque measure. Then the sequence of partitions described above does not seem to be a good approximation of the traders' space (some of them being always given much more weight than others!).

This indicates the way to proceed in order to guarantee the existence of the value. It was first used in *Aumann/Kurz* [1977], by restricting the admissible partitions to have all their elements equal in μ-measure. Here we adopt a slightly more "liberal" approach, requiring the measure of the elements of the partitions to get "close" one to another as $m \to \infty$.

Formally, given a measure μ on (I, C), a sequence $P = \{\Pi_m\}_{m=1}^{\infty}$ of partitions is called μ-*admissible* if it is admissible (i.e., satisfies (2.13) and (2.14)), and furthermore

$$\lim_{m \to \infty} \frac{\max_{B \in \Pi_m} \mu(B)}{\min_{B \in \Pi_m} \mu(B)} = 1. \tag{5.1}$$

The generalized asymptotic value corresponding to the class of all μ-admissible se-quences is called μ-*based value*, or μ-*value*.

In order to guarantee existence of the μ-value for a market game, we need one further assumption, which can be interpreted as an added "competitiveness" require-ment: that the variance of the excess demand, in equilibrium, should be finite. Intuitively, this implies that no arbitrarily "small" group of traders can have an arbitrarily "large" excess demand (recall that the total — hence, average — excess demand in equilibrium is zero). Usually, all allocations will be bounded, therefore this assumption will be surely satisfied.

We can now state our main result. As in Section 4, we will make the simplifying assumption that $L(x - a)$ has full dimension (see Theorem 4.4; in the degenerate case, replace P by proj $_{L(x-a)}P$ and $\mathbf{R}^l$ by $L(x - a)$).

Theorem 5.2: Let v be a market game arising from a non-atomic economy with transferable utility, satisfying (2.1) – (2.5). Let P be the set of all equilibrium prices, and let x be a t.u.c.e. allocation, such that

$$\int_I (z^j)^2 \, d\mu < \infty, \quad \text{for all } 1 \leqslant j \leqslant l, \tag{5.3}$$

and $L(z)$ has full dimension, where $z \equiv x - a$.

Then v has a μ-value ϕv, which coincides with a competitive payoff distribution ν_{p*}. The price vector p^* in P is given by

$$p^* = \int_{\mathbf{R}^l} p(z) \, dN(z), \tag{5.4}$$

where

(5.5) $p(z)$ maximizes $p \cdot z$ over all $p \in P$, for all z in $\mathbf{R}^l$,
   and N is the normal probability measure on $\mathbf{R}^l$ with the same first and second
   moments as z, namely, with expectation vector $0 = \int_I z \, d\mu$, and covariance matrix

$$\Sigma = \left(\int_I z^j \cdot z^k d\mu \right)_{j,k=1}^l. \tag{5.6}$$

Proof: Hart [1978].

For a more detailed discussion of this theorem and its conditions, the reader is referred to *Hart* [1978, Section 3]; it also includes a set of assumptions on a and $\{u_t\}_{t \in I}$ implying (5.3).

The theorem also raises the following interesting question: what is the equilibrium price p^* that corresponds to the μ-value (p^* is called: *value price*)?

The first observation is that in this model, the set P of all competitive prices is determined by *macro-economic considerations only*. Indeed, one needs to know *only* the aggregated utility function[11]) u_I and the *aggregated* (initial) supply $\int_I a$, the

competitive prices being then exactly the set of super-gradients (i.e., supporting hyperplanes) of u_I at $\int_I a$ [see *Hart*, 1978, Corollary 6.19]. All this data is not only "agent

---

[11]) For a in Ω, $u_I(a) = \max \{ \int_I u_t(x(t)) \, d\mu(t) \mid \int_I x = a$ and $x(t) \in \Omega$ for all $t \in I \}$ [cf. *Aumann/Shapley*, p. 213].

free", but also "distribution free" — i.e., the utility function and the income (initial endowment) of any one trader are unspecified; furthermore, not even the distribution of those characteristics in the population need be given. In the case the competitive price is uniquely determined, no problem arises. But which price should be chosen when P is a "large" set? The value considerations point out one such price p^* — as the customary interpretation of this concept indicates, on grounds of "fairness" and "equity". Obviously, additional data — at the *microeconomic* level — will be needed for this purpose.

As in the statement of the theorem, let P be the set of all equilibrium prices, $\underset{\sim}{x}$ a (fixed) competitive allocation, and $\underset{\sim}{z} = \underset{\sim}{x} - \underset{\sim}{a}$.

Let S be a large random sample (coalition), drawn from the set of traders I. If the total supply of S, $\int_S \underset{\sim}{a}$, equals its total demand $\int_S \underset{\sim}{x}$, then every price vector p in P is also an "S-price", namely an equilibrium price for the economy formed by S. In general, however, $\int_S (\underset{\sim}{x} - \underset{\sim}{a}) = \int_S \underset{\sim}{z}$ will be small (by the Law of Large Numbers, since $\int_I \underset{\sim}{z} = 0$), but will *not* vanish. Therefore, the S-prices will be close to P; in fact, they will be close to those p in P maximizing $p \cdot \int_S \underset{\sim}{z}$, i.e., following our notation (5.5), to $p(\int_S \underset{\sim}{z})$. Mathematically, this is an easy consequence of the properties of super-gradients [cf. *Rockafellar*, Theorem 24.6]. Economically, it corresponds to a high price for commodities with a high excess demand, and a low price for those with a low excess demand. It can be also thought of as some kind of an auctioneer's rule in a tâtonnement process [cf. *Arrow/Hahn*, Chapter 11].

Let Z be the distribution of the excess demand $\underset{\sim}{z}$ in the population (i.e., Z is the probability measure on $\mathbf{R}^l$ defined by $Z = \mu \circ \underset{\sim}{z}^{-1}$). Then, if we choose *traders* at random, the distribution of their excess demand will be Z. However, if we choose *samples* (*coalitions*) at random, their aggregated excess demand will no longer be Z-distributed. By the Central Limit Theorem, we will get instead (with the adequate normalization) the normal distribution with the same first and second moments as those of Z — namely, N.

Combining these two results, and noting that the normalization does not matter, since $p(z) = p(\alpha z)$ for all $\alpha > 0$, we finally get: p^*, as defined by (5.4), *is the expected equilibrium price vector of the economy formed by a random subset* (*or, random sample*) *of the agents*.

A close look reveals that this interpretation actually follows from the value considerations. Indeed, let dt be a small trader, then his value, $(\phi v)(dt)$, is the expected incremental worth ("contribution") of dt to a large sample (coalition), picked at random from the population. Let S be such a sample, then the contribution of dt equals the utility of his allocation, minus its net cost, namely

$$[u_t(\underset{\sim}{x}^S(t)) - p^S \cdot (\underset{\sim}{x}^S(t) - \underset{\sim}{a}(t))] \mu(dt), \tag{5.7}$$

where $(p^S, \underset{\sim}{x}^S)$ is a competitive equilibrium in the economy formed by S. By the Law of Large Numbers, S is a "good representation" of the traders space I, hence $\underset{\sim}{x}^S$ will be close to $\underset{\sim}{x}$ (our fixed competitive allocation for the whole economy[12]). Taking expectation in (5.7) we get (in the limit)

$$(\phi v)\,(dt) = [u_t\,(\underset{\sim}{x}\,(t)) - E\,(p^S) \cdot \underset{\sim}{z}\,(t)]\,\mu\,(dt), \tag{5.8}$$

therefore the value payoff distribution is competitive, and the corresponding price p^* is precisely $E\,(p^S)$, i.e., the expected equilibrium price for a random sample (coalition).

References

Arrow, K.J., and *F.H. Hahn*: General Competitive Analysis, San Francisco 1971.

Aumann, R.J.: Markets with a Continuum of Traders. Econometrica 32, 1964, 39–50.

–: Values of Markets with a Continuum of Traders. Econometrica 43, 1975, 611–646.

Aumann, R.J., and *M. Kurz*: Power and Taxes in a Multi-Commodity Economy. Israel Journal of Mathematics 27, 1977, 185–234.

Aumann, R.J., and *M. Perles*: A Variational Problem Arising in Economics. Journal of Mathematical Analysis and Applications 11, 1965, 488–503.

Aumann, R.J., and *L.S. Shapley*: Values of Non-Atomic Games. Princeton 1974.

Champsaur, P.: Cooperation Versus Competition. Journal of Economic Theory 11, 1975, 394–417.

Debreu, G., and *H. Scarf*: A Limit Theorem on the Core of an Economy. International Economic Review 4, 1963, 235–246.

Harsanyi, J.: A Bargaining Model for the Cooperative *n*-Person Game. Annals of Mathematical Studies 40, 1959, 325–355.

Hart, S.: Asymptotic Value of Games with a Continuum of Players. Journal of Mathematical Economics 4, 1977a, 57–80.

–: Values of Non-Differentiable Markets with a Continuum of Traders. Journal of Mathematical Economics 4, 1977b, 103–116.

–: Measure-Based Values of Market Games. TR # 254, Economics Series. Institute for Mathematical Studies in the Social Sciences, Stanford University, 1978, forthcoming in Mathematics of Operations Research.

Hildenbrand, W.: Core and Equilibria in a Large Economy. Princeton 1974.

Kannai, Y.: Values of Games with a Continuum of Players. Israel Journal of Mathematics 4, 1966, 54–58.

Mas-Colell, A.: Competitive and Value Allocations of Large Exchange Economies. Journal of Economic Theory 14, 1977, 419–438.

Rockafellar, R.T.: Convex Analysis. Princeton 1970.

Shapley, L.S.: Values of Large Games VII: A General Exchange Economy with Money. RM-4248-PR, The Rand Corporation, Santa Monica, California, 1964.

–: Utility Comparison and the Theory of Games. La Décision. Paris 1969.

Vind, K.: Edgeworth Allocations in an Exchange Economy with Many Traders. International Economic Review 5, 1964, 165–177.

---

[12]) The non-uniqueness of $\underset{\sim}{x}$ is "inessential": by (2.7), $u_t\,(\underset{\sim}{x}(t)) - p \cdot (\underset{\sim}{x}\,(t) - \underset{\sim}{a}\,(t))$ does not depend on the particular t.u.c.e. allocation $\underset{\sim}{x}$ chosen. Therefore, only the non-uniqueness of the equilibrium price vectors matters in getting (5.8) from (5.7).

Applied Game Theory. 1979 ©Physica-Verlag, Wuerzburg/Germany

Arbitration of Exchange Situations with Public Goods

By *E. Kalai*[1]), Evanston, *A. Postlewaite*[2]), Urbana-Champaign and
J. Roberts[3]), Evanston

Abstract: A small number of players are in a situation involving the possible production of public goods which may benefit all of them. The situation involves a finite number of private goods, owned individually, which can be used for the production of the public goods and for exchanges and sidepayments in an overall compromise. The paper defines a noncooperative game for the participants whose outcomes are compromises in the exchange situations. When played optimally, the game yields outcomes which are in the core of the change situation.

Introduction

The purpose of this paper is to suggest a method of arbitration for exchange situations involving private and public goods. An exchange situation is modeled as follows. We assume that there is a finite set of people, a finite number of private goods and a finite number of public goods. Each individual has a utility function, representing his preferences over bundles of private and public goods, and an initial bundle of private goods. There is a production cone representing the possibilities of production of public goods from private goods. This is available to every coalition in the group. Thus this is a standard model of an economy with public goods.

We incorporate into the exchange situation a graph, whose nodes are the various players, which describes the exchange possibilities for this group. Thus two individuals are linked in the graph if and only if they can trade private goods between themselves. This idea was motivated by *Myerson* [1977] and enables one to model exchange situations with special structures such as the stock market, centralized exchange, and others. It also enables one to model various restrictions to trade which are due to geographical, informational, and legal barriers.

The approach we employ is due to *Hurwicz* [1972], and *Mount/Reiter* [1974]. The idea is that instead of relying on the players to reveal their characteristics to the arbitrator, the arbitrator will design a game whose outcomes will be outcomes in the exchange situation. The arbitrator's goal is to design the game in such a way that the outcome under a game theoretic optimal way of playing will be an "efficient one" from his point of view.

[1]) *Ehud Kalai, John Roberts*, Northwestern University, Graduate School of Management, Nathaniel Leverone Hall, Evanston, Ill. 60201, USA.
[2]) *Andrew Postlewaite*, Department of Economics, University of Illinois at Urbana-Champaign.
[3]) *Roberts'* research was supported by the National Science Foundation under grant No. SOC 76-20953.

The game theoretic solution that we use is the strong Nash equilibrium due to *Aumann* [1967]. The main result that we exhibit is the design of the game whose strong Nash equilibrium will yield outcomes which are in the core of the economy.

In the last section of the paper we give an example of an actual exchange situation in which the application of the arbitration method suggested here seems very reasonable. The reader may find it useful to read the example first if additional motivation for the problem is desirable.

The method suggested in this paper was treated earlier [*Kalai/Roberts*] for the pure private goods case.

The Model of the Environment

Let $N = \{1, 2, \ldots, n\}$ be a set of traders, let l and m be positive integers representing respectively the number of private and public goods in the exchange situation. For every $i \in N$ let $\omega^i \in R_+^l$ be the initial holdings of private goods of individual i and $u_i : R_+^{l+m} \to R$ be his utility function over bundles of goods. Let Z be a convex cone in $R^l \times R_+^m$ containing the origin. Z represents the feasible production possibilities for coalitions in this society. We assume that $(x, y) \in Z$ and $x > \bar{x}$ implies that $(\bar{x}, y) \in Z$, i.e. with more resources you can produce at least as much. Let G be a graph on the set of nodes N. The presence of a link $ij \in G$ means that i and j can trade with one another. We assume throughout the paper that G is connected. An *allocation* for a coalition $S \subseteq N$ is a vector $\{(x^i, y)\}_{i \in S}$ such that $x^i \in R_+^l$ and $y \in R_+^m$. An allocation (x, y) is *feasible* for the coalition S if for every maximal connected subset T of S,
$$\sum_{i \in T} x^i \leqslant \sum_{i \in T} \omega^i \text{ and } (-\sum_{i \in S} (\omega^i - x^i), y) \in Z.$$
An allocation (of the grand coalition N) is *feasible* if it is feasible for N.

To define the core of the exchange situation we first define the notion of blocking by a coalition. This is all done in the usual way but the resulting core will be different from the usual one because it is done relative to the graph g through the definition of feasibility. A coalition S *can block the allocation* z if there is an allocation $\bar{z}$, feasible for S, with $u^i(\bar{z}) \geqslant u^i(z)$ for every $i \in S$, and $u^i(\bar{z}) > u^i(z)$ for some $i \in S$. We define the core of the exchange situation to be all the feasible allocations that cannot be blocked.

On the Use of the Graph

The use of the graph structure on the set of traders permits the modeling of differing forms of market organization. For example, if g is not connected, we are effectively looking at a system of autarkies, while if g is connected but not necessarily complete one obtains models of different forms of economic systems. For example, the complete graph in part a) of Figure 1 represents a system under which all agents are free to exchange directly with one another, while the graph in part b) represents the existence of a middleman through which all trade flows. Figure 2 can be considered as representing two economies, where all foreign trade in the one economy must flow through an export-import agency.

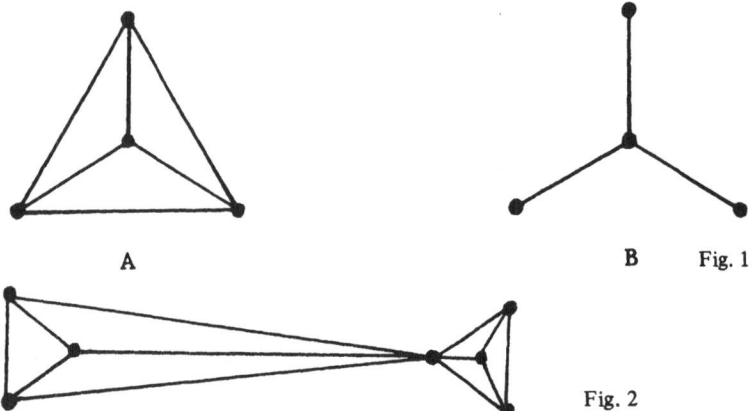

A   B   Fig. 1

Fig. 2

The imaginative reader can easily construct for himself the graphs associated with special markets such as the stock market or trading posts. The comparison of a middle-man market to a free (complete) market was done in detail in *Kalai/Postlewaite/Roberts* [1977].

Implementing the Core by a Strategic Form Game

A designer, facing an environment of the type described earlier, wishes to enforce an outcome which is optimal according to his criterion. Relying on the traders to reveal their true characteristics to him is impractical because the traders may choose to mis-represent themselves with the hope of improving their final outcome. One option open to the designer is to design a game whose outcomes are allocations in the exchange situation and which has the property that under a game theoretic optimal way of play-ing, the players will bring about an outcome which is satisfactory from the designer's point of view.

In this section we will show how the method described above can be used in order to bring about final allocations which are in the core of the exchange situation. In order to do so we define a strategic form game associated with the exchange. To this game we apply the strong Nash equilibrium as the game theoretic solution concept.

A strategy for trader i is a triple $s^i = (\{l_j^i\}_{j \neq i}, c^i, y^i)$ where $c^i \in R_+^l$, $y^i \in R_+^m$ and for $j \in N$ with $j \neq i$ $l_j^i = (r_j^i, g_j^i)$ where $r_j^i \in R_+^l$ and $g_j^i \in R_+^l$. The intuitive meaning is as follows. l_j^i is the exchange that i proposes to j, r_j^i is what i wants to receive from j and g_j^i is what he proposes to give j. c^i is what i proposes to contribute towards the produc-tion of public goods and y^i represents the levels of public goods that he would like to see produced. The quantity of private goods that he is left with is $t^i \equiv \omega^i - \sum_{j \neq i} g_j^i + \sum_{j \neq i} r_j^i - c^i$. We impose the following restrictions on s^i:

1. If $ij \notin G$ then $l_j^i = (0, 0)$.
2. $t^i = \omega^i - \sum_{j \neq i} g_j^i + \sum_{j \neq i} r_j^i - c^i \geq 0$.

In order to define the outcome associated with a strategy n-tuple $s = (s^1, s^2, \ldots, s^n)$ we first define the notion of a consistent coalition. A coalition S is *consistent at* s if for every $i \in S$

1. $(r_j^i, g_j^i) = (g_j^i, r_i^j)$ and $y^i = y^j$ for every $j \in S$,
2. $l_j^i = (0, 0)$ for every $j \notin S$, and
3. $(- \sum_{j \in S} c^j, y^i) \in z$.

Thus a coalition is consistent if there is a complete agreement among its members about exchange and production, it is self-sufficient, and it proposes a feasible production plan. A coalition is *maximal consistent* if no strict superset of itself is consistent. Let $T_1, T_2, \ldots, T_k$ be the distinct maximal consistent sets. It is easy to check that they are pairwise disjoint. Let T_0 be the set of traders that do not belong to any T_i, then $T_0, T_1, \ldots, T_k$ is a partition of the set of traders.

Now we can define the allocation p (s) resulting from the given strategy s. For $j = 1, 2, \ldots, k$ let y^j be the common agreed upon level of public goods in the coalition T_j. Let $y = \sum_{j=1}^{k} y^j$: y will be the levels of public goods produced for all the individuals to enjoy. The private goods will be allocated as follows. If $i \in T_0$ then $x^i = \omega^i$, i.e. no trade. If $i \in T_j$ for some $j > 0$ then $x^i = \omega^i - \sum_{j \neq i} g_j^i + \sum_{j \neq i} r_j^i - c^i$, i.e. the total trade that he has agreed upon with his consistent friends. It is clear that p (s) is a feasible allocation in the economy for every strategy s. Thus we can induce the preferences of the individuals over bundles to obtain their preferences over strategies and the strategic form game is well defined.

The Realtions between the Game and the Market

The following property is obvious from the definition of the game.

Proposition 1: For every strategy n-tuple s, p (s) is a feasible allocation.

Theorem 1: Assume that the utility of every individual is nondecreasing in the public goods. For every coalition S and every S-feasible allocation z there are strategies $\{s^i\}_{i \in S}$ such that, for every n-tuple of strategies $\{t^i\}_{i \in N}$ with $t^i = s^i$ for every $i \in S$, u^i $(p$ $(t)) \geqslant u^i$ (z) for every $i \in S$. In other words, for every feasible allocation for the coalition S the members of S can secure themselves a payoff which is at least as good as the allocation, regardless of the strategies of the others.

Sketch of Proof

Let z be a feasible allocation for S. Let $S^1, S^2, \ldots, S^k$ be the partition into the maximal connected components of S in the restriction of G to S. We define the strategies s^i defining it on one connected component at a time. For example for S^1 choose a spanning tree in the restriction of G to S^1 and let q be one distinct element of it. Define the strategies s^i of $i \in S^1$ to be the ones that transfer all of their ω^i's to q through the tree and transfer the x^i's back through the tree to the individuals. In addition player q will have $c^q = \sum_{i \in S^1} (\omega^i - x^i)$ where all the other $c^i = 0$. Also set $y^i = y$ for every $i \in S^i$. It is clear that this s will have the desired property.

The following theorem is the desired one from the point of view of a designer who wishes to implement the core and believes that players will play strategies which are strong Nash equilibrium. It is also an existence theorem for strong Nash equilibrium whenever the core is not empty.

Given an n-tuple of strategies $s = (s^1, \ldots, s^n)$ we say that coalition T *can improve upon* s if there exist strategies $t^T = \{t^i\}_{i \in T}$ which satisfy the following condition: for the strategy $(s \mid t^T)$ in which i-th strategy is s^i if $i \notin T$ and t^i if $i \in T$,
$u^i (p^i (s \mid t^T)) > u^i (p^i (s))$ for every $i \in T$.

A strategy tuple s is a *strong Nash equilibrium* if there exists no coalition that can improve upon it.

Theorem 2: Suppose the u^i's are increasing in the public goods.
(i) If s is a strong Nash equilibrium then $p (s) \in$ core.
(ii) Conversely, for every core allocation z there exists a strong Nash equilibrium strategy s with $p (s) = z$.

Proof:
(i) If $p (s) \notin$ core then some coalition S has a feasible allocation which they all prefer to $p (s)$. It follows from Theorem 1 then that the coalition S can improve upon s. Thus s is not a strong Nash equilibrium.
(ii) Suppose $z = (x, y) \in$ core. Let s be the strategy defined as in the proof of Theorem 1. Suppose the coalition T can improve on s by using the strategies t^T. Notice that if $i \in C - T$ then the payoff to i is the same for the private goods and possibly larger for the public goods, thus $u^i (p^i (s \mid t^T)) \geqslant u^i (p^i (s))$. Consider the coalition $C \cup T$. $p (s \mid t^T)$ is feasible for it, and thus $C \cup T$ is a blocking coalition for s.

An Example

A doctor (D), a contractor (C), and a landscape architect (L) share an office building. The consider installing a garden in front of the building. They are likely to have different utilities from different size gardens. Also, C and L possess the private goods which are necessary for the building of the garden. In this section we would show how an arbitrator can implement the game described earlier in order to come up with an outcome which would be efficient (in the core sense).

We assume that there are 3 private goods and one public good involved. We think of the first private good as money and denote its quantities by q. All three of them possess enough of it for the production of any reasonable size garden. The second private good is owned by C and we denote its quantities by c while the third one is owned by L and its quantities are denoted by l. We use g to denote the levels of the garden created. Formally we assume that:

$$\omega^D = (M, 0, 0) \qquad u^D\,(p, c, l, g) = p + 0.2\,(g \wedge 10)$$

$$\omega^C = (M, 10, 0) \qquad u^C\,(p, c, l, g) = p + 0.5\,(g \wedge 10) + 0.4\,c$$

$$\omega^L = (M, 0, 10) \qquad u^L\,(p, c, l, g) = p + 0.5\,(g \wedge 10) + 0.4\,l$$

where M is a large positive number and $\wedge$ represents the minimum operator.

The cone of production Z is defined to be generated by the two point $(0, -1, -1, 1)$ and $(-3, 0, 0, 1)$. In other words they could get a unit of garden by paying 3 dollars for it or by investing a unit of C's good and a unit of L's good. From Pareto optimality considerations it is clear that they would build 10 units of garden by using the goods of C and L. Thus a full description of the final outcome set (the core) would be given by determining the money transfers between D, C and L. After computing the core we find that it is characterized to be the convex hull of the following 4 payment schemes: $(-2, 3, -1), (-2, -1, 3)\,(0, -1, 1)$ and $(0, 1, -1)$. In the first two extreme points the doctor contributes 2 dollars which go to the other two. In the first extreme point the contractor gets these two dollars plus an additional 1 dollar from L. In the last two extreme points the doctor contributes nothing and transfers of up to a dollar are possible between C and L.

If we were in a situation where C and L refuse to communicate with one another (in effect making the doctor the middleman) the core payoff set would become bigger and would be generated by the points $(-2, 3, 1)\,(-2, 1, 3)$ and $(2, -1, -1)$; thus the advantage of the doctor is obvious. It is interesting to note that the *Shapley* [1953] value would result in the side payments $(-4/3, 2/3, 2/3)$. *Myerson*'s [1977] modification of the Shapley value as applied to the case of no communication between C and L would result in the sidepayments $(-2/3, 1/3, 1/3)$.

References

Aumann, R.: A Survey of Cooperative Games without Side Payments. Essays in Mathematical Economics in Honor of Oskar Morgenstern. Ed. by M. Shubik. Princeton 1967.

Hurwicz, L.: On Informationally Decentralized Systems. Decision and Organization: A Volume in Honor of Jacob Marschak. Ed. by McGuire-Radner. Amsterdam–New York 1972.

Kalai, E., and *J. Roberts*: A Simple Game of Exchange. Discussion Paper No. 280. The Center for Mathematical Studies in Economics and Management Science, Northwestern University, 1977.

Kalai, E., A. Postlewaite and *J. Roberts*: A Game of Barter with Barriers to Trade. Discussion Paper No. 302, The Center for Mathematical Studies in Economics and Management Science, Northwestern University, 1977, forthcoming in the Journal of Economic Theroy.

Mount, K., and *S. Reiter*: The Informational Size of Message Spaces. Journal of Economics Theory 8, 1974, 161–192.

Myerson, R.: Graphs and Cooperations in Games. Mathematics of Operations Research 2 (3), 1977, 225–229.

Shapley, L.: A Value for *n*-Person Games. Contributions to the Theory of Games II. Ed. by Kuhn-Tucker. Princeton 1953.

Applied Game Theory. 1979 © Physica-Verlag, Wuerzburg/Germany

Efficiency and Altruism in Cooperative Games with Coalition Structures

By *J. Greenberg*, Blacksburg[1])

Abstract: In many economic situations individuals carry out activities as coalitions, and have personal preferences for belonging to specific groups (coalitions). These situations are studied in the framework of cooperative games with coalition structures, by defining for each player a utility function with two arguments, namely his consumption bundle and the coalition to which he belongs.

The optimality analysis brings out a surprising property of the games with hedonic coalitions, namely that transfers among coalitions may be necessary to attain Pareto optimality. Moreover, quite restrictive assumptions are needed to rule out this property.

1. Introduction

In many social and economic situations the optimal solution requires the formation of coalitions that partition the set of agents, since payoffs under such a coalition structure dominate the feasible payoffs for the grand coalition $\{N\}$. (I.e., the game is not superadditive). Examples of such situations are provided by: Segregation due to racial, religious or national differences; local public goods which due to congestion either in production or in consumption, are best provided by local communities rather than by the whole nation; production being carried out in more than one "large" firm; the existence of social and sports clubs, etc.

Cooperative games with coalition structures provide a natural framework for a formal analysis of these situations. In their analysis of such games, *Aumann/Drèze* [1975] impose the condition that each coalition should distribute among its members the total payoff accruing to that coalition. The purpose of this paper is to show that there are situations in which, when modeled as cooperative games with coalition structures, ruling out transfers among coalitions may be inconsistent with the weak and generally accepted criterion of Pareto-optimality. Thus, *transfers among coalitions may be necessary to attain Pareto-optimality*. It is possible that every allocation without such transfers is dominated by an allocation with transfers giving a higher utility level to every individual. Since transfers imply that some coalitions allocate part of their always desired commodity to nonmembers, without receiving any counter-part, (thus, transfers are distinguished from „trade"), the above result could be understood as meaning that "altruism" or "solidarity" may be required to achieve Pareto-optimality (and not only to move from one Pareto optimum to another one).

---

[1]) Prof. *Joseph Greenberg*, Dept. of Economics, Virginia Polytechnic Institute and State University, Blacksburg, VA 24061, USA.

We consider two distinct situations in which transfers are required for Pareto optimality to obtain. In each situation, we constrast two games. In one of them transfers among coalitions are forbidden, while in the other such transfers are allowed. We then investigate the Pareto surfaces of these two games. More specifically, we examine sufficient conditions under which the two Pareto surfaces will, or will not, intersect. In both models we consider a coalition production economy with a single private good, in which preferences are strictly increasing. The main feature of the first model[2]) is that the utility function of each player has two arguments: the amount of the private good he consumes and the coalition to which he belongs. It then seems natural to speak about games with "hedonic coalitions". By its very definition, the hedonic aspect is not transferable and therefore, such situations give rise to cooperative games without side payments. In order to show that altruism may be necessary for efficiency also in games with side payments, we consider another model[3]), (with no hedonic aspect), the main characteristic of which is that while workers (labor) can move costlessly among coalitions, it is infinitely costly to transfer the output across coalitions. (In this case, the transfers are in the form of labor. Since we shall assume that the marginal productivity of labor is positive, transfers imply altruism.)

In both models it is shown that quite restrictive assumptions seem required to guarantee that optimality is compatible with no transfers among coalitions.

2. Notation

There are n individuals, given by the set $N = \{1, \ldots, n\}$. The set of all subsets of N (i.e., the set of all possible coalitions) is denoted by N, and the set of all coalitions containing i by N_i. A generic element of N is denoted S with cardinality $|S|$. We assume that each individual $i \in N$ belongs to one and only one coalition. A partition of N, or coalition structure, is denoted $B = \{B_1, \ldots, B_J\}$ with generic element B_j. The set of all coalition structures is denoted B. Similarly, $B(S)$ denotes the set of all partitions, $B(S)$, of coalition S.

Two games in characteristic function form are defined. In the games $(N; w)$ transfers among coalitions are allowed, while they are excluded in the game $(N; v)$. (Clearly, the precise definitions of v and w depend upon the specific model considered.) To economize notation, we shall omit the subscript on the variable N. Thus, $V = v(N)$, $W = w(N)$, $B = B(N)$, $B = B(N)$, etc.

For $x, y \in R^l$, $x \geqslant y$ means $x_j \geqslant y_j, j = 1, \ldots, l; x \geqslant y$ means $x \geqslant y$ and $x \neq y$; $x > y$ means $x_j > y_j, j = 1, \ldots, l$. Let $A \subset R^l$. The set of maximal elements in A is denoted $\bar{A}$, i.e.,

$$\bar{A} = \{a \in A \mid \nexists \, \bar{a} \in A \text{ with } \bar{a} \geqslant a\}.$$

In particular, $\bar{W}$ (respectively $\bar{V}$) denotes the set of Pareto-optimal allocations in the game $(N; w)$ (respectively $(N; v)$). The core of $(N; w)$ denoted by $Co(N; w)$ is the set of $u \in W$ such that there exists no $S \subset N$ and $a \in W(S)$ with $a \geqslant u_S$, where $u_S = (u_i)_{i \in S}$. Similarly,

---

[2]) This model is based on a work with *Jacques Drèze* [*Drèze/Greenberg*].

[3]) This model is based on a work with *Amoz Kats* [*Greenberg/Kats*].

$$Co\ (N;\ v) = \{u \in v \mid \exists\ S \subset N,\ a \in v\ (S)\ \text{with}\ a \geqslant u_S\}.$$

Finally, note that transfers are necessary for Pareto optimality if and only if $\bar{W} \cap \bar{V} = \emptyset$. Indeed, since every allocation that does not involve transfers can be attained when transfers are allowed (but not made), $v\ (S) \subset W\ (S)$ for all $S \subset N$. In particular, therefore, $W \supset V$, and hence $\bar{W} \supset \bar{V}$. Thus, $\bar{W} \cap \bar{V} = \emptyset$, implies that for all $u \in \bar{V}$ there exists $\bar{u} \in W$ with $\bar{u} \geqslant u$.

3. The Hedonic Model

The domain of the utility functions u_i, $i \in N$, reflects the hedonic aspect. Specifically, $\forall i \in N$, $u_i: R_+ \times N_i \to R$; when i belongs to S and consumes x_i units of the only private good, his utility level is given by $u_i\ (x_i, S)$. We assume that the function u_i is strictly increasing in x_i, but depends upon the coalition in an arbitrary (unrestricted) way. The production function of the private good is given by the correspondence $y: N \to R_+$, where $y\ (S)$ denotes the amount of the private good available to coalition S if and when it forms. Thus, the characteristic functions of the two game $(N;\ v)$ and $(N;\ w)$ are given by:

$$v\ (S) = \{a \in R^S \mid \exists\ B\ (S)\ \text{with}\ a_i \leqslant u_i\ (x_i, B_j),\ i \in B_j \in B\ (S),$$

$$\text{and}\ \forall\ B_j \in B\ (S),\ \sum_{i \in B_j} x_i = y\ (B_j)\},$$

$$w\ (S) = \{a \in R^S \mid \exists\ B\ (S)\ \text{with}\ a_i \leqslant u_i\ (x_i, B_j),\ i \in B_j \in B\ (S)$$

$$\text{and}\ \sum_{i \in S} x_i = \sum_{B_j \in B(S)} y\ (B_j)\}.$$

Thus, in the game $(N;\ v)$ a group of people S can partition itself into subgroups, in each of which the total amount produced is distributed among its own members. In contrast, in $(N;\ w)$, the only constraint on the distribution of the putput is that the total amount consumed by members of S equals the total amount produced in all the subgroups.

Given a coalition structure B and an output distribution x, we denote:
$u\ (x, B) \equiv (u_i\ (x_i, B_j)),\ i \in B_j \in B.$

In the following example, where all individuals are identical, and their preferences are anonymous, i.e., they depend upon the number of members in a coalition but not upon their names, every allocation without transfers is strictly Pareto dominated by an allocation involving transfers, i.e., $\forall u \in \bar{V}, \exists \bar{u} \in W$ such that $\bar{u} > u$.

Example 1:

Let

$$N = \{1, 2, 3\},\ y\ (S) = \begin{cases} 1 & \text{if}\ |S| = 2 \\ 0 & \text{otherwise,} \end{cases}$$

and

$$\forall\ i \in N,\ u_i\,(x_i,\ S) = \begin{cases} 2\,x_i & \text{if } S = \{i\} \\[6pt] x_i & \text{otherwise.} \end{cases}$$

For u in V, there exists $B \in B$ and x such that for all $i = 1, 2, 3\ u_i \leqslant u_i\,(x_i,\ T)$, $i \in T \in B$, and for all $T \in B$, $\sum\limits_{i \in T} x_i = y\,(T)$. For u in $\bar{V}$, B is of the form $(\{i, j\}, \{k\})$; otherwise, $u = (0, 0, 0)$. Without loss of generality, consider the coalition structure $B = (\{1, 2\}, \{3\})$ and the payoff $x = (x_1, x_2, 0)$, $1 > 1 - x_1 \geqslant x_2 \geqslant 0$, so that $u = (x_1, x_2, 0)$. For all such x, u is dominated by $\bar{u} \in W$ defined by means of $\bar{B} = (\{1\}, \{2, 3\})$ and $\bar{x} = (\frac{3}{4}\,x_1, x_2 + \frac{1}{8}\,x_1, \frac{1}{8}\,x_1)$, so that $\bar{u} = (\frac{3}{2}\,x_1, x_2 + \frac{1}{8}\,x_1, \frac{1}{8}\,x_1)$. Note that $\bar{x}$ involves a transfer of $\frac{3}{4}\,x_1$ (which is a positive amount) from $\{2, 3\}$ to $\{1\}$.

One possible explanation one may be tempted to give to this example is that production and consumption preferences conflict. Namely, each individual would rather consume alone but produce together. The following example illustrates that this line of reasoning is wrong, since the production function is additive. That is, each coalition can produce the number of units of x which equals its cordinality. Put differently, the marginal productivity of each individual, in every coalition, is 1. In fact, when production is additive the model can be viewed as a "coalition consumption economy" in which each individual is endowed with 1 unit of x, and no production of x takes place. Even in such cases, it is still possible that $\bar{W} \cap \bar{V} = \emptyset$. Indeed,

Example 2:
Let $N = \{1, 2, 3\}$, $y\,(S) = |\,S\,|$, and let

$$u_i\,(x,\ S) = \begin{cases} 2x - 2 & \text{if } S = \{i\} \\[4pt] x & \text{if } S = \{i, j\} \\[4pt] x - 10 & \text{if otherwise.} \end{cases}$$

Applying the same reasoning as for Example 1 we get $\bar{W} \cap \bar{V} = \emptyset$.

Moreover, even when preferences are separable, that is $u_i\,(x, S) = f_i\,(x) + g_i\,(S)$, and the production correspondence is additive, transfers among coalitions may still be necessary for Pareto optimality. Indeed,

Example 3:
Let $N = \{1, 2, 3\}$, $y\,(S) = |\,S\,|$ and $u_i\,(x, S) = x^2 + g_i\,(S)$ where

$$g_i\,(S) = \begin{cases} 1 & \text{if } |\,S\,| = 2 \\[4pt] 0 & \text{otherwise.} \end{cases}$$

Clearly, $u \in \bar{V}$ implies $u = (x, b)$ with $B = (\{i, j\}, \{k\})$. W.l.o.g. let $B = (\{1, 2\}, \{3\})$ and let $x = (z, 2 - z, 1)$ with $1 \leqslant z \leqslant 2$, i.e., $u = (1 + z^2, 1 + (2 - z)^2, 1)$. Consider the partition $(\{1\}, \{2, 3\})$ and the allocation $(z + \frac{1}{2}, 2\frac{1}{4} - z, \frac{1}{4})$. The corresponding utility levels are: $\hat{u} = ((z + \frac{1}{2})^2, 1 + (2\frac{1}{4} - z)^2, 1 + (\frac{1}{4})^2) \in W$. It is easily verified that $\hat{u} > u$.

Are there general properties of a game (N, v) under which $\bar{V} \cap \bar{W} \neq \emptyset$? It is intuitively clear from the definition, that core allocations never involve transfers among coalitions. Indeed,

Proposition 1: $Co\ (N, w) = Co\ (N, v) \subset \bar{V} \cap \bar{W}$.

Proof:

(i) $Co\ (N, w) \subset Co\ (N, v)$.

Let $u \in Co\ (N, w)$ with $u = u\ (x, B)$. By the monotonicity of preferences in x_i, $u \in Co\ (N, w)$ implies $\Sigma x_i \geq y\ (B_j), \forall B_j \in B$. (Otherwise, B_j would block.) Therefore,
$$\underset{i \in B_j}{}$$
$\Sigma x_i = y\ (B_j), \forall B_j \in B$ and hence $u \in V$. Since $w\ (S) \supset v\ (S), \forall S \subset N$, and $u \in Co\ (N, w)$
$i \in B_j$
we have $u \in Co\ (N, v)$.

(ii) $Co\ (N, v) \subset Co\ (N, w)$.

Suppose to the contrary that $u \in Co\ (N, v) \setminus Co\ (N, w)$, i.e., there exists $S \subset N$, $b_S \in w\ (S)$ with $b_S \geq u_S$. By definition, $b_S \in w\ (S)$ implies that there exists a coalition structure $\bar{B}\ (S)$ such that $\forall i \in S, b_i \leq u_i\ (x_i, T), i \in T \in \bar{B}\ (S)$ and $\Sigma x_i = \underset{T \in \bar{B}(S)}{\Sigma y}\ (T)$.
$i \in S$

Since $b_S \geq u_S$, there exists $T \in \bar{B}\ (S)$ with $b_T \geq u_T$. If $b_T \in v\ (T)$, this would contradict $u \in Co\ (N; v)$. Therefore, $\Sigma \bar{x}_i > y\ (T)$, implying that there exists a coalition R in $\bar{B}\ (S)$
$i \in T$
with $\Sigma \bar{x}_i < y\ (R)$. Therefore, $b_R \in v\ (R)$ which, due to the monotonicity assumption,
$i \in R$
contradicts $u \in Co\ (N; v)$.

(iii) $\bar{V} \cap \bar{W} \supset Co\ (\cdot)$.

Since $Co\ (N, v) \subset \bar{V}$ and $Co\ (N, w) \subset \bar{W}$, we have

$$\bar{V} \cap \bar{W} \supset Co\ (N, v) \cap Co\ (N, w) = Co\ (N, v) = Co\ (N, w).$$

Q.E.D.

Remark: It follows from Proposition 1 that $\bar{V} \cap \bar{W} \neq \emptyset$ whenever the core of the game is non-empty. The converse implication is not true. There exist games with an empty core but with $\bar{V} \cap \bar{W} \neq \emptyset$. Indeed, let

$$N = \{1, 2, 3\}, y\ (S) = \begin{cases} 1 & \text{if } |S| = 2 \\ \\ 0 & \text{otherwise;} \end{cases} \quad \text{and } u_i\ (x_i, S) = x_i \forall i \in N, \forall S \in N;$$

In this game the core is empty while $\bar{V} \subset \bar{W}$ (see Corollary below).

Proposition 2[4]): $u_i\ (x_i, S) = x_i + g_i\ (S), \forall i$, implies $\bar{W} \cap \bar{V} \neq \emptyset$.

---

[4]) Note that this is not sufficient to guarantee non-emptiness of the core. For example, let $N = \{1, 2, 3\}, y\ (S) = |S|, \forall S \subset N$,

and $g_i\ (S) = \begin{cases} 4 & \text{if } |S| = 2 \\ 0 & \text{otherwise.} \end{cases}$

Proof: Let $u = \bar{u}(x, B)$ with $\Sigma x_i = y(B_j)$ for all B_j in B, and B such that
$$\underset{\substack{i \in B_j}}{}$$

$$\underset{\substack{B_j \in B}}{\Sigma} [y(B_j) + \underset{i \in B_j}{\Sigma} g_i(B_j)] \text{ is maximal on } B. \text{ Clearly, } u \in V. \text{ To show that } u \in \bar{W}, \text{ note tha}$$

$\bar{u} = u(\bar{x}, \bar{B}) \geqslant u$ would imply

$$\underset{i \in N}{\Sigma} u_i < \underset{i \in N}{\Sigma} \bar{u}_i = \underset{i \in N}{\Sigma} \bar{x}_i + \underset{\bar{B}_j \in \bar{B}}{\Sigma} \underset{i \in \bar{B}_j}{\Sigma} g_i(\bar{B}_j) = \underset{\bar{B}_j \in \bar{B}}{\Sigma} [y(\bar{B}_j) + \underset{i \in \bar{B}_j}{\Sigma} g_i(\bar{B}_j)],$$

a contradiction.   Q.E.D.

Since the Pareto criterion is invariant under monotonic transformations, we have

Corollary: $u_i(x_i, S) = f^i(x_i) \forall i$ implies $\bar{W} \cap \bar{V} \neq \emptyset$.

Proposition 3[5]): If $y(S)$ is additive, (i.e., $y(S) = \underset{i \in S}{\Sigma} y(\{i\})$, $\forall S \subset N$), and for all

$i \in N$, $u_i(x_i, S) = f_i(x_i) + g_i(S)$, (i.e., u_i is separable), with f_i continuous and concave, then $\bar{W} \cap \bar{V} \neq \emptyset$.

Proof:
1. Define $T = \{t \in R^n \mid \exists x, \underset{i \in N}{\Sigma} x_i \leqslant y(N), t_i \leqslant f_i(x_i), \forall i \in N\}$.

   By concavity of the f_i's, T is a convex set. Let $r \in R^n$ be defined by $r_i = f_i(y(\{i\}))$. By monotonicity, r belongs to the boundary of T and thus there exists $q \geqslant 0$ such that

   (1.1) $q \cdot r \geqslant q \cdot t, \forall t \in T$

   (1.2) $q_i = 0$ implies $y(\{i\}) = 0$ and $r_i \geqslant t_i$ for all t in T such that $q \cdot t = q \cdot r$ (for otherwise reducing x_i and increasing x_j for some j with $q_j > 0$ would increase $q \cdot t$, contradicting (1.1)).

   Let $M = \{i \in N \mid q_i = 0\}$.
2. Define the non-empty set $\bar{B} = \{B \in B \mid \not\exists B' \in B \text{ with } q \cdot g(B') > q \cdot g(B) \text{ or with }$

   $q \cdot g(B') = q \cdot g(B) \text{ and } g_M(B') \geqslant g_M(B)\}$, where $g(B) = (g_i(B_j))_{i \in B_j \in B}$. Then

   for $B \in \bar{B}$, $r + g(B) =_{def} u$ belongs to $\bar{W}$ (and to V, hence to $\bar{W} \cap \bar{V}$). Indeed, there does not exist $\bar{t} + g(\bar{B}) =_{def} \bar{u} \geqslant u$, because:

   (2.1) $q \cdot \bar{u} > q \cdot u$ is ruled out by (1.1) and the definition of u and $\bar{B}$, hence $\bar{u}_{N \setminus M} \geqslant u_{N \setminus M}$ is impossible;

   (2.2) if $q \cdot \bar{u} = q \cdot u$, then $q \cdot \bar{t} = q \cdot r$ and $q \cdot g(\bar{B}) = q \cdot g(B)$; but $q \cdot \bar{t} = q \cdot r$ implies $\bar{t}_i = r_i = f_i(0), \forall i \in M$; and, since $B \in \bar{B}$, $q \cdot g(\bar{B}) = q \cdot g(B)$ rules out $g_M(\bar{B}) \geqslant g_M(B)$; hence $\bar{u}_M \geqslant u_M$ is also impossible.

                                                                                        Q.E.D.

Examples 2 and 3 above together with example 4 below reveal that no pair from the three conditions in Proposition 3 – additivity of $y(S)$, separability of the u_i's and concavity of the f_j's – is sufficient for the result to hold.

---

[5]) See footnote on previous page.

Example 4:

Consider Example 1, with a different representation of the utility functions, namely:

$$u_i(x, S) = \begin{cases} \log x + \log 2 & \text{if } S = \{i\} \\ \\ \log x & \text{otherwise.} \end{cases}$$

Since the sets of Pareto optimal (x, B) are invariant under monotonic transformation of the utility functions, $\bar{W} \cap \bar{V} = \emptyset$, as in Example 1.

4. The Transferable Input-Non Transferable Output Model

We shall now show that altruism may be necessary for Pareto optimality also in situations that give rise to games with side payments. Consider an economy in which each individual is endowed with one unit of labor which is assumed to be a homogeneous commodity. There is a single (private) good which is produced by labor. Specifically, each coalition S is assigned with a production function $y_S(L)$, which is interpreted to mean that if and when coalition S forms, and L hours are worked in that coalition, (not necessarily by its members), the maximum amount of the single commodity that can be produced is $y_S(L)$. We shall assume that for every coalition $S \subset N$ the production function $y_S: [o, n] \to R_+$ is continuous, monotone increasing and concave. Also $y_S(0) = 0$. Thus, within each coalition the marginal product of labor is a nonincreasing function of L, and is always positive.

While labor can be costlessly moved among coalitions, we shall assume that it is infinitely costly to transfer output across coalitions. It therefore follows that the total output produced within a coalition can be distributed only among its own members.

For example, consider an uninhabited island with $2^n - 1$ possible housing locations (one for each coalition, if and when it forms). Clearly, the production functions may depend upon the site. The n individuals are supposed to settle on that island. Each individual can commute costlessly among the different sites (thus, can work in any coalition). Once the houses are built, each individual resides in one and only one community and derives his utility only from the size of his own house. Is it possible that unless some individuals work for free in locations where they themselves do not reside, everyone can be made better off?

In order to answer such questions, we again define the two games $(N; v)$ and $(N; w)$, where,

$$v(S) = \{x \in R_+^S \mid \exists B(S) \text{ such that } \sum_{i \in T} x_i = y_T(|T|), \forall T \in B(S)\}.$$

$$w(S) = \{x \in R_+^S \mid \exists B(S) \text{ such that } \forall T \in B(S), \sum_{i \in T} x_i = y_T(L_T)$$

$$\text{and } \sum_{T \in B(S)} L_T = |S|\}.$$

Thus, under $(N; v)$ members of a given coalition work "full-time" in their own coalition, while under $(N; w)$ labor can be transferred freely among coalitions. In both games, output cannot be transferred. Note that an allocation which belongs to $w(S) \setminus v(S)$, implies that there exists at least one coalition $T \in B(S)$ whose members work, (possibly part time), *for free*, (as output is not transferable), in other coalitions. In that case, since the marginal product of L is always strictly positive, the total amount produced in T could have been increased if all its workers worked full time in their own coalition.

To simplify the analysis even further, let us assume that all individuals are identical. In particular, therefore, the production functions of two coalitions with the same cardinality coincide, i.e., $\forall S, T \subset N$ with $|S| = |T|, y_S(\cdot) = y_T(\cdot)$. However, even under this assumption neither $(N; v)$ nor $(N; w)$ need be games with side payments, since output cannot be transferred across coalitions. We shall therefore impose the additional condition that the game is super-additive, implying that $y_S(|S|) \geqslant \sum_{T \in B(S)} y_T(|T|)$ for all $B(S) \in B(S)$. Under this assumption $(N; v)$ is a game with side payments; its characteristic function v is given by:

$$v(S) = \{x \in R_+^S \mid \sum_{i \in S} x_i = y_S(|S|)\}.$$

Note that the game $(N; w)$ may still be a game without side payments, (see the example below). In view of the assumption that all individuals are identical, we can define: $y_t(\cdot) \equiv y_T(\cdot)$ and denote the average product for a coalition of size t by a_t, namely $a_t = (y_t(t)) / t$.

The following example shows that for $n \geqslant 5$, even when the game is strictly super additive, it is still possible that $\bar{V} \cap \bar{W} = \emptyset$. Let:

$N = \{1, \ldots, n\}$ $n \geqslant 5$ and let $y_{|S|}(L)$ be given by

$$y_1(L) = \begin{cases} 10 L & \text{for } L \leqslant 0.1 \\ \\ 1 + \dfrac{L - 0.1}{10n} & \text{for } L \geqslant 0.1 \end{cases}$$

$$y_j(L) = 1.8 L + \frac{10^{-10}}{n^3}(j - 2)^2 L \qquad\qquad j = 2, \ldots, n - 3$$

$$y_{n-2} = \begin{cases} (4n - 7.8) L & \text{for } L \leqslant \dfrac{1}{2} \\ \\ 2n - 3.9 + \left(\dfrac{0.2}{2n - 5}\right)(L - \dfrac{1}{2}) & \text{for } L \geqslant \dfrac{1}{2} \end{cases}$$

$$y_{n-1}(L) = \left(\frac{2n + 0.1}{n - 0.1}\right) L, \text{ and}$$

$$y_n(L) = 2 L.$$

Since all the five functions defined above are different, we need $n \geqslant 5$.

To check strict super additivity we have to show that for any three positive integers k, l, m such that $k + l = m \leqslant n$,

$$y_k\,(k) + y_l\,(l) < y_m\,(m).$$

It is easy to verify that the above condition holds.

By super additivity, therefore, $\bar{V} = \{x \in R_+^n \mid \Sigma\, x_i = 2n\}$. Let $x \in \bar{V}$.

W.l.o.g. $x_1 \leqslant x_2 \leqslant, \ldots, \leqslant x_n$, and let $A = \sum\limits_{i=1}^{n-2} x_i$. Distinguish between the following two cases:

(i) $A < n - 2$. Then, $x_1 < 1$. Consider the partition $B^1 = (\{1\}, \{2, \ldots, n\})$. Let 1 work for 0.1 hours in $\{1\}$, producing 1 unit, and thus making him better-off. With the remaining $n - 0.1$ hours to be worked in $\{2, \ldots, n\}$ the amount produced is $2n + 0.1$, which is certainly greater than $2n - x_1$.

(ii) $A \geqslant n - 2$. Consider the partition $B^2 = (\{1, \ldots, n - 2\}, \{n - 1, n\})$. Let the first $n - 2$ individuals work in their coalition for $\frac{1}{2}$ hour producing $2n - 3.9$ units. Since $x_i \leqslant x_j$ whenever $1 \leqslant i < j \leqslant n$, $A \leqslant (n - 2)\,(y_n\,(n)/n) = 2n - 4$. Therefore, the first $n - 2$ individuals are made better-off. With the remaining $n - \frac{1}{2}$ hours worked in the coalition $\{n - 1, n\}$, $1.8\,n - 0.9$ units are produced. Now, $y_n\,(n) - A = 2n - A \leqslant 2n - (n - 2) = n + 2 < 1.8\,n - 0.9$. Thus, $n - 1$ and n can also be made better off.

Such situations cannot occur, however, when the number of players does not exceed 4. Indeed,

Proposition 4: Let $n \leqslant 4$. If $(N; v)$ is super additive then $\bar{V} \cap \bar{W} \neq \emptyset$.

Proof:

(i) For $n = 1, 2$ the theorem always holds (regardless of super additivity).

(ii) Let $n = 3$. By super additivity, $\bar{V} = \{x \in R_+^3 \mid \sum\limits_{i \in N} x_i = 3\,a_3\}$. We shall show that either $(3\,a_3, 0, 0) \in \bar{W}$ or $(a_3, a_3, a_3) \in \bar{W}$. To this end, define:

$$f(L) = y_1\,(L) + y_2\,(3 - L), \; 0 \leqslant L \leqslant 3.$$

By super additivity and the concavity of $y_1\,(L)$, we have $y_1\,(3) \leqslant 3\,a_1 \leqslant 3\,a_3$. Thus, if $(3\,a_3, 0, 0) \notin \bar{W}$ then $f(0) > 3\,a_3$. Moreover, $(a_3, a_3, a_3) \notin \bar{W}$ implies that there exists L^* with $y_1\,(L^*) > a_3$ and $y_2\,(3 - L^*) > 2\,a_3$. Since $y_1\,(L)$ is strictly monotone and by super additivity $a_1 \leqslant a_3$, we have $y_1\,(L^*) > a_1$ implying $L^* > 1$. Thus, $(a_3, a_3, a_3) \notin \bar{W}$ implies $f(L^*) > 3\,a_3$ for $L^* > 1$. But by the definition of super additivity $f(1) \leqslant 3\,a_3$, contradicting the fact that f, being a sum of two concave functions, is concave.

(iii) Let $n = 4$. By super additivity, $\bar{V} = \{x \in R_+^4 \mid \sum\limits_{i \in N} x_i = 4\,a_4\}$. As in the case $n = 3$. we shall show that either $(4\,a_4, 0, 0, 0) \in \bar{W}$ or $(a_4, a_4, a_4, a_4) \in \bar{W}$. Define,

$$g(L) = y_1\,(L) + y_3\,(4 - L), \; 0 \leqslant L \leqslant 4,$$

The concavity of the production functions together with super additivity imply that $4\,a_4 \geqslant 4\,a_2 \geqslant y_2\,(4)$ and $4\,a_4 \geqslant 4\,a_1 \geqslant y_1\,(4)$. Therefore if $(4\,a_4, 0, 0, 0) \notin \bar{W}$

then $g(0) > 4a_4$. Moreover, $(a_4, a_4, a_4, a_4) \notin \bar{W}$ implies that $g(L^*) > 4a_4$ for $1 < L^* \leqslant 4$. But, $g(1) \leqslant 4a_4$, contradicting the concavity of g.

Remark: Proposition 1 above is valid for this case, as well. In particular, therefore, if the game is convex, i.e., $a_n \geqslant a_t$ $t = 1, \ldots, n$, then $\bar{W} \cap \bar{V} \neq \emptyset$.

References

Aumann, J.R., and *J.H. Dreze*: Cooperative Games with Coalition Structures. International Journal of Game Theory, 1975.

Dreze, J.H., and *J. Greenberg*: Heconic Coalitions: Optimality and Stability. C.O.R.E. discussion paper number 7735, 1977. To appear in Econometrica, 1979.

Greenberg, J., and *A. Kats*: Unilateral Transfers and Pareto Optimality. Center for Study of Public Choice Discussion Paper, CE-78-4-1. To appear in Econometrica, 1979.

Applied Game Theory. 1979 ©Physica-Verlag, Wuerzburg/Germany

An Approach to the Problem of Efficient Distribution of the Labor Force[1])

By *Y. Kannai*, Rehovot[2]), and *B. Peleg,* Jerusalem[3])

Abstract: We consider three models of economies in which the distribution of the labor force between the private and the public sectors is determined by game-theoretic considerations. More specifically, the only active market in our models is the labor market, and we determine and investigate its Nash equilibrium points. The models vary in the degree of influence that government has in determining the wage of public servants. It should be emphasized that in our models unemployment is possible when the labor market is in equilibrium. Moreover, the rate of unemployment in equilibrium in our models has a simple and natural explanation.

1. Introduction

In the present paper an attempt is made to construct a game-theoretic model describing the choice people make when deciding whether to seek employment in the public (service) sector or in the private (productive) sector. These people are aware of nominal salaries possible in the two sectors, and of the dependence of the output of consumer goods (hence of their real incomes) on the size of the labor force. Here we are faced with a non cooperative *n*-person game and we look for its Nash equilibrium points.

We are dealing essentially with short run behavior. Consequently, we disregard (as customary in economic literature) the role of capital — we assume that our production functions depend only on the size(s) of the labor force. It is the following two observations which shape considerably our models. It appears that the production of goods such as food (and, indeed, the "welfare" of the population) depends significantly on the number of people employed in farming (or more generally in production of tangible consumer's goods) but depends only slightly on the size of the government payroll (or on the number of people working in "public services"). Some authorities [*Parkinson,* 1958, 1960] even claim that most government employees perform no (useful) service at all. But even if we do not adopt such an extremist Parkinsonian point of view, we can still argue, for example, that while a minimal police force, say, is absolutely necessary even for successful farming, the agricultural production of the Soviet Union would

[1]) The research described in this report was carried out while the second author was visiting the Weizmann Institute and was completed while the first author was visiting Bonn University. This visit was made possible by a DAAD grant.

[2]) Prof. *Yakar Kannai*, The Weizmann Institute of Science, Rehovot, Israel

[3]) Prof. *Bezalel Peleg,* The Hebrew University of Jerusalem, Jerusalem, Israel

certainly not go down if most of the K.G.B. was eliminated. It is also apparent that public service employees are at least as wellpaid (if not better paid) than manufacturing or farming workers (this becomes even more evident if factors such as job security and working conditions are included in the considerations). Here, the downward rigidity of wages along with the fact that the marginal contribution to productivity of a government employee is high if the number of public servants is small (as was probably the case in the past) yield a possible (partial) explanation of these phenomena.

In section 2 we describe a simplified static model, where it is assumed, that the wage of a production worker is smaller (by a constant rate) than the wage of a public servant. It turns out that the number of production workers remains bounded even if the population size is unbounded, with disastrous results for all involved. In section 3 and 4 more elaborate dynamic models are presented. Here the government finances its operations by taxing the firm (the employer of production workers) and by printing money. While such printing of money is the direct cause of inflation, it also serves to turn the wheels of the economy. For reasons of computational convenience we choose to consider in sections 3 and 4 a continuum of employment seekers. We emphasize however that this is not an essential feature of the models and that we believe, in fact, that the same qualitative behavior is observed if the population is big enough. (In particular, concepts such as perfect competition play no role in the discussion). The models differ in government commitment. In section 3 (the neo-Parkinsonian model) the government is committed to employ in its services every person seeking a government job (at a nominal wage no smaller than the one at former periods) but does not pay unemployment benefits to people unable to find work in the productive sector, whereas in section 4 the government is required to pay unemployment benefits to all unemployed, but has more freedom in determining the size of the governmental payroll. In this latter model an interesting question arises as to what should constitute a good government policy. Further discussions of the models will be found in section 5.

2. The Static Model

The population in our economy consists of the set of workers $N = \{1, 2, \ldots, n\}$. Every member of N is given the opportunity to decide whether he wants to become a production worker or a services worker. Let p and s denote the number of production workers and services workers, respectively. Thus

$$p + s = n. \tag{2.1}$$

A single consumers' good is produced from labor only according to the production function

$$f(p, s) = p. \tag{2.2}$$

The full output is divided among the n workers. Let β denote the ratio between the salary x of a production worker and the salary y of a services worker. We assume that $0 < \beta < 1$ and that β is given exogenously. Hence

$$x = \beta y \tag{2.3}$$

and

$$px + sy = p \tag{2.4}$$

implying

$$x(p) = \frac{\beta p}{n + p(\beta - 1)} \tag{2.5}$$

$$y(p) = \frac{p}{n + p(\beta - 1)}. \tag{2.6}$$

We regard our economy as a non-cooperative n-person game in normal form in which (i) N is the set of players, (ii) every player has two pure strategies: to become a production worker or to be employed in services, and (iii) the (real) payoff is determined by (2.5) and (2.6). This game is clearly symmetric and the (pure strategies) state is described by the pair $(p, s) = (p, n - p)$. The point (p_*, s_*) is a Nash equilibrium point if and only if

$$y(p_* - 1) \leqslant x(p_*) \tag{2.7}$$

$$x(p_* + 1) \leqslant y(p_*). \tag{2.8}$$

Theorem 2.1: For all n sufficiently large there exists at least one integer p_* such that the pair $(p_*, n - p_*)$ represents a Nash equilibrium point in pure strategies. The integer p_* is either unique or there exist two possible values of p_* differing by the integer one. There exists a constant $C(\beta)$, for $0 < \beta < 1$ such that

$$p_* \leqslant C(\beta) \tag{2.9}$$

independently of n.

Remark 2.2: While the Nash equilibrium points are not uniquely determined (even if p_* is unique) due to the complete symmetry of the game, the equilibrium is indeed unique up to symmetry. We cannot specify the identity of the set of players choosing one of the strategies. Their size, however, is essentially uniquely determined.

Proof: An elementary computation shows that (2.7) is equivalent to the quadratic inequality

$$(1 - \beta)^2 p_*^2 + (n - \beta + 1)(\beta - 1) p_* + n \geqslant 0. \tag{2.10}$$

This inequality is satisfied if and only if either $p_* \geqslant a_2(n, \beta)$ or $p_* \leqslant a_1(n, \beta)$, where

$$a_{1,2}(n, \beta) = \frac{(1 - \beta)(n - \beta + 1) \pm [(1 - \beta)^2 (n - \beta + 1)^2 - 4n(1 - \beta)^2]^{1/2}}{2(1 - \beta)^2}. \tag{2.11}$$

But $\lim\limits_{n\to\infty} (a_2 (n, \beta))/n = 1/(1 - \beta)$. Hence $a_2 (n, \beta) > n$ for n large enough and (2.7) is valid if and only if

$$p_* \leqslant a_1 (n, \beta) = \frac{n - \beta + 1 - [(n - \beta + 1)^2 - 4n]^{1/2}}{2(1 - \beta)}. \qquad (2.12)$$

Similarly, (2.8) is equivalent to the quadratic inequality

$$(1 - \beta)^2 p_*^2 + (n + \beta - 1) (\beta - 1) p_* + \beta n \leqslant 0. \qquad (2.13)$$

This inequality holds if and only if $b_1 (n, \beta) \leqslant p_* \leqslant b_2 (n, \beta)$, where

$$b_{1,2} (n, \beta) = \frac{(1 - \beta) (n + \beta - 1) \pm [(1 - \beta)^2 (n + \beta - 1)^2 - 4\beta n (1 - \beta)^2]^{1/2}}{2(1 - \beta)^2} \qquad (2.14)$$

Here $\lim\limits_{n\to\infty} (b_2 (n, \beta))/n = 1/(1 - \beta)$ implies that $b_2 (n, \beta) > n$ for n large enough and (2.8) is valid if and only if

$$p_* \geqslant b_1 (n, \beta) = \frac{n + \beta - 1 - [(n + \beta - 1)^2 - 4\beta n]^{1/2}}{2(1 - \beta)}. \qquad (2.15)$$

The assertions about existence and uniqueness now follow, once we observe that $a_1 (n, \beta) - b_1 (n, \beta) = 1$, $b_1 (n, \beta) > 0$ and thus the interval $[b_1 (n, \beta), a_1 (n, \beta)]$ contains at least one integer from N. The last statement of the theorem follows from the fact that

$$\lim_{n\to\infty} a_1 (n, \beta) = 1/(1 - \beta). \qquad (2.16)$$

Remark 2.3: We have uniqueness of p_* unless $b_1 (n, \beta)$ is an integer, i.e., we have uniqueness almost always.

Remark 2.4: The estimate (2.9) together with (2.5) and (2.6) imply that $x (p_*)$, $y (p_*) \to 0$ as $n \to \infty$, or that services workers would rather starve to death than switch to production.

The assumption that the production function has the form (2.2) is, of course, a very strong one. For a general production function $f (p, s)$ we can still define a game as above if we keep the assumption (2.3) (discrimination against production workers) intact but replace (2.4) by

$$px + sy = f (p, s) = f (p, n - p). \qquad (2.4')$$

The payoff functions then take the forms

$$x (p) = \frac{\beta f (p, s)}{\beta p + s} = \frac{\beta f (p, n - p)}{n + p (\beta - 1)} \qquad (2.5')$$

$$y(p) = \frac{f(p, s)}{\beta p + s} = \frac{f(p, n - p)}{n + p(\beta - 1)}.$$ (2.6')

The equilibrium conditions are once more (2.7) and (2.8) (where now x and y are given by (2.5'), (2.6')). It is not clear under what general assumptions on f existence of a Nash equilibrium point in pure strategies is guaranteed. If, for example, (2.2) is replaced by

$$f(p, s) = p \min(s, c)$$ (2.2')

where c is a positive constant, then Theorem 2.1 continues to hold (with the same proof) if n is large enough. (The intuitive meaning of (2.2') is that services are very essential up to a certain point, after which the returns are independent of scale). Allowing a more general form of f we can still show that p is bounded in all pure strategies equilibrium points:

Theorem 2.5: Let f be a concave differentiable function with non negative first order partial derivatives such that $f(p, s) > 0$ for $p > 0$ and $f(0, 0) = 0$. There exists a constant C (depending only on f) such that if (p_*, s_*) are numbers of production (services) workers in a pure strategies Nash equilibrium point then

$$p_* \leqslant C.$$ (2.9')

Proof: The inequality (2.7) reads now

$$\frac{\beta f(p_*, s_*)}{\beta p_* + s_*} \geqslant \frac{f(p_* - 1, s_* + 1)}{\beta(p_* - 1) + s_* + 1}$$ (2.17)

or (assuming without loss of generality $p_* > 1$)

$$\frac{f(p_* - 1, s_* + 1)}{f(p_*, s_*)} \leqslant \frac{\beta[\beta(p_* - 1) + s_* + 1]}{\beta p_* + s_*} = \beta\left(1 + \frac{1 - \beta}{\beta p_* + s_*}\right) \leqslant \beta\left(1 + \frac{1 - \beta}{\beta n}\right).$$ (2.18)

Recalling that $0 < \beta < 1$ we obtain from (2.18) the existence of an $\epsilon > 0$ such that

$$\frac{f(p_* - 1, s_* + 1)}{f(p_*, s_*)} \leqslant 1 - \epsilon$$ (2.19)

for n large enough. By concavity and monotonicity,

$$f(p_* - 1, s_* + 1) \geqslant f(p_* - 1, s_*) \geqslant f(p_*, s_*) - f_p(p_* - 1, s_*)$$ (2.20)

(we use subscripts to denote partial derivatives).
    Combining (2.19) and (2.20) we obtain the estimate

$$\frac{f_p(p_* - 1, s_*)}{f(p_*, s_*)} \geqslant \epsilon.$$ (2.21)

By concavity and the assumption $f(0, 0) = 0$,

$$(p_* - 1) f_p (p_* - 1, s_*) + s_* f_s (p_* - 1, s_*) \leqslant f (p_* - 1, s_*). \tag{2.22}$$

The non-negativity of f_s implies that

$$\frac{f_p (p_* - 1, s_*)}{f (p_* - 1, s_*)} \leqslant \frac{1}{p_* - 1}. \tag{2.23}$$

Hence

$$\frac{f_p (p_* - 1, s_*)}{f (p_*, s_*)} \leqslant \frac{1}{p_* - 1}. \tag{2.24}$$

It follows from (2.21) and (2.24) that

$$p_* \leqslant \frac{1}{\epsilon} + 1. \tag{2.25}$$

Remark 2.6: If we add to the assumptions of Theorem 2.5 the additional hypothesis that $f_p > f_s$ then the result shows that the Nash equilibrium is highly inefficient. If we assume that f has a Cobb-Douglas form

$$f (p, s) = A p^a s^{1-a} \tag{2.26}$$

where $1 \geqslant a > 0$, then $x (p_*), y (p_*) \to 0$ (if the pure strategies equilibrium points exist at all). We have the same effect as in Remark 2.4, the severity of the phenomenon being more pronounced, the higher a is.

3. The Neo-Parkinsonian Model

The agents participating in the economy are the workers (or employment seekers), of which there is a continuum $N = [0, n(t)]$, a firm F and the government G. The variable $t = 1, 2, \ldots$, denotes the (discrete) time period, the quantities pertaining to the period $t = 0$ are exogenously given initial conditions. The (positive) numbers $n(t), t = 1, 2, \ldots$, measure the population growth. Let $m(F, t)$ and $m(G, t)$ denote the amounts of money held by the firm and government, respectively, at time t. Let the measurable subsets P, Q and S denote the sets of production workers (employees of F), people seeking employment in production and government employees (public servants) respectively. We assume that $P \subset Q$ and $Q \cup S = N$, i.e., every person seeks employment either in the government or in the firm (but not in both) at time t. The government employs everyone wishing to become a public servant, whereas the firm selects the people who are actually going to be employed from among those who seek productive employment. The Lebesgue measures of P, Q and S are denoted by p, q, and s respectively. Thus

$$q = n - s. \tag{3.1}$$

The monetary (nominal) wages of production workers and government employees are denoted by $x(t)$ and $y(t)$, respectively, unemployed people (members of $Q \setminus P$) getting nothing. In this model (and in the one described in the next section) nominal salaries never go down: $x(t) \geqslant x_*(t-1)$, $y(t) \geqslant y_*(t-1)$. (A subscript-star denotes an equilibrium or an actually observed quantity). There is no saving, nor is the taking of loans allowed, by either individuals, the firm, or the government. The size p of P and $x(t)$ are determined as follows (for $t \geqslant 1$): Set

$$\bar{p} = \frac{m(F, t)}{x_*(t-1)}. \tag{3.2}$$

Then

$$p = \min(\bar{p}, q) \tag{3.3}$$

and

$$x(t) = \frac{m(F, t)}{p}. \tag{3.4}$$

The selection of members of P from Q is done randomly with uniform probability (or at least so people believe). Hence the probability of getting $x(t)$ is p/q and the expected monetary payoff for a member of Q is

$$\xi(q, t) = \frac{m(F, t)}{q}, \quad t \geqslant 1. \tag{3.5}$$

The amount of money held by the government $m(G, t)$ is the sum of the amount $M(G, t)$ the government has (at time t) before printing any money and the quantity of money $\mu(t)$ printed by the government at time t. The government prints money after s becomes known if and only if the government could not otherwise pay all members of S at least $y_*(t-1)$, and does not print more money than necessary. Hence,

$$\mu(t) = \max(0, sy_*(t-1) - M(G, t)), \quad t \geqslant 1 \tag{3.6}$$

and

$$y(t) = \max\left(y_*(t-1), \frac{M(G, t)}{s}\right), t \geqslant 1. \tag{3.7}$$

All employed workers spend all their money to buy the single consumers' good produced by F. Thus the firm gets the amount

$$M(F, t+1) = m(F, t) + m(G, t), \quad t \geqslant 0. \tag{3.8}$$

The firm pays the government income tax whose rate is θ, $0 < \theta < 1$. We assume θ is fixed and cannot be changed. Thus $m(F, t+1)$ and $M(G, t+1)$ denote the amounts of money left to the firm after taxes and the amount of taxes, respectively:

$$m\,(F,\,t+1) = (1-\theta)\,M\,(F,\,t+1) \tag{3.9}$$

$$M\,(G,\,t+1) = \theta M\,(F,\,t+1). \tag{3.10}$$

During every period $t \geqslant 1$, the members of N play a non-cooperative game, every participant having two pure strategies: to become a member of S or a member of Q.

Nash equilibrium points for non-cooperative games with a continuum of players were investigated by *Schmeidler* [1973]. Using a slightly different notation, we define a N-pure strategy to be a (choice of) measurable partition (Q, S) of N. The (expected) payoffs are given by (3.5) and (3.7). We denote this game by

$$\Gamma\,[m\,(F,\,t),\,M\,(G,\,t),\,x_*\,(t-1),\,y_*\,(t-1),\,n\,(t)].$$

The partition (Q, S) is an equilibrium point if nobody can gain by switching. Since p a and s (and the resulting monetary wages, total quantity of money in the economy, total output, price level and hence real wages) are not affected by the action of a single individual, it follows that the condition for equilibrium is $\xi = y$ or

$$\frac{m\,(F,\,t)}{q_*} = \max\left(y_*\,(t-1),\,\frac{M\,(G,\,t)}{s_*}\right) \tag{3.11}$$

Proposition 3.1: If $m\,(F,\,t) > 0$ and $M\,(G,\,t) > 0$ then there exists precisely one value of q_* solving (3.11), and the game $\Gamma\,[m\,(F,\,t),\,M\,(G,\,t),\,x_*\,(t-1),\,y_*\,(t-1),\,n\,(t)]$ has an equilibrium point. The equilibrium point is unique up to a Lebesgue-measure preserving automorphism of $[0,\,n\,(t)]$.

Proof: The function $m\,(F,\,t)/q_*$ is strictly decreasing in q_*, $m\,(F,\,t)/q_* \to \infty$ as $q_* \to 0$. The right hand side $(y\,(s))$ is a non-decreasing positive function of q_*.

From now on we will assume that our systems are, for all $t \geqslant 1$, at an equilibrium of $\Gamma\,[m\,(F,\,t),\,M\,(G,\,t),\,x_*\,(t-1),\,y_*\,(t-1),\,n\,(t)]$.

Example 3.2: Periodic changes in $(p_*\,(t))/(n\,(t))$ and in price level ("business cycles"). Here $n\,(t) = 2^{[t/2]+1}$ (where $[a]$ denotes the greatest integer $\leqslant a$) and $\theta = 1/2$, $x_*\,(0) = y_*\,(0) = m\,(F,\,1) = M\,(G,\,1) = 1$. It can be shown by easy induction on t (for $t = 1$ it is done directly) that $x_*\,(t) = y_*\,(t) = 1$ for all t, whereas $m\,(F,\,t) = M\,(G,\,t) = q_*\,(t) = p_*\,(t) = n\,(t)/2$ if t is odd and $m\,(F,\,t) = M\,(G,\,t) = q_*\,(t) = p_*\,(t) = n\,(t)/4$ if t is even. The price level $n\,(t)/p_*\,(t)$ oscillates between 2 and 4.

Example 3.3: A rapid increase in $n\,(t)$ generates inflation. Here $\theta = 0$, $n\,(t) = (t+1)!$ and $x_*\,(0) = y_*\,(0) = m\,(F,\,1) = M\,(G,\,1) = 1$. In this example $y_*\,(t) = x_*\,(t) = 1$, $q_*\,(t) = p_*\,(t) = m\,(F,\,t) = t!$ but $\mu\,(t) = (t+1)! - t!$. Hence

$$\frac{n\,(t)}{p_*\,(t)} = t + 1$$

i.e., a high inflation rate.

Remark 3.4: Note that in example 3.3 there is no direct (income) taxation. It appears that letting the government determine tax rates for each period would not change the situation.

We turn our attention now to the stationary case.

Definition 3.5: The model is said to be stationary if $n(t) = n$ for all $t = 1, 2, \ldots$ The quadruple $(m(F), M(G), x_*, y_*)$ is said to be stationary if the game $\Gamma[m(F), M(G), x_*, y_*, n]$ at $t = 1$ has an equilibrium point for which $m(F, 2) = = m(F), \mu(1) = 0, M(G, 2) = M(G), x_*(1) = x_*$ and $y_*(1) = y_*$.

Theorem 3.6: The quadruple $(m(F), M(G), x_*, y_*)$ is stationary if and only if

$$m(F) + M(G) = ny_* \tag{3.12}$$

$$\frac{m(F)}{M(G)} = \frac{1 - \theta}{\theta} \tag{3.13}$$

$$x_* \geqslant y_*. \tag{3.14}$$

Proof: Necessity: Let the quadruple $(m(F), M(G), x_*, y_*)$ be stationary, and let q_*, p_*, s_* result from an equilibrium point. Then $\mu(1) = 0$ implies (by (3.6) and (3.7)) that

$$y_* \leqslant \frac{M(G)}{s_*} \leqslant y_*(1), \tag{3.15}$$

Hence $y_* = (M(G))/s_*$. It follows from (3.11) that $(m(F))/q_* = y_*$. But $\xi \leqslant x_*$. Hence $x_* \geqslant y_*$. Also $ny_* = q_* y_* + s_* y_* = m(F) + M(G)$. The equation (3.13) follows from the assumption $\mu(1) = 0$. Conversely setting $q_* = m(F)/y_*$, $s_* = = M(G)/y_*$ we obtain a solution of (3.11), the stationarity of the quantities of money available to the firm and the government following from (3.13) while (3.12) and (3.14) imply that $x_*(1) = x_*, y_*(1) = y_*$.

In the stationary quadruple model the economic system lands at a stationary quadruple after two or three periods, as shown in the following theorem.

Theorem 3.7: If $n(t) = n$ (and $m(F, 1) > 0$ and $M(G, 1) > 0$) then there exist constants $m(F), M(G), x_*, y_*$ such that

$$m(F, t) = m(F), M(G, t) = M(G), x_*(t - 1) = x_*, y_*(t - 1) = y_*$$

$$\text{for all } t \geqslant 3, \tag{3.16}$$

and the quadruple $(m(F), M(G), x_*, y_*)$ is stationary.

Proof: The quantities $x_*(0), y_*(0), m(F, 1)$ and $M(G, 1)$ are given arbitrarily. Nevertheless, (3.9) and (3.10) imply that

$$\frac{m(F, t)}{M(G, t)} = \frac{1 - \theta}{\theta} \tag{3.17}$$

for all $t \geq 2$. Two cases arise: (i) $\mu(2) = 0$. In this case it is clear that
$m(F, t) = m(F, 2), M(G, t) = m(G, t) = M(G, 2)$ for $t \geq 2$ and setting $m(F) =$
$= m(F, 2), M(G) = M(G, 2), x_* = x_*(2), y_* = y_*(2)$ we find from (3.17) and the
equilibrium conditions (3.11) (along with the definitions (3.4) and (3.5)) that the
relations (3.12), (3.13) and (3.14) are satisfied. (ii) $\mu(2) > 0$. It follows from (3.6)
and (3.11) that $m(F, 2)/q_*(2) = y_*(2) = y_*(1)$. Applying (3.9), (3.10) and (3.17)
for $t = 2$ we obtain the relations

$$M(G, 3) = M(G, 2) + \theta\mu(2) \tag{3.18}$$

$$m(F, 3) = m(F, 2) + (1 - \theta)\mu(2). \tag{3.19}$$

We claim that $\mu(3) = 0$. Assume, on the contrary, that $\mu(3) > 0$. Then
$y_*(3) = y_*(2)(= y_*(1))$ and by (3.11) $m(F, 3) = q_*(3) y_*(1)$. It follows from
(3.19) that

$$q_*(3) = \frac{m(F, 2)}{y_*(1)} + \frac{(1 - \theta)\mu(2)}{y_*(1)} = q_*(2) + \frac{(1 - \theta)\mu(2)}{y_*(1)}. \tag{3.20}$$

Hence, by (3.18)

$$\mu(3) = s_*(3) y_*(1) - M(G, 3) = (n - q_*(3)) y_*(1) - M(G, 2) - \theta\mu(2) =$$

$$= (n - q_*(2)) y_*(1) - (1 - \theta)\mu(2) - M(G, 2) - \theta\mu(2)$$

$$= s_*(2) y_*(2) - m(G, 2) = 0$$

a contradiction. Hence $\mu(3) = 0$ and we are in case (i) (for $t = 3$ instead of for $t = 2$).
    There is a certain stabilization (in the sense that there is no inflation) also in geo-
metrically increasing population (Malthus' law).

Theorem 3.8: Let $n(t) = K^t$ for $t = 1, 2, \ldots$ where $K > 1$. Then $y_*(t) = y_*(1)$,
$q_*(t) = (m(F, t))/y_*(1) = (1 - \theta) K^{t-1}$, and price level is constant for $t = 2, 3, \ldots$

Proof: Set $M(F, 2) = a$. Then $m(F, 2) = (1 - \theta)a$ and $M(G, 2) = \theta a$. By (3.11) and
(3.8)

$$a = Ky_*(1). \tag{3.21}$$

The condition for equilibrium at $t = 2$ is

$$\frac{(1 - \theta)a}{q_*(2)} = \max\left(y_*(1), \frac{\theta a}{K^2 - q_*(2)}\right). \tag{3.22}$$

If $y_*(1)$ were less than $\theta a/(K^2 - q_*(2))$ then $q_*(2) = (1 - \theta) K^2$ and
$\theta a/(K^2 - q_*(2)) = a/K^2$. The hypothesis $K > 1$ and (3.21) would then imply that
$a/K^2 < y_*(1)$, a contradiction. Hence the equilibrium condition (3.22) really reads

$$q_*(2) = \frac{(1 - \theta)a}{y_*(1)} = (1 - \theta) K \tag{3.23}$$

and $y_*(2) = y_*(1)$. Continuing inductively we obtain the equations
$y_*(t) = y_*(1), M(F, t) = y_*(1) K^{t-1}$ and $q_*(t) = (1 - \theta) K^{t-1}$ for $t = 1, 2, \ldots$ In
particular (for these values of t) the price level is

$$\frac{M(F, t+1)}{q_*(t)} = \frac{y_*(1) K}{1 - \theta} \tag{3.24}$$

and is independent of t.

Up to now production did not play any role in the neo-Parkinsonian model. In fact,
the passage to a continuum of employment seekers eliminated the need of considering
real (as opposed to nominal) payoffs in the discussion of equilibrium conditions and
of the consequent motion of the economic system. Now we would like to turn our
attention to the question of efficiency and Pareto optimality of the resulting equilibria.
For any production function $f(p, s)$, maximum output is obtained (in our model) at
the maximum of $f(p, n - p)$. (This maximum is assumed at $p = (1 - a) n$ for the Cobb-
Douglas production function $f = p^{1-a} s^a$). Our equilibrium considerations involve only
salaries and money and we should not, in general, expect that at equilibrium output is
efficient. The question remains as to whether our equilibria are at least Pareto-optimal.
If they are not, even priviledged groups of workers (such as members of S) "should"
be better off if they choose to unroll their sleeves. (That they elect not to do so is an
instance of the prisoner's dilemma, as was the case in section 2). The following example
might serve to illustrate the point.

Example 3.9: Let $x_*(0) = 1/3, y_*(0) = 1, n(1) = 2, m(F, 1) = 1$ and $M(G, 1) = 1/2$.
A simple computation shows that at the equilibrium of $\Gamma[1, 1/2, 1/3, 1, 2]$
$q_* = p_* = s_* = 1, x_*(1) = 1, y_*(1) = 1$ and $\mu(1) = 1/2$. Let $f(p, s)$ denote the
production function. Then the price level is $2/f(1, 1)$ and the real wages at the equilibrium points are $x_*(1) = y_*(1) = f(1, 1)/2$. For all $0 < q \leqslant 2, p = q$
$(m(F, 1)/q \geqslant x_*(0))$ and $s = 2 - p, \mu(1) = \max(0, 3/2 - p)$ and the total amount
of money is equal to $3 - p$ if $p = q \leqslant 3/2$. The nominal salaries are $x(1, p) = 1/p$ and
$y(1, p) = 1$ if $p = q \leqslant 3/2$, the real salaries being

$$x(1, p) = \frac{f(p, 2-p)}{p(3-p)} \tag{3.25}$$

and

$$y(1, p) = \frac{f(p, 2-p)}{3-p} \tag{3.26}$$

Hence

$$\frac{dx(1)}{dp} = \frac{f_p - f_s}{p(3-p)} - \frac{f(3-2p)}{p^2(3-p)^2}. \tag{3.27}$$

This function may be positive for many choices of f. If we assume, for instance the
Cobb-Douglas form

$$f(p, s) = p^{1-a} s^a \tag{3.28}$$

for $0 \leqslant a < 1/16$, then an easy computation yields the estimate $2(f_p - f_s) \geqslant f$. Recalling that $p(3-p)/(3-2p) \geqslant 2$ for $p \geqslant 1$ with strict inequality if $p > 1$ we see from (3.27) that $x(1,p)$ (and by (3.26), (3.25) also $y(1,p)$) increases as a function of p, at least in the interval $[1, 3/2]$. In particular the equilibrium point $p = 1$ is not Pareto optimal.

Proposition 3.10: Let (Q, S) denote an N-pure strategy for the game $\Gamma[m(F, 1), M(G, 1), x_*(0), y_*(0), n(1)]$, and let

$$f(p, s) = p. \tag{3.29}$$

Let the strategy (Q', S') be Pareto-preferable to (Q, S). Then $q' > q$.

Proof: Assume on the contrary, that $q' < q$. Then $p' \leqslant p$ while $s' > s$ implies $y(s') \leqslant y(s)$ and $\mu(s') \geqslant \mu(s)$, where, by (3.6) and (3.7), we have a strict inequality in precisely one of the preceding inequalities. Hence the real salary of a government employee has to decrease, contrary to our assumption:

$$y(s') = \frac{y(s')p'}{m(F) + M(G) + \mu(s')} < \frac{y(s)p}{m(F) + M(G) + \mu(s)} = y(s). \tag{3.30}$$

Remark 3.11: If f is differentiable, then an argument similar to the one in example 3.9 shows that the conclusion of proposition 3.10 remains valid if the marginal output f_s of a services worker is less than his real salary $f \cdot y / [m(F) + M(G) + \mu(s)]$. Another proof (and a generalization) of Proposition 3.10 can be found in Proposition 4.2 and its proof.

Theorem 3.12: Let the quadruple $(m(F), M(G), x_*, y_*)$ be stationary and let f satisfy either $f = p$ or the condition stated in Remark 3.11. Then all equilibrium strategies (Q, S) for $\Gamma[m(F), M(G), x_*, y_*, n]$ are Pareto optimal.

Proof: Let (Q_*, S_*) be an equilibrium strategy. If (Q, S) is Pareto-preferable to (Q_*, S_*), then by Proposition 3.10 $q > q_*$. It follows from Theorem 3.6 (especially from the estimate (3.14)) that $p_* = m(F)/x_*$, and therefore $p = p_*$. Hence the expected real salary for a member of Q is less than the expected real salary for a member of Q_*, as price level is the same but the nominal salary is lower for a member of Q.

Remark 3.13: If $f = p$ then it can be shown that the equilibrium strategy (Q_*, S_*) of the game $\Gamma[m(F), M(G), x_*, y_*, n]$ is Pareto-optimal if and only if $q_* \geqslant \min(\bar{p}, n)$ where $\bar{p}$ is given by $\bar{p} = m(F)/x_*$ (compare (3.2)).

4. The Non-Parkinsonian Model

The model described in this section differs from the one described in the preceding section in the following aspects:

a) Every unemploed person gets (at each period) unemployment benefits z, paid for by the government.

b) The government is committed to the payment of unemployment benfits to all unemployed, and may have to print money for this purpose. The government is free to print more money if it desires to do so.

c) The government stands under no obligation to employ in its service everybody seeking government employment.

Formally, the data given at the beginning of the period t are the same as in the neo-Parkinsonian model, except that we are given an additional positive constant z (independent of t), and we assume that

$$z < \min (x_* (t - 1), y_* (t - 1)). \tag{4.1}$$

The government makes sure (at the beginning of the period) that it has enough money to pay unemployment benefits, even though the actual unemployment rate is not yet known. Hence the amount of money printed by the government must satisfy the inequality

$$\mu (t) \geqslant \max (o, n (t) z - M (G, t)) \tag{4.2}$$

(compare and contrast with (3.6)). After learning how much money the government has available (how big is $m (G, t)$) the members of N form the measurable partition (Q, R), where R denotes the set of persons seeking government employment, Q denoting as before the set of persons seeking productive employment. We denote by S and P the subsets of R and Q, respectively, consisting of those people who are actually employed, the selection being random and the probabilities of getting work being p/q and s/r, respectively (r denoting the Lebesgue measure of R). Employment rates and nominal wages (but not expected nominal wages) are determined in the production (firm) sector as in the neo-Parkinsonian model (see (3.2) through (3.4)). A member of Q has the probability $(q - p)/q$ of receiving the unemployment benefit. Hence the expected nominal income of a person seeking employment with the firm (a member of Q) is given by the following counterpart of (3.5):

$$\xi_. (q, t) = \frac{m (F, t) + (q - p) z}{q} = \frac{px (t) + (q - p) z}{q}. \tag{4.3}$$

Let $m_1 (G, t)$ denote the amount of money left for the government to finance its actions in the R sector, after meeting its obligations to the unemployed in Q,

$$m_1 (G, t) = m (G, t) - (q - p) z = M (G, t) + \mu (t) - (q - p) z. \tag{4.4}$$

The expected monetary income for members of R is

$$\eta\,(q,\,t) = \frac{m_1\,(G,\,t)}{r}. \tag{4.5}$$

The nominal wage rate for those people actually employed in the government service is given by

$$y\,(q,\,t) = \max\,(\eta\,(q,\,t),\,y_*\,(t-1)) \tag{4.6}$$

and the size of S is given by

$$s = \min\left(\frac{m_1\,(G,\,t)-rz}{y_*\,(t-1)-z},\,r\right). \tag{4.7}$$

In this manner a non-cooperative game with a continuum of participants $\Gamma\,[m\,(F,\,t),\,m\,(G,\,t),\,x_*\,(t-1),\,y_*\,(t-1),\,z,\,n\,(t)]$ is defined, and the expected payoff for individuals choosing the pure strategies Q and R are given by (4.3) and (4.5), respectively. The partition $(Q,\,R)$ is in equilibrium if

$$\xi\,(q,\,t) = \eta\,(q,\,t). \tag{4.8}$$

Note that (unlike the situation in the neo-Parkinsonian model) the total quantity of money in the economy $(m\,(F,\,t) + m\,(G,\,t))$ is independent of the strategy chosen by the members of N, but depends on government policy.

Existence and uniqueness of equilibria are adequately covered (for the non-trivial cases) by the following.

Proposition 4.1: If $m\,(F,\,t) > 0$ then there exists precisely one value of q solving (4.8), and the game $\Gamma\,[m\,(F,\,t),\,m\,(G,\,t),\,x_*\,(t-1),\,y_*\,(t-1),\,z,\,n\,(t)]$ has an equilibrium point. The equilibrium point is unique up to a Lebesgue measure preserving automosphism of $[0,\,n\,(t)]$.

Proof: Using (3.3) we rewrite (4.3) as

$$\xi\,(q,\,t) = \begin{cases} \dfrac{m\,(F,\,t)}{q} & \text{if } q \leqslant \bar{p} \\[2mm] \dfrac{m\,(F,\,t)-\bar{p}z}{q} + z & \text{if } q > \bar{p}. \end{cases} \tag{4.9}$$

The relations $m\,(F,\,t) > 0$ and $m\,(F,\,t) - \bar{p}z = \bar{p}\,(x_*\,(t-1)-z)$ imply that $\xi\,(q,\,t)$ is a strictly decreasing function of q tending to $+\infty$ as $q \to 0$. Similarly

$$\eta\,(q,\,t) = \begin{cases} \dfrac{m\,(G,\,t)}{n\,(t)-q} & \text{if } q \leqslant \bar{p} \\[2mm] \dfrac{m\,(G,\,t)-(q-\bar{p})\,z}{n\,(t)-q} & \text{if } q > \bar{p} \end{cases} \tag{4.10}$$

Hence $\eta\,(q,\,t)$ is a strictly increasing function of q tending to $+\infty$ as $q \to n\,(t)$.

Characterization of Pareto optimal equilibrium points is similar to (but easier than) the one given in section 3.

Proposition 4.2: If (Q', R') is Pareto-preferred to (Q, R) and $f_p \geqslant f_s$ then $q' > q$.

Proof: Assume, on the contrary, that $q' < q$. Then $p' \leqslant p$ and $s' \geqslant s$. Hence the total production output $f(p', s')$ is not larger than $f(p, s)$, implying that the average real income in (Q', R') is no larger than the one in (Q, R). Thus (Q', R') cannot be Pareto-preferred to (Q, R).

Theorem 4.3: If $f(p, s) = p$ then the equilibrium pair (Q_*, R_*) is pareto-optimal if and only if $q_* \geqslant \min(\bar{p}, n(t))$.

Proof: Necessity: If $q_* < \min(\bar{p}, n(t))$, choose q such that $q_* < q \leqslant \min(\bar{p}, n(t))$. Then $p = q$ and the real income of a member of $P = Q$ is

$$\chi(q, t) = \frac{m(F, t)}{q} \cdot \frac{m(F, t) + m(G, t)}{q} = \frac{m(F, t)}{m(F, t) + m(G, t)} = \chi(q^*, t) \quad (4.11)$$

whereas the expected real income for a member of R is higher than that of a member of R_* as the expected nominal wages are higher while price level is lower. Hence (Q_*, R_*) is not Pareto-optimal.

Sufficiency: If $q_* \geqslant n(t) \geqslant \bar{p}$ then (Q_*, R_*) is Pareto-optimal by Proposition 4.2. Hence we need only to consider the case $n(t) \geqslant q_* \geqslant \bar{p}$ where $n(t) > \bar{p}$. If (Q, R) is Pareto-preferred to (Q_*, R_*) then $q > q_*, p = p_* = \bar{p}$ and the expected real income for a member of Q (in (Q, R)), given by

$$\frac{\xi(q, t)}{[m(F, t) + m(G, t)]/p} = \frac{m(F, t) + (q - \bar{p})z}{m(F, t) + m(G, t)} \cdot \frac{\bar{p}}{q} \quad (4.12)$$

is strictly less than the expected real income for a member of Q_* (in (Q_*, R_*)), given by

$$\frac{\xi(q_*, t)}{[m(F, t) + m(G, t)]/p_*} = \frac{m(F, t) + (q_* - \bar{p})z}{m(F, t) + m(G, t)} \cdot \frac{\bar{p}}{q_*} \quad (4.13)$$

the inequality following from the fact

$$q_*[m(F, t) + (q - \bar{p})z] < q[m(F, t) + (q_* - \bar{p})z] \quad (4.14)$$

implied by the estimate (4.1) and the definition 3.2.

We will discuss from now on only stationary models (compare Definition 3.5). First we will define these models and analyze some properties of their equilibrium points. Afterwards we will elucidate the crucial role played by government actions.

Definition 4.4: The model is said to be stationary if $n(t) = n$ for all $t = 1, 2, \ldots$ The quadruple $(m(F), m(G), x_*, y_*)$ is said to be stationary if the game $\Gamma[m(F), m(G), x_*, y_*, z, n]$ (at $t = 1$) has an equilibrium point for which $m(F, 2) = m(F)$, $M(G, 2) = m(G), x_*(1) = x_*$ and $y_*(1) = y_*$.

Theorem 4.5: The quadruple $(m(F), m(G), x_*, y_*)$ is stationary if and only if

$$z < \min(x_*, y_*) \tag{4.15}$$

$$nz \leqslant m(G) \tag{4.16}$$

$$\frac{m(F)}{m(G)} = \frac{1-\theta}{\theta} \tag{4.17}$$

$$\frac{m(F) + m(G)}{n} \leqslant x_* \tag{4.18}$$

$$m(G) \leqslant (n - q_*) y_* + (q_* - \bar{p}) z \tag{4.19}$$

where

$$q_* = \frac{n(m(F) - \bar{p}z)}{m(F) + m(G) - nz} \tag{4.20}$$

and

$$\bar{p} = \frac{m(F)}{x_*}. \tag{4.21}$$

Proof: The relations (4.15) and (4.16) are necessary and sufficient for the game Γ to be well defined, and (4.17) is clearly necessary for stationarity. In equilibrium.
$x_* \geqslant \xi = \eta$.
Hence

$$m(F) + m(G) = q\xi + r\eta = n\xi \leqslant nx_* \tag{4.22}$$

and $y_* \geqslant \eta$ implies (4.19). We can have $x_*(1) = x_*$ if and only if $q_* \geqslant \bar{p}$; in that case (4.9) and (4.10) imply (4.20). It is easy to see that setting q_* to be equal to the right hand side of (4.20) and putting $r_* = n - q_*$ we obtain an equilibrium point with (4.17) guaranteeing the stationarity of the quantities of money available to the firm and to the government while (4.18) and (4.19) ensure that $x_*(1) = x_*$ and $y_*(1) = y_*$.

Theorem 4.6: If $f_p \geqslant f_s \geqslant 0$ then a stationary equilibrium point is Pareto-optimal.
Proof: Let (Q_*, R_*) be an equilibrium pure strategy for $\Gamma[m(F), m(G), x_*, y_*, z, n]$ where $(m(F), m(G), x_*, y_*)$ form a stationary quadruple. If (Q, R) is Pareto-preferred to (Q_*, R_*), then by Proposition 4.2 $q > q_*$. Hence expected nominal payoff $\xi(q, 1)$ for a member of Q is smaller than $\xi(q_*, 1)$. On the other hand, it was demonstrated in the proof of Theorem 4.5 that in a stationary equilibrium $q_* \geqslant \bar{p}$. Hence $p = p_* = \bar{p}$ and $s \leqslant s_*$ imply $f(p, s) \leqslant f(p_*, s_*)$. Hence the output is no bigger, (while the quantity of money is constant), and the real expected income for a member of Q is strictly less than that for a member of Q_*.

Remark 4.7: In the non-Parkinsonian model, the government can influence convergence properties. If the government so desires, it can bring the system to a stationary state after at most two periods; this can be achieved by the government printing a relatively large quantity of money μ (1) in the first period so that not only the inequality (4.2) shall be satisfied at $t = 1$, but that we shall also have the relation

$$\theta \; [m \; (F, \; 1) + M \; (G, \; 1) + \mu \; (1)] \geqslant nz \tag{4.23}$$

because then the government will not need to print any more money at periods $t \geqslant 2$, and the same equilibrium points will arise (by uniqueness) at each such period (compare Theorem 3.7). If, on the other hand, the government elects to print in every period t the minimum amount possible (i.e., if we have equality in (4.2)) and if μ (1) > 0, then μ $(t) > 0$ for all t and the system never reaches a stationary equilibrium. In fact, if μ $(t) > 0$ and M $(G, \; t)/m$ $(F, \; t) = \theta/(1 - \theta)$, then

$$M \; (G, \; t + 1) \; = \theta \; [m \; (F, \; t) + M \; (G, \; t) + \mu \; (t)] =$$

$$= \theta \; [(1 - \theta) \, M \; (F, \; t) + \theta M \; (F, \; t) + \mu \; (t)] = \theta \; [M \; (F, \; t) + \mu \; (t)]$$

$$= M \; (G, \; t) + \theta \mu \; (t) < M \; (G, \; t) + \mu \; (t) = nz. \tag{4.24}$$

Hence

$$\mu \; (t + 1) > 0.$$

It appears, therefore, that the really interesting problem is what policy should be pursued by the government, and what objective should the government try to achieve. Clearly the choice of objective is open to discussion. We will choose as a gauge for the quality of a government policy a function which involves all periods, not only asymptotic behavior as $t \rightarrow \infty$. We start with the following simply observation.

Proposition 4.8: If $M \; (G, \; t)/m \; (F, \; t) = \theta/(1 - \theta)$ then $p_* \; (t) \leqslant (1 - \theta) \, n$.

Proof: Observe that, at equilibrium,

$$x_* \; (t) \geqslant \xi \; (t) = [m \; (F, \; t) + M \; (G, \; t) + \mu \; (t)]/n \geqslant M \; (F, \; t)/n \tag{4.25}$$

and

$$p_* \; (t) \leqslant m \; (F, \; t)/x_* \; (t) = (1 - \theta) \, M \; (F, \; t)/x_* \; (t) \tag{4.26}$$

and the assertion follows by applying (4.25) to (4.26).

Remark 4.9: The value $p_* \; (t) = (1 - \theta) \, n$ can actually be achieved by the government choosing μ (1) so as to satisfy both (4.23) and the relation

$$x_* \; (0) \leqslant [m \; (F, \; 1) + M \; (G, \; 1) + \mu \; (1)]/n. \tag{4.27}$$

Afterwards to government can choose μ $(t) = 0$ for $t \geqslant 2$. Combining (4.25) with (4.27). we obtain the result.

Definition 4.10: A government policy is an infinite sequence $\{\mu\} = (\mu \; (1),$

μ (2), . . .) such that (4.2) is satisfied.

We assume from now on that $f(p, s) = p$. Motivated by Proposition (4.8), we choose the following criterion for the quality of a government policy.

Definition 4.11: A government policy $\{\mu\}$ is optimal if $d(\{\mu\}) \leqslant d(\{\lambda\})$ for all government policies $\{\lambda\}$, where

$$d(\{\mu\}) = \sum_{t=1}^{\infty} [(1 - \theta) - p_*(t)/n]. \tag{4.28}$$

Note that either $d(\{\mu\}) = +\infty$ or the series converges.

Theorem 4.12: There exists an optimal government policy.

Proof: Similar reasoning to the one described in Remark 4.9 shows the existence of positive constants $A(t)$, $t = 1, 2, \ldots$ such that for every government policy $\{\mu\}$ there exists a policy $\{\lambda\}$ such that $\lambda(t) \leqslant A(t)$ and $p_*(\lambda(t), t) \geqslant p_*(\mu(t), t)$ for all $t \geqslant 1$. Set

$$S = \{\{\mu\}: \mu \text{ is a government policy and } \mu(t) \leqslant A(t) \text{ for all } t \geqslant 1\} \tag{4.29}$$

and

$$d_* = \inf_{\{\mu\} \in S} d(\{\mu\}). \tag{4.30}$$

Then d_* is the infimum of $d(\{\mu\})$ where μ ranges over all governmental policies, and moreover d_* is finite. Let $\{\mu_k\} \in S$ be such that $d(\{\mu_k\}) \leqslant d_* + 1/k$ for $k = 1, 2, \ldots$ For each t the sequence $\{\mu_k(t)\}_{k=1}^{\infty}$ is bounded (by $A(t)$). Using a diagonal argument we may assume, without loss of generality, that $\mu_k(t)$ tends to a limit (as $k \to \infty$) for each t. Denoting this limit by $\mu(t)$, we deduce that for each $T = 1, 2, \ldots$,

$$\sum_{t=1}^{T} [(1 - \theta) - p_*(t)/n] = \lim_{k \to \infty} \sum_{t=1}^{T} [(1 - \theta) - p_{k_*}(t)/n] \leqslant \lim_{k \to \infty} d_* + 1/k = d_*. \tag{4.31}$$

Hence $d(\{\mu\}) \leqslant d_*$ and $\{\mu\}$ is optimal.

We want to study now how the optimal government policy $\{\mu_*\}$ depends on the initial state of the economy. We assume that

$$\frac{M(G, 1)}{m(F, 1)} = \frac{\theta}{1 - \theta}. \tag{4.32}$$

Then

$$m(F, T) = m(F, 1) + (1 - \theta) \sum_{t=1}^{T-1} \mu(t) \tag{4.33}$$

$$M(G, T) = M(G, 1) + \theta \sum_{t=1}^{T-1} \mu(t) \tag{4.34}$$

for all $T \geqslant 1$. If there is no unemployment at the equilibrium point at period t

(i.e., if $q_*(t) \leqslant \bar{p}(t)$) then the equilibrium condition simplifies to

$$\frac{m(G, t)}{n - q_*(t)} = \frac{m(F, t)}{q_*(t)}.$$ (4.35)

Hence

$$q_*(t) = p_*(t) = \frac{nm(F, t)}{m(G, t) + m(F, t)}.$$ (4.36)

Inserting (4.33) and (4.34) in (4.36) we obtain the relation

$$1 - \theta - \frac{p_*(T)}{n} = \frac{(1 - \theta)\mu(T)}{M(F, 1) + \sum\limits_{t=1}^{T} \mu(t)}$$ (4.37)

(where $M(F, 1) = m(F, 1) + M(G, 1)$). Hence

$$d(\{\mu\}) = \sum\limits_{T=1}^{\infty} \frac{(1 - \theta)\mu(T)}{M(F, 1) + \sum\limits_{t=1}^{T} \mu(t)}.$$ (4.38)

Example 4.13: Here we assume, besides (4.32), that

$$x_*(0) \leqslant \frac{m(F, 1) + nz}{n}$$ (4.39)

and

$$M(G, 1) \geqslant nz.$$ (4.40)

In this case $\mu(1) = 0$ is possible. The salary $x_*(0)$ is not bigger than the average income $\xi_*(1) = \eta_*(1)$. Hence there is no unemployment and (4.38) may be used, with the result that $\{\mu_*\} = (0, 0, \dots)$ is an optimal policy, and as a matter of fact is unique, since $\mu(t) \geqslant 0$ for all $\{\mu\}, t$.

Example 4.14: Here we assume, besides (4.32) and (4.39), that

$$M(G, 1) < nz.$$ (4.41)

Hence $\mu(1) > 0$. Once more we can use (4.38). In this case it turns out, however, that $\mu_*(T) > 0$ for all $T \geqslant 1$. We shall prove this by using sort of "Principle of Dynamic Programming." Assume, in fact, that $\mu_*(T + 1) = 0$ but $\mu_*(T) > 0$ for an optimal policy $\{\mu_*\}$. Then $M(G, T + 1) \geqslant nz$ and we get from (4.34) and (4.2) that

$$\sum\limits_{t=1}^{T} \mu_*(t) \geqslant \frac{nz - M(G, 1)}{\theta}.$$ (4.42)

On the other hand $\mu_* (T) > 0$ implies that

$$M (G, T-1) < nz \tag{4.43}$$

(otherwise we would have been in the situation of example 4.13 with $\mu_* (t) = 0$ for $t \geq T$ being the only optimal possibility). Consider now the following minimization problem:

$$\text{Minimize} \ \frac{\mu (T)}{M (F, T) + \mu (T)} + \frac{\mu (T+1)}{M (F, T) + \mu (T) + \mu (T+1)} \tag{4.44}$$

subject to the conditions

$$\mu (T) \geq nz - M (G, T-1) \tag{4.45}$$

$$\mu (T+1) \geq \max \left(0, \frac{nz - M (G, T-1) - \theta \mu (T)}{\theta} \right) \tag{4.46}$$

Denoting the objective function (to be minimized in (4.44)) by $\phi (\mu (T), \mu (T+1))$, and observing that ϕ is a strictly increasing function of $\mu (T+1)$, we infer that at the minimum point equality holds in (4.46). If $\mu (T+1) = 0$ then

$$\mu (T) \geq \frac{nz - M (G, T-1)}{\theta} > nz - M (G, T-1) \tag{4.47}$$

and the function (of a single variable) $\phi (\mu (T)) = \phi (\mu (T), 0)$ has a minimum at a point interior to the domain determined by (4.45). Hence $\phi' (\mu (T)) = 0$ at the minimum. But

$$\phi' (\mu (T)) = \frac{\partial}{\partial \mu (T)} \phi (\mu (T), 0) = \frac{M (F, T)}{[M (F, T) + \mu (T)]^2} \tag{4.48}$$

a contradiction. Hence $\mu (T+1) > 0$ at the minimum. Denoting this minimum of ϕ by $\nu (T), \nu (T+1)$, we claim that the policy $(\mu_* (1), \ldots, \mu_* (T-1), \nu (T), \nu (T+1), 0, \ldots)$ is feasible and yields a smaller value of d than $d (\{\mu_*\})$, a contradiction. Hence $\mu_* (T+1) > 0$ if $\mu_* (T) > 0$, and we obtain inductively the result $\mu_* (T) > 0$ for all T.

Example 4.15: Here we assume (4.32) and (4.40), but unlike example 4.13 we assume that

$$x_* (0) > \frac{m (F, 1) + nz}{n}. \tag{4.49}$$

Here there is bound to be unemployment of a fixed size in the productive sector at the first period $t = 1$, and no government action can decrease $(1 - \theta) - p_* (1)/n$. The government can guarantee, however, that $p_* (t) = (1 - \theta) n$ for all $t \geq 2$ by choosing $\mu (1) = \max (0, nx_* (0) - m (F, 1) - M (G, 1)), \mu (t) = 0$ for $t \geq 2$, and this clearly is the optimal policy.

5. Discussion and Comments on the Model

In this section we collect a number of mostly non-mathematical comments and remarks, having varying degrees of importance, on the models presented in this paper.

1. A number of people might be annoyed by assumptions such as (2.2). It is certainly an oversimplification but we feel there is a certain grain of truth in it, especially if one treat essentially short term phenomena (Definition 4.11 notwithstanding) and one notes that public goods are neither consumed nor payed for in the usual way. We hope to consider in the future more realistic models. One possibility could be to introduce time-delays in the effects of the work performed by public servants (such as, say, teachers), and to make these effects appear additively (rather than multiplicatively) in f. (These ideas were suggested to the authors by W. Neuefeind).

2. We called the model presented in section 3 a neo-Parkinsonian model because the government has to employ every person seeking a civil service career. It is not a Parkinsonian model because the government does not finance its operations (?) by higher and higher direct taxes, but rather by "taxation by inflation". This was done mainly for reasons of technical convenience; we do not think it is essential. The assumption that the tax rate θ is an absolute constant is justified, we feel, on the short run. While the government has no discretion, its actions are nevertheless sometimes useful in pumping money into the economy and enabling the firm to hire more personnel.

3. We ignore any saving or borrowing by either the government, the firm or the public. This is not unreasonable as long as a short term model is concerned. Moreover, we did ignore the role of capital in production. Hence the firm has no incentive to accumulate capital.

4. The rigidity from below of nominal wages is a standard assumption in Macroeconomics. (Note that nominal wages are denoted by x, y and real wages are denoted by x, y).

5. The firm employs as many workers as possible since it desires to maximize its real (not nominal) profits; those increase linearly as a function of p. The real profit is given by $p (1 - (m (F))/(m (F) + m (G)))$, and is maximized subject to the condition $px_* \leqslant m (F)$.

6. If we were to use a continuous model for the static model (of section 2) we would get that at equilibrium $p_* = 0$, $s_* = n$. While this result is "the limit" of the boundedness result (as $n \to \infty$) obtained for p_* in section 2, it is in fact weaker ($p_* (n)/n$ can tend to 0 without $p_* (n)$ being bounded).

7. In the continuous model, the inequalities usually associated with Nash equilibrium points, give way to equations such as (3.11) or (4.8). These equations admit a natural interpretation along classical economic lines of supply and demand considerations for the various sectors. (This was brought to the attention of the authors by C.C. von Weizsäcker.) We wish to add that it was observed recently by several authors in different contexts [Dubey/Shubik; Postlewaite/Schmeidler; Wilson], that the non-cooperative non-optimal Nash-equilibria of games formed in an economic framework tend to price-equilibria (and become more efficient) as the number of participants grows. We prefer to emphasize the Nash-equilibrium nature of our models, as it is this concept which interests us most in the finite case.

8. Our models are highly unrealistic in the assumption that employment depends only on choice by individuals and random choices by the employers, skills, education and seniority playing no role. One can also question the assumption that people are employed for one period only and have to look for a job at every period, and that their previous jobs are irrelevant. One could probably modify the models so as to allow for job participants (newcomers?) have to seek jobs. It was also pointed out by C.C. von Weizsäcker that the productivity of senior workers (of whom it is more difficult to get rid) is lower than that of junior ones, and that unions resist technological innovations, which, if implemented everywhere, would not increase unemployment, and which would certainly increase output. (Another instance of the Prisoners' Dilemma).

9. Note that unemployment can exist even in equilibrium. Thus, in Theorem 3.6 stationary quadruples are characterized. If we have strict inequality in (3.14) (i.e., if production workers insist that the wage rate of employed production workers is strictly larger than that of government employees) then we must have unemployment (the firm simply does not have enough money to pay wages to all members of Q). Otherwise there is full employment (in the stationary case).

10. A-priori, one could introduce a discount factor in (4.28) (Definition 4.11 of an optimal government policy in the stationary non-Parkinsonian model). Theorem 4.12 justifies a-posteriori our leaving out any discount factor, and shows that one need not use other criteria (such as an overtaking principle).

11. Example 4.14 shows that in some cases it is better to perform expansionistic inflationary policies on a piecemeal basis, rather than to perform a painful operation in one period so that there would be no inflation later.

To sum up, it should be evident that we regard this paper as a preliminary draft intended to stimulate discussion. While many assumptions are clearly unrealistic and while there are several oversimplifications, we hope we have focussed on certain aspects of economic life which have not yet received enough attention in the mathematical – analytical economic literature.

Acknowledgements

We are very much indebted to Prof. C.C. von Weizsäcker and to Dr. W. Neuefeind for very interesting discussions, and to Mr. G. Tillmann for carefully reading the manuscript and correcting several errors in it.

References

Dubey, P., and *M. Shubik*: A closed economic system with production and exchange modelled as a game of strategy. Cowles foundation discussion paper 429, Yale University, 1976.

Parkinson Northcote, C.: Parkinson's Law. London 1958.

–: The Law and the Profits. London 1960.

Postlewaite, A., and *D. Schmeidler*: Approximate efficiency of non-Walrasian Nash-equilibria. To appear 1976.

Schmeidler, D.: Equilibrium points of a nonatomic game. J. Stat. Physics 7, 1973, 295–300.

Wilson, R.: A competitive model of exchange, Institute for Mathematical Studies in the Social Sciences. The economics series, Technical report 221, Stanford University, 1976.

Applied Game Theory. 1979 © Physica-Verlag, Wuerzburg/Germany

On Entry Preventing Behavior and Limit Price Models of Entry[1])

By *J. W. Friedman*, Rochester[2])

Abstract: Formally, the market is regarded as a supergame in which one established firm and one potential entrant are players. Both players know all relevant demand and cost functions, and throughout the paper, noncooperative behavior is assumed. The model has two distinct stages: pre- and post entry. In the pre-entry stage, the monopolist chooses his price and capital stock so as to maximize his discounted profits, noting that his investment decision may affect the entry plans of the entrant. Existence of equilibrium is proved, entry preventing behavior is characterized and conditions are shown under which it will be employed.

1. Introduction

Probably the first "limit price" discussions of entry occur in *Bain* [1949] and *Harrod* [1952]. To see what is meant by a limit price, suppose a monopolistic industry into which there is free entry. A price, p^*, is a limit price if no new firm would enter the industry when the monopolist charges p^* or less, and if entry would occur when a price above p^* were used. In deciding what price to adopt, the monopolist compares three per period profit rates, π^m, π^* and π^a. They are, respectively, the highest profit the firm could get as a monopolist (before entry of a new firm), the largest profit attainable when the monopolist keeps his price at or below p^* and the profit which the monopolist believes it would have if entry occurred. If these profit levels are ordered $\pi^m > \pi^* > \pi^a$ then the firm may maximize discounted profits by selecting an entry preventing price and obtaining a profit of π^* per period. Letting $\alpha = 1 / (1 + r)$, where r (≥ 0) is the firm's discount rate, and assuming entry would take place in period t if entry preventing prices were not chosen earlier, then entry prevention is optimal if

$$\frac{1}{1-\alpha} \pi^* > \frac{1-\alpha^t}{1-\alpha} \pi^m + \frac{\alpha^t}{1-\alpha} \pi^a. \tag{1}$$

This model prompts a number of questions. For example, how is p^* determined? It ought to arise as the result of an optimization process undertaken by the potential

---

[1]) I am grateful to Professor Vernon Smith for the opportunity to present an earlier version of this work at the Arizona Conference on Experimental Economics in March, 1977, and to Mordecai Kurz for a suggestion leading to an improvement in exposition and in the proof of theorem 1. I retain responsibility for any remaining errors.

[2]) Prof. *James W. Friedman*, The University of Rochester, College of Arts and Science. Dept of Economics, Rochester, N.Y. 14627, USA.

entrant[3]). How do results change if the market is characterized by oligopoly or by monopolistic competition? How is entry affected if knowledge of future demand and/ or cost conditions is incomplete? Or subject to stochastic elements? I hope to deal with all of these questions in the present paper and in subsequent papers. In the present paper, the assumption is maintained throughout that, prior to the entry of new firms, the market is characterized by monopoly. Also, there are no stochastic elements in the model, and both the established firm and the entrant know the demand and cost functions which prevail before and after entry.

The notion of an entry preventing price has wide appeal; yet in the models examined below an entrant has no direct interest in an established firm's pre-entry price policy. What matters to him is the price pattern which would emerge after he were to come in. That is, he wants to know what equilibrium behavior in the market would be if he were in, and he wants to know what profits he would have under such an equilibrium.

Now, after entry occurs, the established firm and the entrant are in a two person game whose structure and form are entirely independent of the pre-entry price policy of the established firm; hence, whose equilibria are independent of the pre-entry price policy of the established firm. If both participants are fully informed at the outset concerning the profit functions which would prevail after entry, then it is difficult to see the relevance of pre-entry prices to the plans of the entrant.

Though pre-entry price policy has no role in the models examined here, the pre-entry capital policy may be relevant. Imagine that active firms make both price and investment decisions and consider, again, that both firms know in advance what profit functions would prevail after entry. Any equilibria for the model as of the entry time would depend upon the amount of capital held by the established firm just prior to entry. Presumably, before making a commitment to enter, the entrant would check on the size of the established firm's capital, and refrain from entering if the equilibria associated with that capital afforded him negative profits. Thus the established firm may be able to carry out an entry preventing investment policy. The crucial difference between price and investment is, of course, that today's investment decision makes an inevitable change in the future circumstances one will face; but today's price decisions only affect today. In particular, under the assumptions made below, the optimal price for a firm is inversely related to the size of its own capital stock; hence, by choosing a large capital stock, it may force its optimal price to be low. This, in turn, tends to lower the demand of a rival firm.

The model described in section 2, is one in which there is a monopolist and one potential entrant. As long as the potential entrant is not in the market, he uses a criterion, known to the monopolist, upon which to decide whether or not to enter. Once he has come into the market, the model becomes one of duopoly with no possibility of entry by additional firms. Either or both of the firms may leave the market; however, once a firm leaves, he may not enter again. In section 3, the model of section 2 is restricted somewhat, which allows some additional results, and section 4 has a few concluding comments.

---

[3]) To avoid clumsy locutions, I refer to the potential entrant as the entrant, even if he has not decided to come into the industry.

2. Entry and Exit Under Certainty

Imagine a monopolist who chooses his price and the size of his capital stock for each period in a discrete time model. He knows the cost and demand functions which he faces as a monopolist; and he knows what his demand function would be if an entrant came into his market. He also knows the demand and cost functions which the entrant would have. There is one potential entrant whose information is like that of the monopolist. He knows the cost and demand functions he would have if he entered the industry, and he knows the cost and demand functions prevailing for the monopolist both before and after entry. The senses in which the model is characterized by certainty are: a) the information conditions above and b) no stochastic terms enter the cost or demand functions. Throughout the paper, it is assumed that the firms never collude or cooperate in any way.

The model has two distinct stages. These correspond to pre- and post entry. In the pre-entry stage, the monopolist chooses his price and capital stock so as to maximize his discounted profits, taking into account that his investment decision may have a bearing on the plans of the entrant. In the post entry stage, the two firms are in a two person game, with the extra provision that one or both of the firms may leave the market at any time. The entrant will come into the industry if there is any time at which entry would bring positive profits. The time of entry is the one for which discounted profits are maximized. The model which is outlined above is presented in detail in Section 2.1. Following that, in section 2.2, existence and characteristics of equilibrium are discussed.

2.1 A Model of Monopoly with Entry and Exit

The monopolist faces a demand function, $q_{1t} = f(p_{1t})$. p_{1t} is the monopolist's price in period t and q_{1t}, his output and sales level. The entrant's price and output are denoted p_{2t} and q_{2t}. After entry his demand function would be $f_2(p_{1t}, p_{2t}) = f_2(p_t)$ and the (former) monopolist would face $f_1(p_{1t}, p_{2t})$. Both of them know f_1 and f_2. The monopolist is sometimes called "firm 1" and the entrant, "firm 2."

Let $\overset{\circ}{A}_i$ denote the subset of the price space within which f_i is positive[4]):

$$\overset{\circ}{A}_i = \{p \mid p \geqslant 0,\ f_i(p) > 0\}, \quad i = 1, 2. \tag{2}$$

Let $\overset{\circ}{A} = \overset{\circ}{A}_1 \cap \overset{\circ}{A}_2$ and let A be the closure of $\overset{\circ}{A}$ and A_i the closure of $\overset{\circ}{A}_i$. R_+^2 denotes the non-negative orthant of the two dimensional Euclidean space, R^2. The conditions imposed upon the demand functions are given in A1 and A2.

A1. $f_i(p)$ is continuous, bounded and non-negative for all $p \in R_+^2$. For $p \in \overset{\circ}{A}_i$, f_i has continuous second partial derivatives with respet to p_i. For $p \in \overset{\circ}{A}$, f_i is twice continuously differentiable. Where the indicated derivatives exist, the following hold:

---

[4]) The notational conventions for vector inequalities are: $p^0 \geqslant p^1$ means $p_i^0 \geqslant p_i^1$, $i = 1, \ldots, n$. $p^0 > p^1$ means $p^0 \geqslant p^1$ and $p^0 \neq p^1$, and, finally, $p^0 \gg p^1$ means $p_i^0 > p_i^1$, $i = 1, \ldots, n$.

$$f_i^i < 0, f_i^j > 0 \text{ (for } i \neq j), \text{ and } f_i^1 + f_i^2 < 0. \quad i, j = 1, 2. \tag{3}$$

Furthermore, A is bounded.

Under A1, it is known that A contains a unique maximal element, $p^+ = (p_1^+, p_2^+)$. [See *Friedman*, 1977, theorem 3.1]. A1 insures that a firm's sales decline as its own price rises and rise as its rival's price rises. As both prices rise together, both firms suffer a decline in sales. There is a price (p_i^+) for each firm above which nothing is sold, no matter what price is chosen by the rival. It remains to impose a consistency require-ment on f_1 which recognizes that if $p_2 \geq p_2^+$, then firm 1 must necessarily face the same demand function which it faced as a monopolist.

A2. *If $p_2 \geq p_2^+$, then f_1 $(p_1, p_2) = f(p_1)$ for all $p_1 \geq 0$.*

A2 implies that $f(p_1^+) = 0$ and, for any $p_1 < p_1^+, f(p_1) > 0$.

The cost of production in period t is given by $C_i(q_{it}, K_{i,t-1})$ where $K_{i,t-1}$ is the capi-tal of the firm as of the start of the period. Gross investment is denoted I_{it}; hence, total costs of all types for period t are $C_i(q_{it}, K_{i,t-1}) + I_{it}$. The capital stock available at the start of period $t + 1$ is $K_{it} = (1 - \delta) K_{i,t-1} + I_{it}$. The capital stock available at the start of period $t + 1$ is $K_{it} = (1 - \delta) K_{i,t-1} + I_{it}$, where δ is a depreciation rate between zero and one. The initial capital, K_{i0}, equals initial investment, I_{i0}. The full statement of cost assumptions is in A3 and A4.

A3. *The cost of production, $C_i(q_{it}, K_{i,t-1})$, is defined, non-negative and convex on R_+^2, and is twice continuously differentiable on the interior of R_+^2. In addition,*
a) $C_i(0, 0) = 0$, b) $\lim\limits_{q \to 0} C_i^2(q, K) > 0, \text{ when } K > 0,$ c) $C_i^1(q, K) > 0,$
d) $C_i^1(q, K) \to \infty \text{ as } K \to 0,$ e) $C_i^{12}(q, K) < 0, \text{ and f) for } 0 < K_2 < K_1 < \infty \text{ there is }$
$q = \phi_i(K_1, K_2) \text{ such that } C_i(\phi_i(K_1, K_2), K_1) = C_i(\phi_i(K_1, K_2), K_2). \quad i = 1.2.$

A4. $K_{i0} = I_{i0}$. *For $t \geq 1, K_{it} = (1 - \delta) K_{i,t-1} + I_{it}. \delta \in [0, 1]. I_{it} \geq 0$ for $t = 1, 2, \ldots$, and $i = 1, 2. I_{20} \geq I_2' > 0$ [5]).*

The cost function specified in A4 shows positive marginal cost of production (from c)) and increasing marginal cost as output increases, for given capital (from convexity of C_i). At zero output and zero capital, production cost is nil; however, any positive amount of capital has associated with it a level of fixed cost, $C_i(0, K)$, which rises as K increases (from b)). Marginal cost of production goes to infinity as the capital stock of the firm goes to zero (from d)). Condition e) insures that, for a fixed level of output, the marginal cost of production declines as capital increases. Condition f) guarantees that, for any pair of positive capital stock levels, there is an output level above which the total production costs associated with the larger capital stock are less than those associated with the smaller.

For period $t \geq 1$, the firm's profit is

$$\pi_i(p_t, K_{i,t-1}, K_{it}) = p_{it} f_i(p_t) - C_i(f_i(p_i), K_{i,t-1}) - (K_{it} - (1 - \delta) K_{i,t-1}). \tag{4}$$

Let $\alpha_i \in (0, 1)$ be the discount parameter of firm i. The discounted profit stream for a firm, if both are in the market from the outset, is

---

[5]) Of course, if firm 2 does not enter the market until t_0, his investment is zero in every period up to that time.

$$-I_{i0} + \sum_{t=1}^{\infty} \alpha_i^t \, \pi_i \, (p_t, K_{i,t-1}, K_{it}). \tag{5}$$

There is a final assumption made on the π_i.

AS. *For any p and K_i such that $p_i \geqslant C_i^1 \, (f_i \, (p), K_i)$, then*
$2 f_i^i + (p_i - C_i^1 \, (f_i \, (p), K_i)) f_i^{ii} < 0, \quad i = 1, 2.$

Together with convexity of C_i, A5 insures that the functions π_i are concave with respect to (p_i, K_i) in the region of A where the firm's price is no less than its marginal cost.

2.2 Existence of an Entry-Exit Equilibrium

Entry and exit are each treated separately before considering a model in which both are possible. Section 2.2.1 deals with a model in which only entry can occur, in section 2.2.2, an exit model is considered and section 2.2.3 takes up a model in which both entry and exit are possible.

2.2.1 A Model With Entry and Without Exit

First, it is necessary to be precise about the timing of decisions. In time zero, the established firm chooses only K_{10}, his initial capital stock, which is also his initial investment, I_{10}, and is the capital *in place* for period 1. It is in period 1 that he can first produce and sell. Thus, from period 1 onward, he chooses both a price and a capital stock. For the entrant, his decision to enter cannot be made earlier than period 1. If he decides to enter in period $t_0 \geqslant 1$, this means a) that he chooses only an investment level in period t_0, I_{2, t_0}, which is also his initial capital K_{2, t_0} and which is his capital in place in period $t_0 + 1$; b) his *announcement* that he will enter in period t_0 is assumed to be made *before* firm 1 makes his period t_0 choices; and c) firm 2 cannot sell or make price choices until period $t_0 + 1$.

The strategy of the established firm may be thought of as having two major parts. The first consists of a sequence of capital and price decisions which it follows as long as it is alone in the market. The second consists of the capital and price decisions which it follows from the moment of entry of the second firm. The latter necessarily depends upon the capital the estiablished firm possesses when the entrant announces his intention to come in. For the entrant, there is the choice of whether to enter the market. If a date of entry is chosen, then there must be a sequence of capital and price decisions.

The first step of analysis is to look at the subgame which begins at the moment when firm 2 commits himself to enter the market. Most of the present section is devoted to showing that this subgame has a non-cooperative equilibrium. Due to the stationary character of the model, the announcement of entry may be assumed at $t = 1$. Initial capital for the established firm is $K_{10} = I_{10}$. The entrant invests first in period 1 and has sales first in period 2. Thus the two discounted payoff streams for the time period 1 onward are

$$-I_{11} + A* + \sum_{t=2}^{\infty} \alpha_1^{t-1} \pi_1 (p_t, K_{1,t-1}, K_{1t}) \tag{6}$$

$$-I_{21} + \sum_{t=2}^{\infty} \alpha_2^{t-1} \pi_2 (p_t, K_{2,t-1}, K_{2t}) \tag{7}$$

where $A*$ accounts for the effects of period 0 investment and period 1 sales.

For firm 1 a *strategy* may be written $\sigma_1 = (p_{11}, K_{11}, p_{12}, K_{12}, \dots)$ and for firm 2, $\sigma_2 = (K_{21}, p_{22}, K_{22}, p_{23}, K_{23}, \dots)$. A strategy pair is written $\sigma_1 = (\sigma_1, \sigma_2)$. These strategies are *simple* in the sense that the actions taken in any given time period by a firm are not contingent in any way upon the decisions prior to that time. An equilibrium in simple strategies is also an equilibrium in a game in which the strategy sets are enlarged. The payoff stream given by equation (6) may be denoted $G_1 (\sigma)$ and that for equation (7) by $G_2 (\sigma)$.

Let the strategy set for player i, from which σ_i is drawn, be denoted by S_i. S_i is the set of all simple strategies for firm i and is defined precisely below. A non-cooperative equilibrium is defined by: $(\sigma_1^*, \sigma_2^*) = \sigma*$ is a non-cooperative equilibrium if

$$\sigma_i^* \in S_i, \quad i = 1, 2 \tag{8}$$

$$G_i (\sigma*) = \max_{\sigma_i \in S_i} G_i (\sigma_i, \sigma_j^*), \quad i \neq j, \quad i = 1, 2. \tag{9}$$

That the post-entry subgame has a non-cooperative equilibrium is proved below in theorem 1. Prior to theorem 1, the notion of *best reply* is defined: σ_2' is the best reply to σ_1' for firm 2 if

$$G_2 (\sigma_1', \sigma_2') = \max_{\sigma_2} G_2 (\sigma_1', \sigma_2). \tag{10}$$

A non-cooperative equilibrium is also defined by $\sigma* = (\sigma_1^*, \sigma_2^*)$ such that σ_1^* is a best reply to σ_2^* for firm 1 and σ_2^* is a best reply to σ_1^* for firm 2.

Theorem 1: The no entry, no exit two person subgame satisfying A1 − A5, with discounted profits given by equations (6) and (7), and with firm 1 having an initial capital of K_{10} has a non-cooperative equilibrium.

Proof: The steps of the proof are: a) show that the strategy sets of the firms are convex and compact, b) show that the best reply mapping of each firm is defined, single valued and continuous for each strategy of the other, and c) show that a) and b) imply the existence of a non-cooperative equilibrium.

a) Note first that $p_{1t} \in [0, p_1^+]$ and $K_{1t} \in [0, K_1^+]$ where it must now be assured that K_1^+ is finite. Because revenue is bounded, profit per period is bounded by π_2^+, a finite upper bound on revenue per period. Non-negative profit is impossible if I_{1t} exceeds $\pi_1^+/(1 - \alpha_1)$; hence, K_1^+ may be taken as $\pi^+/((1 - \alpha_1)(1 - \delta))$. Let $K_1 = (\{K_{10}, K_{11}, K_{12}, \dots\}), 0 \leqslant K_{10} \leqslant K_1^+$ and $(1 - \delta) K_{1,t-1} \leqslant K_{1t} \leqslant K_1^+,$

$t = \{1, 2, \ldots\}$, $P_1 = \underset{t=1}{\overset{\infty}{\text{X}}} [0, p_1^+]$ and let $S_1 = K_1 \times P_1$. S_1 is the strategy space for firm 1. Both K_1 and P_1 are convex; hence S_1 is convex. Compactness of S_1 may be seen from the following norm for $\sigma_1 \in S_1$.

$$\| \sigma_1 \| = \sum_{t=1}^{\infty} \alpha_1^{t-1} (p_{1t} + K_{1t}) \tag{11}$$

The same argument may be used for S_2, defined in a parallel fashion to S_1.

b) That the best reply mapping of a firm is defined for any strategy of the other firm is immediate because G_i is a continuous function of σ_i and σ_i has a compact set as its domain; hence, G_i takes a maximum. Single valuedness and continuity of one firm's best reply with respect to the strategy of the other remains to be proved. These follow from the concavity conditions imposed in A1 – A5. Whenever the optimal price is so high that demand is zero, the best reply is taken to be the lowest price at which demand is zero. Letting a_1 denote the points in R_+^2 which are in A_1, but are not in $\mathring{A}_1$, the best reply price of firm 1 in period t is given by $\pi_{1t}^1 = 0$, if any price satisfies this condition, and by p_{1t} such that $p_t \in a_1$, if no price satisfies $\pi_{1t}^1 = 0$ [6]). This best reply price depends only on p_{2t} and $K_{1,t-1}$ and may be denoted

$$p_{1t} = \psi_1 (K_{1, t-1}, p_{2t}). \tag{12}$$

$$p_{2t} = \psi_2 (K_{2,t-1}, p_{1t}) \tag{13}$$

may be analogously defined. Note that

$$\frac{\partial \psi_1}{\partial K_{1,t-1}} = \psi_1^1 = \frac{-\pi_1^{13}}{\pi_1^{11}} \text{ if } f_1 > 0$$

$$= 0 \qquad \text{otherwise} \tag{14}$$

If $f_1 > 0$, then $\psi_1^1 < 0$, which means that the larger is the firm's capital, the lower is its optimal price, due to the effect which the larger capital has on lowering marginal cost.

With ψ_1 substituted for p_{1t}, $\pi_1 (p_t, K_{1,t-1}, K_{1t}) = \pi_1 (\psi_1, p_{2t}, K_{1,t-1}, K_{1t})$,

$$\frac{\partial \pi_1}{\partial K_{1,t-1}} = \pi_1^1 \psi_1^1 + \pi_1^3 \tag{15}$$

and

$$\frac{\partial^2 \pi_1}{\partial K_{1,t-1}^2} = \psi_1^1 (\pi_1^{11} \psi_1^1 + \pi_1^{13}) + \pi_1^1 \psi_1^{11} + \pi_1^{31} \psi_1^1 + \pi_1^{33}. \tag{16}$$

---

[6]) Superscripts are used to denote partial derivatives. The "t" subscript denotes the time period to which the demand and cost functions pertain.

The term $\pi_1^1 \, \psi_1^{11} = 0$; for, if $f_1 > 0$, then $\pi_1^1 = 0$ and if $f_1 = 0$, then $\psi_1^{11} = 0$. Equation (16) is clearly strictly negative. Again, if $f_1 = 0$, then $\pi_1^1 = 0$, leaving only π_1^{33}, which is negative. Otherwise, recalling that $\psi_1^1 = -\pi_1^{13}/\pi_1^{11}$, equation (16) reduces to

$$\pi_1^{11} \, (\psi_1^1)^2 + 2\, \pi_1^{13} \, \psi_1^1 + \pi_1^{33} = \frac{\pi_1^{11} \, \pi_1^{33} - (\pi_1^{13})^2}{\pi_1^{11}} < 0. \tag{17}$$

Because the only term of G_1 into which $K_{1,t-1}$ enters non-linearly is $\pi_1 \, (p_t, K_{1,t-1}, K_{1t})$, the strict concavity of π_1 in $K_{1,t-1}$ implies strict concavity of G_1 in $(K_{10}, K_{11}, K_{12}, \ldots)$.

c) Let the best reply functions of the two firms be denoted $\sigma_1 = r_1 \, (\sigma_1)$ and $\sigma_2 = r_2 \, (\sigma_1)$. The two functions, taken together, are a function from $S = S_1 \times S_2$ into S and may be written

$$\sigma' = r\,(\sigma) = (r_1 \, (\sigma_2), r_2 \, (\sigma_1)). \tag{18}$$

Any fixed point of r is a non-cooperative equilibrium and any non-cooperative equilibrium is a fixed pint of r. Thus, if r has a fixed point, the subgame has an equilibrium. By Brouwer's fixed point theorem, the fixed point exists and the theorem is proved.   ☐

The equilibrium of theorem 1 is not, in general, unique; however, in section 3 additional conditions are given under which it is. The lack of uniqueness is an inconvenience which is circumvented by making an arbitrary selection. For each possible value of $K_1 \in [0, K_1^+]$, the capital held by the established firm at the time of entry, choose an arbitrary non-cooperative equilibrium and denote the i-th firm's discounted payoff under the chosen equilibrium, discounted to the announcement period, by $F_i \, (K_1)$.

It is now possible to characterize the entry game in which, if the entrant decides in period t to come in, his payoff for the whole game, discounted to period t is $F_2 \, (K_{1,t-1})$. The payoff to the established firm, given entry in period t, and discounted to period 0 is

$$-I_{10} + \sum_{\tau=1}^{t-1} \alpha_1^\tau \, \pi_1 \, (p_{1\tau}, p_2^+, K_{1,\tau-1}, K_{1\tau}) + \alpha_1^t F_1 \, (K_{1,t-1}). \tag{19}$$

The strategy of firm 1 has been partly specified already — for that part of the game which commences from the entry announcement of the entrant. Firm 1 must also specify a sequence $\{p_{1t}, K_{1,t-1}\}_{t=1}^\infty$ which it follows as long as it is alone in the market. To see how the entrant behaves in equilibrium, imagine an arbitrary price-capital sequence for firm 1 which it follows prior to entry. Clearly, if there is at least one t such that $F_2 \, (K_{1t}) > 0$, then the entrant will come in at some time; however, it need not be in the entrant's best interest to come in at the very first period for which F_2 is positive. Consider the situation as of period 0, and look at the sequence $\{\alpha_2^t F_2 \, (K_{1t})\}$. Firm 2 enters at time t' if the discounted payoff at that time is positive and if further waiting will not increase his discounted payoff. For any given strategy of the established firm, there is a best reply for the entrant, characterized below:

Lemma 1: The entrant comes into the market at time t' if a) $F_2 (K_{1t'}) > 0$,
b) $\alpha_2^{t'} F_1 (K_{1t'}) \geq \alpha_2^t F_2 (K_{1t})$ for all $t > t'$ and c) $\alpha_2^{t'} F_2 (K_{1t'}) > \alpha_2^t F_2 (K_{1t})$ for
all $t < t'$. If there is at least one t such that $F_2 (K_{1t}) > 0$, then there is a unique t'
which satisfies a) – c). If there is no t such that $F_2 (K_{1t}) > 0$, then the entrant never
comes into the market.

Proof: The only thing in the lemma which requires proof is the assertion that if there is at least one t such that $F_2 (K_{1t}) > 0$, then there is a unique t' satisfying a) – c). The conditions a) – c) specify that the entrant comes in at the time at which his discounted profits are maximized, and that, if there are two or more such times, he comes in at the earliest of them; hence, proof reduces to showing that if discounted profits are ever positive, then at some time they are maximized.

By hypothesis there is some t'' such that $F_2 (K_{1t''}) > 0$. Because F_2 is bounded, there is t_0 such that

$$\alpha_2^{t''} F_2 (K_{1t''}) \geq \alpha_2^t F_2 (K_{1t}) \tag{20}$$

for all $t > t^0$. Therefore t' must be a member of the set $\{0, \ldots, t''\}$ which is finite; hence, $\alpha_2^t F_2 (K_{1t})$ takes a maximum on the set. If the maximum is attained for more than one t, then t' is the smallest. Otherwise, it is the time of the unique maximum.

$\square$

For the established firm, only an ϵ-optimal strategy can be assured. That is a strategy such that no other strategy yields a discounted payoff higher by more than an arbitrary positive ϵ. To see this, consider any $\sigma_1 = \{p_{1t}, K_{1,t-1}\}_{t=1}^{\infty}$ specifying the firm's behavior in the absence of entry, and note that the discounted profits of the established firm need not be continuous in σ_1. The set of attainable payoffs is bounded, but not closed; hence, it has a least upper bound which need not be attainable. Thus:

Theorem 2: In the two firm market with one established firm and one entrant, satisfying A1 – A5, with entry allowed, but not exit, there exists an ϵ-equilibrium. This equilibrium is characterized by a pair of strategies such that the entrant's strategy is a best reply to the strategy of the established firm, and no other strategy of the established firm, given that the entrant is always assumed to use a best reply to the choice of the established firm, yields more by at least ϵ.

To see whether the established firm engages in entry preventing behavior, it is necessary to determine the behavior the firm would follow if the entrant did not exist. Let σ_1^* denote a strategy which maximizes the discounted profits of the firm on the assumption that entry is impossible. Let $\{K_{1t}^*\}$ be the associated sequence of capital stocks. If $F_2 (K_{1t}^*) \leq 0$ for all t, then the firm's optimal choice is to follow σ_1^* and entry will not occur. It is also possible that $F_2 (K_{1t}^*) > 0$ for some t; however, no alternative policy for the established firm yields higher profits. In this case too entry preventing behavior would not be followed.

Now consider circumstances in which the established firm will wish to affect or prevent entry. First, assume that there are some values of K_1 for which F_2 is non-positive and others for which it is positive. Assume also that at least one of the K_{1t}^* yields positive F_2. Then if the firm follows a policy such that F_2 is never positive, it is engaging in entry preventing behavior. Such behavior need not be in the firm's best interests; however, under the circumstances sketched above, it is at least possible. It also provides a workable definition of entry preventing behavior.

Now consider an equilibrium in which the entrant does come in at t_0 and the strategy followed by the established firm is not σ_1^*. If entry under σ_1^* would occur at t^* and $t^* \neq t_0$, then the behavior of the established firm is influenced by (the threat of) entry. Even if $t_0 > t^*$, it does not seem reasonable to say that the firm is engaging in entry preventing behavior. It is engaging in entry delaying behavior, but it is not preventing entry.

2.2.2 A Model With Exit and Without Entry

The situation considered here is one in which two firms are in the market at the outset and one or both may leave the market at any time and not return. Assume that both firms make their initial investment in period 0, but that firm 1 may start with an initial capital which is positive. Assume, further, that if either firm left the market, the remaining one would certainly find it profitable to continue. If, for example, firm j leaves the market at the end of period $t - 1$ and, at that time, firm i has a capital stock of $K_{i,t-1}$, then the maximum discounted profit which firm i can earn from t onward, discounted back to t, is denoted $H_i (K_{i,t-1})$. It is immediate that this maximum exists and depends only on the capital which firm i possesses when it becomes a monopolist having no threat of entry.

Exit from the market is handled by assuming that each firm in each period decides on a probability, v_{it}, of remaining in the market in the next period. That is, if firm i is in the market in period t, it chooses p_{it}, K_{it} and v_{it}, where $v_{it} \in [0, 1]$ is the probability that firm i will continue in the market in period $t + 1$, given that it was in the market in period t. Thus, from the time of entry of firm 2, the market is a two player exit game as in *Friedman* [1979]. It remains to define an equilibrium for such a game and to show existence. The equilibrium concept which is employed is the weak non-cooperative equilibrium of *Friedman* [1979]. Letting $V_{it} = \pi_{\tau=1}^{t} v_{i\tau}$, the profit function of the i-th firm is

$$G_i (\sigma) = - I_{i0} + A_i^* + \alpha_i V_{10} V_{20} \pi_i (p_1, K_{i0}, K_{i1})$$

$$+ \sum_{t=2}^{\infty} \alpha_i^t V_{i,t-1} V_{j,t-2} [v_{j,t-1} \pi_i (p_t, K_{i,t-1}, K_{it})$$

$$+ (1 - v_{j,t-1}) H_i (K_{i,t-1})], \; i \neq j. \qquad (21)$$

$$A_1^* = A^*, A_2^* = 0, \text{ and } V_{10} = V_{20} = 1.$$

To picture the strategy of firm i, recall that firm i knows in any period whether its rival is still active and knows that if the rival leaves the market he (firm i) is thenceforth alone. The firm must, then, name a sequence of prices and capital levels. $K_{i0}, p_{i1}, K_{i1}, p_{i2}, K_{i2}, \ldots$ which specify what it chooses as long as the rival is in the market. It must also specify the choices to be followed from period $t + 1$ onward, given that the rival's last period in the market is period t; however, these need not be formally spelled out because the firm would merely face a simple maximization problem. The strategy, σ_i, consists of the sequence $(K_{i0}, p_{i1}, K_{i1}, v_{i1}, p_{i2}, K_{i2}, v_{i2}, \ldots) = (K_i, p_i, v_i)$ together with the implicitly understood behavior to be followed in the event the firm becomes a monopolist.

A particular strategy pair, (σ_1^*, σ_2^*) is a *weak non-cooperative equilibrium* if σ_i^* is in the strategy set of player i ($i = 1, 2$), and

$$G_i(\sigma^*) = \max_{p_i, K_i} G_i(K_i, p_i, v_i^*, \sigma_j^*) = \max_{v_i} G_i(K_i^*, p_i^*, v_i, \sigma_j^*),$$

$$i = 1, 2, \quad i \neq j. \tag{22}$$

That is, neither player can increase his discounted payoff by changing his "survival" probabilitites, v_i, nor can he increase his payoff by changing his price-capital policy. Joint changes of v_i and (p_i, K_i) are not considerd, which is what distinguishes the *weak* non-cooperative equilibrium from a non-cooperative equilibrium. Existence of a non-cooperative equilibrium cannot be assured in this model because the payoff of a firm (G_i) is not, in general, a quasi-concave function of its own strategy (σ_i).

Theorem 3: In any exit model satisfying A1 − A5, the game which commences when the entrant comes into the market has a weak non-cooperative equilibrium.

Proof: For a weak non-cooperative equilibrium in a two player game, there are four *best reply* functions. Each is defined in this way: Let (σ_1, K_2, p_2) be given. Then, for firm 2 there is a v_2 which maximizes its payoff, given the other actions. This v_2 is said to be a *best reply* to (σ_1, K_2, p_2). If the best reply is always unique, then there is a best reply function, and when the best reply need not be unique, the relation is a best reply correspondence. A second best reply relation is that which gives the (set of) optimal (K_2, p_2) for any σ_1 and v_2. Another two best reply relations may be analogously defined for v_1 and (K_1, p_1). These four best reply relations are all defined for all possible strategies and form a mapping, Φ, from the product of the two players' strategy sets to itself. That is

$$\sigma' \quad = (K_1', p_1', v_1', K_2', p_2', v_2') \in \Phi(\sigma) \text{ is defined by}$$

$$\Phi(\sigma) = \Phi(K_1, p_1, v_1, K_2, p_2, v_2) \tag{23}$$

$$= \Phi_1(p_1, K_1, \sigma_2) \times \Phi_2(v_1, \sigma_2) \times \Phi_3(\sigma_1, p_2, K_2) \times \Phi_4(\sigma_1, v_2)$$

where Φ_1, Φ_2, Φ_3 and Φ_4 are the four best reply mappings. Φ_2 and Φ_4 are, in fact functions.

$$\underline{y}'_1 \in \Phi_1 \ (\underline{p}_1, \underline{K}_1, \sigma_2), (\underline{p}'_1, \underline{K}'_1) = \Phi_2 \ (\underline{V}_1, \sigma_2),$$

$$\underline{y}'_2 \in \Phi_3 \ (\sigma_1, \underline{p}_2, \underline{K}_2), \text{ and } (\underline{p}'_2, \underline{K}'_2) = \Phi_4 \ (\sigma_1, \underline{y}_2).$$

(24)

Because the payoffs are linear in the probabilities, the two best reply mappings (σ_1 and Φ_3) which determine $\underline{y}_1$ and $\underline{y}_2$ are upper semi-continuous and they have convex image sets. On this, see *Friedman* [1979]. That the other two best reply mappings are continuous functions is known from the proof of theorem 1; hence, by the Kakutani fixed point theorem, Φ has a fixed point and such a fixed point is a weak non-cooperative equilibrium. $\qquad\qquad\qquad\qquad\qquad\qquad\qquad\qquad\qquad\qquad\qquad\qquad\qquad\square$

2.2.3 Entry and Exit Under Certainty

It remains now to put the two pieces together. Note first that the equilibrium whose existence is proved in theorem 4 is not, in general, unique; hence, one equilibrium is selected for each K_1 whenever more than one exists. Denote the selected equilibrium payoffs by $F_1^*(K_1)$ and $F_2^*(K_1)$.

The situation is parallel to that discussed at the end of section 2.2.1 for the model with entry and no exit. The differences are that the F_i of section 2.2.1 are replaced with the F_i^* and the ϵ-non-cooperative equilibrium with an ϵ-weak non-cooperative equilibrium. Earlier proofs go through, subject to the modifications just given. Thus the following Lemma and theorem are proved.

Lemma 2: The entrant comes into the market at time t' if a) $F_2^*(K_{1t'}) > 0$, b) $\alpha_2^{t'} F_2^*(K_{1t'}) \geqslant \alpha_2^t F_2^*(K_{1t})$ for all $t > t'$ and c) $\alpha_2^{t'} F_2^*(K_{1t'}) > \alpha_2^t F_2^*(K_{1t})$ for all $t < t'$. If there is at least one t such that $F_2^*(K_{1t}) > 0$, then there is a unique t' which satisfies a) – c). If there is no t such that $F_2^*(K_{1t}) > 0$, then the entrant never comes into the market.

Theorem 4: In the two firm market with one established firm and one entrant, satisfying A1 – A5, with both entry and exit allowed, there exists an ϵ-equilibrium. This equilibrium is characterized by: a) The equilibrium strategies yield a weak non-cooperative equilibrium for any subgame commencing with the time the entrant comes into the market. b) Subject to a), the entrant's strategy is a best reply to the strategy of the established firm. c) Given that the entrant always may choose a best reply to the established firm's strategy, the established firm chooses a strategy such that no other strategy could achieve at least ϵ greater discounted profit.

3. Equilibrium in a Special Case of the Model

If some additional assumptions are made, equilibrium is unique in the two firm model with neither entry nor exit. This model is developed in section 3.1 and in section 3.2 some results on entry prevention are obtained, where it is shown that the entrant's equilibrium profit in the market falls as the capital stock of the monopolist, held at the moment of entry, rises. This is what is intuitively expected; however, relaxation of some of the assumptions would results in the contrary.

3.1 Equilibrium in the Two Firm Market with Neither Entry nor Exit

Consider the subgame centered about the choices made in period t. Then

$$\pi_i\,(p_t,\,K_{i,t-1},\,K_{it}) + \alpha_i\,\pi_i\,(p_{t+1},\,K_{it},\,K_{i,t+1}) = \pi_{it} + \alpha_i\,\pi_{i,t+1} \tag{25}$$

is that part of the i-th firm's objective function into which decisions of time t enter. It is possible to think of a single period game in which $(K_{i,t-1},\,p_{i,t+1},\,K_{i,t+1}),\,i = 1, 2,$ are given, the two firms choose $(p_{it},\,K_{it})$, respectively, and equation (25) gives their objective functions. The non-cooperative equilibria of such a game may be denoted by $\phi\,(p_{t-1},\,p_{t+1},\,K_{t-1},\,K_{t+1})$. Such games have been studied in *Friedman* [1977, chapter 9]. If, for each $(p_{t-1},\,p_{t+1},\,K_{t-1},\,K_{t+1})$ there is exactly one equilibrium and the function ϕ is a contraction, then the equilibrium has some special features. Irrespective of the nature of ϕ, a non-cooperative equilibrium for the game (i.e., the original infinite period game) satisfies the function ϕ in each period. Even without further assumptions, ϕ is known to be more specialized than the form given above. In particular,

$$\left.\begin{aligned}
p_{1t} &= \phi_1\,(K_{t-1}),\, p_{2t} = \phi_2\,(K_{t-1}) \\[2mm]
K_{1t} &= \phi_3\,(p_{t+1},\,K_{1,t-1}),\, K_{2t} = \phi_4\,(p_{t+1},\,K_{2,t-1})
\end{aligned}\right\} \tag{26}$$

ϕ_1 and ϕ_2 are derived from equations (12) and (13). $\phi_3\,(p_{t+1},\,K_{1,t-1})$ is the solution to $(\partial\,/\,\partial K_{1t})\,(\pi_{1t} + \alpha_1\,\pi_{1,t+1}) = 0$ if that solution is no smaller than $(1 - \delta)K_{1,t-1}$. When $\phi_3\,(p_{t+1},\,K_{1,t-1}) > (1 - \delta)\,K_{1,t-1}$, then K_{1t} depends only on p_{t+1}. Otherwise, $\phi_3\,(p_{t+1},\,K_{1,t-1}) = (1 - \delta)\,K_{1,t-1}$. Similarly for ϕ_4. Furthermore, if the optimal values of the variables are in the interior of their domains, then the derivatives of ϕ are

$$\left.\begin{aligned}
\frac{\partial p_{1t}}{\partial K_{1,t-1}} &= \frac{-\pi_{1t}^{13}\,\pi_{2t}^{22}}{\pi_{1t}^{11}\,\pi_{2t}^{22} - \pi_{1t}^{12}\,\pi_{2t}^{21}} & \frac{\partial p_{it}}{\partial K_{2,t-1}} &= \frac{\pi_{1t}^{12}\,\pi_{2t}^{23}}{\pi_{1t}^{11}\,\pi_{2t}^{22} - \pi_{1t}^{12}\,\pi_{2t}^{21}} \\[3mm]
\frac{\partial p_{2t}}{\partial K_{1,t-1}} &= \frac{\pi_{1t}^{13}\,\pi_{2t}^{21}}{\pi_{1t}^{11}\,\pi_{2t}^{22} - \pi_{1t}^{12}\,\pi_{2t}^{21}} & \frac{\partial p_{2t}}{\partial K_{2,t-1}} &= \frac{-\pi_{2t}^{23}\,\pi_{1t}^{11}}{\pi_{1t}^{11}\,\pi_{2t}^{22} - \pi_{1t}^{12}\,\pi_{2t}^{21}} \\[3mm]
\frac{\partial K_{1t}}{\partial p_{1,t+1}} &= \frac{-\pi_{1,t+1}^{31}}{\pi_{1,t+1}^{33}} & \frac{\partial K_{1t}}{\partial p_{2,t+1}} &= \frac{-\pi_{1,t+1}^{32}}{\pi_{1,t+1}^{33}} \\[3mm]
\frac{\partial K_{2t}}{\partial p_{1,t+1}} &= \frac{-\pi_{2,t+1}^{31}}{\pi_{2,t+1}^{33}} & \frac{\partial K_{2t}}{\partial p_{2,t+1}} &= \frac{-\pi_{2,t+1}^{32}}{\pi_{2,t+1}^{33}}
\end{aligned}\right\} \tag{27}$$

where

$$\pi_{1t}^{11} = (p_{1t} - C_{1t}^1) f_{1t}^{11} + (2 - C_{1t}^{11} f_{1t}^1) f_{1t}^1 < 0$$

$$\pi_{1t}^{12} = (p_{1t} - C_{1t}^1) f_{1t}^{12} + (1 - C_{1t}^{11} f_{1t}^1) f_{1t}^2 > 0 \tag{28}$$

$$\pi_{1t}^{13} = - C_{1t}^{12} f_{1t}^1 < 0, \; \pi_{1t}^{23} = - C_{1t}^{12} f_{1t}^2 > 0, \; \pi_{1t}^{33} = - C_{1t}^{22} < 0$$

and analogously for firm 2. If the optimal K_{1t} is a boundary value, then
$\partial K_{1t} / \partial p_{1,t+1} = \partial K_{1t} / \partial p_{2,t+1} = 0$. If the optimal value of p_{2t} is high enough that
$f_{2t} = 0, f_{1t}^{12} = f_{1t}^2 = f_{2t}^1 = f_{2t}^2 = f_{2t}^{11} = f_{2t}^{12} = f_{2t}^{22} = 0$ and some of the derivatives in
equation (28) are adjusted accordingly, and the various derivatives must be recalculated.
The derivatives with altered values are

$$\frac{\partial p_{1t}}{\partial K_{2,t-1}} = \frac{\partial p_{2t}}{\partial K_{1,t-1}} = \frac{\partial p_{2t}}{\partial K_{2,t-1}} = 0, \; \frac{\partial p_{1t}}{\partial K_{1,t-1}} = \frac{-\pi_{1t}^{13}}{\pi_{1t}^{11}},$$

$$\frac{\partial K_{1t}}{\partial p_{2,t+1}} = \frac{\partial K_{2t}}{\partial p_{1,t+1}} = \frac{\partial K_{2t}}{\partial p_{2,t+1}} = 0. \tag{29}$$

Should ϕ be a contraction, then from *Friedman* [1977, chap. 9], there is a unique
non-cooperative equilibrium for the game, given the initial capital the firms hold. ϕ is
a contraction if the following conditions hold:

A6. *For p such that f^{ii} and f_i^{12} exist, $f_i^{ii} \leqslant 0$ and $f_i^{ii} + |f_i^{12}| \leqslant 0$. In addition
$C_i^{12} \geqslant -1$ and $- C_i^{12} / C_i^{22} (|f_i^1| + |f_i^2|) < 1, \; i = 1, 2$.*

Lemma 3: *If A1 — A6 hold, then ϕ is a contraction.*

Proof: ϕ is a contraction if $| \phi_j^1 | + | \phi_j^2 | < 1$ for $j = 1, \ldots, 4$. Consider first $j = 1$.
From equation (29) it is seen that $| \phi_1^1 | + | \phi_1^2 | = 0$ if $f_1 = 0$; and $| \phi_1^1 | + | \phi_1^2 | =
= \pi_{1t}^{13} / \pi_{1t}^{11}$ if $f_1 > 0$ and $f_2 = 0$. $\pi_{1t}^{13} / \pi_{1t}^{11} < 1$ is equivalent to

$$(p_{1t} - C_{1t}^1) f_{1t}^{11} + (1 - C_{1t}^{11} f_{1t}^1) f_{!t}^: + (1 - C_{1t}^{12}) f_{1t}^1 < 0 \tag{30}$$

which clearly holds. If f_1 and f_2 are positive, then $| \phi_1^1 | + | \phi_1^2 | < 1$ is equivalent to

$$\pi_{1t}^{13} \pi_{2t}^{22} - \pi_{1t}^{12} \pi_{2t}^{23} < \pi_{1t}^{11} \pi_{2t}^{22} - \pi_{1t}^{12} \pi_{2t}^{21} \tag{31}$$

which, in turn, is equivalent to

$$0 < (\pi_{1t}^{11} - f_{1t}^1) (\pi_{2t}^{22} - f_{2t}^2) - \pi_{1t}^{12} \pi_{2t}^{21} + f_{2t}^2 (\pi_{1t}^{11} - f_{1t}^1 - C_{2t}^{12} \pi_{1t}^{12}) +$$

$$+ f_{1t}^1 \pi_{2t}^{22} (1 + C_{1t}^{12}) \tag{32}$$

That equation (32) holds may be seen by recalling that $-1 \leqslant C_{it}^{12} \leqslant 0$, $\pi_{it}^{12} > 0$ and $\pi_{it}^{ii} - f_{it}^i + \pi_{it}^{12} < 0$. An analogous argument can be made to show that $|\phi_2^1| + |\phi_2^2| < 1$.

Turning now to $|\phi_3^1| + |\phi_3^2| < 1$, when no boundary conditions are in effect, the condition is satisfied when

$$\frac{C_{1t}^{12} f_{1t}^1 - C_{1t}^{12} f_{1t}^2}{C_{1t}^{22}} = \frac{-C_{1t}^{12}}{C_{1t}^{22}} (f_{1t}^2 - f_{1t}^1) < 1 \tag{33}$$

holds. Equation (33) is one of the conditions stated in A6. As to boundary conditions, if $f_{2t} = 0$, then ϕ_3^1 is unchanged and $\phi_3^2 = 0$, so the requirement is still met. If $K_{1,t-1}$ is at a value so high that the choice is zero investment, then $\phi_3^1 = \phi_3^2 = 0$. Again, an analogous argument may be made to show that $|\phi_4^1| + |\phi_4^2| < 1$. $\square$
And now the existence theorem may be given:

Theorem 5: In a two firm market, without entry and exit, which satisfies A1 $-$ A6, *there is a unique non-cooperative equilibrium.*

Proof: Virtually all the elements of proof have been given. It remains to note that the unique $\{p_t, K_t\}$ sequence which satisfies ϕ is clearly a non-cooperative equilibrium. This holds because any equilibrium sequence satsfies ϕ and, due to the concavity of G_i, the given sequence is the only sequence to do so. $\square$

The special structure of the model may be exploited further. The four equations of equation (26) may be condensed into two.

$$\left. \begin{array}{l} K_{1t} = \phi_3 \left(\phi_1 (K_t), \phi_2 (K_t), K_{1,t-1}\right) \\[2mm] K_{2t} = \phi_4 \left(\phi_1 (K_t), \phi_2 (K_t), K_{2,t-1}\right) \end{array} \right\} . \tag{34}$$

The optimal K_t is a fixed point of the system above. Note, however, that if the system of equations is a contraction, the fixed point is unique for a given K_{t-1} and may be written

$$K_{1t} = \psi_3 (K_{t-1}), \quad K_{2t} = \psi_4 (K_{t-1}) \tag{35}$$

Recall that $K_{1,t-1}$ enters ϕ_3 only as a boundary condition which stipulates that the capital stock of the firm cannot fall by a fraction larger than δ in each period. Thus, when the boundary constraints on K_{1t} and K_{2t} are not binding, $\psi_3 (K_{t-1}) = K_1^*$ and $\psi_4 (K_{t-1}) = K_2^*$, a pair of constants. It is surprising that the optimal capital stocks, away from the lower boundary, do not depend on the past history of the market or the length of the firms' horizons, as long as the horizons both extend at least through $t + 1$. Associated with the optimal capital stocks $K^* = (K_1^*, K_2^*)$ are prices $p^* = (\phi_1 (K^*), \phi_2 (K^*))$.

Note what would happen if one firm began in the market with an arbitrary initial capital which exceeded the optimal level by a large amount. Say it were firm 1. Its starting capital is K'_{10} and its subsequent capital levels are $K'_{1t} = (1 - \delta) K'_{1,t-1}$ as long as $(1 - \delta) K'_{1,t-1} \geqslant K^*_1$. For all periods afterward, the optimal level is constant at K^*_1. The capital stock choice of firm 2 and the two firms' prices are found by solving

$$K_{2t} = \phi_4 \left(\phi_1 (K'_{1t}, K_{2t}), \phi_2 (K'_{1t}, K_{2t}), K_{2,t-1} \right) \tag{36}$$

for the fixed point value of K_{2t}, with K'_{1t} and $K_{2,t-1}$ given, and then using ϕ_1 and ϕ_2 to find the prices for period $t + 1$. This process is followed by starting in the last period for which $K'_{1t} > K^*_1$ and working backward to the earliest period in which firm 2 is active.

3.2 Some Observations on Entry Prevention

In section 2.2.1, there is some discussion of entry preventing behavior in which it is noted that if the monopolist can choose an investment policy under which the entrant will never come in, and that policy is not what it would pursue in the absence of an entry threat, then it can engage in entry preventing behavior. It remains to understand whether the equilibrium profits of the entrant, given entry at time t, would rise or fall as K_{1t} increases. Intuition is confirmed, as is seen below:

Theorem 6: In a two firm market with no exit, which satisfies A1 – A6, let firm 2 use a best reply strategy to firm 1's strategy and let firm 1 use prices which are optimal given its capital and the strategy of firm 2. Then, given strategies such that no boundary conditions are in effect, the discounted payoff of firm 2 falls as K_{1t} rises.

Proof: Proof comes down to evaluating the sign of

$$\frac{dF_2}{dK_{1t}} = \frac{\partial F_{2t}}{\partial K_{2t}} \frac{\partial K_{2t}}{\partial K_{1t}} + \sum_{j=1}^{2} \frac{\partial F_2}{\partial p_{j,t+1}} \left[\frac{\partial p_{j,t+1}}{\partial K_{1t}} + \frac{\partial p_{j,t+1}}{\partial K_{2t}} \frac{\partial K_{2t}}{\partial K_{1t}} \right] . \tag{37}$$

As the evaluation is made at a point where firm 2 is in equilibrium and firm 1 is choosing optimal prices, both $\partial F_2 / \partial p_{2,t+1}$ and $\partial F_2 / \partial K_{2t}$ are zero and equation (37) becomes

$$\frac{dF_2}{dK_{1t}} = \frac{\partial F_2}{\partial p_{1,t+1}} \left[\phi^1_1 + \phi^2_1 \frac{dK_2}{dK_1} \right] . \tag{38}$$

As $\partial F_2 / \partial p_{1,t+1} > 0$, equation (38) has the same sign as the term in brackets. dK_2 / dK_1 is obtained by differentiating ϕ_4 and equals

$$\frac{\phi^1_4 \phi^1_1 + \phi^2_4 \phi^1_2}{1 - \phi^1_4 \phi^2_1 - \phi^2_4 \phi^2_2} \tag{39}$$

which makes equation (37) equivalent to

$$\frac{\phi_1^1 - \phi_4^2 \, (\phi_1^1 \, \phi_2^2 - \phi_1^2 \, \phi_2^1)}{1 - \phi_4^1 \, \phi_1^2 - \phi_4^2 \, \phi_2^2} . \tag{40}$$

The numerator of equation (40) is

$$-\frac{\pi_1^{13} \, (\pi_2^{22} \, \pi_2^{33} - \pi_2^{23} \, \pi_2^{23})}{\pi_2^{33} \, (\pi_1^{11} \, \pi_2^{22} - \pi_1^{12} \, \pi_2^{21})} < 0. \tag{41}$$

The denominator is

$$\frac{\pi_1^{11} \, (\pi_2^{22} \, \pi_2^{33} - \pi_2^{23} \, \pi_2^{23}) + \pi_1^{12} \, (\pi_2^{13} \, \pi_2^{23} - \pi_2^{12} \, \pi_2^{33})}{\pi_2^{33} \, (\pi_1^{11} \, \pi_2^{22} - \pi_1^{12} \, \pi_1^{21})} > 0. \tag{42}$$

It is perhaps not transparent that the denominator, equation (42) is positive. That fact may be seen in the following way: Equation (42) is in the form

$$\frac{A_1 \, B_1 + A_2 \, B_2}{E} \tag{43}$$

where A_1 and E are negative and the remaining terms are positive. $|A_1| > A_2$ and $B_1 > B_2$, making $A_1 \, B_2 + A_1 \, B_2 < 0$ and equation (43) positive. Thus, the sign of equation (40) is negative, making dF_2/dK_{1t} negative.   □

In the model it is clear from equations (27) and (28) that the optimal capital for a monopolist who has no fear of entry is larger than the optimal capital K_1^* which would obtain if there were two firms. Letting the former capital be denoted K_1^{**}, it is possible that if the firm had capital of K_1^{**} prior to the entry of firm 2 that, depending on the relative sizes of K_1^{**}, K_1^* and δ, it might take several periods for the firm's capital to fall from K_1^{**} to K_1^*. So it is possible that it would not be profitable for firm 2 to enter the market even though it could make positive profits when the two capital stocks are $K^* = (K_1^*, K_2^*)$.

Another aspect of the model is that if entry preventing behavior is possible, it involves firm 1 choosing its capital stock at or above some critical value, $\bar{K}_1$. Precisely because the high capital stock cannot be reduced at a faster rate than δ per period, the firm is forced to have too much capital during an adjustment interval of several periods. During this interval, it has lower marginal cost at any output level than at K_1^*; hence, it follows a lower price policy and, as a result, lowers the demand of firm 2.

Taking into account both the foregoing remarks and theorem 6, it is clear that entry preventing behavior, or entry forestalling behavior, consists of choosing levels of capital which are above the level which is optimal for the monopolist when he does not take entry into account. The latter level is above the level which would prevail after entry, so the entry preventing intention is unambiguous.

4. Concluding Comments

The most intuitively plausible entry preventing behavior in the present paper is that associated with the entry model of section 3. In it there is the possibility of the established firm preventing entry by choosing its capital so high that an entrant could not make a positive profit. In general it need not be that such a capital level for firm 1 exists, nor, if it does exist, that it is profitable for firm 1 to adopt it. Nonetheless, the means of entry prevention, where it can be practiced, is clear in the model. Note that adopting a very large, entry preventing, capital is like making a credible threat to the entrant. The threat is to use a price so low that he will not make a positive profit, should he come in. What makes the threat credible is that the firm's capital is so high that the "threat price" is simply the ordinary equilibrium price which is natural for the firm to choose after entry and in the light of the capital it would have. This may be contrasted with a model in which there is no capital and the established firm makes a general public announcement that it will switch to some particular (non-equilibrium) low price if any new firm should enter its industry. In this case, after entry, it has no incentive to charge its previously announced price.

The present paper is, I believe, a start on analyzing entry and exit with explicit account taken of the entrants — that is, with the entrant being an active, rational, decision making economic agent. The only earlier work along these lines of which I am aware is *Shubik's* [1959] *firm-in-being*. The present paper is an elaboration of that idea. Whether any of the results here still obtain, and if so, in what form, after the model is further generalized remains to be seen.

References

Bain, J.S.: A Note on Pricing in Monopoly and Oligopoly. Amer. Econ. Rev. 39, 1949, 448–464.
Friedman, J.W.: Oligopoly and the Theory of Games. 1977.
–: Non-cooperative Equilibrium for Exit Supergames. Int. Econ. Rev. 20, forthcoming, 1979.
Harrod, R.F.: Theory of Imperfect Competion Revised. Economic Essays, chapt. 8, London 1952.
Hoggatt, A.C.: Response of Paid Student Subjects to Differential Behaviour of Robots in Bifurcated Duopoly Games. Rev. of Econ. Stud. 36, 1969, 417–432.
Shubik, M.: Strategy and Market Structure. New York 1959.

Applied Game Theory. 1979 © Physica-Verlag, Wuerzburg/Germany

Reinsurance as a Cooperative Game

By *J. Lemaire*, Bruxelles[1])

Abstract: We define axiomatically a concept of value for games without transferable utilities, without introducing the usual symmetry axiom. The model, a generalization of a previous paper extending Nash's bargaining problem, attempts to take into account the affinities between the players, defined by an a priori set of "distances". This new value concept in then applied to compute the value of a reinsurance model. It is shown that the exchange of risks between insurance companies can be formulated as a *n*-person cooperative game without transferable utilities. The determination of an "optimal reinsurance treaty" is then shown to coincide with the computation of the value of the corresponding game. A complete example is given.

1. Reinsurance

Insurance is a process for spreading the strain of misfortune, i.e. the loss resulting from accident, fire, theft, . . . In return for the payment of a premium much less in amount than the possible loss, an insurer accepts to meet the policy holder's losses if they occur. In today's life, however, the insured sums are so enormous (for instance commercial aircrafts, big office buildings) that a single insurer could not bear the possibility of an accident by himself: he has to cede or reinsure part of his risks. The insurer decides on the part of the risk he wishes to support without help (his retention) and arranges with a reinsurer to handle the remainder in compensation of a fixed premium.

The main advantages of reinsurance are

1. to reduce the risk that in the event of a catastrophe, the direct insurer will suffer a net liability in excess of its financial ressources;
2. to stabilize the technical results of the insurer by reducing claim fluctuation. The results will consequently show a more stable progression.
3. to achieve a spread of risks so that the liabilities of the insurer are not too heavily concentrated in any department or any one geographical area. This is one of the reasons why reinsurance is mainly an international business;
4. to enable the direct insurer to handle larger risks than he would otherwise be able to accept. With reinsurance a larger risk can be accepted, which would otherwise have to be refused, with a loss of premium income and prestige;

---

[1]) *Jean Lemaire*, Université Libre de Bruxelles, Campus de la plaine, Institut de Statistique C.P. 210, 50, boulevard du triomphe, B–1050 Bruxelles.

5. if the business of the insurer is growing rapidly, there is likely to be a period during which the reserves are not growing as fast as the risk of an adverse fluctuation in claim experience. Reinsurance might be a cheaper way to achieve security than seeking additional capital.

Naturally the business of reinsurance is a very risky one, since the reinsurer is likely to intervene financially in very few claims, but usually for very large amounts. As on the other hand there is often a lack of statistical evidence to evaluate large risks (earthquakes, explosion of nuclear plants, . . .) the premiums charged by the few professional reinsurance companies are often very high (they are usually expressed as a percentage of the original premium). Most companies then tend to act both as insurer and reinsurer and sign reciprocal treaties with friendly companies, ceding part of their risks and taking charge of part of the other company's risks. Practically one observes even that companies tend to conglomerate in large clusters and sign very developed reciprocal exchange of risks treaties.

This procedure presents a lot of advantages.

1. The more companies in one group, the higher the total retention. Any member of the group can then accept nearly any risk.
2. Moreover, the group as a whole can bear by himself most of his risks and does not have to pay out substantial amounts as reinsurance premiums.
3. The group can even attract outside business, by proposing to the market a "package" reinsurance treaty with a very high retention.
4. The fact that the companies of a group handle the same risks simplifies the administration procedures and reduces some expenses.
5. The companies of a given group can secure technical and legal advices to each other: for instance, one company, specialized in one branch, can provide advices for the rating of a special risk, or the British company of the group can appraise and settle the claim of a Belgian policy-holder, involved in an accident in London, on behalf of their Belgian partner, usually at no charge (although it involves expert's, sollicitor's and administrative costs).
6. The companies can pool their informations, and consequently obtain more reliable statistics, with a gain in ratemaking accuracy and competitiveness.
7. The companies can obtain business in a very cheap way, since only one of them has to perform the ratemaking and the survey of the risks, as well as the appraisal and the settlement of the claims. This implies mutual trustfulness in the ability of each member of the group to accept sound business and to rate correctly.

The groups tend to form according to political or sentimental reasons. As example of such "clusters" we can quote

- the R.C.D. (Regional Cooperation for Development), that groups the societies of Turkey, Iran and Pakistan;
- the forming African Reinsurance Bureau, whose main objective is to slow down the outflow of foreign currencies;
- the International Co-operative Reinsurance Bureau, a section of the International Co-operative alliance, today groups 71 companies in 26 countries spread all over the world.

It is obvious that the companies of the same group have both common (they share the same risks) and antagonistic (they must try to conclude the most favourable treaty) interests. This is characteristic of a game situation, and we shall indeed show in the next sections that the reinsurance market can be modelized as a cooperative game without transferable utilities. It should be obvious from the preceding lines that signing a treaty is not a pure bargaining process in which all considerations besides financial interest are excluded: reinsurance involves respect, confidence, loyalty, common interests and friendship. How can then friendship match with money? We shall in the sequel develop a value concept for non-transferable games that attempts to take into account the affinities between the players.

2. The Model [*Bühlmann*, 1970]

Let $1, \ldots, n$ be n insurance companies wishing to sign an exchange of risks treaty. Let $[S_j, F_j(x_j)]$ be the situation of company j, where $F_j(x_j)$ is the distribution function of its total claim amount, and S_j, the initial surplus, results from the addition of the collected premium P_j to the free reserves R_j. An insurance company certainly does not try to maximize its expected surplus (for otherwise it would never reinsure) but rather attempts to achieve a certain level of benefits and security. We shall assume that each company evaluates its situation by an utility function

$$U_j(x_j) = U_j[S_j, F_j(x_j)] = \int_0^\infty u_j(S_j - x_j) dF_j(x_j),$$

where $u_j(x)$ is the utility of a monetary amount x. As an insurance company is by definition risk-averse, we must have $u_j'(x) \geqslant 0$ and $u_j''(x) \leqslant 0$. The members of the pool will try to improve their situations by concluding a treaty of risk exchanges

$$\bar{y} = [y_1(x_1, \ldots, x_n), \ldots, y_n(x_1, \ldots, x_n)],$$

where $y_j(x_1, \ldots, x_n)$ is the amount j has to pay if the claims for the different companies are respectively $x_1, \ldots, x_n$.

Since all the claims must be indemnified, the $y_j(x_1, \ldots, x_n)$ must satisfy the admissibility condition

$$\sum_{j=1}^n y_j(x_1, \ldots, x_n) = \sum_{j=1}^n x_j = z \tag{1}$$

where z is the total amount of the claims. After the signature of $\bar{y}$, the utility of j becomes

$$U_j(\bar{y}) = \int_\theta u_j[S_j - y_j(\bar{x})] dF(\bar{x}),$$

where θ is the positive orthant of E^n and $F(\bar{x})$ the n-dimensional distribution function of the claims $\bar{x} = (x_1, \ldots, x_n)$.

$\bar{y}$ is Pareto-optimal if there is no $\bar{y}'$, such that $U_j(\bar{y}') \geqslant U_j(\bar{y}) \; \forall j$, with at least one strict inequality. Borch [see *Bühlmann*] has demonstrated that all the Pareto-optimal treaties are characterized by the following relations

$$k_j u'_j [S_j - y_j(\bar{x})] = k_1 u'_1 [S_1 - y_1(\bar{x})] \qquad\qquad k_j \geqslant 0 \; \forall j. \qquad (2)$$

Let $K = \{k_1, \ldots, k_n\}$. The treaty is unique for given K, but there usually exists an infinity of K satisfying (1) and (2).

3. Relationship with Game Theory [*Lemaire*, 1973a]

In the n-dimensional Euclidean space of the companies' utilities, the set of the admissible K forms a n-1-dimensional surface (the Pareto-optimal surface), whose parametric equations are

$$U_j = \int_0^\infty u_j [S_j - y_j(\bar{x})] \, dF(\bar{x}) \qquad\qquad j = 1, \ldots, n.$$

A company will only enter the pool if it increases its initial utility. The Pareto-optimal surface is thus limited by the n hyperplanes $U_j = U_j(x_j)$. The space delimited by the Pareto-optimal surface and the n hyperplanes will be called the space of the game ξ. Note that ξ is closed, bounded and convex, since

$$U_j(\bar{y}) = U_j[\alpha \bar{y}' + (1-\alpha)\bar{y}''] = \alpha U_j(\bar{y}') + (1-\alpha) U_j(\bar{y}'')$$
$$\geqslant \alpha U_j(x_j) + (1-\alpha) U_j(x_j) = U_j(x_j).$$

Definition: A n-player cooperative game without transferable utilities (shortly a non-transferable game) is a triple $[N, v(S), H]$, where

1. $N = \{1, \ldots, n\}$ is the set of the players;
2. $v(S)$, the characteristic function, is defined on all the non-void subsets S of N; the image of each S (the coalitions) is a subset $v(S)$ of $E^{|S|}$, the $|S|$ - dimensional Euclidean space, such that $v(S)$ is non-void, convex, closed and superadditive:
   $\forall S_1, S_2 \subset N \Rightarrow S_1 \cap S_2 = \phi, \; v(S_1 \cup S_2) \supset v(S_1) \times v(S_2)$;
3. H is the set of the "accessible payoffs". More precisely,
   $v(N) = \{x \in E^{|N|} \mid \exists y \in H \Rightarrow y \geqslant x\}$.

Let

1. N be the set of the n companies;
2. $v(S)$ be the space delimited by the projection of the Pareto-optimal surface in $E^{|S|}$;
3. $H = \xi$.

Theorem: The reinsurance market is a non-transferable game $[N, v(S), \xi]$.

Proof: It suffices to show that $v(S)$ satisfies all the requirements. $v(S)$ is non-void since it contains the initial treaty $y_j(\bar{x}) = x_j \ \forall j$; $v(S)$ is convex. In fact, let $\bar{y}'^S$ and $\bar{y}''^S$ be two admissible treaties for coalition S: $\sum\limits_{j \in S} y_j'^S(\bar{x}) = \sum\limits_{j \in S} y_j''^S(\bar{x}) = \sum\limits_{j \in S} x_j$. $\bar{y}^S = \alpha \bar{y}'^S + (1 - \alpha) \bar{y}''^S |$ is also admissible since

$$\sum_{j \in S} y_j^S(\bar{x}) = \alpha \sum_{j \in S} y_j'^S(\bar{x}) + (1 - \alpha) \sum_{j \in S} y_j''^S(\bar{x}) = \alpha \sum_{j \in S} x_j + (1 - \alpha) \sum_{j \in S} x_j = z.$$

$v(S)$ is closed, since the treaties of the Pareto-optimal surface are admissible.

Superadditivity results from the existence of a solution to equations (1) and (2).

The different admissible values of K correspond to the various ways of sharing the benefits of cooperation among the companies. Each has interest to obtain a k_j as high as possible, in order to pay as less as compatible with condition (1). The choice of K thus depends on a further bargaining, in which the interests of each company are conflicting. In other words, we must compute the value of the game, in order to sort out one of the Pareto-optimal treaties. In [1977], we have computed the Shapley value and a generalization of Nash's bargaining point of this game. However, both solutions, as most value concepts of game theory, use a symmetry axiom: every symmetrical game has a symmetrical solution; that is, if the characteristic function of the game is symmetrical with respect to the bissecting line passing through the initial payoffs, the solution grants the same utility increase to each player. If this axiom seems innocuous (it is evident that the final payoff must not depend on a re-numbering of the players), it implies that the game is adequately represented by the characteristic function and that no element outside this function influences the behaviour of the participants and the results of the game. Those models do not take into account the personal affinities and antipathies between the players, and cannot consequently be applied to describe in a realistic way the reinsurance market. We shall thus develop a new value concept that attempts to catch the notion of "affinities" by suppressing the symmetry axiom and introducing "distances" between players. It is a modification of our former [*Lemaire*, 1973b] symmetrical value. We shall then apply this concept to the reinsurance model.

4. Axioms

Axiom 1: Linear invariance. The solution is not affected by a linear transformation performed on the utilities of the players.

Justification: Since utilities are only defined up to a linear transformation, it must obviously be the case for the solution.

Axiom 2: Strong Pareto-optimality. The solution depends on all the sub-treaties relative to all the sub-coalitions (with the exception of the sub-coalitions that form with probability zero – see section 6). Each sub-treaty (and the final treaty) must be Pareto-optimal and satisfy the admissibility condition.

Justification: The axiom expresses the fact that, during a negotiation, the bargaining strength of a player depends on the terms he obtained during the preceding discussions; a player will get more from his partners if he has signed a favourable treaty in a sub-coalition. We thus authorize the formation of any coalition during the bargaining process. Each one may negotiate with a disjoint group in order to unify. During this partial bargain, we suppose that each coalition acts as a single player: no one has the right to disavow his signature and quit his coalition in order to negotiate separately. We also assume that the grand coalition is formed step by step; at each step two coalitions only merge, so that N is obtained after $(n-1)$ steps.[2]) Since the power of a player depends on all the already signed contracts, they must influence the final payoff. Each sub-treaty must of course be Pareto-optimal in the corresponding sub-game, and the admissibility condition must be satisfied.

Axiom 3: Independence of irrelevant alternatives. During each negotiation between two coalitions, exclusion from the prospect space of possible payoffs other than the solution and the disagreement point (the utilities that the players get in case they cannot reach an agreement) does not affect the solution.

Justification: This axiom means that the solution, which by axiom 2 must lie on the upper boundary of the prospect space, only depends on the shape of this boundary in its neighbourhood, and not on distant points. This expresses a structure property of the bargaining process: during the negotiations, the set of the alternatives likely to be selected progressively reduces, so that at the end of the discussion, the solution must only compete with very close points, and not with propositions already eliminated during the prior stages of the bargaining.

Axiom 4: Partial symmetry. If, during a negotiation between two disjoint groups, the prospect space is symmetrical, so must be the treaty signed.

Justification: The classical symmetry axiom is weakened, since we only enforce it for the sets of two players or groups of players. It implies that the affinities between the players do not affect the discussions between two coalitions, which consist of a tough haggling between two groups trying to take as much advantage as they can from the situation. The affinities will intervene in the kind of coalitions that tend to form, in the propensity that some players have to start discussing with a particular group instead of another. In other words, the affinities influence the choice of the groups that enter negotiation, but not their negotiation itself. For example, the recent French political events demonstrate that the fact that the Communists and the Socialists have a strong affinity does not incite them to make concessions to each other: coalition forming and bargaining are two different things.

---

[2]) Those behavioural hypotheses are not very restrictive, since the axiom considers *all* the grouping possibilities. For instance, we prohibit the simultaneous merging of three disjoint groups C_a, C_b, C_c. But the solution will in particular study the grouping of C_a and C_b at one step and the adjunction of C_c during the next step. The two other cases (C_a and C_c unify first then absorb C_b, and C_b and C_c group and join C_a one step later) will also be considered. In the same fashion, some schemes of coalition forming where one player remains isolated until the final step, will intervene in the final treaty.

Therefore, we shall separate the computation of the value of a game in two distinct parts:

1. the coalition forming procedure, which consists of the determination of a set of probabilities $W = \{W_{C_a \cup \overline{C_a}} \; \forall \; C \subset N, \; \forall \; C_a \subset C, \; \overline{C_a} = C \setminus C_a, \; C_a \neq \phi, \; \overline{C_a} \neq \phi\}$, interpreted as "weights associated to orders of formation of the coalitions $C = C_a \cup \overline{C_a}$";

2. the bargaining procedure, which attributes a payoff to each player, given the set W.

5. The Bargaining Procedure: Existence and Unicity Theorem

Let us denote $\overline{y}(C) = \overline{y}(x_i \mid i \in C)$ the treaty signed by a coalition C and $U_i(C) = U_i[y_i(C)]$ the utility $i \in C$ derives from this signature.

Suppose that, at a given moment of the negotiation, a first group C_a of players has reached an agreement and signed a treaty $\overline{y}(C_a)$, allowing to each of its members an utility $U_i(C_a)$, while another group C_b (such that $C_a \cap C_b = \phi$) has concluded a treaty $\overline{y}(C_b)$, giving to each $j \in C_b$ an utility $U_j(C_b)$. Both groups meet in order to conclude a global treaty $\overline{y}(C_a \dot\cup C_b)$ (the symbol $\dot\cup$ has a slightly different meaning than the usual reunion sign. $C_a \dot\cup C_b$ means "C_a joins C_b". The $\dot{}$ is placed to recall that the result not only depends on the set $C_a \cup C_b$, but also on the manner in which this coalition was formed, i.e. on C_a and C_b). If both coalitions cannot agree on a treaty $\overline{y}(C_a \dot\cup C_b)$, they necessarily return to the starting point of the negotiation, awarding to each player $U_i(C_a)$ (if $i \in C_a$) or $U_j(C_b)$ (if $j \in C_b$). For this reason, this point is called the disagreement point.

Lemma: There exists one and only one treaty satisfying the axioms. It can be obtained by maximizing the expression

$$\prod_{i \in C_a} [U_i(C_a \dot\cup C_b) - U_i(C_a)] \prod_{j \in C_b} [U_j(C_a \dot\cup C_b) - U_j(C_b)], \tag{3}$$

providing each term of the product is non-negative.

Proof: See *Lemaire* [1973b].

Theorem: To each set of probabilities W can be associated one and only one treaty $\overline{y}(N)$ satisfying all the axioms. It can be obtained by the recursion

$$y_i(\{i\}) = x_i$$
$$\vdots$$

$$y_i(C) = \begin{cases} \sum_{\substack{C_a \subset C \\ C_a \neq \phi}} W_{C_a \dot\cup \overline{C_a}} \, y_i(C_a \dot\cup \overline{C_a}) & i \in C \quad \begin{cases} \forall \; C \Rightarrow 1 < c < n \\ \overline{C_a} = C \setminus C_a \end{cases} \\[2em] 0 & i \notin C \end{cases}$$

$$\vdots$$

$$y_i\,(N) = \sum_{\substack{C_a \subset N \\ C_a \neq \phi}} W_{C_a \dot{\cup} \overline{C_a}}\, y_i\,(C_a \dot{\cup} \overline{C_a}) \qquad i = 1, \ldots, n \quad \overline{C_a} = N \backslash C_a, \qquad (4)$$

where, at each step, $\sum_{C_a \subset C} W_{C_a \dot{\cup} \overline{C_a}} = 1$ and $W_{C_a \dot{\cup} \overline{C_a}} \geqslant 0$, and $y_i\,(C_a \dot{\cup} \overline{C_a})$ is ob-

tained by maximizing (3), with the disagreement point

$$U_i\,(C_a) \qquad i \in C_a$$
$$U_j\,(\overline{C_a}) \qquad j \in \overline{C_a}\,.$$

Proof:

1. Existence: It is sufficient to verify that $\bar{y}\,(N)$ satisfies all the axioms. This proof is straightforward.
2. Suppose that, for a given set $\{W_{C_a \dot{\cup} \overline{C_a}}\}$, there exist two different optimal solutions

$\bar{y}\,(N)$ and $\bar{y}'\,(N)$, i.e. there exists at least an i such that $y_i\,(N) \neq y_i'\,(N)$.

We shall first show that the two solutions must differ in at least a partial treaty. In other words, it is impossible that $y_i\,(C_a \dot{\cup} \overline{C_a}) = y_i'\,(C_a \dot{\cup} \overline{C_a})$ for all $C_a \subset N$ and that $y_i\,(N) \neq y_i'\,(N)$. (4) expresses that the partial treaties $y_i\,(C_a \dot{\cup} \overline{C_a})$ are summarized by a weighted arithmetic mean. One could of course think of other parameters, like the geometric or the quadratic mean for instance, but the only parameter satisfying the admissibility condition is the weighted arithmetic mean

$$y_i\,(N) = \sum_{\substack{C_a \subset N \\ C_a \neq \phi}} W^i_{C_a \dot{\cup} \overline{C_a}}\, y_i\,(C_a \dot{\cup} \overline{C_a})\,.$$

We shall now show that the admissibility condition also implies that $W^i_{C_a \dot{\cup} \overline{C_a}} = W^1_{C_a \dot{\cup} \overline{C_a}}\;\forall i$. It is sufficient to prove it for $n = 3$. In this case, there are only three ways to form the grand coalition, which we shall note to simplify

$$A = \{12\} \dot{\cup} \{3\}$$
$$B = \{13\} \dot{\cup} \{2\}$$
$$C = \{23\} \dot{\cup} \{1\}\,.$$

Thus $y_1\,(N) = W_A^1\, y_1\,(A) + W_B^1\, y_1\,(B) + W_C^1\, y_1\,(C)$

$$y_2\,(N) = W_A^2\, y_2\,(A) + W_B^2\, y_2\,(B) + W_C^2\, y_2\,(C)$$

$$y_3\,(N) = W_A^3\, y_3\,(A) + W_B^3\, y_3\,(B) + W_C^3\, y_3\,(C).$$

(1) allows us to replace $y_1 (A)$ by $z - y_2 (A) - y_3 (A)$, with similar relations for $y_1 (B)$ and $y_1 (C)$. We obtain

$$y_1 (N) = W_A^1 [z - y_2 (A) - y_3 (A)] + W_B^1 [z - y_2 (B) - y_3 (B)] +$$
$$+ W_C^1 [z - y_2 (C) - y_3 (C)]$$

$$y_2 (N) = W_A^2 y_2 (A) + W_B^2 y_2 (B) + W_C^2 y_2 (C)$$

$$y_3 (N) = W_A^3 y_3 (A) + W_B^3 y_3 (B) + W_C^3 y_3 (C).$$

Summing, and using (1), we get

$$z = y_2 (A) (W_A^2 - W_A^1) + y_3 (A) (W_A^3 - W_A^1) + y_2 (B) (W_B^2 - W_B^1) +$$
$$+ y_3 (B) (W_B^3 - W_B^1) + y_2 (C) (W_C^2 - W_C^1) + y_3 (C) (W_C^3 - W_C^1) +$$
$$+ W_A^1 z + W_B^1 z + W_C^1 z.$$

Since the W's are the coefficients of a weighted arithmetic mean,
$W_A^1 + W_B^1 + W_C^1 = 1$, and the sum

$$y_2 (A) (W_A^2 - W_A^1) + y_2 (B) (W_B^2 - W_B^1) + y_2 (C) (W_C^2 - W_C^1)$$

$$+ y_3 (A) (W_A^3 - W_A^1) + y_3 (B) (W_B^3 - W_B^1) + y_3 (C) (W_C^3 - W_C^1)$$

must be identically equal to zero, $\forall y_2$ and y_3. Thus $W^i = W^1 \ \forall i$.

So there exists a coalition $C_a \subset N$ such that $y_i (C_a \cup \overline{C_a}) \neq y_i' (C_a \cup \overline{C_a})$. Since the solution of the maximization of (3) is unique, this result can only be explained by a difference of the disagreement points $y_i (C_a)$ and $y_i' (C_a)$. Suppose $U_i [y_i (C_a)] <$
$< U_i [y_i' (C_a)]$. There exists a player $j \in C_a$ such that $U_j [y_j (C_a)] > U_j [y_j' (C_a)]$, for otherwise $\bar{y} (C_a)$ would not be Pareto-optimal in the subgame $[C_a, v (C_a'), \xi_{C_a}]$.

The same argument can be repeated iteratively for the coalition C_a: there exists a $C_b \subset C_a$ such that $U_i [y_i (C_b)] < U_i [y_i' (C_b)]$. j must also belong to C_b (or another player j' such that $U_{j'} [y_{j'} (C_b)] > U_{j'} [y_{j'}' (C_b)]$); in fact, if j were a member of $C_a \setminus C_b$, $\bar{y} (C_b)$ would not be Pareto-optimal in $[C_b, v (C_b'), \xi_{C_b}]$ as $\bar{y}' (C_a \setminus C_b)$ in $[C_a \setminus C_b, v (C_b'), \xi_{C_a \setminus C_b}]$ and axiom 2 would be violated.

So we can present a finite succession of coalitions

$$N \supset C_a \supset C_b \supset \ldots \supset C_f \supset \ldots \supset C_F$$

such that, for all $f \leqslant F$:
$$i, j \in C_f;$$
$$U_i [y_i (C_f)] < U_i [y_i' (C_f)];$$
$$U_j [y_j (C_f)] > U_j [y_j' (C_f)].$$

The last term C_F can only be the coalition formed by players i and j (otherwise we could have continued the process). There exists thus two treaties $\bar{y}\,(C_F)$ and $\bar{y}'\,(C_F)$, Pareto-optimal in $[\{ij\}, v\,(C), \xi_{\{ij\}}]$, i.e. such that

$$\max \{U_i\,[y_i\,(\{i, j\})] - U_i\,[y_i\,(\{i\})]\} \cdot \{U_j\,[y_j\,(\{i, j\})] - U_j\,[y_j\,(\{j\})]\}$$

$$= \max \{U_i\,[y_i'\,(\{i, j\})] - U_i\,[y_i'\,(\{i\})]\} \cdot \{U_j\,[y_j'\,(\{i, j\})] - U_j\,[y_j'\,(\{j\})]\} .$$

This contradicts the lemma, applied to the coalitions $C_a = \{i\}$ and $C_b = \{j\}$.

The solution is constructed by induction on the number of players of the coalitions: one must successively compute the value of all the two-player coalitions, then all the three-player sets, . . . to end up finally with the grand coalition. The optimal treaty for a coalition C of c players is obtained by considering the set of its $2^{c-1} - 1$ (strict) sub-coalitions C_a for which there already exists a computed sub-treaty. For each C_a, one computes by (3) a treaty $\bar{y}\,[C_a \cup (C \setminus C_a)]$. The utility granted to a player never diminishes when one or more partners are added to the coalition: (3) always provides a $U_i\,(C_a \cup \overline{C_a})$ greater or equal than $U_i\,(C_a)$. The higher his disagreement point, the higher the utility awarded to a player. The procedure provides $2^{c-1} - 1$ (generally) different partial treaties, which are summed up by a weighted arithmetic mean. The fact that $W^i_{C_a \cup \overline{C_a}}$ does not depend on i allows us to interpret those weights as "probabilities associated to orders of formation of the coalitions".

To sum up, the value concept takes into consideration all the possible orders of formation of the grand coalition, weighted by their respective probabilities; each player allies with other players or sets of players so that after $(n-1)$ junctions N is formed and a treaty concluded. All the grouping possibilities are considered, weighted, and account in the final solution.

For $n = 2$, the value coincides with our unweighted value [*Lemaire*, 1973b] the Nash solution and the Shapley value.

For $n = 3$, the value weights three different partial treaties $\bar{y}\,[\{12\} \cup \{3\}]$, $\bar{y}\,[\{13\} \cup \{2\}]$ and $\bar{y}\,[\{1\} \cup \{23\}]$. Since the disagreement points are computed on the basis of coalitions of one or two persons, the partial treaties are the same as in the symmetrical value. The solution differs generally from the Shapley value.

For $n > 3$, however, the generalization is more than just "adding weights" to the partial treaties, since the disagreement points already take the affinities into account and favour the close partners.

Nothing was said up to now as far as the determination of the weights $W_{C_a \cup \overline{C_a}}$ is concerned. This will be the subject of the next section.

---

[3]) However, the hypotheses of the model imply that they will be forced to cooperate at the final step, since the grand coalition is bound to eventually form. This is a consequence of the fact that we required the value of a n-person game, a value that is useless if we know in advance that N will never form. But, as our theory also provides the value of all the $(n-1)$-person subgames, as well as the probabilities of formation of each sub-coalition, no modification is required when one (or more) of the distances is infinite.

6. Formalization of the Affinity Concept: the Coalition Forming Procedure

We suppose that the affinity between two players can be expressed by a non-negative number, d_{ij}, representing the "distance" (in a broad sense) between i and j: the larger the distance, the lesser the affinity between both players. $d_{ij} = \infty$ means that the antipathy between them is so strong that they will never join together a sub-coalition.[3]) On the other hand, $d_{ij} = 0$ implies that the coalition $\{i, j\}$ will immediately form. This is a relatively uninteresting case, since it amounts to the same thing to consider $\{i, j\}$ as a single player. It is therefore not restrictive to suppose that the (symmetrical) matrix of the distances (the figures of the diagonal are irrelevant) does not contain more than one zero in each row or column (the reunion of three players in a single step is indeed not allowed, although the model could be easily adapted to this case, by introducing as a first stage the merging of the three players with probability one).

Define the "distance" between two coalitions C_a and C_b by

$$d_{C_a, C_b} = \frac{\sum\limits_{i \in C_a} \sum\limits_{j \in C_b} d_{ij}}{|C_a| \cdot |C_b|} .$$

The value of all the two-player coalitions can easily be computed by (3). Suppose, by induction, that we have already computed the solution for all the sets containing at most $(n-1)$ players. It only remains to calculate the value of the grand coalition.

A coalition configuration of order m (shortly a m-configuration) is a vector

$$C^m = (C_1, \ldots, C_m) \quad \begin{cases} C_a \cap C_b = \phi \quad a \neq b \\ \bigcup\limits_{a=1}^{m} C_a = N \\ C_a \neq \phi \; \forall \, a , \end{cases}$$

indicating the coalitions formed after step $(n-m)$. During a negotiation, m successively takes all the integer values, decreasing from n to 1. At the beginning, $n = m$, and $C^n = (\{1\}, \{2\}, \ldots, \{n\})$. After the final junction, $m = 1$ and $C^1 = (\{1, \ldots, n\})$. For $1 < m < n$ there exists several different coalition configurations, denoted by $C_x^m, C_y^m, \ldots$ Let M_m be the set of all the m-configurations. We shall denote $i \sim j$ if i and j belong to the same coalition of C^m, $i \not\sim j$ if they do not.

Each m-configuration C^m generates a number of descendants C^{m-1} obtained by joining two coalitions of C^m. Let D_1 be the set of all the descendants of C^m. Of course, two different m-configurations can produce the same descendant. Let W_{C^m} be the probability that C^m forms during the procedure, and $W_{C^{m-1}|C^m}$ the (conditional) probability that C^m generates C^{m-1}. Naturally, this probability is zero if C^{m-1} cannot be a descendant of C^m.

We must associate to each distance matrix D a set W of probabilities $W_{C_a \cup \overline{C_a}}$, defined $\forall \, C \subset N$, $\forall \, C_a \subset C \Rightarrow \overline{C_a} = C \setminus C_a$, $C_a \neq \phi$, $\overline{C_a} \neq \phi$.

$$D = \{d_{ij}\} \xrightarrow{R} W = \{W_{C_a \dot\cup \overline{C_a}}\}$$

Of course not any rule R that associates a set W to a matrix D is suitable for our problem. A rule will be said *coherent* if it satisfies the following conditions.

Condition 1 (Rules of probability calculus)

1 a) $W_{C^m} \geqslant 0 \quad \forall\, C^m$

1 b) $\displaystyle\sum_{M_m} W_{C^m} = 1 \quad m = 1, \ldots, n$

1 c) $\displaystyle\sum_{D_1} W_{C^{m-1}|C^m} = 1 \quad \forall\, C^m$

1 d) $W_{C^{m-1}} = \displaystyle\sum_{M_m} W_{C^{m-1}|C^m} \cdot W_{C^m} \quad \forall\, C^{m-1}.$

Condition 2 (Relation between affinities and probabilities)

2 a) W_{C^m} is a non-increasing function of $d_{ij} \quad \forall\, C^m \Rightarrow i \sim j$

   W_{C^m} is a non-decreasing function of $d_{ij} \quad \forall\, C^m \Rightarrow i \not\sim j$

2 b) $\displaystyle\lim_{d_{ij} \to 0} W_{C^{n-1}} = 1 \qquad\qquad i \sim j$

2 c) $\displaystyle\lim_{d_{ij} \to \infty} W_{C^m} = 0 \qquad\qquad \forall\, C^m, \; i \sim j$
$\qquad\qquad\qquad\qquad\qquad\qquad\qquad \forall\, m \Rightarrow 1 < m < n.$

Condition 3 (Possible symmetry of two players)

3. If $d_{jl} = d_{il} \quad \forall\, l$, then $W_{C_x^m} = W_{C_y^m}$, where C_y^m is obtained from C_x^m by commuting i and j.

Condition 4 (Relations between successive configurations)

4. If $W_{C_x^m} > W_{C_y^m}$, then $W_{C_x^{m-1}} > W_{C_y^{m-1}} \quad \forall\, m$, if C_x^{m-1} is a descendant of C_x^m and if C_y^{m-1} is the descendant of C_y^m obtained through the same adjunction.

Condition 5 (Relations between configuration probabilities and weights)

5. $W_{C_a \dot\cup \overline{C_a}} = W_{C^2}$, $\forall\, C_a$, where $C^2 = (C_a, \overline{C_a})$.

Condition 6 (Invariance with respect to a similarity)

6. W is not affected by a multiplication of the distances by a positive constant:
if $d'_{ij} = kd_{ij}$ $\forall ij$, $W' = W$.

There exists few coherent rules. In the sequel, we shall use the following rule

$$W_{C^{m-1}|C^m} = \frac{1/d^2_{C_a,C_b}}{\sum\limits_{c=1}^{m} \sum\limits_{d \neq c} 1/d^2_{C_c,C_d}}$$

where $C^{m-1} = (C_1, \ldots, C_a \cup C_b, \ldots, C_m)$ is the descendant of
$C^m = (C_1, \ldots, C_a, \ldots, C_b, \ldots, C_m)$. We thus suppose the attraction between two
coalitions inversely proportional to the square of their distance.

7. Application to Reinsurance

Among all the utility functions, actuaries tend to prefer the exponential utilities, of
the form

$$u_j(x) = \frac{1}{a_j}(1 - e^{-a_j x}) \qquad\qquad j = 1, \ldots, n. \tag{5}$$

They have in fact demonstrated [*Gerber*, 1974a and 1974b; *Leepin*, 1975; *Lemaire*,
1975] that exponential utilities possess very desirable properties for insurers. Their
main argument is that, if the premium P_j is computed by means of the zero-utility
principle

$$\int_0^\infty u_j(R_j + P_j - x)\, dF_j(x_j) = u_j(R_j)$$

the exponential utilities are the only ones for which the principle is additive and iterative, in the sense that

1. the premium for two independent risks is the sum of the individual premiums, and
2. (i)   computing the premium using the distribution function of the losses of the
portfolio, and
   (ii) computing the premium on an individual basis and then integrating with
respect to the structure of the portfolio

lead to the same global income.

It can besides be shown that a_j represents the risk aversion of the company, and
that the ruin probability of j is bounded by $\exp(-a_j R_j)$, if the zero-utility principle is
constantly applied.

Assuming that all the companies of the pool use the form (5), we can solve equations (2). Taking into account condition (1), the solution is

$$y_j(\bar{x}) = q_j z + y_j(0),$$

where

$$q_j = \frac{1/a_j}{\sum\limits_{i=1}^{n} 1/a_i}$$

and

$$y_j(0) = S_j - q_j \sum_{i=1}^{n} \left(S_i + \frac{1}{a_i} \operatorname{Log} \frac{k_i}{k_j} \right).$$

This is a familiar quota-share treaty, very popular in fire, marine and aviation insurance: each company accepts a fixed proportion q_j of the total direct business of each of its partners, and keeps the same proportion of its portfolio. Consequently its liabilities $q_j z$ are the same proportion of the total amount of the claims. The proportion accepted is not subject to bargaining and is inversely proportional to the risk aversion of the insurer. If there are no claims, $z = 0$ and $y_j(\bar{x}) = y_j(0)$. $y_j(0)$ thus represents a side payment, a sum, decreasing with q_j, one has to pay (or to receive) in order to enter the pool. Naturally $\Sigma\, y_j(0) = 0$. The side payment depends on the constants k_j: the companies will have to negotiate the amount of their contribution $y_j(0)$.

As an example, consider three companies willing to sign a reciprocal treaty. We suppose that 1 is a fairly small, recently formed and fast growing company, with a reserve of $S_1 = 2$. Distributions of claim amounts can usually be approximated with good accuracy by a Γ-distribution

$$dF_j(x_j) = \frac{e^{-\tau_j x}\, x^{c_j-1}\, \tau_j^{c_j}}{\Gamma(c_j)}\, dx \qquad\qquad (c_j, \tau_j > 0).$$

Let $m_j = c_j/\tau_j$ and $\sigma_j = c_j/\tau_j^2$ be the means and the standard deviations of those distributions. We suppose that $m_1 = 1.2$ and $\sigma_1 = 1.1$, which means that, due to its recent expansion, 1 does not yet possess sufficient reserves in order to meet the possibility of large claims and desperately needs reinsurance, to keep its ruin probability at a decent level. It therefore chooses a high risk aversion $a_j = .98$.

2 has the following parameters: $S_2 = 4$, $m_2 = 2$, $\sigma_2 = 1.5$, $a_2 = .5$. Its situation is typical of an average sized company: its premium income is steady, the business accepted sound. Due to the extreme skewness of the claims distribution, however, some reinsurance is sought in order to reduce the yearly fluctuations and to reduce the ruin probability.

The parameters of 3 are $S_3 = 12$, $m_3 = 4$, $\sigma_3 = 3$, $a_3 = .17$. It is of course a very large company with large premium income and assets. Its need for reinsurance is not so high, for the company would only be ruined in case of a catastrophe. A risk exchange treaty is then concluded more as a way to attract new business than to increase the safety level. Therefore the risk aversion is kept low.

If we suppose that
- 1 and 2 have a strong affinity (they are both Co-operative companies for instance), and that

– 2 and 3 are already linked by a treaty in another branch,
we can choose as the set of distances $D = (d_{12} = 1, d_{13} = 2.5, d_{23} = 2)$.
The initial utilities are

$$U_1(x_1) = -33.0488$$
$$U_2(x_2) = .8244$$
$$U_3(x_3) = 4.0580.$$

The treaties arising from the two-player coalitions are

1. $\{1\} \dot{\cup} \{2\}$ Quotas $q_1 = .3378$ Side payment $y_1(0) = 2.3418$
   $q_2 = .6622$

   Utilities after reinsurance $U_1[\bar{y}(12)] = -4.2796$
   $U_2[\bar{y}(12)] = 1.6881$

2. $\{1\} \dot{\cup} \{3\}$ Quotas $q_1 = .1478$ Side payment $y_1(0) = 3.0855$
   $q_2 = .8522$

   Utilities after reinsurance $U_1[\bar{y}(13)] = -6.2450$
   $U_3[\bar{y}(13)] = 4.7699$

3. $\{2\} \dot{\cup} \{3\}$ Quotas $q_2 = .2537$ Side payment $y_2(0) = .6430$
   $q_3 = .7463$

   Utilities after reinsurance $U_2[\bar{y}(23)] = 1.1007$
   $U_3[\bar{y}(23)] = 4.2306.$

Adding the third player leads to quotas

$$q_1 = .1146, \quad q_2 = .2247, \quad q_3 = .6607.$$

3, being the least risk averse, takes advantages of this to attract a large proportion of
its partners' portfolios. As a compensation of its increased liabilities, it will naturally
demand a high fixed sum. We obtain the following side payments and utilities.

	Side payments	Utilities
1. $\{12\}\dot{\cup}\{3\}$	$y_1(0) = 2.4458$	$U_1(\bar{y}) = -2.8960$
	$y_2(0) = -1.8673$	$U_2(\bar{y}) = 1.7362$
	$y_3(0) = -.5785$	$U_3(\bar{y}) = 4.1634$
2. $\{13\}\dot{\cup}\{2\}$	$y_1(0) = 2.7476$	$U_1(\bar{y}) = -4.2439$
	$y_2(0) = .7666$	$U_2(\bar{y}) = 1.0153$
	$y_3(0) = -3.5142$	$U_3(\bar{y}) = 4.8388$
3. $\{1\}\dot{\cup}\{23\}$	$y_1(0) = 3.5005$	$U_1(\bar{y}) = -9.9894$
	$y_2(0) = -.8645$	$U_2(\bar{y}) = 1.5644$
	$y_3(0) = -2.6360$	$U_3(\bar{y}) = 4.6708.$

We notice that the last company to enter the bargaining has a solid disadvantage.

Using the rule defined in section 6, we find the weights of the three different coalition formations

$$W = (.7092, \quad .1135, \quad .1773).$$

The final solution is then

$$y_1 (0) = \quad 2.6670 \qquad U_1 (\overline{y}) = -3.8441$$
$$y_2 (0) = -1.3905 \qquad U_2 (\overline{y}) = \quad 1.6551$$
$$y_3 (0) = -1.2765 \qquad U_3 (\overline{y}) = \quad 4.3557.$$

1 and 2 take advantage of their vicinity to reduce their side payments to 3.

As the initial utilities correspond to side payments of $(4.6532, 1.1209, -0.2286)$, the final solution achieves the same utility increase as a gain in capital of $(1.9862, 2.5114, 1.0479)$.

We can notice that

– 1 has been growing too rapidly and is therefore compelled to reinsure nearly all of its portfolio, acting more as a broker than as an insurance company. Despite the levelling of the claims achieved by the treaty, the balance sheet of the year is bound to show a deficit. All reinsurance can achieve is to guarantee that the loss will not be too large;
– 2 has improved its situation since its premium income has risen sharply while its risks are now spread over the three companies;
– 3 has succeeded in its goal to enlarge its business without suffering any diminution of its secureness.

References

Bühlmann, H.: Mathematical Methods in Risk Theory. Berlin 1970.

Gerber, H.: On additive premium calculation principles. ASTIN Bull. VII, 1974a, 215–222.

– : On iterative premium calculation principles. Mitt. der Vereinigung Schweizerischer Versicherungsmath. 74, 1974b, 163–172.

Leepin, P.: Über die Wahl von Nutzenfunktionen für die Bestimmung von Versicherungsprämien. Mitt. der Vereinigung Schweizerischer Versicherungsmath. 75, 1975, 27–45.

Lemaire, J.: Optimalité d'un contrat d'échange de risques entre assureurs. Cahiers du C.E.R.O. 15, 1973a, 139–156.

– : A new value for games without transferable utilities. International J. of Game Theory 2, 1973b, 205–213.

– : Sur l'emploi des fonctions d'utilité en assurance. Bull. A.R.A.B. 70, 1975, 64–73.

– : Echange de risques entre assureurs et théorie des jeux. ASTIN Bull. IX, 1977, 155–179.

Applied Game Theory. 1979 ©Physica-Verlag, Wuerzburg/Germany

Tariff Games[1])

By *D.L. Allen*, Virginia[2])

Abstract: Previous empirical studies of tariff reductions have regarded such events as unilateral, loss-minimizing situations. This paper develops the "trade power index" (TPI), which is similar in computation but not in interpretation to the Shapley-Shubik power index, to analyze the tariff reductions made by the US, the UK, and the EEC at the Kennedy Round as multilateral, benefit-maximizing actions. Regression results indicate that the TPI is significantly related to the tariff reductions made in the Round. The evidence of this paper, then, supports the hypothesis that bargaining for benefit-maximization has some role in explaining the conduct of tariff setters. Further, the TPI demonstrates the possibility of entering real world data into the calculation of power indices in other than voting situations.

1. Introduction

The Kennedy Round of tariff negotiations, conducted from May, 1963 to June 30, 1967 under the auspices of the GATT,[3]) is still considered to have been one of the most significant events in recent tariff history. Its major contributions may be categorized under two general headings: the scope and depth of the resultant tariff concessions and the adoption of new techniques for achieving these reductions. Concessions were made on over $ 40 billion of trade, this being eight times the corresponding value for the previous Dillon Round [GATT, p. 17]. Reductions on industrial commodities averaged 35 percent [*Preeg*, chapter 13]. The methods employed in this round were clearly related to its outcome. The need was seen for a multilateral framework under which bilateral balance might not be achieved, but overall balance in concessions should be received by all participants.[4])

For these reasons the Kennedy Round affords a rare opportunity for substantive empirical research concerning tariffs and tariff policies. Most of the research of this type pre-dates the Kennedy Round dealt with analysis of the tariff structure of a particular country. Attempts were made to associate the inter-industry variation in levels of protection to variations in industrial characteristics, particularly labor-related ones [see, for example, *Balassa; Ball*]. But, a country's tariff structure at any point in time is the product of the changing political pressures and industrial traits over many years. Therefore, analysing a tariff structure with respect to the industrial or political charac-

---

[1]) The author is indebted to John Dutton, Andy Stoeckel, Ed Tower, and Roy Weintraub for invaluable assistance on previous versions of this paper. Sole responsibility for remaining errors goes to the author. This paper was extracted from the author's Duke University Ph.D. dissertation.

[2]) Prof. *Deborah L. Allen*, Dept. of Economics, VPI, Blacksburg, Virginia 24061, U.S.A.

[3]) General Agreement on Tariffs and Trade.

[4]) LDC's and special conditions countries were largely exempt from extending full reciprocity.

teristics prevailing at one point in time should not be expected to yield clear-cut or even meaningful results [*Cheh*]. However, the occurance of a major change in a country's tariff structure at one fairly discrete point in time (where a five year span may be considered as such a point) makes it possible to conduct analysis with respect to the levels of variables current at the time of the change.[5])

2. Recent Analysis

Recently, *Cheh* [1974], *Bale* [1977], and *Reidel* [1977] have taken advantage of this opportunity. *Cheh* demonstrated that the U.S. tended to minimize the labor adjustment costs that would be expected to be incurred as a result of the Round. *Bale* presented further evidence supporting *Cheh*'s hypothesis for the U.S. *Reidel* attempted to duplicate *Cheh*'s study for Germany but reported that "no evidence is found to supported a similar conclusion with regard to nominal tariff concessions granted by the EEC as they pertain to the German Federal Republic" [*Reidel*, 1977].

The reasoning behind the above hypothesis is probably similar to the following. Countries, recognizing that there are gains to be realized from freer trade, agree to reduce their tariffs, on average, by approximately the same amount, in percentage terms. Each country attempts to meet the average reduction effected by the other countries by making reductions of different amounts in its own various tariffs. *Cheh*'s hypothesis is that these differences in reductions may be at least partially explained by differences in labor adjustment cost. That is, each country will meet the necessary average reduction with a minimum of labor adjustment cost. This theory seems to have been borne out by *Cheh*'s and *Bale*'s studies, if not by *Reidel*'s and appears reasonable as far as it goes.

3. An Alternative Form of Analysis

There is more to a multilateral round of negotiations than is reflected in the above line of reasoning. The above theory centers on constrained loss-minimization. However, the main purpose of the Round was to increase trade so that each participant might realize gains. Some inclusion of a benefit side thus seems appropriate in analysing the actions of countries in the Round. Further, this round was a truly multilateral round. A 50 percent reduction on all tariffs, with a bare minimum of exceptions, was agreed to by the countries as a working hypothesis. It was then up to each country to draw up an exceptions list, this being a list of industries on which it desired to reduce tariffs by less than 50 percent. But, these lists were not automatically accepted. Each item had to be defended as to why it should be granted exception before a panel of countries who had a trading interest in each particular good. Simultaneously, other countries would be withdrawing old offers and making new offers in an attempt to

---

[5]) *Finger* [1974]. In this article *Finger* analyzed tariff changes rather than structures, but the limited nature of the changes that took place in the Dillon Round make this round somewhat less amenable to study.

achieve reciprocity for themselves vis-a-vis their trading partners taken together. Further, offers in several major industrial categories were contingent upon the actions of other countries. In fact, industries where such conditional offers were made comprised 30 percent of all imports of manufacturers by the participating countries [*Preeg*, p.93]. The Kennedy Round, then, was a bargaining situation. This, too, should be accounted for in analysing the differences in the amounts of the reductions made by a country in the tariffs protecting its various industries.

Therefore, while previous papers have regarded the Kennedy Round as a situation of unilateral, loss-minimization, the present paper analyses this round as a multilateral bargaining situation of constrained benefit-maximization. Each country enters the Round desiring to emerge with the maximum possible benefit (the valuation of which is to be discussed in the next section). Recognizing that benefits to itself are created by the reduction of its partners' tariffs rather than by its own actions, a country would realize that it must bestow benefits in order to receive them. In other words, tariff concessions had to be bargained for.

The feasible region of gains would be restricted by the extent of interdependence among the countries, that is by the export supply and import demand conditions prevailing among the countries, as well as by the political and other considerations which partially determine how much concession each country is willing to make. This paper attempts to answer the question: within the realm of concessions that were made by a country, is there any relation between the sizes of the concessions made on the several commodities and the degree of bargaining power that the country held on each commodity vis-a-vis its trading partners?

4. Bargaining Tools and Goals

The preliminary questions to be answered are: what were the bargaining instruments and what were the targets of the negotiators? Next, an appropriate measure must be found which incorporates these bargaining and payoff concepts so that they may be compared to the tariff concessions that were made by each country.

Various sources provide strikingly similar answers to the preliminary questions. In regard to targets, the fundamental purpose of the round was to increase the volume of world trade by reducing impediments to its conduct. But the goal of each negotiator was to obtain the most concession in return for the least possible concessions on the part of his own country. Still, each country had to be satisfied that it had achieved reciprocity before the rounds could be concluded. Some idea of how reciprocity was to be judged should provide an indication of the bases on which concessions were valued.

Four measures of the reciprocity of concessions have been suggested. They are: 1. the average depth of tariff cuts, 2. the trade volume offered for concession, 3. the loss of duties collectable and 4. the projected trade impact. Of these, "the impact of tariff and other concessions on future trade might be considered the main criterion of reciprocity" [*Preeg*, p. 132–133]. On this point, Michael Blumenthal, deputy U.S. negotiator, stated in speeches on two separate occasions,

"We can only look at each offer from the viewpoint of our exports back home . . . the yard-stick of increased export opportunity must still remain the basic criterion in judging the value of proposed offers" [*Preeg*, p. 140].

The payoffs to negotiations are then to be denominated in increments to export production. Bargaining power should thus be defined as the ability to allow for additional export production.[6])

Calculation of potential export expansion depends on the amount of the tariff cut, the volume of trade upon which reductions can be offered, and the relevant demand and supply elasticities (see the Appendix for a mathematical statement). While the latter two factors are out of the control of the negotiators, the first two are not. For simplification, it will be assumed that partial offers were not made. That is to say that offers were made on entire commodity groups. This assumption has the effect of removing the volume of trade from the list of negotiating instruments. However, this assumption clearly does not remove this factor from the list of determinants of the potential for export creation, or, bargaining power. Thus, the instrument variable of negotiators, through which they might exercise the power given them by this and the other three factors, is the level of tariff reduction.

5. Definition and Calculation of the Trade Power Indices

Multilateral tariff reductions increase the efficiency of the allocation of resources worldwide by reducing the relative price distortions caused by tariffs. Where unemployed factors exist, increased aggregate demand stimulated by tariff reductions (and the concomitant reductions in prices faced by demanders) can evoke increases in aggregate production. However, the expansion of a particular country's export production occasioned by such a round is a function of the tariff reductions made by its trading partners, rather than the reductions made by itself (second round effects aside). Therefore, the payoff to each country is determined by the combination of the actions of all of the other participants. Further, the actions taken by any given country affect the payoffs to accrue to the others. Where reciprocity of concessions is to be considered, then, the total payoff to the group can be maximized through cooperation among the members.[7]) In other words, countries behaving in a benefit-maximizing manner would be expected to bargain across goods for concessions.[8])

---

[6]) The concomitant reduction in import-competing production should also play a role here. Negotiators would be concerned with the increase in export production only to the extent that it added to total production. Any decreases in home production due to tariff reductions should then be subtracted from the increases in export production to derive the net production increase. However, to simplify the modeling of both the bargaining position of the countries and the trade among them, the assumption of 0 elasticity of substitution between imports and import-competing goods is made.

[7]) It is the guarantee that each country was to be satisfied that it had achieved reciprocity before the Rounds could be concluded and the enforcement of this reciprocity through the GATT after the conclusion of the Rounds that makes this a cooperative situation. Otherwise the various incentives for and methods of cheating would have to be included in the analysis.

[8]) *Finger* [1974] hypothesizes that a similar type of trading took place to a limited extent in the Dillon Rounds. Industrialized countries tended to make larger reductions on those goods which were supplied by other industrialized countries since this group would have more to offer in return than would the LDC's.

In order to test this bargaining hypothesis, some measure is needed for the value of a particular country's cooperation on each good. The Shapley Value solution to the game-theoretic bargaining situation provides one such measure. For an n-person, non-constant-sum game the Shapley Value solution singles out a particular payoff vector (imputation) which distributes among the players the joint gains which would be realized by the group should all players cooperate. This solution can be considered to be equitable in the sense that is assigns payoffs based upon the average marginal contribution each player would make by joining the group. In this same sense the Shapley Value assigned to a particular player, in relation to the Shapley Values of the other players has been used to capture the concept of power in various contexts [see *Shapley/Shubik; Miller*]. The measure of bargaining power to be used in the tariff negotiation framework is similar. Before presenting the exact measure to be used to represent the bargaining power of countries in the tariff games, a short digression of the calculation and application of the Shapley Value solution would be instructive.

The Shapley Value solution to a game necessitates the modeling of only the third level form of a game, this being the characteristic function form. The characteristic function of a 3-person game may be specified in general as follows:

$$v(\bar{1}) = V_1, \quad v(\bar{2}) = V_2, \quad v(\bar{3}) = V_3$$

$$v(\overline{12}) = V_{12}, \quad v(\overline{13}) = V_{13}, \quad v(\overline{23}) = V_{23}$$

$$v(\overline{123}) = V_{123}$$

where the V's represent the largest payoff that the coalition to which they pertain could guarantee to itself. A modeling of all of the possible strategies each player may use and the payoffs accruing to each as a result of every strategy combination among players is not required. Instead, modeling may be restricted to determining the best possible outcome that each coalition could guarantee for itself. The Shapley Value solution then designates one specific imputation as "the" solution to the game. This imputation assigns to each player his average marginal contribution to the coalitions he might join. For the general, 3-person game the Shapley Value solution is calculated as follows:

$$SV_1 = 1/3\,[v(\overline{123}) - v(\overline{23})] + 1/6\,[v(\overline{12}) - v(\bar{2})] + 1/6\,[v(\overline{13}) - v(\bar{3})] + 1/3\,[v(\bar{1})]$$

$$SV_2 = 1/3\,[v(\overline{123}) - v(\overline{13})] + 1/6\,[v(\overline{12}) - v(\bar{1})] + 1/6\,[v(\overline{23}) - v(\bar{3})] + 1/3\,[v(\bar{2})]$$

$$SV_3 = 1/3\,[v(\overline{123}) - v(\overline{12})] + 1/6\,[v(\overline{13}) - v(\bar{1})] + 1/6\,[v(\overline{23}) - v(\bar{2})] + 1/3\,[v(\bar{3})].$$

Therefore, to calculate the bargaining positions of countries in the Kennedy Round, the tariff negotiations are to be modeled in characteristic function form for each of the many commodities on which negotiations were conducted. This modeling will deal, in particular, with the negotiations concerning the US, the UK, and the EEC. While other countries, notably the Nordic countries, Canada and Japan had substantial influence in certain commodity negotiations, it was the first three blocs mentioned above that played major roles throughout the round. At the present stage of analysis, it was decided that the addition of more countries would add more to the complexity of the modeling effort than to the results. Thus, this paper handles only the three blocs, while realizing that eventual expansion would be desirable.

To see how this modeling might be done consider the case of three countries (A, B, C) engaged in negotiations for tariff reductions on a good, j, where each good is assumed to be differentiable from the others by some classification scheme. Each country may or may not produce some of each good. Unless the most finely detailed of classification schemes is used, each good will be composed of several sub-goods (i.e., chemicals may be considered as a good whose sub-goods are benzenoids, glycerine, and amino acids, etc.). Therefore, each country may be both an importer and an exporter of a given good. Each country is here assumed to trade at least some of the good with each of the other two countries. The characteristic function for this game might then be specified as follows:

$$v^j(A) = v^j(B) = v^j(C) = 0$$

$$v^j(AB) = X^a_{jb} + X^b_{ja}$$

$$v^j(AC) = X^a_{jc} + X^c_{ja}$$

$$v^j(BC) = X^b_{jc} + X^c_{jb}$$

$$v(ABC) = v(AB) + v(AC) + v(BC) -$$

$$-{_a}\sigma_{bc} - {_b}\sigma_{ac} - {_c}\sigma_{ab} - {_a}\sigma_{cb} - {_b}\sigma_{ca} - {_c}\sigma_{ba}$$

where X^i_{jk} is the value of country k's exports to country i of good j that would be stimulated should countries i and k reduce their tariffs to each other on good j. ${_i}\sigma_{kl}$ is the value of the increase in i's imports from k, when they reduce their tariffs on good j to each other, that is due to substitution away from imports from country l. The characteristic function specified above may be considered as the general characteristic function to be used in this model for the tariff game between countries A, B, and C on any commodity group.

Shapley Values calculated from these characteristic functions (using the definition given above) capture, at least to some extent, each of the elements discussed in Section 4 as being determinants of bargaining power in tariff games. This game represents the "as if" case of – if each pair of countries could enter into a bilateral agreement, what gains could they generate. A country's power will then be a function of the contributions it could make by joining each possible coalition. Thus, the higher are the values of the additional exports that would be evoked from your partners by your tariff reductions, the greater will be your power value. Similarly, the more your partners' tariff reductions would stimulate your exports to them the greater will be your power value. These two influences reflect your power as derived from your dual position as both an importer and an exporter. To the extent that you can increase the exports of your partner via your tariff reductions you should be able to extract concession from him. Similarly, a strong position as a potential exporter means that you are the one adding to the value of the group's payoff and should be so rewarded.

Thus, relatively high supply price elasticities for export production would contribute to a higher power value as they would allow a country to make larger contribution to the group. Similarly, relatively high tariff rates and import demand elasticities would also increase the contributions the country could make to each of its partners and, thus, would lead to a higher power value. Further, large traders (either importers or exporters) should have relatively large power values because the percent changes in exports or imports brought about by its partners' and its own tariff reductions, respectively, will translate into larger absolute amounts of trade creation. The own conditions are basically in keeping with the traditional elasticity approach to tariff theory as concerns the ability of countries to affect their terms of trade. The divergence of the desirable partner properties is due to the bargaining approach used here, as opposed to the unilateral approach of traditional theory. The difference between the properties that would lead to a more advantageous bilateral agreement in this model and those that are generally accepted as conducive to more beneficial customs union formations are due to the differences in the arguments posited for the utility functions of countries in the two approaches. Traditional customs union theory "has been confined mainly to a study of the effects of customs unions on welfare rather than, for example, on the level of economic activity ..." [Lipsey]. Thus, changes in production efficiency and consumer welfare as brought about by relative price changes have been the usual focuses of attention. This model posits different arguments for countries' utility functions, these being the levels of export production, since it seems probable that these factors would be the major concerns of tariff negotiators, as discussed in Section 4. Therefore, the differences in the categorizations of "properties conducive to beneficial customs unions" and "properties conducive to beneficial bilateral agreements" should not be seen as contradictory, but rather as different in kind.

One further calculation is necessary to put these power values into the form required for the present application. The values have meaning only when compared across countries within a given good. Thus, each value for a country is to be divided by the sum of the values across countries for that good. The resultant index, here termed the trade power index, will be defined by

$$TPI_{ij} = \frac{SV_{ij}}{\sum_i SV_{ij}}$$

where SV refers to Shapley Value, i refers to country and j to goods.[9]

In order to calculate values for the characteristic functions of the several tariff games, it is necessary to develop a model of the trade between the negotiating countries. This model is presented in the Appendix. The resulting equations to be used for calculation of the changes in trade that would be expected to occur due to tariff reductions are as follows:

$$EX_{jk}^i = \alpha_{jk} \frac{\beta_{jk}^i}{\gamma_{jk}^i} ET_{jk}^i + \alpha_{jk} \frac{\delta_{jl}^i}{\gamma_{jk}^i} ET_{jl}^i \qquad (1)$$

---

[9] It may be noted that the calculation of this index is similar to that of the Shapley-Shubik power index. However, the interpretation is quite different.

where

EX^i_{jk} = the percentage change in exports of good j from country k to country i

ET^i_{jk} = the percentage change in the tariff factor $(1 + t^i_{jk})$ affecting imports of good j from country k into country i (t^i_{jk} is the ad valorem rate of duty imposed on good j coming from country k into country i)

ET^i_{jl} = the percentage change in the tariff factor affecting imports of good j from country l into country i

α_{jk} = the export supply elasticity of good j in country k

$\beta^i_{jk}, \gamma^i_{jk}$ and δ^i_{jl} are as defined in the Appendix. Briefly, they are functions of the suppliers' export supply elasticities for good j, the market shares of the various suppliers in the importer's market for good j, the demand elasticity of the importer for good j, and the importer's elasticity of substitution between import sources.

In the case of an infinite export supply elasticity on the part of country k, the appropriate equation is:

$$EX^i_{jk} = (c_l + \eta_{ij})\, ET^i_{jk} + c_l\, ET^i_{jl} \qquad (2)$$

where η_{ij} is the price elasticity of import demand for good j in country i and c_l is a function of η_{ij}, country i's elasticity of substitution between import sources and the market shares of the import sources of good j into country i.

To fill in the values for the characteristic functions, the appropriate EX^i_{jk} is multiplied by the pre-Round level of imports of good j into country i from country k. 1964 prices are to be used, where the usual indexing problems are recognized. The calculations for each good are conducted independently. Following *Balassa* [1965], and *Cooper* [1965], the assumption of no cross elasticities is made. The values for the EX^i_{jk} and X^i_{jk} for each country and each good and resultant TPI's are presented in Table 1.

Com-modity	EXUSUK	EXUSEEC	EXUSUKl	EXUSEECl	EXEECUS	EXEECUK	EXUKEEC
1	.00889	.00951	.00618	.00922	.07934	.04987	.01175
2	.04011	.03616	.04314	.03625	.07639	.04703	.04223
3	.0	.0	.0	.0	.08717	.05867	.08266
4	.07595	.07595	.06646	.07228	.12084	.07458	.07601
5	.08071	.08071	.08514	.08554	.01747	.01122	.02570
6	.17843	.18614	.1278	.16374	.18695	.11582	.14657
7	.33046	.36652	.16908	.33253	.45533	.25979	.27768
8	.03511	.03425	.03275	.03361	.05649	.03513	.03017
9	.05575	.05688	.04241	.05464	.10189	.06287	.07071
10	.14473	.14699	.11652	.13585	.12075	.07409	.08387
11	.11225	.11494	.08771	.10896	.13390	.08171	.10199
12	.05238	.04900	.05504	.04992	.05499	.03449	.03074
13	.19710	.20876	.13025	.19277	.20390	.12198	.13311
14	.08552	.08673	.08552	.08673	.08945	.05902	.06216
15	.04002	.03560	.05697	.03876	.04891	.02994	.02363

D.L. Allen

Commodity	EXUSUK	EXUSEEC	EXUSUKI	EXUSEECI	EXEECUS	EXEECUK	EXUKEEC
16	.02334	.02101	.02547	.02128	.0	.0	.02928
17	.10277	.09949	.09434	.09694	.07245	.04508	.07511
18	.19274	.20678	.10809	.19279	.30598	.17304	.21521
19	.20456	.22277	.10710	.20091	.29823	.17973	.22596
20	.09803	.09728	.08606	.08848	.13180	.08220	.09884
21	.13441	.14077	.07071	.11919	.39834	.23664	.29138
22	.11347	.10830	.10789	.10667	.09332	.05721	.08834

Commodity	EXUKUS	EXEECUSI	EXEECUKI	EXUKEECI	EXUKUSI	XUKUS	XUKEEC
1	.01663	.07934	.04987	.01214	.01718	5248	5334
2	.06525	.09395	.08724	.05913	.11755	11356	22794
3	.12136	.08650	.04997	.07537	.11979	26345	3852
4	.11816	.13222	.09397	.09905	.13502	50790	23868
5	.04029	.03099	.01563	.04210	.06727	37966	39862
6	.22532	.14139	.09629	.14291	.19806	25740	125188
7	.42856	.34913	.18780	.26270	.24717	11898	91220
8	.04583	.07737	.05342	.04189	.05965	36572	28334
9	.10356	.11794	.07018	.07407	.11692	31859	86594
10	.12995	.12491	.07907	.09487	.13584	41286	14283
11	.15486	.14007	.08625	.10884	.17180	25475	41594
12	.04722	.07522	.05765	.04453	.08092	16450	26183
13	.20148	.18197	.10922	.12908	.17516	8745	37717
14	.09811	.08985	.05959	.08408	.15529	155724	256791
15	.03890	.08619	.08096	.05812	.08783	16253	14242
16	.04554	.0	.0	.04754	.07473	11232	11569
17	.11471	.09247	.05569	.08506	.16957	5023	18112
18	.32585	.20640	.16740	.21433	.22414	346	25409
19	.32143	.22718	.12108	.17173	.22636	268564	330599
20	.15397	.14338	.09359	.11385	.17745	61509	61700
21	.43083	.30916	.20043	.25625	.25194	42199	144972
22	.14411	.11994	.07723	.12125	.19695	59956	59055

Commodity	XUSUK	XUSEEC	XEECUS	XEECUK	TPIUS	TPIUK	TPIEEC
1	29	278	38854	823	.48411	.02943	.48646
2	436	14069	6981	1919	.30373	.36459	.33169
3	341	393	65856	3403	.47314	.19626	.33061
4	3951	10191	66638	24126	.39919	.26802	.33278
5	25981	23800	20866	41042	.39662	.37210	.23128
6	50378	113185	70207	101293	.30137	.29557	.40306
7	33690	160146	57759	48521	.35244	.20740	.44015
8	8256	30264	122206	86616	.34891	.18014	.47096
9	41693	248796	61380	86344	.33946	.22814	.43241
10	23718	60001	110151	56578	.41883	.19736	.38381
11	28702	117876	83467	69063	.37507	.20690	.41803
12	3094	8967	46966	25700	.33883	.22095	.44022
13	21849	91645	34950	36978	.37649	.19406	.42945
14	49422	206197	552033	253507	.36519	.23441	.40040
15	5288	28346	84044	37584	.38563	.08451	.52986
16	1680	13613	48865	8630	.34352	.45424	.20224

Com-modity	XUSUK	XUSEEC	XEECUS	XEECUK	TPIUS	TPIUK	TPIEEC
17	15961	52918	23829	27995	.38252	.20358	.41389
18	10387	67926	484	4843	.34771	.19279	.45950
19	106318	522896	646497	471766	.36086	.24121	.39794
20	37877	55521	179034	113489	.36070	.24849	.39081
21	209905	621222	213593	312358	.31644	.25627	.42728
22	36649	125570	110667	90087	.38357	.25033	.36610

$EXik$ indicates the percent change of imports into country i from country k when i reduces
           its tariffs to both countries k and l by 50 percent.

$EXik1$ indicates the percent change in imports into country i from country k when i reduces
           its tariff to country k only by 50 percent.

Xik is the 1964 value of imports into country i from country k in $ 1000 units.

Tab. 1

6. Testing the Hypothesis

6.1 The Model

The hypothesis of this paper is that the bargaining power of a country on a particular good, as defined above, will have some relation to the tariff reductions that were made by the country on that good. The dependent variable for each country is, thus, the percentage changes in the nominal tariff rates imposed by the country (DT). As discussed by *Cheh* [1967], the notion of effective protection was not prominent at the time of the Kennedy Round. The change in the nominal rates is thus seen to be the appropriate dependent variable. Data for this variable are taken from *Preeg* [1970, p. 208–210] who has calculated the average percentage reductions made by the US, the UK, and the EEC on each of 22 commodity groups, as detailed in Table 2. The trade power index (TPI) has been entered as one explanatory variable. The 1964 level of the nominal tariffs (T) has also been entered as an explanatory variable to control for the fact that equal percentage changes in tariffs would translate into different amounts of absolute reductions for different initial levels of tariffs (data for this variable was also taken from *Preeg*). The equation to be estimated for each country, i, is:

$$DT^i_j = \beta_{0i} + \beta_{1i} (TPI)^i_j + \beta_{2i} (T)^i_j + \epsilon^i_j \qquad (3)$$

where $i = 1, 2, 3$ (Countries US, UK, EEC) and $j = 1, \ldots, 22$ (Commodities). The commodity groups to be included in the analysis are groups of 2-digit BTN categories[10]) where an average reduction of less than 50 percent took place. To the extent that these reductions were below 50 percent, some negotiations must have impacted on the final reductions made in these categories. Such categories are, thus, considered, appropriate for the present analysis.

Initially, separate equations are to be estimated for each of the three blocs. Then a model restricted to $\beta_{11} = \beta_{12} = \beta_{13}$ will be estimated. The standard F test on Theil's U statistic will be used to test the validity of this restriction. Should this restriction not be rejected, further credence would be given to the power value modeling since it would be expected that a measure so derived should enter similarly into the functions

[10]) BTN stands for Brussels Tariff Nomenclature.

Goods Group	Brussels Tariff – Nomenclature Categories
1. Wood, natural cork	4401–05, 4501–02
2. Articles of wood, cork	44–46 minus categories comprising group 1
3. Pulp	47
4. Paper	48
5. Natural fiber and waste	5001–03, 5301–05, 5401–02, 5501–04, 5701–04
6. Yarn and basic fabrics	50–57 minus categories comprising group 5
7. Special fabrics, apparel, other	58–63
8. Unwrought base metal, pig iron, scrap	7301–03, 7401, 7501, 7601, 7701, 7801, 7901, 8001
9. Steel	7304–20
10. Other base metals	7402–12, 7502–04, 7602–07, 7702, 7802–05, 7902–04, 8002–05
11. Articles of base metal, miscellaneous	73–83 minus categories comprising groups 8, 9 and 10
12. Mineral Products	25–26
13. Stone, ceramic and glass	68–70
14. Chemical products	28–39
15. Rubber products	40
16. Raw hides, skins, fur	4101, 4301
17. Articles of leather, fur	41–43 minus 4101, 4301
18. Footwear and headwear	64–65
19. Nonelectrical machinery	84
20. Electrical machinery	85
21. Transportation equipment	86–89
22. Precision instruments	90–92

Tab. 2

determining all of the tariff changes. Initially, estimations of the above equations will be done using Zellner's method of seemingly unrelated equations. This is believed to be appropriate due to the expectation of correlation of errors across countries for individual observations. Error component analysis will be used should other types of error correlation be evidenced.

6.2 Expected Results

It is anticipated that the sign of the β_{2i}'s will be negative. As mentioned above, a given percentage reduction would translate into a larger absolute reduction for a larger initial tariff. It is expected that the sign of the β_{1i}'s should be positive. Should the variable prove to be significant with a positive sign, this would indicate that countries make benefit-maximizing, inter-commodity trades for concessions. That is, a country with a strong bargaining position on one good would make a relatively large concession on that good in order to induce its partners to make concessions on other goods that would be particularly advantageous to itself. This type of bargaining would be implied by the benefit-maximization goal of negotiators posited earlier, since it would allow each country to come away from the Rounds with the maximum possible amount of concessions from its partners compatible with overall reciprocity.

7. Econometric Results

The regression results for the three equations (3), $i = 1, 2, 3$, are presented in Tables 3, 4 and 5. These equations were first estimated using Zellner's technique of seemingly unrelated equations, as planned. However, due to the low calculated covariances across equations, the results obtained via this technique were nearly identical to those produced by ordinary least squares (OLS) estimation. Therefore, the OLS results are reported here for the single country equations. The restricted model was also estimated with OLS.

All of the estimated coefficients have the expected sign. Further, each of the coefficients for (T) and (TPI) in the UK and EEC equations are significantly different from 0 at at least the 2.5 percent level. Further, in the UK and EEC equations, approximately 33 and 59 percent respectively of the variations in tariff reductions have been explained by these variables. The F-statistics for these equations indicate that all of the coefficients are significantly different from 0 at the 1 percent level of significance for the EEC and at the 5 percent level for the UK equations.

Equation 1: Dependent variable: DTUS

Independent Variables	Estimated Coefficient	Standard Error	T-Statistic
C	0.271	0.374	0.724
TPIUS	0.492	0.936	0.525
TUS	-0.572	0.617	-0.926

R-squared = 0.0691
Sum of Squared Residuals = 0.793374
F-Statistic (2., 19.) = 0.705099

DTUS   = Percent change in the ad valorem nominal tariffs of the US
TPIUS   = Trade Power Indices of the US
TUS   = Initial (1964) level of the ad valorem nominal tariffs of the US

Tab. 3

Equation 2: Dependent variable: DTEEC

Independent Variables	Estimated Coefficient	Standard Error	T-Statistic
C	-0.363	0.150	-2.423
TPIEEC	2.293	0.439	5.217
TEEC	-1.611	0.697	-2.311

R-squared = 0.59
Sum of Squared Residuals = 0.317701
F-Statistic (2., 19.) = 13.79

DTEEC   = Percent change in the ad valorem nominal tariffs of the EEC
TPIEEC   = Trade Power Indices of the EEC
TEEC   = Initial (1964) level of the ad valorem nominal tariffs of the EEC

Tab. 4

Equation 3: Dependent variable: DTUK

Independent Variables	Estimated Coefficient	Standard Error	T-Statistic
C	0.417	0.120	3.459
TPIUK	0.853	0.381	2.237
TUK	− 1.340	0.565	− 2.371

R-squared = .334
Sum of Squared Residuals = 0.453251
F-Statistic (2., 19.) = 4.7668

DTUK = Percent change in the ad valorem nominal tariffs of the UK
TPIUK = Trade Power Indices of the UK
TUK   = Initial (1964) level of the ad valorem nominal tariffs of the UK

Tab. 5

While the estimated coefficients are not significantly different from 0 in the US equation, the uniformity of the estimated signs lends some additional support to the hypothesis. The lack of significance of the *TPI* coefficient for the.US might be attributed to the lack of substantial variation in the US's TPI's across goods. The lack of significance of the (*T*) coefficient in the US equation may be due to the particular form of aggregation used. Alternatively, it may indicate that the US made no attempt to either keep formerly relatively high tariffs high or to equalize protection levels across goods.

The restricted model, constrained so that $\beta_{11} = \beta_{12} = \beta_{13}$ produced slightly different coefficient estimates and explained 27 percent of the total variance in the tariff reductions (see Tab. 6). However, the results of the F-test were such that the constraint could not be rejected at the 1 percent level of significance. This lends further support to the bargaining hypothesis. Additionally, the results from all of these equations were robust to several alternative import demand and export supply elasticity assumptions (these assumptions are discussed in the appendix).

8. Conclusions

The results of this paper support the hypothesis that tariff negotiators bargain for concessions, at least to some extent, in order to maximize the benefits to accrue to each country. These results both are consistent with and add to the results obtained by *Cheh* and *Reidel*.

Cheh's results indicate that approximately 50 percent of the variation in the tariff reductions granted by the US were explained by the policy of minimizing labor adjustment costs. The conditions in 1964 were such that the US had virtually equal bargaining power on each of the goods analyzed in the present study so that the TPI had no significant explanatory value over the tariff reductions. However, the overall results of this paper indicate that should different conditions prevail in the next tariff round so that there was substantial variation in this variable, it could be expected to add to the explanation of the variation in tariff reductions.

Equation 4 (Restricted Model)[1]): Dependent variable: DT

Independent Variables	Estimated Coefficient	Standard Error	T-Statistic
CUS	$-$ .015	.140	$-$ .110
TUS	$-$ .461	.516	$-$ .893
CUK	.339	.116	2.913
TUK	-1.403	.604	-2.322
CEEC	$-$ .045	.130	$-$ .349
TEEC	$-$ .699	.796	$-$ .878
TPI	1.228	.306	4.011

R-squared = .279
Sum of Squared Residuals = 1.618
Number of observations = 66

[1]) This model was estimated by stacking the observations of each variable for all of the
countries as follows:

$$\begin{pmatrix} \overline{DTUS} \\ \overline{DTUK} \\ \overline{DTEEC} \end{pmatrix} = \begin{bmatrix} \overline{1} & \overline{0} & \overline{0} & \overline{TPIUS} & \overline{TUS} & \overline{0} & \overline{0} \\ \overline{0} & \overline{1} & \overline{0} & \overline{TPIUK} & \overline{0} & \overline{TUK} & \overline{0} \\ \overline{0} & \overline{0} & \overline{1} & \overline{TPIEEC} & \overline{0} & \overline{0} & \overline{TPIEEC} \end{bmatrix} \times \begin{bmatrix} CUS \\ CUK \\ CEEC \\ \beta_1 \\ \beta_{21} \\ \beta_{22} \\ \beta_{23} \end{bmatrix} + \epsilon$$

where each vector (as denoted by a bar above the name) is of dimension 22×1.

This formulation is equivalent to estimating the 3 equations separately with the restriction
$\beta_{11} = \beta_{12} = \beta_{13}$ being imposed across equations.

CUS, CUK, and CEEC represent the constant terms for the respective countries' equations.

Tab. 6

Thereas *Reidel*'s study found virtually no relation between tariff reductions and
labor adjustment costs for Germany, the present study indicates that approximately
59 percent of the variation in the reductions granted by the EEC may be explained by
the bargaining hypothesis.

In summary, then, it appears that both the theory of unilateral loss-maximization
and the theory of bargaining for benefit-maximization have some role in explaining the
conduct of tariff negotiators.

References

Allingham, M.G.: Economic Power and the Value of Games. Zeitschrift für Nationalökonomie
XXXV, 1975, 293- 299.
Armington, P.: A Theory of Demand for Products Distinguished by Place of Production. IMF Staff
Papers XVI, 1969, 159 178.
Balassa, B.: Tariff Protection in Industrial Countries: An Evaluation. Journal of Political Economy
LXXIII, 1965, 573 594

Bale, M.: United States Concessions in the Kennedy Round and Short-Run Labor Adjustment Costs: Further Evidence. Journal of International Economics **VII**, May 1977.

Ball, D.: United States Effective Tariffs and Labor's Share. Journal of Political Economy **LXXV**, 1967, 183–187.

Brams, St.: Game Theory and Politics. New York 1975.

Cheh, J.: US Concessions in the Kennedy Round and Short-Run Labor Adjustment Costs. Journal of International Economics **IV**, 1974, 323–340.

Cooper, R.N.: Tariff Dispersion and Trade Negotiations. Journal of Political Economy **LXXII**, 1964, 597–606.

Fieleke, N.S.: The Pattern of United States Tariffs: The Myth and the Reality. New England Economic Review, July/August, 1974, 15–18.

Finger, J.M.: GATT Tariff Concessions and the Exports of Developing Countries – US Concessions at the Dillon Round. Economic Journal **LXXXIV**, 1974, 566–575.

GATT: Press Release no. 1008, November 8, 1967, p. 17.

Hnyilicza, E., and *R. Pindyck*: Pricing Policies for a Two-Part Exhaustible Resource Cartel: The Case of OPEC. European Economic Review **VIII**, 1976, 139–155.

Lipsey, R.: Theory of Customs Unions: A General Survey. Economic Journal **LXX**, 1960, 496–513.

Miller, D.R.: A Shapley Value Analysis of the Proposed Canadian Constitutional Amendment Scheme. Canadian Journal of Political Science **I**, 1973, p. 140.

Preeg, E.H.: Traders and Diplomats. Washington, D.C., 1970.

Rapoport, A.: N-Person Game Theory. Ann Arbor 1970.

Reidel, J.: Tariff Concessions in the Kennedy Round and the Structure of Protection in West Germany: An Econometric Assessment. Journal of International Economics **VII**, May 1977.

Riker, W.: Some Ambiguities in the Notion of Power. American Political Science Review **LVIII**, 1964, 341–349.

Shapley, L.: A Comparison of Power Indices and a Non-Symmetric Generalization. Rand Series, 1977.

Shapley, L., and *M. Shubik*: A Method for Evaluating the Distribution of Power in a Committee System. American Political Science Review **XLVII**, 1954, p. 787.

Applied Game Theory. 1979 ©Physica-Verlag, Wuerzburg/Germany

A Game-Theoretic Approach
to International Monetary Confrontations

By *K. Hamada*, Tokyo[1])[2])

Abstract: International monetary confrontations can be regarded as a two-stage game. The first stage is a game situation to agree on a set of monetary rules, i.e., to choose an international monetary regime, and the second stage is a game situation of policy interplays given a set of rules.

The first stage of the game for agreeing on a monetary regime can be characterized as the battle of the sexes; the second stage of the game of policy interplay can be characterized as a variant of the prisoner's dilemma. If the pay-off of a country is the social welfare that is dependent on the rate of inflation and the balance of payments, then the Cournot-Nash equilibrium as well as the Stackelberg leadership solution lies on the inflationary side of the set of von Neumann solutions under the dollar standard system. This result can be generalized in a dynamic model of world inflation and the balance of payments, by a simple formulation of a differential game.

1. Introduction

The more integrated the world economy is through international trade and capital movements, the more interdependent become the national economic policies. In many examples of international economic confrontations, such as trade wars and international monetary disputes, national policy authorities react with each other while taking account of the mutual interdependence of their economic policies. Thus the field of economics potentially contains many interesting problems that can be most appropriate studied by the tool of game theory.

In this paper, I shall discuss application of game theory to monetary aspects of international economy. One of the eminent features of international monetary confrontations is that national policy authorities are not only concerned with policy interactions between countries, but also with the set of rules under which these policy interplays take place. Depending on the nature of monetary rules of the world economy, that is, depending on what kind of monetary regimes are chosen, economic consequences of a given combination of national economic policies may differ a great deal. For example, depending on whether a world economy is under the gold standard, the

---

[1]) Prof. *Koichi Hamada*, University of Tokyo, Dept. of Economics, Bunkyo-ku, Tokyo 113.

[2]) I have written this work whilst visiting the London School of Economics and Political Science from the University of Tokyo, under a grant from the Japan Foundation, to which I am very grateful. I thank Professor Takashi Takayama for his valuable comments.

dollar standard, the SDR standard, or the flexible exchange regime, a given combination of national monetary policies may exert different impacts on economic activities
of constituent countries. Therefore one needs to study the structure of conflicts concerning the choice of an international monetary regime as well as concerning policy
interplays under a given monetary regime.

The second feature of international monetary confrontations worth mentioning
here is the following. There are many academic and practical proposals for the ideal
situation of the world monetary regime. In these ideal proposals, economists discuss
various desirable properties of the system on the assumption that a particular international monetary were actually adopted. Most of these discussions, however, neglect the
study of the benefit-cost structures that a particular system gives to participating countries, and therefore refrain from studying the political feasibility of these ideal plans.
This reminds us of one of the Aesop's fables. Once mice were to succeed in putting a
bell to a cat, then everything is all right. How to put the bell to the cat, however, remains to be the most difficult problem.

Thus the purpose of this essay is to apply tools of game theory to clarify the benefit-cost structure of agreeing on a monetary regime, as well as that of policy interplays
under given monetary regimes. Before going into our main discussion, let us ask why
these strategic or political economy aspects of international monetary relations have
been neglected.

First, as mentioned in the beginning, traditional general equilibrium analysis becomes less powerful when an individual agent has a non-negligible influence on the
economy. This methodological limitation has probably led economists to evade strategic consideration of world monetary relations where few large countries have substantial influences. Here we have every reason why the introduction of game theory to this
area is most welcomed.

In economics, however, even tools of game theory have been utilised most effectively to analyze situations where the number of participants is very large and where
each participant has negligible influence. The highly elegant study of the core and its
relationship to the competitive equilibrium typically illustrates this. If one aims at obtaining some new insight into economic reality by the use of game theory, rather than
just supporting or elaborating on properties of existence and efficiency of competitive
equilibrium, which was established by the general equilibrium analysis, one must develop some apparatus that is suited for the analysis of interdependence among few participants.

The second reason why strategic aspects of international monetary relations have
been neglected lies in the following attitude of economists. Some economists, even
though they are potentially representing the interest of a particular country or a group,
find themselves to sound more persuasive when they advocate a plan as if it were an
ideal plan for the general interest rather than analyzing its benefits and costs to various
countries. Many economists, needless to say, feel at home and secured, if they confine
their attentions to purely economic aspects. It is often argued that since economists
are laymen in political science, it would be dangerous to be involved excessively in
political considerations. Too much concern with political feasibility would make ideal
plans discarded. This view is well expressed in the following concluding remarks in a

textbook by *Yeager* [1968, 148–149]. "For at least two reasons, 'political impossibility' should not rule a proposal out of consideration. First, discussing how a proposed change is likely to work can be a useful pedagogical device . . . Secondly, 'political impossibility' is not an inherent operative property of an economic arrangement. It may be overcome . . . If an economist, concerned with his reputation for practicality and reasonableness, makes amateur assessments of political feasibility and accordingly recommends policies other than those he truly considers best, he is shirking the responsibilities of the expert he claims to be."

Indeed, economists should not give up consideration of an ideal system just because of the difficulty of achieving it. But I believe, at the same time, that economists, as social scientists, should not shy away from analyzing of economic variables related to be feasibility of plans. In order to obtain a positive theory to explain negotiation processes and policy interplays as well as a normative theory to help the feasible design of better international monetary regimes, we need a theoretical framework to analyze the benefit-cost structure of interdependent world.

In the next section, I shall formulate a model of international monetary confrontations as a two-stage game: the first stage being a game of agreeing on a set of monetary rules, that is, of choosing an international monetary regime; the second stage being a game of policy interplays under a given set of monetary rules. After the discussion of the relationship between these two stages, it will be shown that the first stage of agreeing on a monetary regime can be characterized as a game similar to the battle of the sexes.

In section 3, the second stage, the game of policy interplays will be characterized as a variant of the prisoner's dilemma. Taking the dollar standard as an example, a game of monetary policy interplays is formulated, where the pay-off of a country is assumed to depend on the rate of inflation and the balance of payments, and where the strategies are monetary expansions of participating countries. Then it will be shown that the Cournot-Nash solution as well as the Stackelberg leadership solution tends to lie on the inflationary side of the set of Pareto efficient configurations, that is the set of von Neumann solutions.

In section 4, I shall study the same problem in a dynamic context and formulate a simple model of a differential game with quadratic welfare functions. It will be shown that the qualitative nature of the results obtained in the static formulation in section 3 can be extended to a dynamic framework.

2. The Game for Choosing a Monetary System

When one considers the strategic structure of an international monetary confrontation it is helpful to conceive them as game-like situations with two distinct layers or stages. The first stage consists of choosing or agreeing on a set of rules, namely an international monetary regime. The second stage consists of playing with economic policies given a set of rules, namely, under a given monetary regime.

By a set of rules, or the same thing, by an international monetary regime, I mean such systems as the gold standard, the fixed exchange rate system, including the dollar standard, the flexible exchange rate system, and the managed floating rate system. The rules of the game specify the set of players, permissible strategies, and pay-off functions corresponding to combinations of strategies.

Under the fixed-exchange-rate regime the rule of the game is that monetary authorities except that of the reserve currency country are assigned to intervene in the foreign exchange market in such a way as to keep the exchange rates constant. Under the flexible-exchange-rate regime, the rule requires that monetary authorities do not intervene in the exchange market. In these regimes it is quite clear what are the rules of the game, and what are policy interplays under a given set of rules. In some cases, the distinction is more subtle. Under the managed float system, monetary authorities have at least a certain range of freedom regarding how often and how much they intervene in the exchange market, so that the degree of intervention itself becomes a component of strategies.

Even though I have formulated a monetary confrontation as a two-stage game, this does not mean that the two stages are consecutively played. These two stages or layers of games are often played at the same time. Moreover, these two stages are closely related to each other. Poor economic performances under a given set of rules reveal their defects and lead to a change in the rules themselves.

Figure 1 symbolizes this two-stage game in the form of a game tree. If one finds the combination of strategies, for the second stage of the game, that are optimal under a certain criterion, then one can reduce the whole game to a single stage game by the principle of optimality. Suppose, for example, p_1 is the vector of pay-offs for participants that will be realized by the combination of optimal strategies under the first set of rules, and similarly p_2 is the vector of pay-offs with the same property under the second set of rules. Then the whole game is reduced to a single stage game to choose between p_1 and p_2. If the combination of optimal strategies to be played in the second stage is known to each participant, the ultimate pay-offs will be determined as soon as the players agree on the choice of a regime.

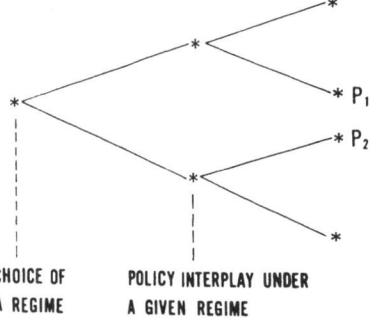

CHOICE OF   POLICY INTERPLAY UNDER
A REGIME   A GIVEN REGIME   Fig. 1

As will be shown in the next section, it is by no means an easy question to settle as to how to determine the combination of optimal strategies, or the most reasonable outcome, and which solution concepts of the game should be used in the second stage.

For the moment, let us discuss the property of the first stage of the game, that is a game of choosing a regime, under the assumption that the pay-offs resulting from the combination of optimal policy interplays are given.

An eminent characteristic at this stage of the game is that the ongoing set of rules prevails until some agreement on a change of rules is reached by all the participants, or at least by a considerable number of participants [Hamada, 1974]. This property gives this stage a pay-off structure similar to the battle of the sexes. The battle of the sexes is a name of the game with a pay-off structure attached to the following scenario. A boy and a girl, each has a choice for an evening's entertainment, a prize fight or to a ballet. Suppose if one is allowed to use a conventional male chauvinistic example, the boy prefers the fight and the girl the ballet. However, if they reach no agreement, both must stay solitary at home. (See the pay-off matrix in Table 1). Unless a new agreement for the monetary reform is reached, the current system − for example, the managed float system at the present moment − will continue. If we normalize the pay-offs to give the present system the values of the origin, the pay-off matrix for adopting two alternative systems, say, return to the Gold standard, or adoption of fixed exchange rates based on dollars, could be illustrated as in Table 2. Suppose the gold standard is preferred by Europe and the Dollar standard is preferred by the United States. Then we could assign the values of parameters such that $a < b$, and $c > d$. If b and c are positive, then this game of choosing a regime becomes a game analogous to the battle of the sexes. However, there is no assurance that a, b, c, d, are positive, because it is not always certain that pay-offs from the new system are better for the participants. The probability of negative values for these parameters becomes higher when we consider the adjustment cost for each country to adapt itself to any changes in the international monetary regime.

boy \\ girl	fight	ballet
fight	(3, 1)	(0, 0)
ballet	(0, 0)	(1, 3)

Tab. 1

U. S. \\ Europe	Gold Standard	Dollar Standard
Gold Standard	(a, b)	(0, 0)
Dollar Standard	(0, 0)	(c, d)

Tab. 2

Let us summarize some of the properties of the battle of the sexes [Luce/Raiffa, 91−92]. There are two equilibrium pairs in the sense of Nash, as indicated by the two diagonal pairs in Table 1, since each strategy in one of the pairs is best against the other in the same pair. However, of combined differently, these strategies do not give equilibrium pairs in the sense of Nash. Moreover, the two equilibria on the diagonal yield different returns to players. In the terminology of game theory the two equilibria are neither interchangeable nor equivalent [Luce/Raiffa, p. 106].

Mixed strategy combinations in the non-cooperative form do not improve the outcome. Only preplay communication and co-ordination between the players will realize the alternating solution of the diagonal situations. An equitable solution is obtained if they alternate to go to a fight or to a ballet together, or to toss a coin to decide to go jointly to one of the entertainments [*Luce/Raiffa*, p. 94].

From these observations, one can draw at least two conclusions on the first stage of choosing a monetary regime.

First, in order that a new system be adopted by a group of participants, the pay-off structure should take the form of a genuine battle of the sexes. More precisely, at least one of the pairs (a, b) or (c, d) should be a pair of positive numbers. This could be achieved, of course, if an attractive plan is proposed. More realistically, however, this would be achieved if the existing system shows serious defects which are well recognized by participating countries. The deterioration of the pay-offs in the ongoing regime mean the relative increase in payoffs in new regimes such as (a, b) and (c, d) because the origin $(0, 0)$ is used to express the *status quo* situation only as a normalizing device.

Thus international economic crisis might as well lead to fundamental changes in the international monetary regime. As *Samuelson* [1967, p. 699] puts it, "One does not have to be cynical, but merely realistic to guess that if fundamental changes are to come, they will come in the wake of some international crisis rather than as a result of predetermined planning and agreement."

Second, it can be concluded that a compromising combination or a hybrid of basic regimes, rather than a pure form of one regime, is more likely to be agreed upon. A reasonable solution to the game of the battle of the sexes is a coordinated alternation of the choice between fight and ballet. It is difficult to conceive a randomized solution for the choice of a regime, but it is quite common to devise a system that has a mixture of characteristics of pure regimes. The most notable example is the adjustable-peg system in the Bretton Woods regime, which can be regarded as a convex combination of two pure regimes, fixed exchange rates and flexible exchange rates. Most of the time, the adjustable-peg system is operated in a manner similar to the fixed-rate system, but under the circumstances of "fundamental disequilibrium"[3]) in the balance of payments, exchange rates become adjustable so that the elements of flexible rate system are brought in.

It is not a mere coincidence that the Bretton Woods system is described as a hybrid of originally thoroughbred plans by *Keynes*. Comparing plans to species of dogs, he observes "(T)he loss of the dog (his plan of Clearing Union) we need not too much regret, though I still think that it was a more thoroughbred animal than what has now come out from a mixed marriage of ideas (the actually agreed International Monetary Fund)". However, he was right also in recognizing, "(Y)et perhaps, as sometimes occurs, this dog of mixed origin is a sturdier and more serviceable animal and will prove not less loyal and faithful to the purposes for which it has been bred [*Keynes*, p. 369].

---

[3]) It is well-known that since the IMF charter does not have exact definition of "fundamental disequilibrium" – the rule was incomplete – the interpretation of this phrase created a lot of confusion.

This framework for the choice of a regime can be effectively applied to the problem of the monetary union. The incentive for participating in a monetary union where exchange rates are more or less fixed among the currencies of participating countries depends on the benefit-cost of pay-offs of creating a monetary union. By some historical studies of the unification of currencies in Germany, Italy, and historical studies of example of monetary union such as Latin Monetary Union, Scandinavian Monetary Union, one can argue [*Hamada*, 1977] that it is difficult to achieve and to sustain a monetary union without political unification, because in most cases benefits can be obtained only in the long run while costs are immediate.

3. The Game of Interplay of Economic Policies

Let us now turn to the second stage, the game of policy interplays. I shall illustrate how a game approach can be applied to analyze this policy game, by taking up the example of the interplays of monetary policies under the regime of fixed exchange rates, particularly under its contemporary variant, the dollar standard system. In order to study the interdependence of monetary policies, I shall utilize an extremely simple model of the world economy that is developed by *Johnson* [1976].

Suppose there are n countries in the system under fixed exchange rates. Goods are assumed to be mobile enough to allow the assumption of a common price level p. Then we have

$$D_i + R_i = M_i, \qquad i = i, \ldots, n,$$

where D_i, R_i, M_i are respectively domestic credit, international reserves, and money outstanding in country i. The demand for money is expressed as

$$M_i^d = p \, L^i (Q_i), \; Q_i \text{ being real output.}$$

We shall neglect the effect of changes in interest rates, in order to focus on the strategic aspect of policy interplays.

Equating the demand and supply of money and differentiating the above equation by time and normalizing by M_i, we get (η_i being income elasticity of the demand for money, and dot indicating time derivative)

$$\dot{D}_i/M_i + \dot{R}_i/M_i = \dot{p}/p + \eta_i (\dot{Q}_i/Q_i). \tag{1}$$

Define excess credit expansion x_i and the normalized balance of payments z_i by

$$x_i = \dot{D}_i/M_i - \eta_i (\dot{Q}_i/Q_i), \; \text{ and } \; z_i = \dot{R}_i/M_i.$$

Then writing the common rate of inflation $\dot{p}/p$ as π, one obtains from (1)

$$z_i = \pi - x_i, \qquad i = 1, \ldots, n. \tag{2}$$

Defining $w_i = M_i/M$, where $M = \sum_{i=1}^{n} M_i$, and notating $\sum_{i=1}^{n} w_i z_i = (\sum_{i=1}^{n} \dot{R}_i)/M = G_R$, we get by summing equations (2),

$$\pi = \sum_{i=1}^{n} w_i x_i + G_R. \tag{3}$$

Equations (2) and (3) look much simpler, but they are essentially equivalent to the formulae derived by *Johnson* [1976] with simplifying notations.

Thus, in this simplified world the rate of inflation depends on the weighted average of excess credit expansions and the relative increase in outside money for the world G_R, while the balance of payments depends on the difference between this weighted average and a country's excess credit expansion. Here we see the public goods (bads) nature of credit expansions.

In order to construct a game-theoretic model, we need pay-off or welfare functions of the countries. Let the utility function of a monetary authority u^1 (π, z_i) be a strictly concave function of the rate of inflation π and the balance of payments z_i with a satiation coordinate (a_i, b_i) such that

$$u_1^i = \frac{\partial u^i}{\partial \pi} \gtreqless 0 \qquad \text{according as} \qquad \pi \gtreqless a_i$$

$$u_2^i = \frac{\partial u^i}{\partial z_i} \gtreqless 0 \qquad \text{according as} \qquad z_i \gtreqless b_i.$$

In other words, a monetary authority is assumed to have a most desired rate of inflation (or deflation) and a most desired increase (or decrease) in foreign reserves. Usually, a lower rate of inflation is preferred. But if the rate of inflation reduces below some specific value further reduction of inflation, or further increase in the degree of deflation may not be desirable. Similarly, a surplus in the balance of payments beyond a certain level means the reduction of consumption-investment opportunity for a nation. Accordingly, it is assumed that there is a most preferred value for the balance of payments for a country.

The game situation we have here is the interplay of monetary policies, and strategies are the excess credit expansion x_i's of monetary authorities. The reaction curve for country i is obtained by maximizing u^i (π, z_i) with respect to x_i, given the assumption that all other x_j's $(j \neq i)$ are given. Differentiating

$$u^i (\sum_{i=1}^{n} w_j x_j + G_R, \sum_{i=1}^{n} w_j x_j + G_R - x_i)$$

with respect to x_i, one obtains

$$w_i u_1^i + (w_i - 1) u_2^i = 0, \qquad\qquad i = 1, \ldots, n \qquad\qquad (4)$$

or

$$-\frac{d\pi}{dz_i} \bigg|_{u_i \text{ const.}} = u_2^i / u_1^i = w_i/(1 - w_i). \qquad\qquad (5)$$

In Figure 2, the indifference map and the reaction curve for which $a_i > 0$ and $b_i < 0$ are shown. From equation (5), the reaction curve is seen to be the locus of tangency points of indifference curves to the lines with slope of $w_i/(1 - w_i)$. It is also clear that the smaller a country is, the more deviates its reaction curve from the most desired inflation line $\pi = a_i$. The point that a country can choose on its reaction curve of course depends on the strategies taken by other countries. The Cournot-Nash solution is given by the intersection of these reaction curves.

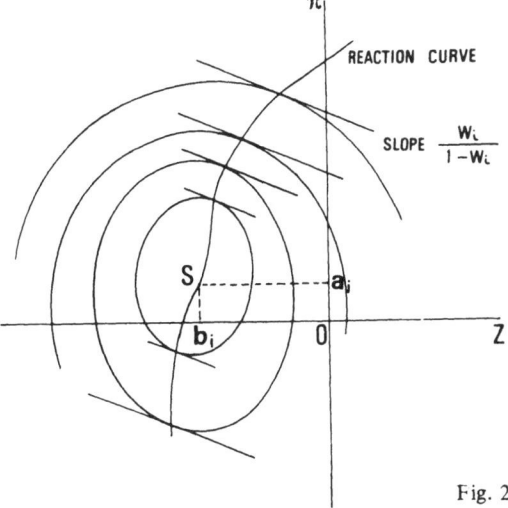

Fig. 2

On the other hand, the Pareto optimum configuration, or the contract surface, is given by maximizing the following with respect to all x_i's:

$$\sum_{k=1}^{n} \beta_k u^k \, (\pi, z_k) = \sum_{k=1}^{n} \beta_k u^k \, (\sum_{i=1}^{n} w_i x_i + G_R \, , \sum_{i=1}^{n} w_i x_i + G_R - x_k), \sum_{k=1}^{n} \beta_k = 1, \beta_k > 0,$$

where β_k is the index of relative bargaining power of country k. The optimum conditions are

$$\sum_{k=1}^{n} \beta_k u_1^k w_i + \sum_{k=1}^{n} \beta_k u_2^k w_i - \beta_i u_2^i = 0, \qquad i = 1, \ldots, n . \tag{6}$$

Summing by index i and noting $\sum_{i=1}^{n} w_i = 1$,

$$\sum_{k=1}^{n} \beta_k u_1^k + \sum_{k=1}^{n} \beta_k u_2^k - \sum_{i=1}^{n} \beta_i u_2^i = 0$$

or

$$\sum_{k=1}^{n} \beta_k u_1^k = 0. \tag{7}$$

Thus the Pareto optimal configuration is achieved if the weighted average of the marginal costs of inflation is equated to zero.

The Cournot-Nash solution implies, however,

$$\sum_{i=1}^{n} w_i u_1^i = \sum_{i=1}^{n} (1 - w_i) u_2^i \tag{8}$$

as is obtained by summing equations (4) with respect to i. Comparing (7) with (8), one can see that the Cournot-Nash solution does not lie on the Paretian configuration, that is, the set of the von Neumann solution. If both sides of (8) are positive, international reserves and credit expansion takes on the nature of public goods, and the Cournot-Nash solution has a deflationary bias. Each country welcomes a higher rate of increase in total world money supply, but each country tries to expand its own money supply at a slower rate than others because it is eager to accumulate more international reserves. Conversely, if both sides of (8) are negative, international reserves and credit expansion takes on the nature of public bads, and the Cournot-Nash solution has an inflational bias. Each country tries to expand its money supply at a higher rate than others in order to get rid of excessive international reserves.

Suppose the most desired rate of inflation is identical for every country, that is, $a_i = a$ for any i. Then the Paretian configuration is naturally $\pi = a$. Because π is common for all countries, u_1^i has the same sign for all i, and because on the reaction curve u_1^i and u_2^i have the same sign, one obtains

$$\pi \gtreqless a, \qquad \text{according as } z_i - b_i \gtreqless 0.$$

By summing with a weight w_i for each equation, $z_i - b_i \gtreqless 0$ implies

$$\sum_{i=1}^{n} w_i z_i - \sum_{i=1}^{n} w_i b_i \gtreqless 0$$

or

$$G_R - \sum_{i=1}^{n} w_i b_i \gtreqless 0. \tag{9}$$

Thus the Cournot-Nash solution gives a rate of inflation higher (lower) than the most desired rate, if and only if, the relative increase in outside international reserves is larger (smaller) than the weighted average of desired increases in reserves by these countries.

The Stackelberg leadership solution can also be defined so long as a single country, say country 1, behaves as the leader and the others as followers. The leader maximizes $u^1(\pi, z_1)$, subject to the reaction curves of the others

$$w_i u_1^i + (w_i - 1) u_2^i = 0, \qquad i = 2, \dots, n.$$

The leadership solution, which is of special interest in the dollar standard with the U.S. as the leader, does not in general lie on the Paretian surface.

The above argument can be illustrated by the Stackelberg diagram of two country case in Figure 3, where two countries are of equal size ($w_1 = w_2 = 1/2$). They prefer most price stability ($a_1 = a_2 = 0$) and some balance of payments deficits ($b_1 < 0, b_2 < 0$). Also it is assumed that there is no exogenous growth of international reserves for the world ($G_R = 0$).

In the diagram, values of excess credit expansions are taken along both axes. The line $x_1 = x_2$ indicates the combination of monetary policies that would maintain the equilibrium in the balance of payments; $x_1 + x_2 = 0$ indicates their combination for

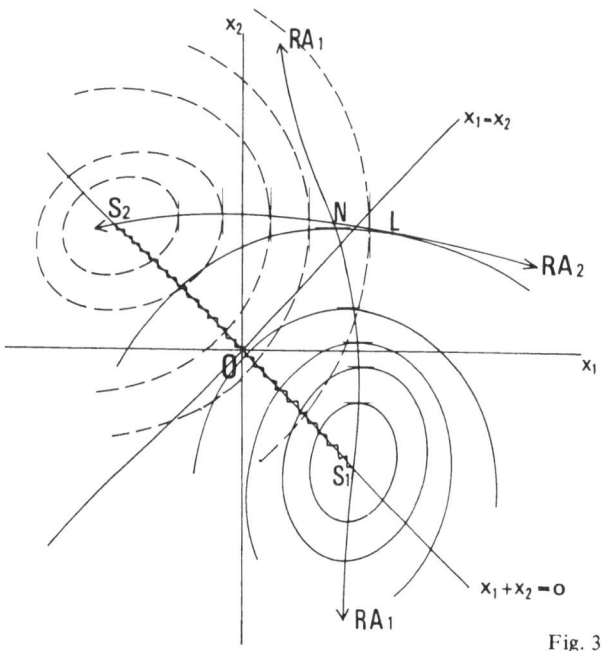

Fig. 3

price stability. The indifference curves of country 1 are drawn with solid curves, those of country 2 with dotted curves. The π axis in (z_i, π) plane in Figure 2 coincides with the line $x_1 = x_2$, and the z_i axis coincides with the line $x_1 + x_2 = 0$ with inverse direction for country 1 and with the same direction for country 2. The reaction curve for country 1 is the locus of points where indifference curves of country 1 have horizontal tangent lines; that for country 2 is the locus of points where indifference curves of country 2 have vertical tangent lines. The Cournot-Nash solution N is given by the intersection of these reaction curves. The Stackelberg leadership solution L with country 1 as the leader is the best point for country 1 along the reaction curve of country 2. The Pareto-efficient frontier, or the contract curve, is given in this example as the line between S_1 and S_2. In Figure 3, one can note that neither the Cournot-Nash solution nor the Stackelberg leadership solution lies on the contract curve.[4])

It can be seen that the benefit-cost structure of this game resembles that of the prisoner's dilemma. The cooperative strategies for players are to choose a point on the contract curve. The non-cooperative strategies can be viewed as expressed in the reaction curves. If both cooperate, they end up in a certain point on the contract curve. If both do not cooperate, the Cournot-Nash equilibrium will be achieved. If one cooperates, and the other does not, then the latter will achieve its first best situation.

As is shown by equation (9), it is possible to let the Cournot-Nash solution lie on the contract curve by manipulating the value of increase in outside world money G_R. The role of international organization, or agreement, is to change the pay-off structure

---

[4]) For a more detailed analysis of this game of policy interplay, particularly of the effect of the sizes of countries and the number of countries, see *Hamada* [1976].

from that of the prisoner's dilemma to that of such a game in which the Nash solution is included in the Paretian surface, that is, to restore the incentive compability in a broad sense of the word. This is in fact what the peace research in political science attempts to do.

Next, let us briefly look at the benefit cost structure of the flexible exchange rate system. It is believed that at least in the long run monetary authorities can choose their desired values of inflation, since interdependence through the balance of payments is blocked under flexible exchange rates. In the short run, monetary interdependence still seems to exist through conflicts on the levels of interest rates [Mundell], or through the effect of the terms of trade [Hamada/Sakurai]. However, interdependence under flexible rates are more subtle than under fixed rates, and the regime of flexible exchange rates offers a certain degree of monetary interdependence. Of course, if a country adopts the flexible exchange rate system by abandoning its intervention in the exchange market to fix exchange rates is not without costs. The country will lose the benefits attached to a wider circulation of a common currency [Hamada, 1977]. However, one can safely say that the adoption of a flexible exchange rate system by a country can be regarded as a kind of max-min strategy for a country that tries to create a currency area by itself.

4. A Dynamic Formulation of the World Money Game

Some readers might feel the analysis of policy interplays under fixed rates given in the previous section too static, because monetary authorities should have preference on the level of reserves rather than on the change in reserves, that is, the balance of payments. To formulate these conflicting situation in a dynamic model requires in general the method of differential game. In this section I shall present a simple example to illustrate how the results obtained in previous sections could be extended to a dynamic context.

Consider the case of two countries of identical size ($w_1 = w_2 = 1/2$). Let me use in this section the symbol z as the *level* of international reserves of country 2. Thus in terms of x_1 and z_2 in the previous section, $\dot{z} = -z_1 = z_2$ ($\dot{z}$ being time derivative of z).

The basic differential equation of the system is then

$$\dot{z}(t) = (x_1(t) - x_2(t))/2. \tag{10}$$

Also we have

$$\pi(t) = (x_1(t) + x_2(t))/2. \tag{11}$$

Country 1 is the reserve currency country, and I shall assume that it is concerned only with the rate of inflation π and levels of consumption flows over time. Thus the welfare function for each instant will be specified as

$$u^1(t) = -\pi(t)^2 + (c\dot{z}(t) - h\dot{z}(t)^2), \qquad c, h > 0. \tag{12}$$

Underlying the welfare function (12) is the assumption that increased consumption due to the balance of payments deficit has a marginal utility as long as $\dot{z}(t)$ is small, but that the marginal utility is decreasing. Country 1 maximizes the discounted utility integral, ρ being the rate of discount,

$$J^1 = \int_0^\infty \{-\pi(t)^2 + c\dot{z}(t) - h\dot{z}(t)^2\} e^{-\rho t} dt, \qquad \rho > 0. \tag{13}$$

subject to the differential equation (10) and an initial condition $z(0) = z^0$.

Country 2, the non-reserve country, is assumed to have the following instantaneous utility function, because it needs some level of international reserves to accommodate trade transactions.

$$u^2(t) = -\pi(t)^2 + (-c\dot{z}(t) - h\dot{z}(t)^2) - m(z(t) - \bar{b})^2, \qquad m > 0. \tag{14}$$

Here for simplicity coefficients c, h are assumed to be identical with those of country 1. Since a deficit of the balance of payments of country 2, namely a negative value of $\dot{z}(t)$, corresponds to an increased flow of instantaneous consumption flow, the coefficient on $\dot{z}(t)$ is negative. This country is assumed to have a loss function concerning the level of international reserves, which is proportional to the distance between actual and the most desired level of reserves that is denoted by $\bar{b}$. Thus country 2 is assumed to maximize the utility integral:

$$J^2 = \int_0^\infty \{-\pi(t)^2 - c\dot{z}(t) - h\dot{z}(t)^2 - m(z(t) - \bar{b})^2\} e^{-\rho t} dt,$$

given the differential equation (10), and an initial level of reserves $z(0) = z^0$.

Then this situation of policy interplays is readily to be interpreted as a differential game. In fact, because of the simplified structure of this game, one can easily obtain the Nash solution of this non-zero-sum differential game.

One can distinguish two types of the Nash solution to a differential game. One is the open-loop solution in which each player is assumed to neglect the possibility that the other player's strategy may depend on the values of state variables, here on the value of $z(t)$. The other is the closed-loop solution in which each player does take account of the dependence of the other player's path on the state variable.

Formally, the Nash solution $\hat{x}_1(t)$, $\hat{x}_2(t)$ of the open-loop type is defined such that for any (piece-wise continuous) feasible control $x_1(t)$ and $x_2(t)$ [see, e.g. *Simaan/Takayama*],

$$J^1(\hat{x}_1(t), \hat{x}_2(t)) \geqslant J^1(x_1(t), \hat{x}_2(t)),$$

$$J^2(\hat{x}_1(t), \hat{x}_2(t)) \geqslant J^2(\hat{x}_1(t), x_2(t)).$$

The Nash solution $\hat{x}_1(t, z(t))$, $\hat{x}_2(t, z(t))$ of the closed-loop type is defined as the paths satisfying for any (piece-wise continuous) feasible control $x_1(t, z(t))$ and $x_2(t, z(t))$

$$J^1(\hat{x}_1(t, z(t)), \hat{x}_2(t, z(t))) \geqslant J^1(x_1(t, z(t)), \hat{x}_2(t, z(t)))$$

$$J^2(\hat{x}_1(t, z(t)), \hat{x}_2(t, z(t))) \geqslant J^2(\hat{x}_1(t, z(t)), x_2(t, z(t))).$$

Here I shall examine the property of the simples solution, that is, the Nash solution of open-loop type by using a variational approach [Berkovitz; Simaan/Takayama].

The Hamiltonians can be expressed for the two countries (omitting the time notation):

$$H^1 = -\pi^2 + c\dot{z} - h\dot{z}^2,$$

$$H^2 = -\pi^2 - c\dot{z} - h\dot{z}^2 - m(z - \bar{b})^2 + \lambda(x_1 - x_2)/2,$$

where $\pi = (x_1 + x_2)/2$, $\dot{z} = (x_1 - x_2)/2$, and $\lambda (\equiv \lambda(t))$ is an auxiliary variable that can be interpreted as the undiscounted shadow value attached to the level of reserves for country 2. The open-loop Nash solution is given by the solution satisfying the following conditions in addition to (10):

For country 1

$$\frac{\partial H^1}{\partial x_1} = -\frac{x_1 + x_2}{2} + \frac{c}{2} - \frac{h}{2}(x_1 - x_2) = 0. \tag{15}$$

For country 2

$$\frac{\partial H^2}{\partial x_2} = -\frac{x_1 + x_2}{2} + \frac{c}{2} + \frac{h}{2}(x_1 - x_2) - \frac{\lambda}{2} = 0 \tag{16}$$

$$\dot{\lambda} - p\lambda = -\frac{\partial H^2}{\partial z} = 2m(z - \bar{b}) \tag{17}$$

$$\lim_{t \to \infty} \lambda e^{-pt} = 0. \tag{18}$$

By simple substitutions one obtains the following pair of differential equations in terms of z and λ.

$$\dot{z} = \lambda/4h \tag{19}$$

$$\dot{\lambda} = 2m(z - \bar{b}) + p\lambda. \tag{20}$$

By the use of a phase diagram that is familiar to economists, the optimal trajectory is found to be stable branches of the saddle point, because all the other paths do not satisfy the transversality condition (18) (Figure 4). It is easy to notice that at the equilibrium P, the rate of inflation is positive.

$$\pi = (x_1 + x_2)/2 = c/2 > 0. \tag{21}$$

Thus the open-loop Nash solution converges to an inflationary solution, so long as the marginal utility of increasing consumption by running a deficit in the balance of payments is positive. The characteristic equation of (19) and (20) is

$$(\rho - y)y + m/(2h) = 0,$$

and the root corresponding to stable branches is

$$y = \rho/2 - 1/2\sqrt{\rho^2 + 2(m/h)}.$$

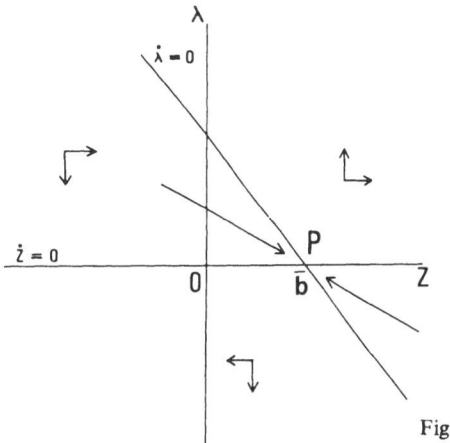

Fig. 4

Thus it is easy to check that increase in m/h will speed up the process of convergence.

Next let us consider the Paretian or von Neumann solution by a coordinated behaviour of the two monetary authorities. Suppose they coordinate to maximize $\beta J^1 + (1 - \beta) J^2$, where β is the weight of bargaining power attached to country 1. Then the problem is reduced to a usual variational problem in which the Hamiltonian is expressed

$$H = -\pi^2 + (2\beta - 1)\, c\dot{z} - h\dot{z}^2 - (1 - \beta)\, m\, (z - \bar{b})^2 + \lambda\, (x_1 - x_2)/2.$$

Thus one gets the optimal conditions:

$$\frac{\partial H}{\partial x_1} = -\frac{x_1 + x_2}{2} + \frac{2\beta - 1}{2}\, c - \frac{x_1 - x_2}{2}\, h + \frac{\lambda}{2} = 0 \tag{22}$$

$$\frac{\partial H}{\partial x_2} = -\frac{x_1 + x_2}{2} - \frac{2\beta - 1}{2}\, c + \frac{x_1 - x_2}{2}\, h - \frac{\lambda}{2} = 0 \tag{23}$$

$$\dot{\lambda} - \rho\lambda = -\frac{\partial H}{\partial z} = 2\, (1 - \beta)\, m\, (z - \bar{b}) \tag{24}$$

$$\lim_{t \to \infty} \lambda\, e^{-\rho t} = 0. \tag{25}$$

It is easy to see that all along any path satisfying (22) and (23),

$$\pi = (x_1 + x_2)/2 = 0. \tag{26}$$

Thus the trajectory of the von Neumann solution gives price stability during all the time of adjustment process. The optimal path is the stable branches of the following differential equations: (Figure 5).

$$\dot{z} = \frac{1}{2h}\, \lambda + \frac{2\beta - 1}{2h}\, c \tag{27}$$

$$\dot{\lambda} = 2(1 - \beta)m(z - \bar{b}) + \rho\lambda. \tag{28}$$

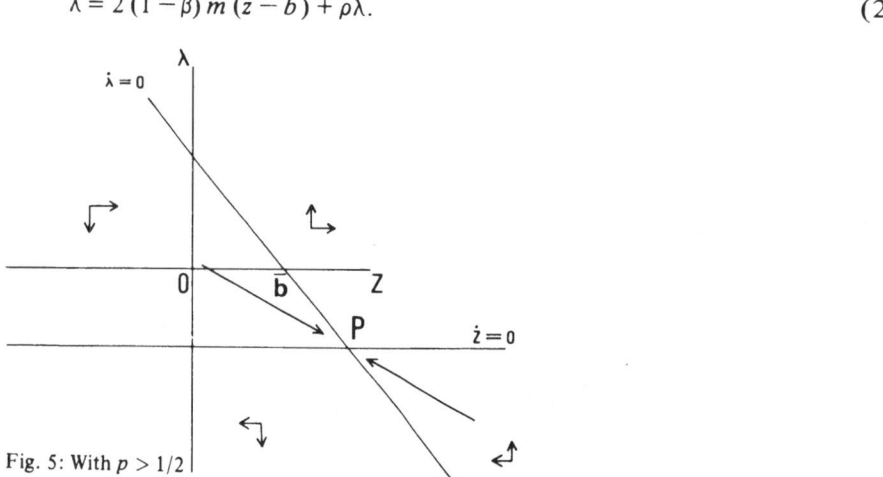

Fig. 5: With $p > 1/2$

If $\beta > 1/2$, then the equilibrium situation corresponds to a larger value of z than $\bar{b}$.

It may be slightly more realistic to assume that u^2 has the following form because a larger level of reserves than the equired level $\bar{b}$ may not do any harm, and the benefit from possible decumulation of reserves is already counted in the term of consumption increases.

$$u^2 = -\pi(t)^2 + (-c\dot{z}(t) - h\dot{z}(t)^2) - m(z(t) - \bar{b})^2, \quad \text{if } z(t) \leq \bar{b}$$

$$= -\pi(t)^2 + (-c\dot{z}(t) - h\dot{z}(t)^2), \qquad \text{if } z(t) > \bar{b}. \tag{14'}$$

The resulting pair of differential equations for the Nash solution can be written

$$\left. \begin{array}{l} \dot{z} = \lambda/4h \\[2mm] \dot{\lambda} = 2m(z - \bar{b}) + \rho\lambda \end{array} \right\} \quad \text{if } z \leq \bar{b}$$

$$\left. \begin{array}{l} \dot{z} = 0 \\[2mm] \dot{\lambda} = \rho\lambda \end{array} \right\} \quad \text{if } z > \bar{b}.$$

In this particular example, equilibrium point P is seen to be extended to a line segment on the horizontal axis from $z = \bar{b}$. The qualitative characteristics will not be different from the main discussion. (The reader will notice that triangular area locating in the south-east of P and containing the stable branch in Figure 4 is compressed to a single line segment. See Figure 6.)

There still remain many questions to ask. What are the effects of inclusion of the balance of payments objective for country 1, or those of asymmetrical utility function, or the relative size? What are properties of the closed-loop solution or the Stackelberg solution?

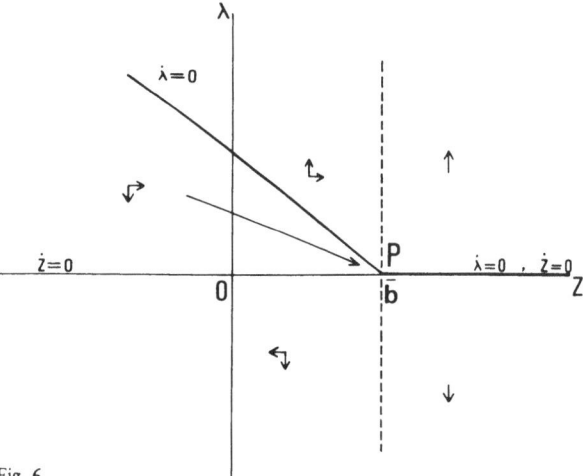

Fig. 6

We hope, however, that our dynamic analysis suggests that the results we have obtained in our static formulation are quite likely to be generalized to a dynamic situation. If the rules of the game are not properly set, the non-cooperative interplay of monetary policies would result in a situation outside the von Neumann solution frontier.

5. Concluding Remarks

Game theory, at the current stage of its development, may not be able to offer national policy authorities the optimal strategy, or the best prescription to follow when they are engaging in monetary negotiations or in the interplay of monetary policies. It seems to require also further research in order to find out whether game theory can provide testable hypotheses for policy behaviour of various countries. We have argued, however, that game theory does help to clarify the benefit-cost structure of monetary conflicts among national economies, and that it provides some insight into the optimal design of monetary rules that are desirable, and, at the same time, politically feasible.

The practical applicability of the above analysis is limited, of course, by several simplifying assumptions. Let me just mention an important limitation that has resulted from the assumption concerning the decision unit. I took a nation as a decision unit, which to most readers may seem to be a natural choice. In many circumstances, however, it is hard to define a uniform national interest. Individuals in the same country do not necessarily share common benefits and costs. The interest of the rich and that of the poor may differ. The interests of exporters do often conflict with those of importers; those of consumers conflict with those of producers. In fact, the majority of trade and monetary disputes in the present world stem from the difference of international interests as well as that of domestic interests. We would need more stratified approaches to cope with this complex combination of international and international conflicts of interest, which I hope to discuss on some other occasion.

References

Berkovitz, L.: Lectures on Differential Games. Differential Games and Related Topics. Ed. by
    Kuhn and Szego. Amsterdam 1971.
Hamada, K.: Alternative Exchange Rate Systems and the Interdependence of Monetary Policies.
    National Monetary Policies and the International Monetary System. Ed. by R.Z. Aiber. Chicago
    1974.
– : A Strategic Analysis of Monetary Interdependence. Journal of Political Economy **83** (4), 1976.
– : On the Political Economy of Monetary Integration. The Political Economy of Monetary Re-
    form. Ed. by R.Z. Aliber. New York 1977.
Hamada, K., and *M. Sakurai*: Transmission of Stagflation under Fixed and Flexible Exchange
    Rates. Journal of political Economy **85** (5), 1978.
Johnson, H.G.: The Monetary Approach to Balance of Payments Theory. The Monetary Approach
    to the Balance of Payments. Ed. by J. Frenkel and H.G. Johnson. London 1976.
Keynes, J.M.: The International Monetary Fund. Chapter 27, in S.E. Harris, ed. 1947.
Luce, R.D., and *H. Raiffa*: Games and Decisions. New York 1975.
Mundell, R.A.: International Economics. New York 1968.
Samuelson, P.A.: Economics: An Introductory Analysis, 7th ed. New York 1967.
Simaan, M., and *T. Takayama*: Dynamic Duopoly Game: Differential Game Theoretic Approach.
    Faculty Working Paper #155, University of Illinois, Urbana–Champaign 1974.
Yeager, L.B.: The International Monetary Mechanism. New York 1968.

Applied Game Theory. 1979 ©Physica-Verlag, Wuerzburg/Germany

OPEC Pricing and Output Decisions
A Partition Function Approach to OPEC Stability

By *D. Gately*, New York[1])

Abstract: This paper is concerned with the stability of OPEC as a cartel, in particular the bargaining within OPEC over price and output shares for its member countries.

Although many of the questions concerning OPEC behavior are inherently game-theoretical, there have been few formal attempts to employ game theory. In this paper, the OPEC cartel is modeled as a game in partition function form.

Given the estimated partition function, we utilize various solution concepts, such as the set of imputations and the core, as modified for the partition function form, to examine questions about OPEC stability and the prospects for oil prices and output shares. We would hope to achieve a better understanding of the bargaining within OPEC over price and output shares and to predict the likely partition and rule-of-thumb strategy selected.

1. Introduction

Since the quadrupling of oil prices in 1973/74 there have been numerous opinions and a few serious modeling efforts concerned with the future of oil prices. The assessments and conclusions have varied widely. Critical surveys appear in *Fischer/Gately/ Kyle* [1975] and *Gately* [1977].

Many of the questions concerning world oil and the behavior of OPEC are inherently game theoretical. The most prominent perhaps is the question of OPEC stability — whether OPEC can agree on a strategy to set prices and restrict output to the degree necessary to support those prices. Opinions on this question have varied. Milton Friedman represents one extreme:

"... in order to keep prices up the Arabs would have to curtail their output by ever larger amounts. But even if they cut their output to zero, they could not for long keep the world price of crude at $ 10.00 per barrel. Well before that point the cartel would collapse ... World oil prices are weakening. They will soon tumble." [Newsweek, March 4, 1974.]

Others have asserted that OPEC is an exception to the rule that all cartels are unstable and that OPEC will be able to maintain its control over world oil prices for the foreseeable future.

---

[1]) Prof. Dr. *Dermot Gately*, Dept. of Economics, New York University, 40 W 4th, New York, N.Y. 10003, USA.

An intermediate position is taken by *Osborne* [1976] and *Danielsen* [1976]. *Osborne* [1976, p. 843], who had initially predicted that OPEC would collapse within a year after the price was quadrupled, writes in a recent article discussing the various types of problems facing cartels:

> "A cartel is inherently unstable only if it faces inherently insoluble problems . . .
> We considered [three of] these problems only in enough detail to see that, while
> difficul't they are not inherently insoluble . . . The two remaining problems . . .
> to determine quotas and to deter cheating . . . are not inherently insoluble either.
>
> From this it does not follow that a cartel is stable. Though it can solve its prob-
> lems in principle, it might nevertheless fail to solve them in practice . . . So much
> depends on the particular features of their environments that no general prediction
> about the durability of cartels is justified."

Recently, some interesting analyses of OPEC behavior and OPEC stability have appeared, such as *Eckbo* [1976] and *Moran* [1978]. However, few formal attempts to apply game theory to such questions have been made. *Hnyilicza/Pindyck* [1976] employ the Nash cooperative solution to analyze pricing and output-share decisions for a cartel consisting of two groups, the "savers" like Saudi Arabia and the "spenders" like Iran. *Sampson* [1977] estimates a characteristic function for OPEC to determine the existence of a core and various other solution concepts.

Considering a different question about OPEC behavior, the *Gately/Kyle/Fischer* [1977] analysis of OPEC's selection of a rule-of-thumb pricing strategy can be viewed as a game against nature. On account of unavoidable uncertainty about the true values of certain parameters characterizing the future of the world oil market (the long-run responsiveness of demand and non-OPEC supply to increases in OPEC prices), the pricing problem facing OPEC is not the selection of *the* optimal price-path. Instead, it is to select a market-signalled rule-of-thumb pricing strategy that will serve OPEC relatively well regardless of the parameter values actually characterizing the world oil market. Attached as Appendix A is Table IV—2 from that paper. It lists the payoffs to a monolithic OPEC of 8 rule-of-thumb pricing strategies under 12 "states of nature." A relatively cautious pricing rule (*B*) is found to yield OPEC relatively good results, almost regardless of the true parameter values.

In the present paper we utilize the same basic model of the world oil market: the price of oil is set by OPEC, or a subset of OPEC countries, and the demand for OPEC oil is determined as the difference between world energy demand and non-OPEC energy supply. We explore questions of OPEC stability, particularly the bargaining within OPEC over price and output shares for the member countries.

2. The Model

The negotiations within OPEC over price and output shares are modeled as a game in partition function [see *Thrall/Lucas; Owen*] form without side payments. A partition function approach is used, rather than a characteristic function, because the payoff to a particular oil exporter (e.g., one which has split off from OPEC) depends on how the other oil exporters are partitioned – whether they're acting together or independently. Side payments among OPEC members in monetary terms, while possible in principle, are not considered as a realistic possibility in view of the institutional setup of OPEC. More appropriate would be side payments in the form of output shares over time. However, rather than modeling such transfers explicitly as side payments, these alternative arrangements for output shares over time will be modeled as components of alternative pricing-prorationing strategies to be selected by negotiations within OPEC.

Following *Eckbo* [1976], OPEC is assumed to consist of three players: the "moderates," consisting of Saudi Arabia, Kuwait, UAE, Qatar, and Libya; the "price-pushers" – Iran, Venezuela and Algeria; and the "expansionist fringe," which includes Indonesia, Nigeria, Iraq, Ecuador and Gabon. The shares of the estimated 900 billion barrels of OPEC oil reserves by the moderates, price-pushers, and expansionist fringe are, respectively, 64 %, 19 %, and 17 %; the shares of OPEC productive capacity (now about 37 million barrels per day) are, respectively, 52 %, 28 %, and 20 %. The moderates have relatively small populations, substantial financial reserves and have the least need for additional current revenue. They tend to be relatively cautious about further price increases, because the burden of output restriction most likely will fall on them. In addition, their disproportionately large share of oil reserves gives them an interest in satisfying world oil demand that will endure long after many other OPEC members have exhausted their reserves. In contrast, the price-pushers, with much greater populations and ambitious spending plans, have a greater need for additional current revenue and have generally advocated much higher prices. The expansionist fringe countries are the poorest, and have enormous populations relative to their oil resources. Being in most need of additional revenue, they are least likely to be willing to restrict output.

Three partitions of the three countries are investigated:

I.   {(moderates, price-pushers, fringe)},
     all three groups act together as a coalition and jointly set price and determine output shares;
II. {(moderates, price-pushers), (fringe)},
     the moderates and price-pushers act together to set price and their own output shares, while the fringe acts independently as a price-taker and output maximizer;
III. {(moderates), (price-pushers), (fringe)},
     the three groups act independently, with the moderates setting the price and (perhaps) restricting output, while the others act as price-takers and output maximizers.

No other partitions were considered likely and so were ignored.

The behavioral assumptions are simple. The price-taker players in partitions II and III are assumed to maximize output. When they do otherwise, i.e., restrict output below capacity, they are acting, at least implicitly, as part of the supply-restricting, price-setting coalition. All players and coalitions of players are assumed to act to maximize their own self-interest; this is what *Rosenthal* [1971] calls individual-and-group-rational behavior.

Under Partitions I and II the price-setting coalition can choose from a variety of strategies, each of which involves decisions about current and future prices and about output shares over time for the members of the supply-restricting group. Under Partition III, the moderates alone decide about current and future prices; no decisions about output shares within the supply-restricting group are necessary because it consists of only one player.

The agreement of all players in a supply-restricting coalition is necessary for a particular strategy to be adopted. Hence the fringe can veto any or all of the strategies corresponding to Partition I. But then, having given up any influence it may have had on pricing and prorationing, it must accept its payoff under the strategy selected under Partition II or III. Similarly, the price-pushers can reject any or all of the strategies under Partitions I and II, but then must accept its payoff under the strategy selected by the moderates under Partition III.

The payoff to a particular player from a particular strategy is the resulting time-stream of future profits, reduced to a scalar by calculating the discounted present value of the profit stream. Since the different players have different time preferences between current and future profits, different discount rates are used: 3 % for the moderates, 10 % for the price-pushers, and 15 % for the expansionist fringe. There are, of course, problems in reducing everything to a scalar, but they will not be dealt with here.[2])

3. The Case with Certainty About Oil Market Parameter Values

As was noted in *Gately* et al. [1977], the true functional specifications and parameter values for the world oil market are not known to the degree of certainty required. Among these are the short-run and long-run elasticities (and lag structures) for demand and non-OPEC supply, the future rate of growth of world income, the income elasticity of demand, and the price expectations mechanisms for demanders and non-OPEC suppliers. Still further uncertainty exists with respect to technological change and the discovery of additional oil reserves, but this will be ignored.

Assume for the moment, however, that the true functional specifications and parameter values are known with certainty by all participants. The payoffs to the various players can then be presented in a simple matrix. The columns correspond to the three players and the rows represent the various pricing-prorationing strategies available to the supply-restricting group under each of the three partitions. Each entry in the matrix, X_{ij}, measures the discounted present value of the time-stream of future profits accruing to player j under the pricing-prorationing strategy i.

Specifically, suppose we assume that the true parameter values for the world oil market are given by those listed in row 4 of Appendix A. The corresponding matrix of payoffs under a variety of pricing prorationing strategies is presented in Table 1.

---

[2]) As *Bobrow/Kudrle* [1976, p. 34] note, "Because payoffs and decision assets are multidimensional and intertemporal and actors have different discount rates, calculation problems are formidable". Some of the problems with using the discounted present value of profits are discussed in *Gately* et al. [1977]. If, instead, we used the discounted present value of the utility of profits, where each group's estimated utility function displayed diminishing marginal utility for profits within each time period, some of those problems might be alleviated. In particular, it would more adequately assess the value of a profit stream with high variance over time, i.e., enormous profits in some years and sharp declines (and cash-flow problems) in other years.

Partition Number	Strategy Number	Pricing Rule[2]	Marginal Increases[3]			Marginal Decreases[3]			Moderates	Price-Pushers	Expansionist Fringe
I	1	J	.05	.2	.75	.75	.2	.05	2940	820	381
	2	J	.1	.5	.4	.6	.3	.1	3023	803	376
	3	B	.1	.5	.4	.6	.3	.1	2450	685	331
	4	B	.4	.35	.25	.4	.35	.25	2756	656	306
	5	D	.1	.5	.4	.6	.3	.1	2077	626	336
	6	D	.4	.35	.25	.4	.35	.25	2416	587	276
II	7	J	.3	.7	.0	.7	.3	.0	2967	803	389
	8	J	.65	.35	.0	.65	.35	.0	3062	774	389
	9	B	.65	.35	.0	.65	.35	.0	2493	649	349
	10	D	.65	.35	.0	.65	.35	.0	2114	565	374
	11	K	.3	.7	.0	.7	.3	.0	2571	660	328
III	12	K	1	.0	.0	1	.0	.0	2527	686	328
	13	B	1	.0	.0	1	.0	.0	2236	753	348
	14	D	1	.0	.0	1	.0	.0	1544	739	364

Payoffs to[1]

[1]) Payoffs measured in billion $, net present value.

[2]) The various pricing rules are described in Appendix B.

[3]) Such shares are for the moderates, price-pushers, and fringe respectively.

Tab. 1: Payoffs to each group under the various pricing-prorationing strategies for the three partitions. (Set of Parameter Values No. 4 from Appendix A)

Each row represents a different pricing-prorationing strategy, defined as market-signalled pricing rule[3]) together with a prorationing agreement for sharing changes in demand. The prorationing shares for marginal increases describe how much of a one-unit increase in demand for the supply-restricting coalition's output (over the previous year's demand) shall be supplied by the various members. Similarly, the shares for marginal decreases measure the respective cutbacks resulting from a one-unit decrease in demand for their output. For Partitions II and III, of course, those players not restricting supply do not share in any changes in demand because they are producing at capacity.

Strategies can differ by employing different pricing rules and/or different prorationing agreements. Strategies with a common pricing rule have the same price-path over time, but the time paths of output levels for supply restricting players differ according to the prorationing agreement. For example, under Partition II, strategies 7 and 8 employ the same pricing rule, but the prorationing agreement of strategy 7 is more generous to the price-pushers (a bigger share of output increases and a smaller share of output cutbacks) than that of strategy 8. This results in a somewhat bigger payoff (803 vs. 774) to the price-pushers and a smaller one (2967 vs. 3062) to the moderates. (The payoff to the fringe is unchanged at 389 because its output level and the price-path are the same under the two strategies.)

[3]) Such pricing rules are described in Appendix B.

The various pricing-prorationing strategies listed in Table 1 are, of course, only a small number of those possible. But supposing that they are the only strategies available, which will be chosen? Before answering this question, we shall first explain some game-theoretic terms such as the players' guaranteed minimum payoffs and dominance of one payoff vector by another.

The minimum payoffs occur under Partition III, when there is no cooperation among the players in setting price and restricting output. Under that partition the moderates alone determine the choice of strategy. Assuming they would act to maximize their own self-interest, the moderates' guaranteed minimum payoff is their maximum payoff under any Partition III strategy. Also determined by that Partition III strategy choice of the moderates are the resulting payoffs to the price-pushers and the fringe, which are their respective guaranteed minimum payoffs.

Given two strategies and the respective payoff vectors X and X^* (not necessarily corresponding to the same partition), we say that X dominates X^* if each member of the X vector's supply-restricting coalition is better off under X than under X^*. Clearly, there will always be one Partition III payoff vector that dominates all other Partition III payoff vectors (and perhaps some other payoff vectors from Partitions I and II as well). It will not be the case, however, that a single Partition II (or Partition I) payoff vector would dominate all other Partition II (or Partition I) payoff vectors, since different prorationing agreements can always be worked out to shift profits from one coalition member to another.

Returning to Table 1, we see that strategy 12 dominates the other Partition III strategies, as well as dominating strategies 3, 5, 6, 9, and 10, since the moderates receive a larger payoff under 12. However, it is dominated (along with strategies 4 and 11) by strategies 1, 2, 7, and 8. All these undominated strategies (1, 2, 7, 8) involve the same pricing rule (J) but have different output shares over time.

If we assume that the 14 strategies in Table 1 are the only ones available, then the output ought to be one of the undominated strategies. We could be fairly sure that Partition III would not come about and that a coalition of the moderates and price-pushers (perhaps with the fringe) would be fairly stable. Of course, if other strategies are analyzed, this conclusion may be modified.

Such results, of course, are relevant only for the particular set of parameter values assumed. Under different assumptions we would expect different results, certainly for price and output levels over time and perhaps also for the degree of cooperation within OPEC. For example, using a set of parameter values that are more pessimistic from OPEC's viewpoint, listed in row 10 of Appendix A, we get the payoff matrix presented in Table 2.

In this case we see that strategy 13 dominates 2, 3, 6, 10, 11, 12, 14, and 15. Strategy 9 dominates numbers 4 and 8. The undominated strategies are 1, 5, 7, 9, and 13. However, the guaranteed minimum payoffs, which are determined by strategy 13, ensure the price-pushers and the fringe more than they would receive under strategies 1, 5, 7, and 9. Hence, they would refuse to cooperate with the moderates, putting the sole burden of supply restriction on the moderates. Of course, if additional strategies (or monetary side payments[4]) were considered, perhaps involving prorationing agreements more generous to the price-pushers and fringe, this conclusion might be modified.

Partition Number	Strategy Number	Pricing Rule[2])	Prorationing Shares For:						Payoffs to[1])		
			Marginal Increases[3])			Marginal Decreases[3])			Moderates	Price-Pushers	Expansionis Fringe
I	1	J	.05	.2	.75	.75	.2	.05	1749	442	208
	2	D	.1	.5	.4	.6	.3	.1	1259	447	224
	3	D	.4	.35	.25	.4	.35	.25	1454	419	197
	4	B	.1	.5	.4	.6	.3	.1	1696	445	212
	5	B	.4	.35	.25	.4	.35	.25	1770	436	202
	6	K	.1	.5	.4	.6	.3	.1	1474	450	222
II	7	B	.1	.9	.0	.9	.1	.0	1640	456	228
	8	B	.65	.35	.0	.65	.35	.0	1708	428	228
	9	J	.1	.9	.0	.9	.1	.0	1722	446	217
	10	D	.2	.8	.0	.8	.2	.0	1143	465	251
	11	D	.65	.35	.0	.65	.35	.0	1295	403	251
	12	K	.2	.8	.0	.8	.2	.0	1396	468	252
III	13	B	1	0	0	1	0	0	1617	469	228
	14	J	1	0	0	1	0	0	1191	481	231
	15	K	1	0	0	1	0	0	1356	503	252

[1]) Payoffs measured in billion $, net present value.
[2]) The various pricing rules are described in Appendix B.
[3]) Such shares are for the moderates, price-pushers, and fringe respectively.

Tab. 2: Payoffs to each group under various pricing-prorationing strategies for the three partitions. (Set of Parameter Values No. 10 from Appendix A)

4. A Framework for the Case with Uncertainty About Oil Market Parameter Values

One of the main themes of *Gately* et al. [1977] was how a monolithic OPEC ought to select its pricing strategy in the face of uncertainty about the true parameter values for the world oil market. This was noted above.

For a non-monolithic OPEC facing similar uncertainty the framework is much the same. Each group within OPEC would like to see a pricing-prorationing strategy that will yield it a payoff greater than that from any other strategy, regardless of the true parameter values for the world oil market. However, it is doubtful that a single strategy would dominate, consistently serving all groups as well as possible under the circumstances. Not only do the three groups have different discount rates and marginal

---

[4]) We would normally expect side payments to be made in terms of output shares. At times, however, this may not be possible and money payments (direct or indirect) would be necessary. For example, suppose that under Partition II the price-pushers are producing at capacity but in the future will be required to reduce output when OPEC demand fails. To secure their cooperation in such future cutbacks, the moderates could make a monetary side payment. They might do this, rather than raising price, which might have undesired long-term effects on world demand and non-OPEC supply.

utilities for current income but their oil reserves differ dramatically, ranging from as little as 20 years (at current output levels) for some countries to more than 75 years for countries such as Saudi Arabia. Consequently, those with relatively small oil reserves much prefer an aggressive pricing strategy even if it jeopardizes the long-run market for OPEC oil, while those with substantial reserves would be much more cautious. Under a set of parameter values favorable to OPEC, an aggressive strategy might benefit all groups more than a cautious strategy. Under a different set of parameter values, with little price-responsiveness over the short-term (or medium-term) but dramatic response after ten or fifteen years, an aggressive pricing strategy could still serve those countries with a high discount rate and small oil reserves better than a cautious pricing strategy, but it would be a disaster for the other countries with longer-term interests in the world oil market.

It is relatively easy to make a formal statement of the problem involved under uncertainty. Let X_{ijk} measure the payoff to group j when pricing-prorationing strategy i has been followed (under one of the three partitions) and the assumed state of nature is described by the functional specifications and set of parameter values given by case k. Given these payoffs, it would then be necessary to estimate each player's utility from each strategy, taking account of each player's risk aversion in its preference function. With this information one could then begin to analyze the bargaining within OPEC over price and output shares.

It is, however, considerably more difficult to actually solve such a problem. For one thing, it is very time consuming to estimate the payoffs for a large number of pricing-prorationing strategies under a wide range of alternative sets of parameter values. Although we shall not carry out such an analysis, we shall make a few conjectures about the effects of differing perceptions about the true state of nature among the different groups:

i)   Shared perceptions about the true state of nature, or shared subjective probability distributions for the likelihood of the various states of nature, will simplify the negotiation process and enhance the prospects for cooperation.
ii)   If the moderates are more pessimistic than the others about the true state of nature, then agreement will be more difficult to achieve, especially on a relatively aggressive strategy.
iii) Conversely, if the moderates are more optimistic than the others, then an agreement is more likely, particularly on a more aggressive pricing strategy.

5. Conclusions

The numerical estimates of this paper are primarily illustrative and no conclusive statements about OPEC stability are made. Yet we do feel confident about certain issues.

First, the question of cartel stability cannot be resolved axiomatically. The circumstances of particular cartels must be examined in detail, within a framework such as ours, before any conclusions are reached.

Second, the appropriate framework for analyzing OPEC stability follows the lines of this paper: n-person cooperative game theory, using a partition function (rather than a characteristic function), in which members of a supply-restricting coalition control price and their own output levels, while non-members act as price-taking output maximizers. Side payments between OPEC members typically would take the form of changed output shares over time, but utility is not conserved when such transfers are made because of differing marginal utilities and discount rates. It's possible, but not entirely satisfactory, to reduce to a scalar the payoffs from a particular pricing-prorationing strategy.

Third, our analysis supports the conclusion of *Gately* et al. [1977] that OPEC ought to pursue a relatively cautious pricing strategy in the face of uncertainty about the true set of parameter values describing the world oil market. We feel that such a strategy is even more likely when OPEC is assumed to be non-monolithic. Given the moderates' low discount rate, low marginal utility for additional current revenue, large reserves of oil, and their disproportionate power over prices, we would expect them to pursue a strategy which is fairly generous with respect to output shares for the rest of OPEC.

Fourth, our analysis supports some ideas of M. Olsen's theory of collective action, as noted in *Bobrow/Kudrle* [1976]. Specifically, we expect some "exploitation of the large by the small," the burden of output restriction being borne disproportionately by the moderates, with much of the benefits going to the others. Moreover, we observe that the variation in size of OPEC members, taken together with discount rates inversely correlated with size, is conducive to a stable cartel, with output restricted primarily by the moderates.

References

Bobrow, D.B., and *R.T. Kudrle*: Theory, Policy, and Resource Cartels: The Case of OPEC. Journal of Conflict Resolution 20 (1), 1976, 3–56.

Danielsen, A.L.: Cartel Rivalry and the World Price of Oil. Southern Economic Journal, January 1976, 407–415.

Eckbo, P.L.: The Future of World Oil. Cambridge 1976.

Fischer, D.: A Dynamic Model to Calculate Optimal Price Strategies for Associations of Raw Material Exporting Countries with Special Reference to OPEC. Dissertation, New York University, 1975.

Fischer, D., D. Gately, and *J.F. Kyle*: The Prospects for OPEC: A Critical Study of Models of the World Oil Market. Journal of Development Economics 2, 1975, 363–386.

Gately, D.: The Possibility of Major, Abrupt Increases in World Oil Prices by 1990. Brookhaven National Laboratory, December 1977, mimeo.

Gately, D., and *J.F. Kyle* in association with *D. Fischer*: Strategies for OPEC's Pricing Decisions. European Economic Review 10, December 1977, 209–230.

Hnyilicza, E., and *R.S. Pindyck*: Pricing Policies for a Two-Part Exhaustible Resource Cartel: The Case of OPEC. European Economic Review 8 (2), 1976, 136–154.

Moran, T.H.: Oil Prices and the Future of OPEC. Resources for the Future Research Paper No. R–8, Washington, D.C., 1978.

Osborne, D.K.: Cartel Problems. American Economic Review 66 (5), 1976, 835–844.

Owen, G.: Game Theory. London 1968.

Rapoport, A.: N-Person Game Theory. Ann Arbor 1970.

Rosenthal, R.: External Economics and Cores. Journal of Economic Theory 3, 1971, 182–188.

Sampson, M.: Analysis of an International Policy Coalition: A Game Theoretic Study of OPEC. Research Paper, Indiana University, 1977.

Thrall, R.M., and *W.F. Lucas*: N-Person Games in Partition Function Form. Naval Research Logistics Quarterly 10, 1963, 281–298.

Appendix A

Objective function values (billion $, NPV) for eight rules-of-thumb using twelve sets of Parameter values for a monolithic OPEC [from *Gately/Kyle/Fischer*]

	Sets of Parameter Values					Rules-of-Thumb							
#	Demand Elasticity[1])	Lags[2])	Demand Growth[2])	Non-OPEC Supply Growth[2])	OPEC Capacity by mid–80's	A	B	C	D	E	F	G	H
1	−.16	longer	4%–6%	1/2%–2%	45	5128	5369	5072	5125	5199	4966	5021	4946
2	−.16	longer	4%–6%	1/2%–2%	60	4581	3867	4651	4839	5003	4892	4800	4888
3	−.16	longer	2%–6%	1/2%–2%	45	2579	2811	3025	2868	2950	3074	2897	3062
4	−.16	long	2%–6%	1/2%–2%	45	1623	2498	2207	2077	1888	1589	1697	1522
5	−.16	longer	2%–6%	2%–4%	45	2111	2195	2226	2369	2203	2360	2167	2240
6	−.33	longer	2%–6%	1/2%–2%	45	1800	2015	1893	1771	1772	1298	1703	1560
7	−.16	long	2%–6%	2%–4%	45	1474	1987	1799	1697	1643	1073	1391	1255
8	−.33	long	2%–6%	1/2%–2%	45	1671	1854	1699	1649	1665	1037	1565	1147
9	−.33	longer	2%–6%	2%–4%	45	1486	1728	1569	1565	1535	1087	1454	1149
10	−.33	long	2%–6%	2%–4%	45	1440	1597	1454	1384	1399	819	1230	949
11	−.33	long	1%–4%	2%–4%	45	1102	1253	1171	1151	1141	692	920	805
12	−.33	long	1%–4%	2%–4%	60	1170	1313	1211	1188	1194	694	938	845

[1]) Point elasticity at $ 12.
[2]) Explanation and details in *Gately/Kyle/Fischer* [1977].

Appendix B

Pricing rules used in pricing-prorationing strategies

	Pricing Rules[1])			
	B	D	J	K[2])
Annual price change which is "regularly scheduled (unless one of the two possibilities described below occurs):	− 5%	none	− 10 %	none
Unless spare capacity exceeds:	25%	30%	20%	
and (output)$_{t-1}$ < c (output)$_{t-2}$ where c = ...	1.00	.95	1.00	
then cut price by	20%	30%	30%	
or unless spare capacity is less than:	15%	15%	15%	
and (output)$_{t-1}$ > b (output)$_{t-2}$ where b = ...	1.02	1.00	1.00	
then increase price by:	15%	60%	10%	

[1]) Note that price is a real price and not a nominal price. Hence, holding real price constant means increasing the nominal price at the world's rate of inflation; decreasing the real price by 5% would mean, assuming a 5% rate of inflation, that the nominal price is held constant.
[2]) Rule K holds price constant at $ 12 as long as possible. If OPEC demand would exceed OPEC capacity, of if OPEC demand would be negative, a market clearing price would be calculated.

Applied Game Theory. 1979 © Physica-Verlag, Wuerzburg/Germany

An Application of the Entrepreneurial Theory of Games

By *S.C. Littlechild*, Birmingham[1])

Abstract: In a previous paper, I have argued that the inadequacies of game theory stem from its failure to incorporate an explicit account of the bargaining proceess, a process which necessarily takes place over time. To study this process requires emphasis on the different personalities of the players, their different and initially conflicting perceptions and expectations, the choices which they perceive as open to them, the formation of their plans, and the subsequent revision of expectations and plans as bargaining takes place. The present paper will attempt to apply these ideas to the game centered on the COMSAT communications satellite, as described by McDonald in the Game of Business.

1. Introduction

Game theory is almost universally considered a plaything of mathematicians. Despite the numerous conferences and workshops on game theory and the potential insights which this theory seems to have in a vast range of spheres, it has taken over 30 years to set up a conference on *applications* of game theory. Why is this? Why has game theory not been applied with the same enthusiasm and success as linear programming?

I have elsewhere suggested an answer to these questions [*Littlechild*]. It is not only that game theory is more complex because it involves multiple objectives of the several players in the game. More important is the fact that these players will generally have *different perceptions* of the nature and rules of the game and of the intentions of other players. Furthermore, the essence of game theory lies in the interaction between players which can only take place over *time*. What this means is that a central feature of game theory must be an account of the *learning and bargaining process* which leads up to the final outcome. Such an account has been lacking from most (but not all) expositions of game theory.

The present paper begins with a review of this argument, giving an illustration using the "Walt Disney game" described by *McDonald* [1975] in his illuminating volume *The Game of Business*. The next section appraises a recent development in game theory which involves a bargaining process, but of a different kind than the one I envisage. Section 5 introduces the „COMSAT communications satellite game", again as described by McDonald. The entrepreneurial model itself is formulated mathematically in section 6, with a numerical illustration using the COMSAT game in section 7. There is a brief concluding section with suggestions for further work.

---

[1]) *S.C. Littlechild*, University of Birmingham, North Ring Road, Birmingham, England.

2. An Entrepreneurial Theory of Games

Almost all the work on game theory to date has assumed *complete information*, that is, all the participants are assumed to know the rules of the game and to be able to compute optimal strategies. Only recently have games of *incomplete information* been studied. Even here an ingenious technique has been applied to transform the game from one of incomplete information to one of complete but *imperfect information* [e.g. *Harsanyi*]. In effect, this trick enables the players to learn the rules of the game as play proceeds, but it does so by removing the element of *surprise* from the game. The players may not know everything, but at least they know what they don't know!

It would probably not be denied that in real life, as opposed to classroom situations, most game involve incomplete information. Typically a player does not know completely who else might be in the game, what kinds of actions might be taken, what kinds of payoffs are available, etc. It is only in the course of actually playing the game that he learns these things — and then only some of them. For this reason he may make what, in retrospect, are mistakes; he may overlook opportunities, he may accept too low a payoff or hold out for too high a payoff.

What I am suggesting, then, is the need for a theory which emphasises the *subjective* nature of the game, as seen through the eyes of the different participants, and which studies the process by which the players come to understand and play the game. Such an approach stands in contrast to the usual formulation in which the game is specified *objectively*, and known to all players. In this latter formulation, interest centres on the nature of the *final solution*, which may be required to have properties of stability, rationality, equity, etc. In the subjective formulation, by contrast, interest centres to a greater extent on the properties of the *process* by which a solution (if any) is reached.

The term "entrepreneurial" theory of games is derived from recent usage in economics. Writers of the so-called "Austrian School" of Economics, notably C. Menger, L. Mises, J. Schumpeter, F.A. Hayek and the late O. Morgenstern, have always emphasised the role of incomplete knowledge in economic and social affairs [see also *Shackle,* esp. chapt. 36; *Simon*, esp. chapt. 14–16]. The term "entrepreneurship" is used by some of these writers, especially *Kirzner* [1973], to denote the element of alertness to new opportunities which must be involved in a world of incomplete knowledge. It is entrepreneurial activity which sets and keeps the market process in motion, and which in principle explains how an economic equilibrium can be reached.

Two points of clarification should be made. First, an entrepreneurial theory of games should not necessarily by seen as an *alternative* to established theories and devices, such as the characteristic function, the Shapley value, the core, etc. It is rather intended to *complement* them, to provide a framework within which they may perhaps be fitted.

Second, an entrepreneurial theory is not so much a specific model as a way of looking at situations; it is an *approach* to modelling. The appropriate entrepreneurial model for one situation will quite likely be very different from that for another situation. In some situations the task is to learn who is playing the game; in other situations the players are specified but the rules or the characteristic function values are unknown.

Finally, as I tried to show in my earlier paper, even where all these parameters are known, there is still a need for an account of the learning and bargaining process which results from conflicting expectations and plans of different players.

3. The Walt Disney Game

The following simple illustration is taken from *McDonald* [1975]. In 1940–41, Walt Disney got into financial difficulties with his new experimental film Fantasia. His traditional banker, the Bank of America, was concerned about the extension of his loan. For the bank, no game was involved, or at most a two-person game. It was purely a question of bargaining over the terms of the loan.

Walt Disney, on the other hand, took the view that if the Bank of America refused to lend him funds, another bank would be available. For Disney, it was a three-person game. The question for him was whether the new bank would collude with the Bank of America against him, or whether he could play off one bank against the other.

Evidently there was a fundamental difference of perception here about the nature of the game. Two of the players had different beliefs, and neither of them — at this stage — could objectively be said to be right or wrong.

However, such a conflict in perceptions could not continue indefinitely. As it happened, Walt Disney was persuaded by his advisers that no alternative bank was available. In effect, he accepted that his initial perception was mistaken. An amicable agreement was eventually reached between Disney and the Bank of America.

This simple episode may be thought of as a two-stage entrepreneurial game. At the first stage the players had different opinions about the number of players in the game and, hence, about the game tree and the characteristic function. Consequently, they held different views about what payoff vector would be reasonable. A change in perceptions on the part of one player led to a second stage, in which the perceptions of both players concerning the game tree and the characteristic function were essentially the same, and a solution was soon agreed upon.

4. Convergent Transfer Sequences

The conventional n-person game theory model is defined as follows. Let $N = (1, 2, \ldots, n)$ be a set of players and let $v: 2^N \to R$ be a characteristic function. Then (N, v) is a game in characteristic function form. A payoff vector is written $x = (x_1, x_2, \ldots, x_n)$. In such games, interest centres on the properties of the payoff vector which constitutes a solution. It may be individually or coalitionally rational, stable, fair, etc. There is no need to define these concepts here.

Recently, several writers have proposed an alternative approach (*Stearns; Billera; Kalai/Maschler/Owen*]. Let $x(1), x(2), \ldots, x(t), x(t + 1), \ldots$ be a sequence of payoff vectors in which $x(t + 1)$ is derived from $x(t)$ by means of some "transfer function". Interest then centres on the existence (if any) and stability properties of limit points to these sequences, depending upon the transfer mechanism chosen. These models thus shift attention from the outcome itself to the process which generates that outcome. This is in the spirit of the approach we have proposed. However, there is an essential difference. The transfer sequence models involve a mechanical procedure based upon the characteristic function alone. If the resulting process can be given an interpretation in terms of the perceptions and actions of the players, it is that they agree upon the characteristic function itself, and the bargaining is an iterative procedure for calculating the relative strengths of each player which the characteristic function implies.

In the entrepreneurial model, by contrast, the players do *not* agree upon the characteristic function or, in some cases, upon the set of players. Bargaining is largely a process of discovering what that function is. In practice, and in the entrepreneurial model, not only the proposed payoff vectors, but also the images of the characteristic function held by players, change over time. One is, nevertheless, interested in whether these payoff vectors and the beliefs about the characteristic functions converge over time, and if so what properties they have. In this sense, the entrepreneurial model may be considered an extension of the transfer sequence models.

5. The COMSAT Game

The following account and analysis is based upon the description given by *McDonald* [1975] of the game played around the communications satellite in the U.S.A. during 1970–1971. Only the briefest of details can be given here.

In March 1970 the Federal Comunications Commission invited interested parties to apply for authorisation to construct and operate domestic satellite systems, either alone or jointly. Ten parties expressed interest, and each of these had different objectives. Four of the parties already had substantial traffic, namely

American Telephone and Telegraph (A)
Western Union (W)
General Telephone and Telegraph (G)
Television Networks (N)

These parties preferred to rent satellite facilities, and were looking to co-operate with a potential owner who would launch and maintain the satellities.

Three of the parties had only a small amount traffic, namely

RCA Alaska (R)
MCI Lockheed (M)
Western Telecommunications (T).

Broadly speaking, these there parties were prepared to put up their own satellites, but preferred one or more partners, especially the ones with traffic.

The final three parties had no traffic, namely

COMSAT (C)
Hughes Aircraft (H)
Fairchild (F)

These parties were interested in the satellite hardware which they wished to rent to partners with traffic.

The interests of the parties were thus partly complementary but they were also partly conflicting. For example, the satellite manufactures and owners were competing with each other for the parties with traffic, while the latter were competing with each other for the source of this traffic (the ultimate customers).

The parties concerned were initially interested in only a limited number of potential coalitions, half-a-dozen each at most, which they ranked in order of preference. Not surprisingly, these preferences were not consistent between the parties, i.e. one party often ranked highly a certain coalition which the other potential members did not rank highly, or even did not consider at all. One reason for this was that the parties undoubtedly had different information about the objectives and bargaining strengths of each other, and consequently formed expectations which were to some degree bound to be disappointed.

Negotiations between the players proceeded for one year. At the end of this time certain coalitions had been formed with agreed payoffs. We do not have information about the detailed negotiations between the parties, nor about the precise manner in which their expectations and plans changed in the course of these negotiations, as their knowledge about each other increased. We do know, however, that the coalitions and payoff vectors finally agreed upon were substantially different from those initially contemplated.

6. A Formal Entrepreneurial Model

The following model is just one of many which might be constructed in the spirit of the entrepreneurial approach. It is designed with a view to the COMSAT application, though it is necessarily too simple to capture correctly all the interesting features of that situation.

We shall assume that the set of players $N = \{1, 2, \ldots, n\}$ is given and known to all players. We also assume that there exists a real-valued characteristic function $v: 2^N \to R$ defined over all possible subsets of N, called coalitions. However, we do *not* assume that each player knows all these characteristic function values. At any time t, we assume that player i knows only a subset $S_i^t \subseteq 2^N$ of the characteristic function values. He must, therefore, make his plans in ignorance of some of the opportunities available in the game, as well as in ignorance of the knowledge and intentions of the other players.

We suppose that at time t, each player i puts forward a (non-negative) claim $\hat{x}_i^t$ to a payoff. The resulting proposed payoff vector $\hat{x}^t = (\hat{x}_1^t, \hat{x}_2^t, \ldots, \hat{x}_n^t)$ is not necessarily feasible with respect to the true characteristic function. However, we suppose that the players gradually acquire more knowledge over time, and in any case change their plans, thereby generating a sequence of claims $\hat{x}^1, \hat{x}^2, \ldots, \hat{x}^t, \ldots$ over a discrete sequence of time periods. The game is assumed to finish as soon as a payoff vector $\hat{x}^t$ is proposed which is actually feasible, i.e. which satisfies

$$\sum_{i=1}^{n} \hat{x}_i^t \leqslant v(N). \tag{1}$$

We shall present sufficient conditions on the process by which the sequence of claims is generated to ensure that a feasible payoff vector is eventually obtained. No suggestion is made, however, that these are the weakest or most interesting conditions. The purpose of the present paper is mainly illustrative. Mathematical properties of entrepreneurial processes are best dealt with elsewhere.

Let us denote by $v_i^t \colon 2^N \to R$ player i's subjective perception of the characteristic function at time t. His subjective game at time t will be denoted (N, v_i^t). Denote by $C(N, v)$ the core of the game (N, v) that is

$$C(N, v) = \{x \colon \sum_{i \in S} x_i \geqslant v(S), S \subset N, \sum_{i=1}^{n} x_i = v(N)\}. \tag{2}$$

We shall assume that player i at time t puts forward a claim $\hat{x}_i^t$ which is compatible with a payoff vector lying in the core $C(N, v_i^t)$ of his subjective game (N, v_i^t). To avoid complications, we shall assume that the cores of the objective game and all the subjective games are always non-empty.

Recall we have assumed that at time t player i knows only the subset $S_i^t \subseteq N$ of characteristic function values. Since he is aware of no further opportunities, he believes that the value of any coalition S not in S_i^t is merely the superadditive cover generated by sub-coalitions in S_i^t. For example, if he knows $v(1, 2) = 20$ and $v(3, 4) = 10$, but knows no other values of the true characteristic function, he assumes $v(1, 2, 3, 4) = 30$, $v(1, 2, 3) = v(1, 2, 4) = 20$, $v(1, 3, 4) = v(2, 3, 4) = 10$, $v(1, 3) = v(1, 4) = v(2, 3) = v(2,4) = 0$, etc. Formally, player i's subjective characteristic function at time t is given by

$$v_i^t(S) = \max_{y} \sum_{T \subseteq N} y(T) v(T) \tag{3}$$

$$\text{subject to } \sum_{T \ni j} y(T) = \begin{cases} 1 & j \in S \\ 0 & j \notin S \end{cases}$$

$$y(T) = \begin{cases} 0 \text{ or } 1 & T \subseteq S_i^t \\ 0 & T \subseteq 2^N - S_i^t. \end{cases}$$

Now define $\bar{x}_i^t$ as the maximum payoff to him which player i at time t believes is consistent with some payoff vector lying in the core of his subjective game. That is

$$\bar{x}_i^t = \max \{x_i^t \colon x^t \in C(N, v_i^t)\}. \tag{4}$$

Similarly, define $\underline{x}_i^t$ as the minimum payoff

$$\underline{x}_i^t \min \{x_i^t \colon x^t \in C(N, v_i^t)\}. \tag{5}$$

We suppose that player i claims a payoff $\hat{x}_i^t$ lying somewhere between these two extremes

$$\hat{x}_i^t = \theta \, \bar{x}_i^t + (1 - \theta_i^t) \, \underline{x}_i^t \tag{6}$$

where $0 \leqslant \theta_i^t \leqslant 1$.

Different players will take different views as to how much they should concede at any time, and how their bargaining policy should develop over time. We shall assume that whatever policy they adopt, they are *ultimately* prepared to make *any* concession necessary to achieve agreement, thus

$$\lim_{t \to \infty} \theta_i^t = 0. \tag{7}$$

It remains only to define how knowledge about the characteristic function changes over time. Each player has an incentive to discover characteristic function values hitherto unknown to him, for several reasons: (i) to improve the opportunities open to potential coalitions of which he will be a member; (ii) to improve his bargaining power within such coalitions by reference to alternative available coalitions; (iii) to prevent him claiming an unrealistic payoff in ignorance of opportunities open to other players.

We shall assume that each player never forgets any characteristic function values, and has a positive probability p_i, constant over time, of discovering an hitherto-unknown value by the next period. We may thus calculate a transition probability matrix by using the binomial theorem. For example, suppose a player in a four-person game knows 7 of the 16 coalition values. The probability that he will discover any further 3 of the remaining 9 values by the next period is $p^3 (1-p)^6$. Formally, the probability of transition from S_i^t to S_i^{t+1} is given by

$$P(S_i^t, S_i^{t+1}) = \begin{cases} p_i^{|S_i^{t+1}|-|S_i^t|} (1-p_i)^{|2^N|-|S_i^{t+1}|} \\ \qquad \text{for } S_i^t \subseteq S_i^{t+1} \subseteq 2^N \\ 0 \qquad \text{otherwise.} \end{cases} \tag{8}$$

In fact, however, we do not need to be so specific about the transition matrix which describes each player's acquisition of knowledge. For purposes of this paper, all we need to assume, which is implied by (8), is that each player eventually learns all the characteristic function values with probability arbitrarily close to unity, i.e.

$$\lim_{t \to \infty} \text{Prob } \{S_i^t = 2^N\} = 1. \tag{9}$$

We may now prove the following property of this entrepreneurial game.

Proposition: A feasible global payoff vector is eventually attained with probability 1, i.e.

$$\lim_{t \to \infty} \Pr \left\{ \sum_{i=1}^{n} \hat{x}_i^t \leqslant V(N) \right\} = 1. \tag{10}$$

Proof: Recall that

$$\hat{x}_i^t = \theta_i^t \bar{x}_i^t + (1 - \theta_i^t) \underline{x}_i^t. \tag{6}$$

Now $\bar{x}_i^t$ takes at most 2^N different values, each of them necessarily finite by construction of v_i^t. It follows from (7) that

$$\lim_{t \to \infty} \theta_i^t \bar{x}_i^t = 0. \tag{11}$$

Let $M_i \geq 0$ denote the (finite) value which $\underline{x}_i^t$ attains when $S_i^t = 2^N$. If again follows from (7) that

$$\lim_{t \to \infty} \Pr \{ \sum_{i=1}^n (1 - \theta_i^t) \underline{x}_i^t = \sum_{i=1}^n M_i \} = 1. \tag{12}$$

Since by (5) each $\underline{x}_i^t$ is a member of a payoff vector lying in the core of $C(N, v)$, we have necessarily

$$\sum_{i=1}^n M_i \leq v(N). \tag{13}$$

Combining (11), (12) and (13) yields

$$\lim_{t \to \infty} \Pr \{ \sum_{i=1}^n [\theta_i^t \bar{x}_i^t + (1 - \theta_i^t) \underline{x}_i^t] \leq v(N) \} = 1 \tag{14}$$

which is equivalent to (10).

7. Interpretation of the COMSAT Game in Terms of the Entrepreneurial Model

McDonald [1975] lists and ranks some three dozen coalitions considered by the 10 players. He asks how the players managed to evaluate and reject the other thousand or so possible coalitions. The answer surely is that the majority of these were never considered in the first place because there was no prospect of their viability, or at least nothing to make them stand out compared to the more favourable coalitions initially investigated. However, each player may well have overlooked some potential coalitions which begin to look more favourable after initial negotiations prove unsuccessful. He may also have overlooked coalitions of advantage to other players which explain the reluctance of these players to accept his own proposed terms.

The coalitions initially envisaged by the players are listed in order of preference [as conjectured by *McDonald*] in Table 1. Note that each player initially explores only those coalitions of which he is a member. The characteristic function values have been chosen to be broadly consistent with *McDonald*'s account. The two right-hand columns contain the maximum and minimum individual payoffs consistent with membership in each individual player's subjective core. We assume that the players make initial claims lying somewhere between these limits. As a result, the first global payoff vector will have a toatl claim somewhere between 44 and 80.

Player code	Values of coalitions considered initially	Payoffs perceived as possible	
		Maximum $\bar{x}_i$	Minimum $\underline{x}_i$
A	AC = 5, A = 2	5	2
C	ACGMNTRW = 16, CN = 2, C = 1	16	1
F	AFGMNRTW = 14, FN = 3, F = 1	14	1
G	CG = 5, GH = 8, GRW = 7, GW = 6, GHW = 7, G = 1	8	7
H	GH = 8, HN = 8, HW = 8, GHW = 7, H = 2	8	8
M	AM = 3, MN = 2, M = 1	3	2
N	AN = 4, CN = 3, FN = 3, HN = 8, MN = 3,		
	RN = 3, TN = 3, NW = 4, N = 1	8	8
R	NR = 3, GRW = 7, RW = 5, R = 1	7	7
T	NT = 3, HT = 3, RT = 2, T = 1	3	2
W	NW = 6, GW = 6, HW = 8, GHW = 7, RW = 5,		
	CW = 5, W = 3	8	6
	Global Totals	80	44

Tab. 1: Initial Stage of the COMSAT Game

A claim of this magnitude, even as low as 44, is not feasible with respect to the true (objective) characteristic function, as we shall see in a moment. We therefore assume that, in the course of negotiations, the players gain a better knowledge of the characteristic function by discovering the values of certain other relevant coalitions. This new information is the kind which the players would be most likely to learn as they attempted, and generally failed, to form the coalition of their first choice. The values thus learned are shown in Table 2, together with the effect on the maximum and mini-

Player Code	Values of additional coalitions considered subsequently	Payoffs perceived as possible	
		Maximum $\bar{x}_i$	Minimum $\underline{x}_i$
A	CN = 3, C = 1	2	2
C	AFGMNRT = 14, GH = 8, AC = 5, MT = 2,		
	A = 2, N = 1	2	1
F	ACGMNRTW = 16, GH = 8, CN = 3, AC = 5, R = 1	1	1
G	AC = 5, HW = 8, W = 3, R = 1	3	3
H	GW = 6	8	5
M	AC = 5, AN = 4, A = 2, N = 1	1	1
N	GH = 8, W = 3	4	3
R	GW = 6, AN = 4, W = 3	1	1
T	AN = 4, HN = 8	2	1
W	GH = 8	6	5
	Global Totals	30	23

Tab. 2: Second Stage of COMSAT Game

mum payoffs subjectively perceived as possible. Subjective payoff vectors upon which these entries are based are set out in Table 3.

Perceived by Player	Payoff to players										Coalitions forming basis for superadditive cover $v_i(N)$
	A	C	F	G	H	M	N	R	T	W	
A	2	3									AC = 5
C	3 3	2 1	8 8			1 2			2 2		ACGMNRTW = 16
F	4	1	1	8			2	1			ACGMNRTW + F = = 16 + 1 = 17
G	3	2		3	5			1		3	GH + AC + W + R = 8 + 5 + 3 + 1 = 17
H				3 3	8 5		0 3			3 3	HN + GW = 8 + 6 = 14
M	4	1				1	1				AC + M + N = = 5 + 1 + 1 = 7
N	0 1		4 5	4 3			4 4			3 3	AN + GH + W = = 4 + 8 + 3 = 15
R	2		2				2	1	4		AN + GW + R = 4 + 6 + 1 = 11
T					4 4		4 4	0 1	2 1		HN + RT = 8 + 2 = 10
W				6 4	2 4		0 1			6 5	GH + NW = 8 + 6 = 14
Objective core	2	3	1	2	6	1	2	1	1	4	HN+GW+AC+F+ +R+M+T=8+6+ +5+1+1+1+1 GH+WR+AC+ +FN+M+T 8+5+5+3+1+1 ≈ 23

Tab. 3: Calculations underlying Table 2. Payoff vectors achieving specified maximum and minimum values $\bar{x}_i$ and $\underline{x}_i$ at 2nd stage. (Note: these vectors are not necessarily unique)

Most of the subjective payoffs, both maximum and minimum, have decreased from the first stage shown in Table 1. For example, COMSAT (player C) and Fairchild are each assumed to realise that the other could launch satellites to provide equivalent communications facilities. These alternatives sharply reduce their maximum perceived payoffs. Similarly, the Television Networks (N) realise that Hughes (H) can make a deal with General Telephone and Telegraph (G) for traffic. The Networks' maximum and minimum payoffs are both reduced.

At this second stage the maximum and minimum subjective vectors total 20 and 23, respectively. The views of the players are now more realistic, i.e. more consistent with what the game "really" offers, and there is less uncertainty about the scope for bargaining.

The final row in Table 3 shows the unique payoff vector which lies in the core of the game (at least, in the core so far as it is defined by the characteristic function values so far revealed). The maximum global payoff actually available is 23. Thus, if the players at the second stage were all willing to accept their minimum estimated payoff, their claims would be globally feasible. This is true even though some players, such as COMSAT, currently have a minimum estimate below their true core value, while other players, such as Western Electric, have a minimum estimate which is above.

Since it is unlikely that all players at this second stage would limit their claims to their minimum perceived payoffs, it seems necessary for the game to continue somewhat longer. One would expect that, as a result of further learning, the perceived maximum and minimum payoff vectors would come closer together, tending in the limit to the unique core payoff vector. At the same time the bargaining fractions θ would tend to zero, i.e. players would become more conservative in their claims.

Summary and Concluding Remarks

I have argued that game theory models, as conventionally presented, focus exclusively upon the final outcome of the game. It is true that a few writers have recently presented theories of transfer sequences leading up to such an outcome, but these sequences have been generated in a somewhat mechanical way, reflecting bargaining behaviour rather than learning behaviour.

In the present paper I have suggested an alternative kind of process. It assumes that participants have different knowledge about the characteristic function, and consequently envisage different payoff vectors. Increasing knowledge acquired over time is likely, under appropriate assumptions, to lead to a convergence of these vectors to a sinle equilibrium outcome. Such a process was illustrated very cursorily with the 10-person COMSAT game.

In retrospect, the COMSAT game was too large a game for a pioneering illustration; a 3 or 4-person game would have been easier to handle. To analyse such a game in detail would be a useful next step. At the same time, it is necessary to explore the formalisation of the model, and to state and prove more interesting propositions about convergence (or otherwise) of the sequence of sujective payoff vectors. Despite its inadequacies, it is hoped that the present paper will form a useful basis for such further discussion.

Appendix

The ideas put forward in this paper may be used to shed further light on S.J. Brams' excellent analysis of "Faith versus Rationality in the Bible" [page 430–445 in this book]

By taking into account the rewards of man's actions in these games, and the entrepreneurial learning process over time which these rewards may induce, we are led to a slightly different interpretation of God's motives, one in which His actions in the two situations studied may be seen not as contradictory but as consistent.

Brams begins his analysis with the two games as somehow "given" — he does not explain how they came to be initiated. In fact, the two situations are quite different in this respect. In the first case, God initiates the game by suddenly requesting Abraham to sacrifice his son. In the second case, Jephthah initiates the game by requesting a favour of God and promising a sacrifice in return for satisfactory performance.

What lessons would man form from the outcome of these two situations? In the first case he learns that if God requests him to do something, he should unquestioningly obey, and God will see him all right. In the second case he learns that he should try to solve his own problems without invoking God's help or rather, that if he does invoke God's help, he should not lay down the terms of the deal.

Is it not possible that God wished man to draw precisely these conclusions? In other words, God wished to induce in man a learning process about the nature of God and his relation to man. This learning would evidently have significance for man's subsequent behaviour, beyond the particular context of these two situations. In effect, God and man are players in a multi-stage entrepreneural game, of which these two episodes are merely two stages. Only in context of the series of games as a whole can God's actions be properly appreciated.

If this interpretation is true, then it is not necessary to conclude that "a less forgiving god than in Abraham's case must be posited to explain the choices of Jephthah and God in the game played over the sacrifice of Jephthah's daughter". [Brams, pp. 442ff.]. The interpretation given above allows us to interpret God's actions as consistent with each other and part of a single plan. In the one case He is rewarding an appropriate response, in the other case He is discouraging an inappropriately phrased initiative.

References

Billera, L.J.: Global Stability in *n*-Person Games. Trans. Am. Math. Soc. **172**, 1972, 45–56.
Harsanyi, J.C.: Games with Incomplete Information Played by 'Bayesian' Players, Parts I–III. Management Science **14** (3), (5), (7), 1967.
Kalai, G.M., M. Maschler, and *G. Owen*: Asymptotic and other Properties of Trajectories and Transfer Sequences leading to the Bargaining Set. Int. J. Game Theory **4** (4), 1975, 193–213.
Kirzner, I.M.: Competition and Entrepreneurship. Chicago 1973.
Littlechild, S.C.: An Entrepreneurial Theory of Games. Paper presented at the International Symposium on Extremal Methods and Systems Analysis in honour of A. Charnes, University of Texas at Austin, September 13–15, 1977, reprinted as Discussion Paper Series A No. 212. Faculty of Commerce and Social Science, University of Birmingham.
McDonald, J.: The Game of Business. New York 1975.
Shackle, G.L.S.: Epistemics and Economics. Cambridge 1972.
Simon, H.A.: Models of Man. New York 1957.
Stearns, R.E.: Convergent Transfer Schemes for *n*-Person Games. Trans. Am. Math. Soc. **134**, 1968, 449–59.

Applied Game Theory. 1979 ©Physica-Verlag, Wuerzburg/Germany

Fair Subsidies for Urban Transportation Systems

By *H. Diaz* and *G. Owen*, Bogota[1])

Abstract: Urban transportation systems are frequently subsidized by the governmental unit concerned. The problem of determining a fair rate of return for these systems is considered here from the point of view of the Shapley value (for non-atomic games). This allows us to obtain a "fair" rate of subsidy which can be made to depend on the particular route traveled, the number of passengers on the route, general frequency of service on that route, and physical condition of the cars used.

1. Introduction

Urban transportation systems are frequently subsidized by the local government (municipality, township, etc.) involved. There are various justifications given for this. In the first place, it is generally felt that the users of mass transportation (bus, subways, etc.) are normally poorer than those who drive their own automobiles; thus, in effect, subsidies from tax monies represent a transfer payment from the richer to the more disadvantaged classes. Secondly, even the person who normally drives his own automobile will sometimes have an occasion to use mass transit (his car may be undergoing repairs or otherwise unavailable); thus this person should be willing to pay (through taxes) for the availability of this service when he needs it. Finally, a mass transit system helps to integrate the city as an economic entity, and therefore helps all the inhabitants (not just the system's users) to benefit from the city's economic well-being.

For whatever reasons, the local government frequently finds it politic to maintain low transit fares, which are usually insufficient to cover costs. The operators therefore ask for subsidies, which must be negotiated with the government, and renegotiated every time that costs are significantly changed – as, e.g., by a new wage agreement with labor, or by an increase in the price of fuel. Eventually, agreements are reached, which generally guarantee to cover the operators' costs, minus fares, plus a "fair profit" – anything between 6 % and 20 % per annum – on investment. There is substantial acrimony, and very little incentive for efficiency in managing the system.

---

[1]) *Hugo Diaz* and Prof. *Guillermo Owen*, Instituto SER de Investigaciones, Calle 26, # 25-50 Bogota 1, D.E., Colombia.

The present paper will attempt to solve this problem by an application of game theory. It is to be understood that, based as it is on the Shapley value, we cannot expect people to accept it unless and until they have been convinced of the "reasonableness" and fairness of Shapley's axioms. Apart from this, there may be serious political objections to this approach; the transport companies can paralyze a city such as New York, Paris, or Bogota by striking, and political decision-makers will think twice before trying to limit their revenues. We feel nonetheless that this is a fair model, which could well be used for the purpose of determining "fair" returns.

2. The Mathematical Game

We shall assume here that there is a large number, m^*, of passengers, divided into n classes. Two passengers belong to the same class if they have essentially the same transportation needs, i.e., they must travel from the same origin to the same destination. There are m_i passengers in the i-th class ($i = 1, \ldots, n$), with

$$m^* = \sum_{i=1}^{n} m_i. \tag{1}$$

On the other hand, we assume a total of p transportation routes; frequency of service on the j-th route is f_j. We shall assume that the utility, for a passenger of class i, of a space on route j, is c_{ij} if he can get this without waiting. Since he must in general wait, however, before a bus passes, his utility on route j is rather given by the expression

$$c_{ij} (1 - e^{-af_j}) \tag{2}$$

where a is a parameter which represents, in some sense, the importance of quick service. The expression (2) is approximately equal to c_{ij} if f_j is large, whereas, if f_j is small, (2) is approximately equal to $ac_{ij}f_j$. Thus, when frequency is low, doubling the frequency will approximately double the utility, while, for high frequencies, any further increase is nearly worthless.

For the sake of simplicity, we shall further assume that each bus can handle a total of K passengers. If so, we find that an "efficient" social system will assign passengers to routes so as to maximize utility. Thus, the maximum utility to be expected is given by

$$w = \text{Max} \sum_i \sum_j c_{ij} (1 - e^{-af_j}) x_{ij} - \sum_j f_j r_j \tag{LP–1}$$

$$\text{subject to} \quad \sum_j x_{ij} \leqslant m_i \qquad i = 1, \ldots, n$$

$$\sum_i x_{ij} \leqslant K f_j \qquad j = 1, \ldots, p$$

$$x_{ij} \geqslant 0$$

where x_{ij} is the number of passengers of class i on route j, while r_j is the cost of operating a vehicle on route j. It may be seen that this is, essentially, a classical transportation problem [see, for example, *Owen*, 158–167] and that algorithms exist to solve such problems with as many as 50 sources and 1000 destinations. In our case, this. means that a moderate-sized computer can be used to solve such a program with as many as 50 routes and 1000 classes of passenger.

To transform this into a game, we shall assume that there are two types of player: the passengers, and the frequencies on the different routes. (In this treatment, a bus company becomes a coalition of frequencies.) Generally, a coalition can be represented by a pair

$$(S; G) = (s_1, \ldots, s_n; g_1, \ldots, g_p)$$

where s_i is the number of passengers of type i, and g_j the frequency of service on route j, within the coalition. We then obtain a characteristic function

$$w(S; G) = q(S; G) - \sum_j r_j g_j \tag{3}$$

where q is given by the linear program

$$q(S; G) = \max \sum_i \sum_j c_{ij} (1 - e^{-ag_j}) x_{ij} \tag{LP-2}$$

subject to
$$\sum_j x_{ij} \leqslant s_i$$

$$\sum_i x_{ij} \leqslant K g_j$$

$$x_{ij} \geqslant 0.$$

Also of interest is the dual linear program

$$q(S; G) = \min \sum_i s_i u_i + K \sum_j g_j v_j \tag{LP-3}$$

subject to
$$u_i + v_j \geqslant (1 - e^{-ag_j}) c_{ij}$$

$$u_i, v_j \geqslant 0$$

which, of course, can also be used to define the characteristic function $w(S; G)$ of the game.

In this analysis, the g_j are clearly continuous variables. The s_i are discrete, but, for a large city, they tend to be of the order of 100 or 1000 and so can be treated as though they were continuous. We shall therefore treat this as a non-atomic game, and proceed to find its Shapley value. As is known from *Aumann/Shapley* [1974] and *Billera/ Heath/Raanan* [1978], this value can be found from the integrals

$$\alpha_i = \int_0^1 w^i(tm_1, \ldots, tm_n; tf_1, \ldots, tf_p) \, dt \tag{4}$$

$$\gamma_j = \int_0^1 w_j(tm_1, \ldots, tm_n; tf_1, \ldots, tf_p) \, dt \tag{5}$$

where w^i is the partial derivative of w with respect to s_i, and w_j is the partial derivative with respect to g_j. This will give us the value function

$$\varphi [w] (S; G) = \sum_{i=1}^{n} \alpha_i s_i + \sum_{j=1}^{p} \gamma_j g_j. \tag{6}$$

The interpretation of this is that if a bus company runs buses at a frequency g_j on the several routes, then its profit should be $\Sigma \gamma_j g_j$. Now, we have $w = q - \Sigma r_j g_j$, and thus it is easy to see that

$$w^i = q^i$$

$$w_j = q_j - r_j$$

where, again, q^i and q_j are partial derivatives of q with respect to s_i and g_j. Thus, equations (4) – (5) take the form

$$\alpha_i = \int_0^1 q^i (tM; tF) \, dt \tag{7}$$

$$\gamma_j = \int_0^1 q_j (tM; tF) \, dt - r_j \tag{8}$$

where $M = (m_1, \ldots, m_n)$ and $F = (f_1, \ldots, f_p)$. If, now, we set

$$\beta_j = \int_0^1 q_j (tM; tF) \, dt \tag{9}$$

we will have $\gamma_j = \beta_j - r_j$. Thus the profit to a bus company should be

$$\Sigma \gamma_j g_j = \Sigma \beta_j g_j - \Sigma r_j g_j.$$

Since its expenses are $\Sigma r_j g_j$, and profit equals revenue minus expenses, we conclude that the line's revenues should be equal to $\Sigma \beta_j g_j$. In other words, bus companies should receive a revenue of β_j for each run on the j-th route.

3. Determination of the Parameters

The purely mathematical part of this analysis is very straightforward. Somewhat more complicated is the determination of the several numbers and parameters, m_i, c_{ij}, a which go into the defining programs. Presumably, the frequencies f_j are known in advance, or can somehow be controlled by the government. As for the m_i, these can presumably be obtained from the census figures (if these can be trusted) or from a poll, though there is implicit here a problem in cluster analysis if the n classes of passenger are to be defined in efficient fashion.

A real problem arises in the determination of the c_{ij}. As mentioned above, c_{ij} represents the utility to a passenger of class i of riding on a bus on route j. Actually, to be more realistic, this should really represent the total utility (to society), which, as mentioned in Section 1 above, may be greater than the utility to the user himself. It is clear that even a rough evaluation of c_{ij} is going to be difficult, though it should be added that government is constantly doing this sort of thing: every social program involves, at least implicitly, an evaluation of social utility. We may put an upper bound on the c_{ij}: it can be no more, say, than the cost of a taxi ride from the passenger's origin to his destination; it may well be considerably less since the bus is usually less comfortable, and, besides, the passenger might feel that he would never bother to make this trip unless the bus was available. Moreover, it is clear that, starting from this maximum, c_{ij} must decrease as the distance which the passenger must walk – to and from the bus stops – increases. At the same time, c_{ij} must be somehow related to the general operating costs of transportation along a given route: other things being equal, there must be more utility to a long bus ride over difficult terrain than to a short ride over level ground – which the passenger could perhaps forego by walking. Probably, c_{ij} should take into account the cost – to the state – of providing alternative transportation, as well as the economic cost if the passenger were to stay home instead of travelling.

Similar questions must also be asked about the parameter a which, as mentioned above, represents somehow the need for prompt transportation. As such, it might be expected to vary from person to person; for the sake of simplicity, we assume it constant. In a sense, it is related to the c_{ij}, inasmuch as we seem to be comparing, somehow, the inconvenience of waiting as against the inconvenience of walking. In any case, it is clear that the evaluation of both a and the c_{ij} must, to a certain extent at least, be subjective in nature. An opinion poll, asking people whether they would rather walk six blocks, with no waiting, or wait fifteen minutes for door-to-door service, would probably be necessary.

4. Computation of the Value

It remains now to compute the value of this game, or, more precisely, the numbers α_i and β_j. We consider the two dual linear programs (LP–2) and (LP–3), and note that s_i appears in the objective function of (LP–3) but not in its constant terms. Thus, if we think of q $(S; G)$ as determined by (LP–3), a straight-forward differentiation yields

$$q^i = u_i^* \tag{10}$$

where $u^* = (u_i^*)$ is the minimizing u-vector in (LP–3). On the other hand, we note that g_j appears, in both programs, both among the constant terms and in the objective function. The derivative of q with respect to g_j is therefore made up of two parts: the contribution due to the appearance of g_j in the objective function of (LP–2) (and thus among the constant terms of (LP–3)), and that due to its appearance in the objective function of (LP–3) (and thus among the constants of (LP–2)). The first of these, obtained by differentiation of (LP–2), is $ae^{-agj} \sum\limits_{i} c_{ij} x^*_{ij}$, and the second, obtained by differentiating (LP–3), is Kv^*_j. Thus we obtain

$$q_j = ae^{-agj} \sum_i c_{ij} x^*_{ij} + Kv^*_j \tag{11}$$

where $x^* = (x^*_{ij})$ is the maximizing vector in (LP–2), and $v^* = (v^*_j)$ is the minimizing v-vector in (LP–3).

Remembering that the integrands in (7) – (9) are to be evaluated at the points $(tM; tF)$, we note that the program (LP–2) takes the form

$$q\,(tM;\,tF) = \max \sum_i \sum_j (1 - e^{-atfj})\, c_{ij} x_{ij}$$

subject to
$$\sum_j x_{ij} \leqslant tm_i$$

$$\sum_i x_{ij} \leqslant Ktf_j$$

$$x_{ij} \geqslant 0$$

or, if we make the substitution $x_{ij} = ty_{ij}$,

$$q\,(tM;\,tF) = t \max \sum_i \sum_j c_{ij}\,(1 - e^{-atfj})\, y_{ij} \tag{LP–4}$$

subject to
$$\sum_j y_{ij} \leqslant m_i$$

$$\sum_i y_{ij} \leqslant Kf_j$$

$$y_{ij} \geqslant 0.$$

In this case, we see that, as t varies, the objective function varies (continuously) with t, but the constraint set, in the variables y_{ij}, remains the same at all times. As a straightforward application of parametric linear programming, we conclude that the interval $0 \leqslant t \leqslant 1$ will be split into finitely many subintervals, in each of which the maximizing vector (y^*_{ij}) is constant. At the end-points of the sub-intervals, the maximizing vector is not unique. This will, however, happen only at the end-points, unless the several functions e^{-afjt} are linearly dependent, and this in turn happens only in the degenerate case that two or more of the frequencies are equal.

Finally, looking at the dual program

$$q\,(tM;\,tF) = t \min \sum_i m_i u_i + K \sum_j f_j v_j \tag{LP–5}$$

subject to
$$u_i + v_j \geqslant c_{ij}\,(1 - e^{-atfj})$$

$$u_i, v_j \geqslant 0$$

we see that, in each of the sub-intervals where (y^*_{ij}) is constant, the minimizing vectors

(u_i^*) and (v_j^*) are determined by the fact that, from duality theory (slack complemen-tarity) certain of the inequalities must hold exactly as equations. Then, in each such sub-interval, u_i^* and v_j^* must be linear functions of the right-hand terms in the con-straints. Thus, they are linear functions of the exponential terms e^{-afjt}. In other words, we will obtain a partition

$$0 = t_0 < t_1 < \ldots < t_k = 1 \tag{12}$$

where, for $t_{h-1} \leqslant t \leqslant t_h$,

$$y_{ij}^* = y_{ij}^*(h) \tag{13}$$

$$x_{ij}^* = ty_{ij}^*(h) \tag{14}$$

$$u_i^* = L_{ih}(\exp(-atf_1), \ldots, \exp(-atf_p)) \tag{15}$$

$$v_j^* = L'_{jh}(\exp(-atf_1), \ldots, \exp(-atf_p)) \tag{16}$$

where L_{ih} and L'_{jh} represent linear functions. Then q^i and q_j, given by $(10) - (11)$, can be integrated, with no trouble, in each of the sub-intervals. Determination of the several sub-intervals, together with the corresponding values of y^*, u^* and v^*, is of course, a relatively simple exercise in parametric linear programming. (This assumes non-degeneracy.)

5. Implementation

The above analysis allows us to determine the revenue that a bus company should receive. Under a market situation, of course, such revenue would come entirely from passenger fares, but in our treatment it is understood that only part will come from fares; the rest will come through government subsidy.

We have already discussed the rationale for government subsidy. In practice, the set-ting of fares is very much a political matter and we may imagine that the government generally has very little latitude in this. We will however make a recommendation on this as well. It would probably be too complicated to establish a system of differential fares. Then, if the fare throughout the system is b, we find that a passenger of type i, in using a bus on line j obtains a net utility gain equal to

$$\theta_{ij} = c_{ij}(1 - e^{-afj}) - b. \tag{17}$$

On the other hand, our analysis states that such a passenger should receive a net gain equal to α_i. Thus we must have $\theta_{ij} \geqslant \alpha_i$, or

$$b \leqslant \min\{c_{ij}(1 - e^{-afj}) - \alpha_i\} \tag{18}$$

where the minimum is taken over all pairs (i, j) such that $x_{ij}^* > 0$, and x^* is the maximizing vector for (LP−1). This gives us an upper bound on the fare, and it is clear that, at this fare, some of the passengers may well be receiving a net gain of more than their "value", α_i. (This represents a political gain for the decision-makers.)

Once the fare has been set, it remains to fix the subsidy so as to let each line have the desired revenue. In fact, the total number of passengers on route j is

$$z_j^* = \sum_i x_{ij}^* \tag{19}$$

and the average number of passengers on a bus on that route is z_j^*/f_j. The average receipts, then, from fares, are bz_j^*/f_j, and we obtain a subsidy

$$\sigma_j = \beta_j - \frac{bz_j^*}{f_j} \tag{20}$$

for each run of a bus on route j.

6. Discussion

The above gives us a method of determining "fair" revenue levels for public transportation systems as well as of deciding what portions of this revenue should be provided by fares and by subsidies. Admittedly the method is imperfect; much will depend on subjective evaluations. Worse yet, the general process of determining bus fares and/or subsidies is largely tied up in municipal politics, and it is probably more than a bit optimistic to expect mathematical analysis to replace the usual give-and-take process of politics. Nevertheless, our model has some virtues.

For one thing, the general effect of the value is to decrease revenues on routes which have more buses than necessary, and to increase it on others that need more buses. Thus periodic adaptation might well lead to a very efficient equilibrium. For another, we note that the operating expenses, r_j, do not implicitly enter the calculation of the "fair revenue" γ_j. Of course c_{ij} is closely related to r_j, and thus r_j will eventually influence γ_j. Nonetheless, this influence is by some "ideal" value of r_j − given perhaps by general transportation costs − and thus an inefficient company would be unable to ask for an increase in subsidies to cover some inordinately high costs; increase would, rather, come if the general costs of all companies were to go up. Finally, a general acceptance of this model − if such a thing is possible − would do much to alleviate the unfortunate and recurrent threats of transit strike which are so repeatedly voiced throughout the world.

Finally, we note certain possible modifications of the model. For example, it would not be too difficult to include the possibility of buses of varying capacity on the several routes; similarly, we might treat two different times on the same route as being, essentially, different routes. Buses of different degrees of comfort could also be handled. On each of these modifications, of course, the number of variables would increase, and there is some danger of reaching the point where our computers would be unable to handle the relevant linear programs. At the same time, we note that the linear programs can, very likely, prove amenable to partition: for example, the east-west passengers probably have no use for north-south lines, and so we might be able to treat east-west and north-south transit as two independent problems.

References

Aumann, R.J., and *L.S. Shapley*: Values of Non-Atomic Games. Princeton 1974.

Billera, L.J., D.C. Heath, and *J. Raanan*: Internal Telephone Billing Rates: a Novel Application of Non-Atomic Game Theory. To appear in Operations Research, 1978.

Dantzig, G.B.: Linear Programming and Extensions. Princeton 1963.

Owen, G.: Finite Mathematics. Philadelphia 1970.

Applied Game Theory. 1979 © Physica-Verlag, Wuerzburg/Germany

Costs and Their Assessment to Users of a Medical Library:
A Game-Theoretic Method for Allocating Joint Fixed Costs[1])

By *E. Bres*, Austin[2]), *A. Charnes*, Austin[2]), *D. Cole Eckels*, Houston[3]), *S. Hitt*, Houston[3]), *R. Lyders*, Houston[3]), *J. Rousseau*, Austin[2]), *K. Russell*, Houston[3]), *M. Schoeman*, Austin[2])

Abstract: This paper is part of a larger study which identifies how costs arise in the formation and operation of a service institution like a library that provides a variety of related and interrelated services to a heterogeneous mix of users.

These users belong to different institutions (members of the library system) with diverse characteristics and functions and which are assessed fees to recover these costs, i.e., to support the annual operating budget.

Fair and efficient fees that recover the fixed portion of the annual joint costs are determined via construction and solution of a cooperative game model in which the players are the institutions.

The procedure is illustrated by the example of an independent medical liberary serving over 10, 000 individuals in 22 academic, research and clinical institutions.

1. Introduction

The present paper is part of a larger study, "Costs and Their Assessment to Users of a Medical Library: Parts I–IV," which identifies how costs arise in the formation and operation of a service institution like a library that provides a variety of related and interrelated services to a heterogeneous mix of users. We may distinguish two general categories of users: 1. Institutional or departmental users and 2. individual users within the institutions or departments. It is possible, then, to construct a means for recovering these costs (i.e., supporting the library's annual operating budget) through assessments to the institutions housing the users. Even if direct institutional assessment or subcription is not necessary because the library is supported through other means (e.g., as part of a general operating budget for a university), such a cost allocation process is useful to monitor performance and establish patterns of growth and expanding needs for resources.

[1]) This research was partly supported by Grant Number LM02333 from the National Library of Medicine, Department of Health, Education and Welfare, and by ONR Contract Number N00014-75-C-0569. Reproduction in whole or in part is permitted for any purpose of the United States Government.

[2]) *E. Bres, A. Charnes, J. Rousseau*, and *M. Schoeman*, Center for Cybernetic Studies, Business Economic Building, 203 E, The University of Texas at Austin, Austin, TX 78712, USA.

[3]) *D. Cole Eckels, S. Hitt, R. Lyders*, and *K. Russell*, Houston Academy of Medicine - Texas Medical Center Library, Houston, TX.

A reasonable division of the annual operating costs of a library distinguishes those attributable to each specific direct service provided (called out-of-pocket costs) and those essential to general library operations as an integrated facility (called support costs).

Out-of-pocket costs are variable in nature, responding to the usage experience of each service; whereas support costs, which may be relatively substantial in some libraries, are characteristically more complex to analyze[4]).

A large portion of support costs are related to physical inputs, such as books and journals purchased or plant (space) and equipment leased. Also to be considered are those personnel and their activities that contribute to library operations overall, but at levels not determined by the quantity of direct library usages. One aspect of support costs, then, arises from the necessity to establish and maintain a basic facility with the requisite capacity to provide a predetermined set of services for its users. In a sense, these costs are relatively fixed with respect to usage once having defined the type of services to be offered. However, qualitative considerations of service may require changes in supportive facilities and activities as usage varies. Thus, some portion of support costs are related to direct service usage.

The total operating costs of a library might, therefore, be described as having the following three components: direct costs for each service provided; variable support costs related to usage of services; and fixed support costs related to certain characteristics and functions of the institutions using the library.

To justify the budget to the participating institutions and maintain cooperative support for the library, the fees assessed to the institutions must be both equitable and efficient. That is, in addition to recovering total costs, the fees should reflect so far as possible the way in which costs are actually generated. Thus, institutional fees should be determined in a way that directly incorporates both individual user behavior and variation among institutions in informational requirements.

A reasonable and equitable charging mechanism, therefore, might recover the direct costs and variable portion of support costs through usage based fees, and the fixed support costs through institutionally based (or "membership") fees. Consequently, an institutional member of a library system would pay a fee which is partially based on the costs generated through its members' use of services and partially based on the costs of maintaining that type and quality of library that the institution would need to have available whenever its members wished to use it.

A methodology for determining usage based fees has been presented elsewhere [E. Bres/A. Charnes/D. Cole Eckels, Part II]. The present paper describes a method for determining institutionally based fees which incorporates notions from the mathematical theory of games. The model will be described both algebraically and in its library context. An illustration using data compiled at the HAM-TMC Library will show what results can be expected and how they may be interpreted.

---

[4]) The following discussion is given in great detail in Bres, et.al [1977, Parts II and III].

2. Joint Fixed Cost Allocation as an *n*-person Game

The fixed portion of the annual support costs of operating a library typically include costs associated with library acquisitions (such as purchases of books and serials, etc.); costs of different types of personnel; certain administrative costs (such as legal and professional fees, supplies, etc.); and expenditures on various types of equipment.

Such non-usage related costs must be incurred to accommodate the demands of all the member institutions as a group. However, distinct differences in informational requirements and quality and type of facilities demanded may exist among the member institutions. Such differences are determined by the various (non-usage related) characteristics peculiar to different institutions, for example, the set of academic programs offered, accreditation standards and potential user population.

Given these circumstances how might a set of fees (one per institution) be assessed that is *efficient* in that total annual fixed support costs are recovered yet *equitable* in the sense that an institution's fee closely reflects that institution's contribution to costs?

The notions of game theory enable us to provide an analytic formulation of such a "fairness plus efficiency" approach to cost allocation. We can construct an *n*-person game in which the "players" are the institutions served by the library (i.e., the member institutions of the library system) and whose characteristic function reflects the fixed portion of annual support costs that would be generated by each subset of institutions acting alone. Then, any of the concepts of solution to this game can be employed as a basis for allocating total fixed annual support costs to member institutions.

A reasonable approximation to total fixed annual support costs can be obtained (admittedly, in a somewhat round-about fashion) by subtracting estimates of usage related and certain other costs from the total annual operating budget. However, it is exceptionally difficult to calculate characteristic function values for all proper subsets of the "players" in our game for two reasons. First, the various items that contribute to fixed support costs are themselves difficult to isolate with any precision. Second, even if an exhaustive list of contributors were obtained, there is little or no quantitive evidence that systematically relates their levels (and thus, costs) back to the programs, accreditation standards and other institutional characteristics which generate these costs.

Therefore we make the following observation and assumption. The relevant costs are those which must be incurred in order to *maintain* a certain quality and type of library as dictated by the member institutions' diverse requirements, irrespective of library usage levels. The initial set-up costs of the library are the fixed joint costs of *constructing and equipping* this quality and type of facility in the first place, and are determined by this same set of (non-usage related) institutional characteristics. On this basis we assume (as a first approximation) that, for any group of institutions, the fixed portion of annual support costs of operating a library to accommodate their joint requirements would be proportional to the fixed joint set-up costs of that library; and that the constant of proportionality would be the same for all groups of institutions. In this way the *relative* ranking of institutions would remain the same for either initial set-up costs or fixed annual support costs.

Since data on initial set-up costs are more easily obtainable, we can then construct a game in which the "players" are again the institutions served by the library but whose characteristic function now reflects the initial set-up costs that would be generated by any group of institutions if a library were to be constructed to meet that group's requirements. A solution to this game would then determine an allocation of joint set-up costs to member institutions, from which we can determine a hierarchy of the institutions based on their individual cost allocations. We can then apply this hierarchy to the fixed portion of annual support costs of operating the library system to determine the institutionally based fees.

The following two sections develop a general characteristic function for the game which reflects initial set-up costs, and indicate how a hierarchy of institutions can be derived. The final section illustrates this procedure with a particular example.

3. Development of a General Characteristic Function

The initial costs which must be incurred for the provision of any library facility are here taken to be 3.1 purchases of books and journals, 3.2 purchases of equipment and furnishings, and 3.3 the construction cost of the building. The level of these costs will be determined by such factors as a) the size of the potential user population, b) the set of programs to be accommodated, and c) accreditation standards for member institutions.

Let $N \equiv \{j: j = 1, \ldots, n\}$ be the set of the participating institutions and $M \equiv \{i: i = 1, \ldots, m\}$ be the set of programs offered by all institutions. Further, let M_j denote the set of programs conducted by institution j, $j \in N$, and $M_S \equiv \underset{j \in S}{\cup} M_j$

denote the set of programs conducted by any subset S of institutions. We allow for institutions to have programs in common, so that in general $M_j \cap M_k \neq \emptyset$, for $j, k \in N$. Finally let x_j denoted the total potential membership of institution j, $j \in N$.

3.1 Purchases of Books and Journals

We assume that a program must be supported by a specific minimum core collection of books and journals if that program is to be conducted at all, irrespective of the number of people engaged in that program or of the institution to which they belong.

Let $Y_i =$ set of books and journals required for a minimum core collection to support program i, $i \in M$.

We allow for different programs to have overlapping core collection requirements, so that in general $Y_i \cap Y_l \neq \emptyset$, $i, l \in M$.

Let $Y_S \neq \underset{i \in M_S}{\cup} Y_i =$ set of books and journals required for a minimum core collection to support all programs conducted by subset S of institutions $s, S \subseteq N$.

$c_1 =$ average unit cost of a book or journal. Then, letting $g_1(S)$ denote the cost to subset S of core collections to support all its programs, we can write

$$g_1(S) = c_1\, y_S, \quad S \subseteq N \tag{1}$$

where $y_S \equiv |\, Y_S\,| = $ size (in number of books and journals) of Y_S.

3.2 Purchases of Equipment and Furniture

The costs of certain types of furnishings such as carpets, drapes, and decorative material may be included as part of the costs of construction of the library building. The relevant items for our purposes here are book shelves, user study facilities (such as chairs, tables and lamps), and employee work facilities (such as desks, chairs, filing cabinets, etc.). All of these can meaningfully be related to potential user population, size of collections and personnel requirements (which, in turn, are determined by user population and collection size).

Let $b_1 = $ average number of books and journals that can be accommodated per square foot of shelf space,

$c_2 = $ cost per square foot of shelf material,

$g_2(S) = $ cost to subset S of providing shelf materials to accommodate its collections.

Then we have that

$$g_2(S) = c_2 \cdot \frac{y_S}{b_1}, \quad S \subseteq N, \tag{2}$$

We assume that seating and study facilities must be available for a certain proportion, p_j, of the potential user population of institution j, $j \in N$. We may expect this proportion to vary according to the type of institution under consideration.

Let $c_3 = $ average cost of study facilities (e.g., table, lamp, chair and/or lounge chair).

$g_3(S) = $ cost to subset S of providing study facilities for its membership.

We then have that

$$g_3(S) = c_3 \sum_{j \in S} p_j x_j, \quad S \subseteq N. \tag{3}$$

The amount of staff work facilities (such as desks, chairs, filing cabinets, etc.) to be provided will depend upon the number of personnel required to operate the library. Library personnel may be categorized under three general headings according to the functions they perform: public services, technical services, and administration.

For any subset S of institutions let $Z^P(S), Z^T(S), Z^A(S)$ denote the required number of public, technical and administrative services personnel respectively.

Public services typically include answering reference questions, providing general information and help, reshelving material used in the library and check-out of material to be used outside the library. Personnel requirements for such services are then determined by the expected levels of these activities, in particular the number of circulations, which in turn can be related to either the size of the potential user population or that proportion of the population for which seating accommodation is to be provided. Following the latter we can write the desired number of public services personnel for any subset S of institutions as

$$Z^P(S) = f^P(\sum_{j \in S} p_j x_j), \qquad S \subseteq N. \tag{4}$$

Technical services such as ordering, cataloging and binding are related to books; so that requirements for technical services personnel are more accurately defined as being a function of the size and type of collections rather than the number of users. Accordingly, we may write

$$Z^T(S) = f^T(y_S), \qquad S \subseteq N. \tag{5}$$

The number of administrative personnel required is primarily determined by the number of other staff needed to operate the library rather than by collections or user population, so that we write

$$Z^A(S) = f^A[Z^P(S), Z^T(S)], \qquad S \subseteq N \tag{6}$$

i.e.,

$$Z^A(S) = f^A[f^P(\sum_{j \in S} p_j x_j), f^T(y_S)] \qquad S \subseteq N. \tag{7}$$

Let c_4, c_5, c_6 denote the average costs of providing work facilities per public, technical and administrative services employee, respectively. Then, for any subset S of institutions, the cost of providing work facilities may be denoted $g_4(S)$ and is given by

$$g_4(S) = c_4 Z^P(S) + c_5 Z^T(S) + c_6 Z^A(S), \quad S \subseteq N \tag{8}$$

i.e.,

$$g_4(S) = c_4 f^P(\sum_{j \in S} p_j x_j) + c_5 f^T(y_S) + c_6 f^A[f^P(\sum_{j \in S} p_j x_j), f^T(y_S)]$$
$$S \subseteq N. \tag{9}$$

3.3 Building Construction Cost

Essentially, we have three determinants of the size of library building required: space for collections of books and serials, study and browsing space for users, and work space for library personnel. Let b_2 denote the average number of books and journals that can be housed per square foot of library space (with due allowance for walk-ways between stacks of shelves). Then for subset S the square footage required to house its collections is given by

$$\frac{1}{b_2} \cdot y_S, \qquad S \subseteq N. \tag{10}$$

Let k denote the area (in square feet) required per study or seating facility. Then for subset S the area required for úser study and seating facilities is given by

$$k \sum_{j \in S} p_j x_j, \quad S \subseteq N. \tag{11}$$

Finally, let k^P, k^T, k^A denote the average number of square feet of work space required per public, technical, and administrative services employee respectively. Then for subset S total employee work space is given by

$$k^P Z^P (S) + k^T Z^T (S) + k^A Z^A (S), \quad S \subseteq N. \tag{12}$$

The total area (in square feet) of a library to accommodate subset S of institutions may be denoted by K_S and is given by

$$K_S = \frac{1}{b_2} \cdot y_S + k \sum_{j \in S} p_j x_j + k^P Z^P (S) + k^T Z^T (S) + k^A Z^A (S),$$
$$S \subseteq N. \tag{13}$$

In general the construction cost function may exhibit increasing returns to scale (decreasing average costs) over a broad range of capacities. Writing $g_5 (S)$ as the construction cost for a library for subset S we then have that

$$g_5 (S) = g (K_S), \quad S \subseteq N \tag{14}$$

where $g' > 0$ and $g'' \leqslant 0$.

Let $C (S)$ denote the initial set-up costs of building and equipping a library to meet the requirements of subset S of institutions. These costs are the costs of collections, shelf materials, study and employee work facilities, and construction of the building. We thus have

$$C (S) = g_1 (S) + g_2 (S) + g_3 (S) + g_4 (S) + g_5 (S), \quad S \subseteq N. \tag{15}$$

That is, for $S \subseteq N$,

$$C (S) = c_1 y_s + \frac{c_2}{b_1} \cdot y_S + c_3 \sum_{j \in S} p_j x_j + c_4 Z^P (S) + c_5 Z^T (S) +$$
$$+ c_6 Z^A (S) + g (K_S) \tag{16}$$

where

$$Z^P (S) = f^P \left(\sum_{j \in S} p_j x_j \right)$$

$$Z^T (S) = f^T (y_S)$$

$$Z^A (S) = f^A [f^P \left(\sum_{j \in S} p_j x_j \right), f^T (y_S)]$$

$$K_S = \frac{1}{b_2} \cdot y_S + k \sum_{j \in S} p_j x_j + k^P Z^P (S) + k^T Z^T (S) + k^A Z^A (S).$$

We have derived a general expression for the initial costs of building and equipping a library to meet the requirements of any group of institutions. This cost function can then be used as the characteristic function for the "game" of allocating initial set-up costs to member institutions. By making some further assumptions about the functions $f^P, f^T, f^A, z^P, z^T, z^A$ and g we are then in a position to calculate $C(S)$ for any subset S of N and thus derive the characteristic function values for the game.

4. Determination of a Hierarchy of Institutions

We have derived an n-person game (N, C), a solution to which allocates the initial set-up costs of constructing and equipping a library to the institutions served by the library.

Let $(c_1, \ldots, c_n)$ be a vector of optimal allocations (by whatever criteria) to the member institutions. We can rank these allocations in decreasing order of magnitude to obtain (relabelling if necessary)

$$c_1 \geqslant c_2 \geqslant \ldots \geqslant c_n.$$

We may derive a set of indexes $I_1, \ldots, I_n$ in which the index for the j-th institution is the ratio of the cost allocation to the j-th institution to that of the largest cost allocation to all institutions. Thus,

$$I_j = \frac{c_j}{c_1}, \quad j = 1, \ldots, n$$

and clearly $I_1 = 1.0$. We then have a hierarchy of institutions based upon the set of indexes in which

$$1 = I_1 \geqslant I_2 \geqslant \ldots \geqslant I_n > 0.$$

The fixed portion of annual support costs, denoted F, is then allocated according to this hierarchy. Thus, the institutionally based fee for institution j, denoted b_j, is given by

$$b_j = \frac{F}{\sum\limits_{k=1}^{n} I_k} \cdot I_j \qquad j = 1, \ldots, n.$$

5. An Application: The HAM-TMC Library Experience

The Houston Academy of Medicine-Texas Medical Center (HAM-TMS) Library is an independent medical library that jointly serves and is supported by 22 academic, research and clinical institutions. Serving over 10,000 individuals, the library ranks second in total expenditures of all medical libraries in the U.S. and Canada, with an annual budget in excess of $1.25 million which is recovered through fees assessed to the supporting institutions.

The 22 institutions can be categorized under six general headings.

Medical Schools:	Baylor
	University of Texas (UTMS)
Medical Associations:	Houston Academy of Medicine (HAM)
Nursing Schools:	Texas Women's University (TWU)
	U.T. System School of Nursing (UTSSN)
	Prairie View A&M University School of Nursing (PVAMUSN)
Graduate Medical and	
Health related Schools:	U.T. Graduate School of Biomedical Sciences (GSBS)
	U.T. School of Public Health (SPH)
	U.T. School of Allied Health (SAH)
	U.T. Dental Branch (DB)
Specialized Research	
Institutes:	U.T. System Cancer Center M.D. Anderson
	Hospital and Tumor Institute (UTMDA)
	Texas Institute of Rehabilitation and Research (TIRR)
	Texas Research Institute of Mental Sciences (TRIMS)
	Speech and Hearing Institute (SHI)
	Institute of Religion (IR)
Teaching and Research	
Hospitals:	Methodist Hospital
	St. Luke's Episcopal/Texas Children's
	Hospital (SL/TCH)
	Harris County Hospital District (HCHD)
	Hermann Hospital
	St. Joseph Hospital
	Memorial Hospital
	Veteran's Administration Hospital (VA)

The categories are necessarily broad with, in some cases, substantial variation in size and emphasis of institutions under a particular heading. In addition, the nature of the institutions is such that more than one classification is possible. For example, UTMDA might be classified as a graduate medical school, a research hospital, or a specialized research institute; SHI could also be considered as a graduate college; TIRR, TRIMS and IR are small in comparison with UTMDA and probably more closely resemble research hospitals than spezialized research institutes; even the hospitals themselves vary as to their emphases on teaching and research.

In addition, Baylor Medical School includes smaller departments with programs in Allied Health and Biological Sciences, but data are not available on them in any dis-aggregated fashion, whereas they are available under the U.T. System. In the interests of comparability, therefore, we absorb SAH and GSBS into the U.T. Medical School, so that Baylor and U.T.M.S. may be considered to conduct essentially the same programs.

5.1 Some Practical Simplifications

To make the model operational we require further assumptions about the various functions and parameters involved in the general expression (16) for the characteristic function values.

We assume constant ratios between the required number of public services personnel and planned seating and study capacity, and between the required number of technical services personnel and the size of collections. We further assume that the number of administrative personnel required is a constant fraction of the total number of other staff required to operate the library irrespective of their distribution over public and technical services. In this way we can rewrite equations (4), (5), (7) and (13) as

$$Z^P(S) = f^P \sum_{j \in S} p_j x_j, \qquad\qquad S \subseteq N \qquad (17)$$

$$Z^T(S) = f^T y_S, \qquad\qquad S \subseteq N \qquad (18)$$

$$Z^A(S) = f^A [f^P \sum_{j \in S} p_j x_j + f^T y_S] \qquad\qquad S \subseteq N \qquad (19)$$

$$K_S \quad = [\frac{1}{b_2} + k^T f^T + k^A f^A f^T] y_S +$$

$$+ [k + k^P f^P + k^A f^A f^P] \sum_{j \in S} p_j x_j, \qquad\qquad S \subseteq N \qquad (20)$$

where f^P, f^T and f^A are now redefined as constants.

The construction cost function, g, can be approximated linearly by assuming a constant average cost, γ, per square foot of space. Equation (14) thus becomes

$$g_S(S) = \gamma K_S, \qquad\qquad S \subseteq N \qquad (21)$$

Substituting equations (17) through (21) into equation (16) and collecting terms we obtain

$$C(S) = \alpha y_S + \beta \sum_{j \in S} p_j x_j, \qquad\qquad S \subseteq N \qquad (22)$$

where

$$\alpha = [c_1 + \frac{c_2}{b_1} + c_5 f^T + c_6 f^A f^T + \frac{\gamma}{b_2} + \alpha k^T f^T + \gamma k^A f^A f^T] \qquad (23)$$

and

$$\beta = [c_3 + c_4 f^P + c_6 f^A f^P + \gamma k + \gamma k^P f^P + \gamma k^A f^A f^P]. \qquad (24)$$

Note that α and β are constants so that the characteristic function of the game is now reduced to a simple linear expression in two variables. That is, the cost to any group of institutions of building and equipping a library to meet that group's require-ments is a linear function of a) the size and type of collections of books and serials needed to support that group's programs and b) the amount of seating and study capacity which that group needs to have available at all times. Moreover, these two variables reflect the essential (non-usage related) differences between institutions, namely, differences in the amount, quality and type of informational requirements and facilities demanded. The coefficients α and β reflect both the direct and indirect effects of these variables on cost.

5.2 Institutional Collection Requirements[5])

The HAM-TMC Library collection consists of approximately 133, 000 books and serials, encompassing all fields of medicine, nursing and some special areas of research, different amounts of which are required by the different institutions to support their programs.

We proceed on the basis that all institutions require a core collection consisting minimally of a certain amount of literature on general medicine and nursing. Differences in informational requirements between institutions are then reflected in the extent of this collection and the amount of other specific research items needed.

Institution	No.Books and Serials	Medium-Term Average Membership	Proportion to be Accommodated	Medium-Term Average Membership to be Accommodated
Baylor	123, 000	3, 030	1/4	758
HAM	61, 500	2, 000	1/100	20
TWU	10, 000	950	1/4	238
UTMS (Plus SAH, GSBS, SHI, UTSSN)	133, 000	3, 450	1/4	863
DB*	2, 460	750	1/8	94
SPH*	6, 150	400	1/8	50
UTMDA*	24, 600	450	1/8	56
PVAMUSN*	6, 150	130	1/8	16
TIRR	9, 000	25	1/5	5
TRIMS*	6, 750	35	1/8	4
IR	9, 000	65	1/10	7
HCHC	9, 000	40	1/10	4
Hermann	9, 000	20	1/10	2
Methodist	9, 000	90	1/10	9
SL/TCH	9, 000	80	1/10	8
Memorial	9, 000	50	1/10	5
St. Joseph	9, 000	35	1/10	3
VA*	2, 250	36	1/20	2

Tab. 1: Institutional Requirements of HAM-TMC Collection and Membership of Institutions
*) Own Library

[5]) A fuller treatment is given in *Bres*, et. al [1977, Part III].

Library personnel have estimated the number of books and serials in the HAM-TMC collection that each institution would require if it were to provide its own library in the absence of the HAM-TMC facilities; these figures are listed in Table 1. Six of the institutions already have their own collections (so indicated by an asterisk) consisting either of specialized materials relevant to their particular fields or smaller quantities of much the same type of literature as that at the HAM-TMC Library. These institutions essentially use the HAM-TMC collections as supplementary literature for back-up purposes only, although to varying degress; so that the relevant figure is the additional material to their current collection that would be required in the absense of the HAM-TMC Library.

Note that Table 1 shows a combined figure for UTMS, GSBS, SAH, SHI and UTSSN. This grouping of institutions presents no real administrative problems since their fees traditionally come from the same source. If necessary, subdivision of the composite fee may be determined by a simple small game along the lines presented here.

The advantage of this grouping is that the entire HAM-TMC collection is now required by this composite institution and every other institution requires only some proportion of this total in its core collection. In this way, since the figures in Table 1 refer to the same general type of literature, we can now approximate the core collection requirements for any group of institutions by the requirements of the "largest" individual member of that group.

Although this one hundred percent overlap of institutional requirements may not be a strictly accurate description for a few of the coalitions, it certainly is the case for the overwhelming majority of the one-quarter million-plus involved, and it is to be expected that any error introduced will be relatively slight. Moreover, as we shall see later, this approximation considerably simplifies calculation of the characteristic function values and the solution procedure for our game.

5.3 Institutional Membership

Table 1 also lists estimates of the membership of each institution at planned capacity for the medium-term future, the fraction of membership to be provided with seating and study facilities, and the resulting number of users to be accommodated. The proportions are those supplied by HAM-TMC personnel.

Those institutions with their own libraries are assumed to have existing seating and study facilities equal to fifty percent of their total needs, so that the figures in Table 1 represent the necessary additional facilities were they to go it alone. Consequently, the relevant proportions are one-half of what they would be if these institutions had no other library.

5.4 A Solution

Several notions of solution may be appropriate for the game (N, C), in which the characteristic function is defined by equation (22), one which is the *Shapley* value [1953]. For the general n-person game (N, C) the Shapley value for player j, φ_j, is given by

$$\varphi_j = \sum_{\substack{S \subseteq N \\ S \ni j}} \frac{(n-s)!\,(s-1)!}{n!} [C(S) - C(S - \{j\})], \qquad j \in N. \tag{25}$$

The summation ranges over all the subsets of N to which player j belongs, and $s \equiv |S|$ denotes the number of members of subset S.

In the present context the Shapley value calculates the increase in cost generated by each institution for each possible combination of institutions, and takes an average.

We can consider the game (N, C) to be a composite one comprised of the following two games:

$$(N, u) \text{ where } u(S) = \alpha y_S, \qquad\qquad\qquad S \subseteq N \tag{26}$$

$$(N, v) \text{ where } v(S) = \beta \sum_{j \in S} p_j x_j, \qquad\qquad S \subseteq N. \tag{27}$$

The game (N, u) is that of allocating to member institutions all initial set-up costs associated either directly or indirectly with providing the size, quality and type of collections of books and serials necessary to support all programs. The game (N, v) is that of allocating all set-up costs incurred either directly or indirectly through the provision of seating and study facilities for library users.

Paraphrasing an axiom of *Shapley* [1953]: if the j-th institution is a "player" in both games, then the Shapley value solution for the cost allocation to that institution in the composite game is equal to the sum of the Shapley value solutions for the cost allocations to that institution in the separate games; that is,

$$\varphi_j(C) \equiv \varphi_j(u + v) = \varphi_j(u) + \varphi_j(u) + \varphi_j(v).$$

Now note that the game (N, v) is inessential, that is,

$$\sum_{j \in N} v(\{j\} = \beta \sum_{j \in N} p_j x_j = v(N),$$

so that the core consists of the single point

$$(\beta p_1 x_1, \dots, \beta p_n x_n).$$

This unique vector is also the Shapley value solution for the game (N, v), that is,

$$\varphi_j(v) = \beta p_j x_j, \qquad j \in N. \tag{28}$$

Now consider the game (N, u). A general result due to *Littlechild/Owen* [1973] is that a simple and easily calculated expression may be obtained for the Shapley value whenever the characteristic function is a "cost" function with the property that the cost of any subset of players is equal to the cost of the "largest" player in that subset.

As already noted in section 5.2, we can approximate the core collection requirements for any group of institutions by the requirements of the "largest" individual member of that group.

Following *Littlechild/Owen* we subdivide the set N of players according to type. Let N_i denote the set of players of type i, $i = 1, \dots, m$, and let $n_i > 0$ denote the number of players of type i.

Thus

$$N = \bigcup_{i=1}^{m} N_i \text{ and } n = \sum_{i=1}^{m} n_i.$$

The "cost" associated with player type i is given by αy_i ($i = 1, \ldots, m$), where y_i denotes the size of collections required to support all programs of institutions type i. We can order the players such that

$$0 = y_0 < y_1 < y_2 < \ldots < y_m.$$

Define the game on N by the subadditive cost function

$$u(\emptyset) = 0, \ u(S) = \max \{\alpha y_i\} \tag{29}$$

with the maximization taken over all i such that $N_i \cap S \neq \emptyset$.

Now define

$$R_k = \bigcup_{i=k}^{m} N_i \text{ and } r_k = \sum_{i=k}^{m} n_i, \qquad k = 1, \ldots, m.$$

Note that $R_1 = N$ and $r_1 = n$.

With the characteristic function of the game (N, u) defined by (29), the expression for the Shapley value may be simplified to [see *Littlechild/Owen*]

$$\varphi_j(u) = \alpha \sum_{k=1}^{i} \frac{y_k - y_{k-1}}{r_k}, \quad \text{for } j \in N_i, \ i = 1, \ldots, m. \tag{30}$$

Table 2 contains the essential calculations for determining the Shapley values.

Expression (30) can be interpreted as the following rule: Divide the cost of catering for the smallest type of institution equally among all institutions. Divide the incremental cost of catering for the second smallest type of institution (above the cost of the smallest type) equally among all institutions except the smallest type, etc.

Two remarks are worth noting:
a) If we set $\varphi_j(u) = \Phi_i$ for $j \in N_i$, then $\Phi_i = \Phi_{i-1} + \alpha (y_i - y_{i-1})/r_i$ \hfill (31)

b) If $n_i = 1$ for all i, that is, there is only one player of each type, then

$$\Phi_i = \Phi_{i-1} + \alpha (y_i - y_{i-1}) / (n - i + 1). \tag{32}$$

The Shapley value of the original game (N, C) can now be conveniently calculated as

$$\varphi_j(C) = \varphi_j(u) + \varphi_j(v)$$

$$= \alpha \sum_{k=1}^{i} \frac{y_k - y_{k-1}}{r_k} + \beta p_j x_j, \text{ for } j \in N_i, \quad i = 1, \ldots, m. \tag{33}$$

Type i	Institutions	No. of Institutions n_i	No. of Volumes y_i	$y_i - y_{i-1}$	r_i	$\dfrac{y_i - y_{i-1}}{r_i}$	$\sum\limits_{k=1}^{i} \dfrac{y_k - y_{k-1}}{r_k}$
1	VA*	1	2, 250	2,250	18	125.00	125.00
2	DB*	1	2, 460	210	17	12.35	137.35
3	SPH*	2	6, 150	3,690	16	230.63	367.98
	PVAMUSN*						
4	TRIMS*	1	6, 750	, 600	14	42. 86	410. 84
5	TIRR	8	9, 000	2, 250	13	173. 08	583. 92
	IR						
	HCHD						
	HERMANN						
	METHODIST						
	SL / TCH						
	MEMORIAL						
	ST. JOSEPH						
6	TWU	1	10, 000	1, 000	5	200. 00	783. 92
7	UTMDA*	1	24, 600	14, 600	4	3,650. 00	4,433.92
8	HAM	1	61, 500	36, 900	3	12,300. 00	16,733. 92
9	BAYLOR	1	123, 000	61, 500	2	30,750. 00	47,483. 92
10	UTMS	1	133, 000	10, 000	1	10,000. 00	57,483. 92
	(Plus SAH, GCBS, SHI, UTSSN)						

Tab. 2: Calculations of Shapley Values
*) own library

A solution to our game, and hence a hierarchy of institutions, can be obtained in a straightforward manner once values for the coefficients α and β have been determined. This hierarchy can then be applied to the fixed portion of annual support costs to derive the institutionally based fees.

However, we can eliminate the problems which inevitably arise in obtaining reliable estimates for all the various factors that constitute α and β by the following procedure.

Since α is a constant, the hierarchy of institutions based solely on the allocation of those set-up costs associated either directly or indirectly with collections will be independent of α, and is completely determined by the data in the final column of Table 2. This hierarchy is thus given by a set of weights $\{w_j^1\}$ in which the weight for any institution is given by the ratio of the Shapley value for that institution to that for the "largest" institution, i.e.,

$$w_j^1 = \frac{\varphi_j(u)}{\max\limits_{k}\{\varphi_k(u)\}} = \frac{\sum\limits_{k=1}^{i} \dfrac{y_k - y_{k-1}}{r_k}}{\sum\limits_{k=1}^{m} \dfrac{y_k - y_{k-1}}{r_k}}, \qquad j \in N_j, \; i = 1, \ldots, m$$

which is independent of α.

If we can isolate that fraction, f, of annual fixed support costs, F, associated with collections, we can then use this set of weights directly to obtain a first set of institutional allocations as follows:

$$b_j^1 = \frac{w_j^1}{\sum\limits_k w_k^1} f F, \qquad j \in N$$

where b_j^1 = allocation to institution j of annual fixed support costs associated with collections.

Similarly, since β is a constant, the hierarchy of institutions based solely on the allocation of those set-up costs associated either directly or indirectly with providing seating and study facilities will be independent of β, and is completely determined by the data in the final column of Table 1.

The corresponding set of weights $\{w_j^2\}$ in this second hierarchy is given by

$$w_j^2 = \frac{\varphi_j(v)}{\max\limits_k \{\varphi_k(v)\}} = \frac{p_j x_j}{\max\limits_k \{p_k x_k\}}, \qquad j \in N$$

which is independent of β. We thus obtain a second set of allocations as follows:

$$b_j^2 = \frac{w_j^2}{\sum\limits_k w_k^2} (1-f) F, \qquad j \in N$$

where b_j^2 = allocation to institution j of annual fixed support costs associated with seating and study facilities.

The sum of these two allocations b_j^1 and b_j^2 to institution j will then be that institution's contribution to annual fixed support costs; that is, the institutionally based fee for institution j is given by

$$b_j = b_j^1 + b_j^2, \qquad j \in N.$$

We note here that in some cases it may be desirable to adjust the weights as calculated above in order to reflect particular policy prescriptions or any important but less tangible institutional differences. This may be accomplished by multiplying the weights by appropriate factors either less than or greater than unity. Alternatively, one can attempt to incorporate such considerations directly in the calculation of the original weights. In either case the rest of the procedure is unchanged.

Approximately fifty-five percent of the fixed support costs of the HAM-TMC Library for 1976 were estimated as attributable to annual acquisitions of books and serials and related activities such as binding and cataloging. Projected fixed support costs for 1977 were $579, 065. Using the figure of 55 % as a typical yearly average then yields an amount of $318, 486 as projected fixed support costs associated with collections and the remaining $260, 579 is thereby attributed dircetly and indirectly to the provision of study facilities. The appropriate weights and corresponding cost allocations are given in Table 3.

Institution	Collections		Study Facilities		Base Fee
j	Weight w_j^1	Cost Allocation b_j^1 ($)	Weight w_j^2	Cost Allocation b_j^2 ($)	b_j ($)
Baylor	0.826038	93,500	0.878331	92,126	185,626
HAM	0.291106	32,950	0.023175	2,431	35,382
TWU	0.013637	1,542	0.275782	28,926	30,468
UTMS (Plus SAH, GSBS, SHI, UTSSN)	1.500000	169,788	1.000000	104,888	274,676
DB*	0.002389	270	0.108922	11,425	11,695
SPH*	0.006401	725	0.057937	6,077	6,802
UTMDA*	0.077133	8,731	0.064890	6,806	15,537
PVAMUSN*	0.006401	725	0.018540	1,945	2,670
TIRR	0.010158	1,150	0.005794	608	1,758
TRIMS*	0.007147	809	0.004635	486	1,295
IR	0.010158	1,150	0.008111	851	2,001
HCHD	0.010158	1,150	0.004635	486	1,636
Hermann	0.010158	1,150	0.002317	243	1,393
Methodist	0.010158	1,150	0.010429	1,093	2,243
SL/TCH	0.010158	1,150	0.009270	972	2,122
Memorial	0.010158	1,150	0.005794	608	1,758
St. Joseph	0.010158	1,150	0.003476	365	1,515
VA*	0.002175	246	0.002317	243	489
	2.813691	318,486	2.484355	260,579	579,065

Tab. 3: Institutional Weights and Cost Allocations[1])
[1]) All figures rounded to nearest dollar.
*) own library

Note that the U.T. Medical School weight for "collections" is 1.5 as opposed to 1. 0. This is so because U.T. Medical School is a relatively new institution in the HAM-TMC Library system and consequently has not been charged in the past for the early years' acquisitions which other institutions, particularly Baylor, have gradually accumulated. Therefore, on equity grounds it was decided to "phase in" U.T. Medical School over five years, so that after that period its weight would be 1. 0.

The figures in Table 3 show that Baylor contributes almost one-third of the annual fixed support costs, whereas U.T. Medical School plus other U.T. graduate institutions are allocated over one-half of the costs. As might be expected, the hospitals and small research institutes each contribute less than one-half of one percent of annual fixed support costs.

As an aside we noted that knowing the fraction f defined above we can then determine the base fees directly by deriving a composite "weight" or, rather, allocation factor for each institution in an obvious way.

$$b_j = b_j^1 + b_j^2 = \left[f \frac{w_j^1}{\Sigma w_k^1} + (1-f) \frac{w_j^2}{\sum\limits_k w_k^2} \right] F$$

$$= \left[\frac{f w_j^1 \sum\limits_k w_k^2 + (1-f) w_j^2 \sum\limits_k w_k^1}{\sum\limits_k w_k^1 \cdot \sum\limits_k w_k^2} \right] F$$

so that

$$b_j = w_j F, \qquad j \in N$$

where

$$w_j = \left[\frac{f w_j^1 \sum\limits_k w_k^2 + (1-f) w_j^2 \sum\limits_k w_k}{\sum\limits_k w_k^1 \cdot \sum\limits_k w_k^1} \right]$$

and

$$\sum_j w_j = 1.$$

Thus the composite weights $\{w_j\}$ are allocation factors that give directly the proportion of fixed operating costs to be allocated to each institution.

References

Bres, E., A. Charnes, D. Cole Eckels, S. Hitt, R. Lyders, J.J. Rousseau, K. Russell, and *M. Schoeman*: Costs and Their Assessment to Users of a Medical Library. Parts I–IV. CCS Reports 301–304. Center for Cybernetic Studies, Austin 1977.

Littlechild, S.C., and *G. Owen*: A Simple Expression for the Shapley Value in a Special Case. Management Science 20 (3), 1973, 370–372.

Shapley, L.S.: A Value for N-Person Games. Contributions to the Theory of Games 2, Princeton 1953.

Applied Game Theory. 1979 ©Physica-Verlag, Wuerzburg/Germany

Efficiency / Equity Analysis of Environmental Problems –
A Game Theoretic Perspective

By *J.P. Heaney*, Gainesville[1])

Abstract: Procedures for efficiency/equity analysis in U.S. water resources planning are compared to related concepts in cooperative *N*-person game theory. The intent is to recommend procedures for allocating and sharing the costs of water pollution control programs. The selected solution concept is two part tariff which is a proration between the minimum and maximum costs a player can be charged and still retain stable coalitions for all sub groups. A single linear program is solved 2nd times to obtain the required lower and upper bounds.

Introduction

Efficiency analysis is herein defined to be determination of the economically efficient solution to the problem under consideration. Procedures range from simple benefit-cost analysis to sophisticated nonlinear optimization. The objective is to maximize the positive difference between benefits and costs to whomsoever they may accrue.

Equity analysis is herein defined as the evaluation of how the costs of the economically efficient solution should be apportioned among the affected purposes (cost allocation) and/or among the affected groups (cost sharing). The procedures are applicable to both the cost allocation and sharing problems. For notational simplicity, the equity analysis will be referenced to the allocation problem only. Procedures for analyzing other aspects of the problem, e.g., multiple objectives, environmental impacts, socio-political factors, are not examined. It is recognized that the results of these other studies may significantly alter the solution(s) obtained from this efficiency/equity analysis.

Emergence of Efficiency/Equity Analysis in Environmental Management

Prior to major federal (U.S.) initiatives in the pollution control field in the 1960's, there was relatively little interest in efficiency/equity analysis. Where pollution controls were prescribed, the conventional engineering analysis was usually restricted to evaluating one or a very few alternatives. The selected single purpose control served a single political jurisdiction so there was little concern about equity aspects.

Early federal involvement was restricted to providing construction grants to municipalities to partially subsidize the cost of installing controls. This program was criticized because it sometimes encouraged construction of small, inefficient control

---

[1]) Prof. *James P. Heaney*, University of Florida, Environmental Engineering Sciences, A.P. Black Hall, Gainesville, Florida 32611, USA.

units. Significant savings could often we realized if several cities joined together and built one larger control unit with the waste being transported to that point. Savings resulted from a) economies of scale in constructing larger control facilities; and/or b) making fuller use of the available assimilative capacity of the air, land, and/or water [see *Chacko; Dorfman* et al.; *Biswas; Thrall* et al. for examples]. The optimal strategy typically called for widely varying control levels among polluters. Such solutions seemed to be "unfair" unless some sort of transfer payments among polluters were permitted.

Continued refinement of cost-effectiveness procedures has led to the promulgation of relatively complex mixes of control of sewage, combined sewer overflows, urban and agricultural runoff, and other sources using a wide variety of control techniques, many of which are multi-purpose [see *Heaney/Nix* for example]. Indeed, this study, financed by the U.S. Environmental Protection Agency, is a result of the need to develop cost allocation techniques to support this very large program (about 5×10^9/yr).

In contrast with this recent interest in environmental analysis, the related field of water resources has a long history of efficiency/equity studies. The next section reviews these studies. Then, a review of attempts apply cooperative game theory to the cost allocation problem is presented. Based on these reviews, the main unresolved problems are identified. Lastly, proposed solutions to these problems are explained.

Efficiency/Equity Studies in Water Resources

TVA *Studies*

The original requirements for cost allocation in the Tennessee Valley Authority project in the southeastern U.S. were promulgated in 1935 [*Ransmeier*]. Allocation of costs was required among 1. flood control, 2. navigation, 3. fertilizer production, 4. national defense, and 5. development of power. The most controversial aspect of this problem dealt with the allocation of cost to power development. The outcome of this allocation would have a major impact on the cost of publicly produced power as compared to privately produced power which was the primary power producer at that time.

Ransmeier and other prominent authorities were brought in to evaluate the numerous alternatives for allocating costs. They formulated a set of five criteria which are listed below [*Ransmeier*].

1. The method should have a reasonable logical basis. It should not result in charging any objective with a greater investment than the fair capitalized value of the annual benefit of this objective to the consumer. It should not result in charging any objective with a greater investment than would suffice for its development at an alternative single purpose site. Finally, it should not charge any two or more objectives with a greater investment than would suffice for alternate dual purpose or multiple purpose improvement.
2. The method should not be unduly complex. In a democracy public agencies should be subject to the understanding scrutiny of the citizen body to the maximum possible extent. The goal should always be presentation of problems and programs for public consideration in clear and simple terms.

3. The method should be workable. It should not be dependent upon estimates which are difficult to prepare, and it should not require the use of data as to actual operation of the project which will not become available until after the passage of a considerable period of time.

4. The method should be flexible. It should be equally applicable to an isolated project or a co-ordinated system. In the case of a system it should be applicable to both the initial and ultimate stages of development. It should be readily adaptable to changing conditions over plant or system life.

A final criterion which sometimes has been suggested but which we put forward here only for the sake of discussion is:

5. The method should apportion to all purposes present at a multiple purpose enterprise a share in the overall economy of the operation. This point has two implications. First, no purpose should be assessed costs as great as would suffice for alternate single purpose stream improvement. Second, no purpose should be assessed costs as great as its capitalized benefit value.

The conditions described in these five criteria can be summarized as follows:

a) The cost allocated to purpose i should not exceed its benefits, i.e.,

$$x(i) \leqslant B(i) \qquad\qquad \forall i \in N \tag{1}$$

where

$x(i) =$ cost allocated to purpose i,

$B(i) =$ benefit of purpose i, and

$N =$ set of all purposes $= \{1, 2, \ldots, i, \ldots, n\}$.

b) The cost allocated to purpose i should not exceed its alternative single purpose cost, i.e.,

$$x(i) \leqslant C(i) \qquad\qquad \forall i \in N \tag{2}$$

where

$C(i) =$ alternative cost if purpose i acts independently.

c) The cost allocated to any two or more purposes should not exceed their alternative cost acting as a subgroup of s of the n purposes, i.e.,

$$\sum_{i \in S} x(i) \leqslant C(S) \qquad\qquad \forall S \subset N \tag{3}$$

where

$C(S) =$ alternative cost if the subset of s purposes acts independently; and

$S =$ any subset of the master set, N.

The first two criteria can be combined into a single condition,

$$x(i) \leqslant \min\ [B(i),\ C(i)] \quad \forall i \in N \tag{4}$$

In retrospect, it is remarkable that condition (3), so critical in specifying the core of a game, was articulated about 40 years ago by analysts wrestling with this problem. Unfortunately, its implications are not discussed elsewhere by *Ransmeier* [1942]. Criteria 2, 3, and 4 are not directly convertible to axioms. Their primary message is that the method should be simple, workable, and flexible. The first sentence of criterion (5) implies that the savings should be apportioned among all purposes. This criterion is certainly compatible with game theoretic approaches. The provisions stated in the last two sentences are contained in the above condition (4).

Notably absent from this list of criteria is the condition that the sum of the costs allocated to the various purposes equals the total cost of the project. Most of the standard methods of cost allocation require that this condition be met. However, *Ransmeier* [1942] concluded, that, based on economic theory, it is not possible to meaningfully allocate joint costs. The crux of his argument is that the criterion for economic efficiency requires that the marginal benefits of a purpose be compared with the marginal costs. *Ransmeier* viewed marginal costs as being equivalent to direct costs, i.e., the costs of the elements of the project which are used solely by that purpose. Accordingly, the joint costs are viewed as sunk costs which need not be recovered.

Other Agency Studies

With the rapid growth of federal involvement in water resources projects during the 1930's and 1940's, inter-agency committees were established to standardize techniques for performing economic and cost allocation analysis. Three reports [Federal Inter-Agency Committee; Inter-Agency Committee; Water Resource Council], present the following accepted guidelines:

1. The total of all benefits exceeds the total of all costs.
2. The scale of development is such as to provide maximum total benefits over total costs.
3. The benefits of adding each purpose as the last increment at least equal the separable costs of the added capacity.
   With regard to cost allocation the committee recommended the separable costs, remaining benefits method which assigns to each purpose [Inter-Agency Committee]:
4. its separable costs; and
5. a share of the residual or remaining joint costs in proportion to the remaining benefits, i.e., the benefits (as limited by alternative costs) less the separable costs.

Before discussing the criteria, it is necessary to define "separable costs". They are defined as the difference between the cost of the multiple-purpose project and the cost of the project with the purpose omitted [Inter-Agency Committee]. Separable costs include direct costs and the incremental costs of changing the size of the multi-purpose cost element. Thus, the use of separable costs represents a significant change from *Ransmeier's* [1942] procedure which used direct costs. Separable costs are simple to define in familiar game theoretic terminology, i.e.,

$$SC\,(i) = C\,(N) - C\,[(N) - \{i\}] \qquad \forall\, i \in N \qquad\qquad (5)$$

where

$SC(i)$ = separable cost to purpose i,

$C(N)$ = total cost for the grand coalition of n purposes, and

$C[(N) - \{i\}]$ = total cost for the grand coalition with purpose i excluded.

These additional five criteria can be represented as follows:

a) $B(N) \geqslant C(N)$ (6)

where

$B(N)$ = total benefits for the grand coalition of n purposes.

b) $B(N) - C(N) \geqslant B(W) - C(W) \quad \forall \text{ sets, } W$ (7)

where

$B(W)$ = total benefits for a coalition of w purposes, $w \neq n$, and

$C(W)$ = total costs for a coalition of w purposes, $w \neq n$; and

c) $B(i) \geqslant SC(i)$ $\qquad\qquad \forall i \in N$ (8)

d) $x(i) \geqslant SC(i)$ $\qquad\qquad \forall i \in N$ (9)

e) i) $NSC = C(N) - \sum_{i \in N} SC(i)$ (10)

where

NSC = nonseparable (joint)costs of the multipurpose project.

ii) $\beta(i) = \dfrac{[\min(B(i), C(i))] - SC(i)}{\sum\limits_{i \in N} [\min(B(i), C(i)) - SC(i)]} \quad \forall i \in N$ (11)

where

$\beta(i)$ = proportion of nonseparable costs assigned to purpose i.

Thus, the total cost assigned to purpose i is

$x(i) = SC(i) + \beta(i)(NSC) \qquad \forall i \in N$ (12)

Implicit in equation (12) is the requirement that

$$\sum_{i \in N} x\,(i) = C\,(N) \tag{13}$$

A recent article by *Loughlin* [1977] indicates that the above procedures are still being used.

Other Studies

Loehman et al. [1971, 1973, 1974] have used the Shapley value as a cost allocation device. The 1973 study included an application to river pollution caused by three cities. *Sorenson* [1972], a student of *A*. Charnes, applied game theory to problems of regional water allocation. He also presented alternative definitions of the characteristic function. This evaluation helps examine questions of ownership of common resources and indicates whether a game will be competitive or cooperative. *Giglio/Wrightington* [1972] introduce game theoretic approaches to evaluate cost allocation problems in the water resources field. Their analysis indicates potential problems which might arise, e.g., games without cores. They recommend the separable costs, remaining benefits method or another method which prorates savings based on the measure of pollution. *Gately* [1974] applies several different solution notions to the problem of regional cooperation among Tamil Nadu, Andhra Pradesh, and Kerala-Mysore. *Heaney* et al. [1975a, 1975b] apply game theory to the problem of efficiently and equitably allocating a common storage unit for storm drainage or pollution control among competing users. Solutions are presented for cases where ownership of the common storage unit is known or unknown. *Carbone/Sweigart* [1976] present an optimization model which addresses the efficiency/equity questions associated with a regional air pollution problem.

Littlechild/Thompson [1977] applied game theory to the problem of allocating the cost of an airport runway among several different sizes of aircraft. They obtain a very simple allocation formula for this special case of a convex game. *Billera* et al. [1977] applied non-atomic game theory to set up internal telephone billing rates at Cornell University. This procedure can be extended to water, sewer, and electric rate making where continuous recording meters are used. In the accounting field, game theoretic approaches are beginning to be used, e.g., the work of *Hamlen* et al. [1977] and *Jensen* [1977]. A thorough discussion of conventional methods of cost allocation in accounting is presented by *Thomas* [1974] who concludes that it is futile to allocate joint costs. The *Shapley* value [1953] and the nucleolus [*Schmeidler*] have merged as the two most popular solution notions in cooperative game theory. The Shapley value is based on the argument that each player should receive the expected value over all coalition formation sequences of the incremental value he brings to each coalition, or

$$\Phi_i = \sum_{S \subseteq N} \frac{(s-1)\,!\,(n-s)\,!}{n\,!}\,[v\,(s) - v\,(s-(i))]. \tag{14}$$

The nucleolus is that solution which maximizies the minimum savings accruing to any coalition. For an N person game, the nucleolus is found by solving at most $n-1$ linear programs. This paper compares conventional practices to the Shapley value. A similar comparison with the nucleolus is underway [*Heaney*, 1979].

Conventional and Game Theoretic Criteria

Summary of Conventional Criteria

Workable procedures for examing the efficiency/equity aspects of environmental problems in the U.S. are needed to support the current multi-billion dollar per year control program. Based on extensive experience in the water resources and related fields, an acceptable procedure should satisfy the following conditions. It remains to be considered whether any procedure satisfies all of these conditions and is still "simple, workable, and flexible" (see page 3).

A) Economic Efficiency
  1. Total benefits equal or exceed total costs, i.e.,

$$B(N) \geqslant C(N) \tag{15}$$

  2. The scale of development is such as to provide maximum total benefits minus total costs, i.e.,

$$B(N) - C(N) > B(W) - C(W) \quad \forall \text{ sets, } W \tag{16}$$

  3. The benefits of adding each purpose as the last increment at least equal the separable costs of the added capacity, i.e.,

$$B(i) \geqslant SC(i) \qquad\qquad \forall i \in N \tag{17}$$

B) Equity Criteria
  1. The cost allocated to purpose i should not exceed its benefits nor its alternative single purpose cost, i.e.,

$$x(i) \leqslant \min[B(i), C(i)] \qquad \forall i \in N \tag{18}$$

  2. The costs allocated to any two or more purposes should not exceed their alternative cost acting as a subgroup of s out of a total of n purposes, i.e.,

$$\sum_{i \in S} x(i) \leqslant C(S) \qquad\qquad \forall S \subset N \tag{19}$$

  3. Each purpose i pays at least its separable costs, i.e.,

$$x(i) \geqslant SC(i) \qquad\qquad \forall i \in N \tag{20}$$

4. Each purpose pays a share of the residual joint (nonseparable) costs in proportion to the remaining benefits (as limited by alternative costs) less the separable costs. The nonseparable costs (NSC) and apportioning factor, $\beta(i)$, need to be defined.

$$NSC = C(N) - \sum_{i \in N} SC(i), \tag{21}$$

$$\beta(i) = \frac{[\min(B(i), C(i))] - SC(i)}{\sum\limits_{i \in N} [\min(B(i), C(i)) - SC(i)]} \qquad \forall i \in N \tag{22}$$

where

$$\sum_{i \in N} \beta(i) = 1.0 \tag{23}$$

This condition can be stated as

$$x(i) = SC(i) + \beta(i)(NSC) \qquad \forall i \in N \tag{24}$$

5. The sum of the costs allocated to all purposes equals the total cost, i.e.,

$$\sum_{i \in N} x(i) = C(N). \tag{25}$$

From equation 10,

$$C(N) = NSC + \sum_{i \in N} SC(i). \tag{26}$$

Substituting eqn. (21) into eqn. (26) yields

$$C(N) = NSC + \sum_{i \in N} [x(i) - \beta(i)(NSC)].$$

From eqn. (23), if $\sum\limits_{i \in N} \beta(i) = 1$, then eqn. (25) is automatically satisfied.

Relation to Game Theory

Overall, seven conditions need to be satisfied. The reader familiar with cooperative n-person game theory will recognize that conditions (4), (3), and (25) constitute the core of a cost game. However, there is one defect in condition (3). Condition (4) recognizes the possibility that, for a single purpose, $B(i)$ could be less than $C(i)$. Thus, it is proper to use the minimum of $B(i)$ and $C(i)$ as the maximum justifiable alternative investment. The same argument should apply to condition (3) since it is possible for $B(S) < C(S)$ for any $S \subset N$. However, because of condition (6), $B(N) \geqslant C(N)$. Thus, condition (3) needs to be modified to

$$\sum_{i \in S} x(i) \leqslant \min[B(S), C(S)] \qquad \forall S \subset N \tag{27}$$

to render conditions (4), (27), and (25) equivalent to conditiond for the core of a cost game. Condition (16) is automatically satisfied if the game is subadditive. It does serve as a reminder that explicit proof is needed that the n purpose coalition is optimal. The similarity of conventional criteria and the core is striking. Of course, the vital link is criterion (3) which was stated by *Ransmeier* in 1942 but which has never been used subsequently in the water resources filed until the recent introduction of core theory.

Finding the Minimum Cost Allocation to Purpose *i*

The next question to be resolved is whether to deduct direct or separable costs prior to allocating the remaining costs. Accepting criteria (17) implies that separable costs must be deducted first. This condition is disturbing to many practitioners who routinely use direct costing techniques. The defect in direct costing is that fails to incorporate the external benefits (or costs) caused by the project elements. As a very simple example consider two cities of equal size located along a river. Two wastewater control options are being considered: two treatment plants, or a single large plant at downstream city 2 with a pipe from city 1 to city 2. The costs are as follows:

Alternative 1
Treatment plant at city 1   = $10
Treatment plant at city 2   = $10
                               $20

Alternative 2
Pipe from city 1 to city 2   = $ 4
Treatment plant at city 2   = $14
                               $18

Using the direct cost method, city 1 pays for the pipe and a prorated share of the treatment plant, or

$$x(1) = 4 + 1/2 (14) = 11 > C(1).$$

He objects and drops out. Thus, using direct costs may violate conditions (4) and/or (27).

The use of separable costs evolves from the condition (17) for economic efficiency wherein additional purposes should be added as long as their incremental benefits are at least equal to the incremental costs of serving them. Accepted procedures define the separable cost to purpose i as the increment in cost given that this purpose joins last. The apparent reasoning is that this is the most favored order in which to join and thereby represents the minimum cost that this purpose should pay. However, in game theory, it is known that

$$C(N) - C(N - \{i\}) \leq C(S) - C[(S) - \{i\}] \qquad \forall S \subset N \qquad (28)$$

only if the game is convex [*Shapley*, 1971]. Figure 1 depicts the standard presentation of optimality conditions. The game is convex in the range $0 \leqslant Q \leqslant Q_1$, i.e., marginal costs are nonincreasing. However, the optimal solution dictates expansion to Q^*. Thus, in this case, and quite often, the lowest incremental cost is not $C(N) - C[(N) - \{i\}]$. Consequently, the economic efficiency criteria that $MB(i) \geqslant SC(i)$ is too restrictive if $SC(i)$ is defined as $C(N) - C(N - \{i\})$. Then, what is the minimum cost assignable to purpose i?

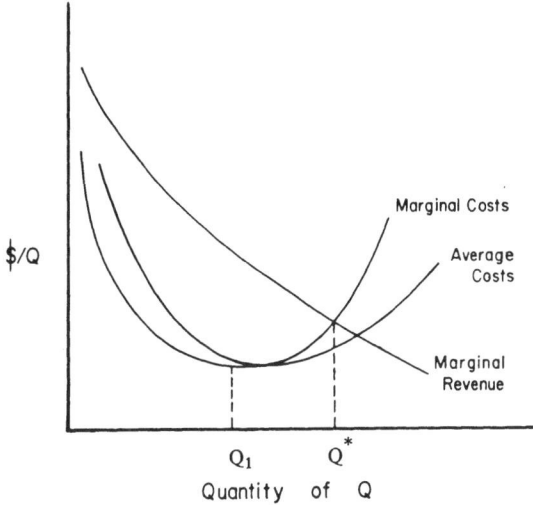

Fig. 1: Conventional Determination of Optimal Level of Output, Q^*

Consider the following simple example of a three city cost game.

$$B(1) = 6, B(2) = 7, B(3) = 6$$
$$C(1) = 2, C(2) = 4, C(3) = 6$$
$$C(12) = 5, C(13) = 6, C(23) = 6$$
$$C(123) = 6$$

Inspection of this game indicates that $B(i) \geqslant C(i)$ for all i and $B(S) \geqslant C(S)$ for all $S \subset N$. Thus, there is no further need to consider benefits. For this game to be subadditive

$$C(1) + C(2) \geqslant C(12), \tag{29}$$

$$C(1) + C(3) \geqslant C(13), \tag{30}$$

$$C(2) + C(3) \geqslant C(23), \tag{31}$$

$$C(1) + C(23) \geqslant C(123), \tag{32}$$

$$C(2) + C(13) \geqslant C(123), \text{ and} \tag{33}$$

$$C(3) + C(12) \geqslant C(123). \tag{34}$$

For this example, the maximum value of $C(123) = 8.0$ because of condition (32). At that value of $C(123)$ the coalition is inessential, i.e., $C(1) + C(23) = C(123)$. If $C(123)$ exceeds 8.0 it would be economically inefficient to form this coalition.

This subadditive game has a core since

$$C(12) + C(13) + C(23) \geqslant 2\, C(123). \tag{35}$$

Furthermore, since

$$C(12) + C(13) \geqslant C(123) + C(1), \tag{36}$$

$$C(12) + C(23) \geqslant C(123) + C(2), \text{ and} \tag{37}$$

$$C(13) + C(23) \geqslant C(123) + C(3) \tag{38}$$

the game is convex if $C(123) \leqslant 6.0$. For this special case

$$C(N) - C(N - \{i\}) \leqslant C(S) - C(S - \{i\}) \qquad\qquad \forall\, S \subset N. \tag{39}$$

Accordingly, separable costs, as conventionally defined, would be the minimum incremental costs, as confirmed below in Table 1 for this example. However, for $C(123) = 8.0$, the minimum incremental cost occurs if some of the players join second, not last. In this case, the economic criterion would incorrectly compare incremental benefits with incremental costs which were not the lowest. Figures 2a and 2b show the core of the original convex game with $C(123) = 6$ and a modified game with $C(123) = 8.0$.

		Incremental Cost of Joining			
		Second After Player		Third	With $C(123) =$
Player i	First	j	k	$6.0^1)$	$8.0^2)$
1	2	1	0	0	2.0
2	4	3	0	0	2.0
3	6	4	2	1	3.0

[1]) convex game

[2]) nonconvex game with a core

Tab. 1: Incremental Costs for Three Person Game with Variable Costs for Grand Coalition

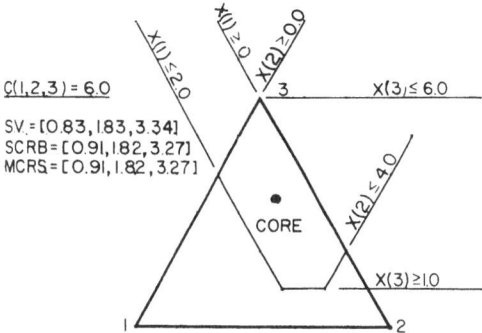

Fig. 2a: Convex Cost Game

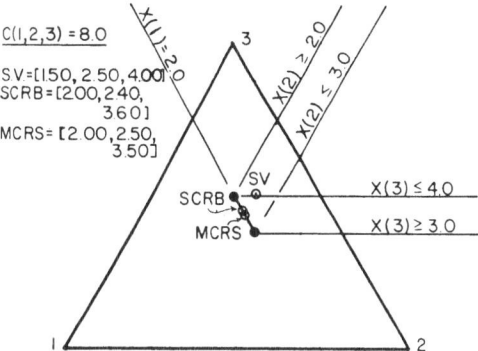

Fig. 2b: Nonconvex Cost Game with Core

For games with cores, the minimum feasible (within the core) cost which can be assigned to each player is found by solving the following linear programming problem for player i [*Charnes/Kortanek*]:

minimize $x(i)$

subject to $x(i) \leqslant \min [B(i), C(i)]$ $\forall i \in N,$

$\displaystyle \sum_{i \in S} x(i) \leqslant \min [B(S), C(S)]$ $\forall S \subset N,$

$\displaystyle \sum_{i \in N} x(i) \geqslant C(N),$ and   (40)

$x(i) \geqslant 0$ $\forall i \in N.$

The constraint set simply delimits the core of the game so that the minimum value of $x(i)$ is the lowest feasible cost which can be assigned to player i without violating the core conditions. For the three person game, it is the minimum costs shown on Figures 2 a) and 2 b).

This result is important for several reasons:

1. It unambiguously defines the minimum *feasible* cost which can be assigned to player i thereby eliminating the problem of what is meant by marginal or incremental costs. Thus, $x(i)_{min}$ should replace $SC(i)$ in the economic criteria.
2. Because it is a linear program, it is very easy to calculate, even for large games, if information on the values for all coalitions are available, or to approximate, using values which are available. The separable costs, remaining benefits method only includes the costs of going alone, $C(i)$, being next to last to join, $C(N - \{i\})$, and the cost of the grand coalition. The costs of all other coalitions are not included.
3. Because it is a feasible solution, the calculated value of $x(i)$ can be constrained to be not less than $x(i)_{min}$.

If the game has no core, and problem (40) is solved, then $\sum_{i \in N} x(i) > C(N)$. A way

to satisfy $\sum_{i \in N} x(i) = C(N)$ is to make the conditions on the intermediate coalitions

less restrictive. Since it is impossible to satisfy the intermediate constraints, it seems reasonable to relax the constraints on the sub-coalitions until a core appears. If there are any savings then the set of imputations is nonempty. It is the added constraints on the sub-coalitions, S, that eliminate the core. However, a proportionate relaxation of these constraints will yield a non empty core. Thus, if no core exists, solve the following problem:

$$\text{minimize} \quad \theta$$

$$
\begin{aligned}
\text{subject to} \quad & x(i) \leqslant \min[B(i), C(i)] && \forall i \in N \\
& \sum_{i \in S} x(i) - \theta \{\min[B(S), C(S)]\} \leqslant \min[B(S), C(S)] && \forall S \subset N \quad (41) \\
& \sum_{i \in N} x(i) = C(N), \text{ and} \\
& x(i) \geqslant 0 && \forall i \in N.
\end{aligned}
$$

The optimal solution to this problem tells the minimum value, θ^*, which yields a nonempty core. The results of this analysis are inserted into the following formulation to find the minimum cost for player i, i.e.,

$$\text{minimize} \quad x(i)$$

$$
\begin{aligned}
\text{subject to} \quad & x(i) \leqslant \min[B(i), C(i)] && \forall i \in N \\
& \sum_{i \in S} x(i) \leqslant \min[B(S), C(S)](1 + \theta^*) && \forall S \subset N \quad (42) \\
& \sum_{i \in S} x(i) = C(N), \text{ and} \\
& x(i) \geqslant 0 && \forall i \in N.
\end{aligned}
$$

As an example, consider the following subadditive cost game without a core

$$C(1) = 4, C(2) = 4, C(3) = 5$$

$$C(12) = 5, C(13) = 8, C(23) = 6$$

$$C(123) = 10.$$

These characteristic function values imply the following bounds on the charges to the three players: $x(1) = 4$, $2 \leqslant x(2) \leqslant 4$, and $x(3) = 5$. Given unique values of $x(1) = 4$ and $x(3) = 5$, then $x(2) = 1.0$ which is impossible since $2 \leqslant x(2) \leqslant 4$.

Thus, the values on the player coalitions are relaxed by solving problem (41). For this example, $\theta^* = 0.0526$ and the resulting θ core is the unique point [3.68, 1.58, 4.74]. The Shapley value is [3.33, 2.33, 4.34]. The separable costs, remaining benefits method runs into problems when the game has no core. The separable costs are [4, 2, 5]. The nonseparable costs are -1. For this case, the nonseparable cost weighting factors, B_i, are not clearly defined since some of the values are zero. The SCRB solution is [4, 1, 5] which allots all of the savings to Player 2, not a very reasonable compromise solution.

Finding the Maximum Cost Allocation to Purpose i

In the previous section, it was argued that the minimum assignable cost should be a feasible solution to the game, i.e., within the core. Correspondingly, the maximum assignable cost should also be feasible. From condition (4),

$$x(i) \leqslant \min[B(i), C(i)] \qquad \forall i \in N. \tag{4}$$

But, for games with cores, is this solution always feasible? The answer is no as shown by the example with $C(123) = 8.0$ Referring to Figure 2 b), the maximum feasible cost allocations are

$$x_{max} = [2.0, 3.0, 4.0].$$

In this case, $x(i)_{max} \leqslant C(i) \forall (i)$. From before, the minimum values of $x(i)$ are $x_{min} = [2.0, 2.0, 3.0]$. Assume that all other players j are allotted their minimum cost. How much would player i play? He pays

$$x(i)_{max} = \min[B(i), C(i), C(N) - \sum_{\substack{j=1 \\ j \neq i}}^{n} x(j)_{min}]. \tag{43}$$

In this case

$$x(1)_{max} = \min[6, 2, 8.0 - 2.0 - 3.0] = 2.0$$

$$x(2)_{\max} = \min [7, 4, 8.0 - 2.0 - 3.0] = 3.0$$

$$x(3)_{\max} = \min [6, 6, 8.0 - 2.0 - 2.0] = 4.0$$

as shown in Figure 2b as the upper bound on the core. In general, the maximum cost can be found by solving the linear programm,

maximize $x(i)$

subject to $x(i) \leqslant \min [B(i), C(i)]$ $\forall i \in N,$

$\displaystyle\sum_{i \in S} x(i) \leqslant \min [B(S), C(S)](1 + \theta^*)$ $\forall S \subset N,$ (44)

$\displaystyle\sum_{i \in N} x(i) = C(N),$ and

$x(i) \geqslant 0$ $\forall i \in N.$

For games without cores, θ^* is greater than zero.

Solution Concepts

The conventional method uses the separable costs remaining benefits method to find a "fair" solution. Analogously, the Shapley value and other solution concepts are used in game theory. Referring to the example, the SCRB and Shapley value solutions are shown in Figures 2a – 2b. The two solutions are almost identical for the convex game, $C(123) = 6.0$. The Shapley value lies outside the core for the nonconvex game, $C(123) = 8.0$. For the example of a game without a core, the separable costs, remaining benefits method does not work well since some of the remaining benefits are zero.

In my opinion, the fairest solution is the one which falls in the center of the core. If the game is convex, the Shapley value and the SCRB approximation are acceptable. The unresolved problem occurs when the game is not convex since these methods are now taking weighted averages of infeasible solutions. The proposed remedy outlined below can be expressed as a generalization of the SCRB method.

Step 1.   Find the minimum $(x(i)_{\min})$ and maximum $(x(i)_{\max})$ costs by solving linear programs (29) or (3), and (32) respectively.

Step 2.   Find the proration of the nonseparable cost as before, where

$$\beta(i) = \frac{x(i)_{\max} - x(i)_{\min}}{\displaystyle\sum_{i \in N} [x(i)_{\max} - x(i)_{\min}]} \qquad \forall i \in N \qquad (22)$$

and

$$NSC = C(N) - \sum_{i \in N} x(i)_{\min} \qquad (21)$$

Step 3. Find the fair solution for each purpose using

$$x (i) = x (i)_{min} + \beta (i) (NSC).$$ (45)

For lack of a better name, this method is called the minimum cost, remaining savings method (MCRS).

Summary and Conclusions

Procedures for efficiency/equity analysis have evolved over the past 40–50 years in the water resources field. The specific motivation for this study is to recommend procedures for environmental analysis to assist the U.S. Environmental Protection Agency in managing a 5×10^9/yr water pollution control effort.

In the water resources field, intensive studies of the cost allocation problem have been made in response to controversy surrounding the early development in the 1930's of the Tennessee Valley Authority in the Southeastern U.S. The growth of federal involvement in U.S. water projects during the 1930's and 1940's stimulated the development of generalized procedures for efficiency/equity analysis. These methods are still widely used. These efficiency/equity criteria are used to evaluate the suitability of the proposed procedures.

Related criteria were developed by game theorists over a similar time span. Of specific relevance is core theory and solution notions such as the Shapley value. During the past several years, investigators have attempted to demonstrate the relevance of these methods to water resources and environmental problems. Remarkably, *Ransmeier* [1942] has stated some of the key assumptions of core theory as related to the TVA cost allocation problem. Unfortunately, no subsequent work was done on this subject. One correction in the conventional theory is needed to have conditions equivalent to the core. The characteristic function for an *s* member coalition needs to be defined as the minimum of the benefits and costs since it is possible for subcoalitions to not satisfy the economic efficiency criterion that $B (S) \geqslant C (S)$. With that change, the two approaches are equivalent.

The next major task was to determine the minimum costs which a purpose must be assessed. Separable costs are defined as the incremental costs of adding a purpose as the last purpose or

$$SC (i) = C (N) - C (N - \{i\}) \qquad\qquad \forall i \in N.$$ (46)

It has been assumed that this is the minimum cost which purpose *i* could be assessed and is thereby the relevant one to compare to its marginal benefits. Another way of defining these minimum costs is to determine the direct costs, i.e., the costs of those project elements serving only one purpose.

The direct cost method is shown to be incorrect since it fails to recognize the net change in total cost each project element causes. It is the total change in cost which is relevant.

Separable costs, as defined, are minimum incremental costs only if the game is convex; often it is not. When the game is nonconvex the purpose would have a lower incremental cost if it joined earlier. Thus, accepted practices for economic/equity analysis may be incorrect when the game is nonconvex.

A general method for finding the minimum costs for purpose i is to solve the linear program which minimizes x (i) subject to the core constraints. This method is conceptually sound and computationally simple since efficient computer codes are widely available. An exact solution can be found or an approximation obtained using available characteristic function data. If the game does not have a core, the inequalities on the s member coalition constraints are relaxed and the minimum costs for that problem are found. Analogous procedures are used to find the maximum feasible costs.

Knowing the minimum and maximum feasible costs for each player, a generalization of the separable costs, remaining benefits method is proposed as the solution procedure. Each purpose is assigned its minimum costs (instead of its separable costs) and a proportionate share of the remaining costs base on the difference between its minimum and maximum assignable costs. Maximum feasible costs are used instead of benefits. The resulting formula is identical to accpected pratice if x $(i)_{min}$ and x $(i)_{max}$ are used in place of separable costs and benefits. The solution to this game is guaranteed to be in the core since it is a convex combination of two feasible solutions. Also, because it is directly related to accepted practice, it should be much easier to implement than solutions which are unfamiliar to practitioners.

References

Billera, L.J., D.C. Heath, and *J. Raanan*: Internal Telephone Billing Rates – A Novel Application of Non-Atomic Game Theory. T.R. No. 331, School of Operations Research and Industrial Engineering. Ithaca, N.Y. 1977.

Biswas, A.K. (ed.): Systems Approach to Water Management. New York 1976.

Carbone, R., and *J.R. Sweigart*: Equity and Selective Pollution Abatement Procedures. Management Science **23** (4), 1976.

Chacko, G.K. (ed.): Systems Approach to Environmental Pollution. ORSA Health Applications Section. Arlington, Va., 1972.

Charnes, A., and *K. Kortanek*: On Balanced Sets, Cores and Linear Programming, Technical Report No. 12, Dept. of Industrial Engineering and Operations Research, Cornell University, Ithaca, N.Y. 1966.

Dorfman, R., et al. (ed.): Models for Managing Regional Water Quality. Cambridge, Mass., 1972.

Federal Inter-Agency River Basin Committee: Proposed Practices for Economic Analysis of River Basin Projects. U.S.G.P.O., Washington, D.C., 1950.

Gately, D.: Sharing the Gains from Regional Cooperation: A Game Theoretic Application to Planning Investment in Electric Power. International Economic Review **14** (1), 1974.

Giglio, R.J., and *R. Wrightington*: Methods for Apportioning Costs Among Participants in Regional Systems. Water Resources Research **8** (5), 1972.

Hamlen, S.S., W.A. Hamlen, Jr., and *J.R. Tschirhart*: The Use of Core Theory in Evaluating Joint Cost Allocation Schemes. The Accounting Review, July 1977.

Heaney, J.P.: Economic/Financial Analysis of Urban Water Quality Management Problem, US Environmental Protection Agency, Cincinnati, OH, 1979.

Heaney, J.P., and *S.J. Nix*: Storm Water Management Model: Level I – Comparative Evaluation of Storage-Treatment and Other Management Practices. Environmental Protection Technology Series EPA-600/2-77-083, U.S. Environmental Protection Agency. Cincinnati, OH., 1977.

Heaney, J.P., and *H. Sheikh*: Game Theoretic Approach to Equitable Regional Environmental Quality Management. Mathematical Analysis of Decision Problems in Ecology. Ed. by A. Charnes and W.R. Lynn. Berlin 1975a.

Heaney, J.P., et al: Urban Stormwater Management Modelling and Decision-Making. Environmental Protection Technology Series EPA-670/2-75-022, U.S. Environmental Protection Agency. Cincinnati, OH, 1975b.

Inter-Agency Committee on Water Resources: Proposed Practices for Economic Analysis of River Basin Projects, U.S. Government Printing Office, Washington, D.C. 1958.

Jensen, D.L.: A Class of Mutually Satisfactory Allocations. The Accounting Review, Oct. 1977.

Littlechild, S.C., and *G.E. Thompson*: Aircraft Landing Fees: A Game Theory Approach. The Bell Jour. of Economics 8 (1), 1977.

Loehman, E., and *A. Whinston*: A New Theory of Pricing and Decision-Making in Public Investments. The Bell Journal of Economics and Management Sciences 2 (2), 1971.

–: An Axiomatic Approach to Cost Allocation for Public Investment. Public Finance Quarterly 2 (2), 1974.

Loehmann, E., E.D. Pingry, and *A. Whinston*: Cost Allocation for a Regional Pollution Treatment System. Economics and Decision Making for Environmental Quality. Ed. by J.R. Conner and E. Loehman. Gainesville 1973.

Loughlin, J.C.: The Efficiency and Equity of Cost Allocation Methods for Multipurpose Water Projects. Water Resources Research 13 (1), 1977.

Ransmeier, J.S.: The Tennessee Valley Authority. Nashville, Tenn., 1942.

Schmeidler, D.: The Nucleolus of a Characteristic Function Game. SIAM J. Appl. Math. 17, 1969. 1162–1170.

Shapley, L.S.: A Value for n-Person Games. Annals of Mathematics Study 28, 1953, 307–317.

–: Cores of Convex Games. Rand Report P–4620, Santa Monica, CA., 1971.

Sorenson, S.W.: A Mathematical Theory of Coalitions and Competition in Resource Development. Ph. D. Dissertation, U. of Texas at Austin, 1972.

Thomas, A.L.: The Allocation Problem: Part Two, Studies in Accounting Research, #9. American Accounting Association, Sarasota, Fl., 1974.

Thrall, R.M. et al. (ed.): Economic Modeling for Water Policy Evaluation. Amsterdam 1976.

Water Resources Council: Policies, Standards and Procedures in the Formulation, Evaluation, and Review of Plans for Use and Development of Water and Related Land Resource. 87th Cong., 2d sess., Senate Doc. 97, 1962.

Models of Control and Confrontation

Applied Game Theory. 1979 ©Physica-Verlag, Wuerzburg/Germany

Dynamic Standard Setting for Carbon Dioxide

By *E. Höpfinger*, Karlsruhe[1])

Abstract: Under the assumption that a continuous increase in atmospheric carbon dioxide beyond a critical value, caused by the combustion of fossil fuel, will lead to irreversible and large changes of the climate of the earth, the problem of limiting CO_2 emission becomes an urgent concern. The subject of how to determine and adapt an emission standard for carbon dioxide is treated as a three-person infinite stage game, the players of which are the decision units of regulators, producers, and population. After the description of the model solutions are derived for several solution concepts and discussed. In special cases the solutions differ substanially from each other.

Introduction

The emission of carbon dioxide into the atmosphere resulting from fossil fuel use has been increasing at an exponential rate for more than one century. If this expansion continues, the concentration of carbon dioxide in the atmosphere may be doubled in about the next 60 years according to *Rotty* [1977]. The effects on the global climate may well appear suddenly and could get out of control before remedial actions become effective.

Since easily accessible fossil fuels contain such big amounts of carbon there is a strong tendency to use them as a source of energy that could last for nearly two more centuries. This is much more so since the competing nuclear energy meets increasing resistance by citizen groups. But it is the vastness of this carbon reserve that causes deep concern within the climatological community. The amount of carbon in recoverable fossil reserves is ten times the amount now contained as carbon dioxide in the entire global atmosphere.

As these reserves are being used, the concentration of carbon dioxide in the atmosphere will surely increase; and because carbon dioxide absorbs a portion of the infrared radiation emitted by the earth, it is generally believed that a higher atmospheric temperature will result ("greenhouse effect"). Although it is uncertain how much warming is produced by a given increase, the increased atmospheric carbon dioxide could have a considerable impact on man's environment.

Significant physical effects that may be expected with high fossil use are the melting of polar sea ice and/or decreasing precipitation in mid-latitude regions. Major socio-

---

[1]) Dr. *E. Höpfinger*, Inst. für Wirtschaftstheorie und Operations Research der Universität Karlsruhe, Kaiserstr. 12, D–7500 Karlsruhe 1

political impacts could plausibly attend a substantial increase of carbon dioxide, for example:

— large and persistent fluctuations in global food supply, due to repeated crop failures in various regions of the world which are caused by chronic and severe weather variability;
— increasingly regulated demographic migration between regions and across national borders, due to a climate-related collapse of selected webs in regional economies; shifts in the power balance among nations due to physical effects stimulating the economic and cultural decline in some regions and stimulating increased growth and prosperity elsewhere.

At the present time the physical processes causing variations of temperature are poorly understood [see *Williams* and *Augustsson* et al.], and changes due to atmospheric carbon dioxide increases are impossible to detect since there is no accurate knowledge of the natural variability of the global average temperature. As outlined by *Markley* et al. [1977], and *Rotty* [1977], the other physical and sociopolitical effects are also highly uncertain.

Although a large part of the climatological community shares the opinion that mankind needs and can afford a time window between five and ten years for vigorous research and planning in order to narrow the uncertainties sufficiently so as to justify a major change in energy policies, the model analyzed in this paper excludes an increase of relevant knowledge about the physical effects. Thus the model deals with the pessimistic view of the climatic aspects of carbon dioxide. It is global in character because the global effects seem to dominate the local or regional ones.

Given these substantial uncertainties about the development of climate, the problem of what energy policies governments should choose, becomes important. This problem is approached as a conflict situation among the groups of governments, producers emitting carbon dioxide, and population. In order to work out the global aspects this conflict situation has been formalized as a multistage three person game, the players of which are called regulator, producer, and impactee. Thus we neglect conflicting interests among governments, producers, and different groups of populations, such as of developed and developing countries. The regulator stands for an international agency, the producer for an organization of all producers, and the impactee for the community of people possibly affected by the carbon dioxide problem.

The paper is based on the assumption that a continuous increase of atmospheric carbon dioxide beyond a critical value will lead to irreversible and large changes of the climate which are regarded as a catastrophe. All three players have their subjective probability of the level of the critical value. Since, by assumption, there is no increase of knowledge about the climatological process, the regulator can only be concerned about the reactions of the producer and especially of the impactee.

After the specification of the model the results for several solution concepts are derived. These are quite different in general but can all be interpreted in terms of fair play or power. Given that the model allows prescriptive answers although it is primarily descriptive.

Since data are often unknown or scarcely available or arbitrary – as in the case of the regulator where the utility function may be conceived of as reflecting a trade-off between the interests of producer and impactee – solutions are derived as functions of the parameters. Hence parameter analysis can reveal the crucial parameters. For the purpose of illustration a small numerical example is added.

The Model

The conflict situation is described by a three-person dynamic or multistage game in extensive form [see *Owen*, or *McKinsey*] which resembles stochastic games. At each stage a component game of perfect information is played which is completely specified by a state. The players' choices control not only the payoffs but also the transition probabilities governing the game to be played at the next stage. Each player has his own subjective estimate of the transition probability due to his subjective probability of the "true critical value".

The set of states of the game is

$$S = \{(C, L) \mid C_p \geqslant C \geqslant 0, L \geqslant 0\} \cup \{k \geqslant 0\}$$

C being the amount of carbon dioxide in the atmosphere;
C_p the maximal amount of carbon dioxide if all fossil fuel is burnt;
L the upper bound of carbon dioxide emission during a period;
k the ciritical value for a catastrophe.

Let (C^1, L^1) denote the first state. Then C^1 can be assigned the present amount of atmosphere carbon dioxide, and L^1 the present maximal emission of CO_2 or some multiple of it.

The perfect information of the component games is specified as follows:
For state (C, L) the regulator's set of choices is

$$M_R (C, L) = \{l \mid 0 \leqslant l \leqslant L\}$$

where l denotes the upper bound of the emission of carbon dioxide by the producer.

Then the producer chooses the amount of carbon dioxide to be emitted. His set of choices or measures equals

$$M_P (C, L, l) = \{a \mid 0 \leqslant a \leqslant l, a \leqslant (C_p - C) / \beta\};$$

$0 < \beta < 1$ is defined below. The impactee's set of measures equals

$$M_I (C, L, l, a) = \{p \mid 0 \leqslant p \leqslant 1\}.$$

Knowing the choices l and a he chooses the degree p of the pressure he wants to exert on the regulator. p can denote the probability of a vote to suspend the government or of an aggression against institutions.

The sets of measures in the case of k, i.e. a catastrophe has occureed at amount k of carbon dioxide in the atmosphere, equal

$$M_R(k) = \{0\};$$

$$M_P(k, 0) = \{0\};$$

$$M_I(k, 0, 0) = \{0\};$$

which means that there is no pressure.

Given state (C, L) and the choices (l, a, p) the following states are possible at the next stage:

$$(C + \beta a, L), \quad (C + \beta a, L/2), \quad \{k \geqslant C\}.$$

The first component of the first and second states indicates that the constant share βa of emitted carbon dioxide is added to the amount of carbon dioxide in the atmosphere. This is consistent with results of box models for the CO_2 cycle of the earth [*Avenhaus/Fenyi/Frick*] if a is emitted at a constant rate during the time period. The estimates for β range between 0.01 and 0.5. Amount $(1 - \beta) a$ is assumed to disappear into the biosphere, the upper mixed layer of the sea, and the deep sea. The second components express that the old upper bound either remains or is reduced by half. It is assumed that there is a probability $p\nu$ that L is replaced by $L/2$, where $0 < \nu < 1$ is a parameter provided that the catastrophe will not occur. $k \geqslant C$ denotes the amount of carbon dioxide in the atmosphere at which the catastrophe occurs.

All three players are assumed to have subjective probabilities relating to the critical amount k of carbon dioxide. They characterize the transition probabilities. For simplification of the model we assume that the subjective probabilities concentrate on points denoted by C_R, C_P, and C_I for regulator, producer, and impactee. We assume $C_R < C_P$, $C_I < C_P$ thus allowing the producer to neglect a possible catastrophe.

The subjective probabilities P_R, P_P, P_I for the transition from (C, L) to the possible new states are

New state t	$P_R(t \mid C, L, l, a, p)$	$P_P(t \mid C, L, l, a, p)$	$P_I(t \mid C, L, l, a, p)$
$(C + \beta a, L)$	0 if $C \leqslant CR < C + \beta a$ or $C_R < C < C + \beta a$ $1 - p\nu$ if $C + \beta a \leqslant C_R$ or $C_R < C = C + \beta a$	$1 - p\nu$	0 if $C \lessdot C_I < C + \beta a$ or $C_I < C < C + \beta a$ $1 - p\nu$ if $C + \beta a \leqslant C_I$ or $C_I < C + \beta a$
$(C + \beta a, \frac{L}{2})$	0 if $C \leqslant C_R < C + \beta a$ or $C_R < C < C + \beta a$ $p\nu$ if $C + \beta a \leqslant C_R$ or $C_R < C = C + \beta a$	$p\nu$	0 if $C \leqslant C_I < C + \beta a$ or $C_I < C < C + \beta a$ $p\nu$ if $C + \beta a \leqslant C_I$ or $C_I < C = C + \beta a$
C_R	1 if $C \leqslant C_R < C + \beta a$ 0 else	0	1 if $C \leqslant C_R = C_I < C + \beta a$ 0 else
C_I	1 if $C \leqslant C_I = C_R < C + \beta a$ 0 else	0	1 if $C \leqslant C_I < C + \beta a$ 0 else

If the inequality $C \leqslant C_j < C + \beta a$ holds, player j thinks that with probability 1 catastrophe C_j will occur since with the scheduled emission a the critical threshold is passed. The probability for $C_j < C < C + \beta a$ is only defined so that the scope of the definition covers all possible states and choices. Nevertheless, the probability is defined such as to express the idea of player j that although C_j has turned out as a view too pessimistic, $C_j < C$ and any further increase $C < C + \beta a$ will result in a catastrophe. From the results below it is obvious that the specific definition of $C_R < C$ has no consequence.

State k cannot be changed: $P_j (k \mid k, 0, 0, 0) = 1$ $(j = R, P, I)$. Since no utility functions are known for the three players, we start with linear ones which are simplest to assess. Let the transition from state s and measures (l, a, p) to state t have the utility $U_j (s; l, a, p, t)$ for players $j = R, P, I$.

$$U_R (C, L; l, a, p; C + \beta a, M) = c_1 l + c_2 a + c_3 p, \quad (M = L, L/2);$$

$$U_R (C, L; l, a, p; k) = c_1 l + c_2 \frac{k - C}{\beta} + c_3 p + c_R;$$

$$U_R (k; 0, 0, 0; k) = 0;$$

$$U_P (C, L; l, a, p; C + \beta a, M) = c_4 a, \quad (M = L, L/2);$$

$$U_P (C, L; l, a, p; k) = c_4 \frac{k - C}{\beta} + c_P;$$

$$U_P (k; 0, 0, 0; k) = 0;$$

$$U_I (C, L; l, a, p; C + \beta a, M) = c_5 a + c_6 p, \quad (M = L, L/2);$$

$$U_I (C, L; l, a, p; k) = c_5 \frac{k - C}{\beta} + c_6 p + c_I;$$

$$U_I (k; 0, 0, 0; k) = 0.$$

The parameters are assumed to have the signs $c_1 \geqslant 0, c_2 > 0, c_3 < 0, c_4 > 0$, $c_5 > 0, c_6 < 0$. c_j $(j = R, P, I)$ is the additional payoff to player j due to catastrophe and therefore regarded as largely negative. $c_1 \geqslant 0$ reflects the regulator's internal difficulties in setting small standards, $c_2 > 0, c_4 > 0, c_5 > 0$ the benefits of energy production; $c_3 < 0$ the damage to the regulator due to pressure exerted on him; and $c_6 < 0$ the burden of organization. The term $(k - C)/\beta$ expresses that energy production is only valuable up to the critical amount. Thus the idea is excluded that in the case of a slowly developing catastrophe energy production by combustion of fossil fuel may give additional benefits during the initial stages of the catastrophe.

A play π of the game is given by an infinite sequence

$$\pi = (s^1, l^1, a^1, p^1; s^2, l^2, a^2, p^2; \ldots)$$

of states, measures of the regulator, producer, and impactee, respectively. According to the list of transition probabilities, there are only sequences where

$$C^1 \leqslant C^i \leqslant C_p \text{ and } L^i \in \{L^1, L^1/2, L^1/4, \ldots\},$$

and

$$a^i = \frac{C^{i+1} - C^i}{\beta} \text{ if } s^{i+1} = (C^{i+1}, L^{i+1}).$$

Furthermore if $s^i = k$ then $s^m = k$ for $m > i$. As a first approach we define the utility of a play as the undiscounted infinite sum of the transition utilities:

$$\underline{U}_j(\pi) = \sum_{i=1}^{\infty} U_j(s^i, l^i, a^i, s^{i+1}).$$

Since the summed-up internal utilities $\Sigma c_1 \, l^i$ can become infinite we omit them by specifying $c_1 = 0$. Let $(s^1, l^1, a^1, p^1, \ldots)$ denote a play where $s^i = (C^i, L^i)$ and $s^{i+1} = k$.

Then

$$\underline{U}_R(s^1 \ldots) = \sum_{j=1}^{i} (c_2 a^j + c_3 p^j) + c_2 \frac{k - C^i}{\beta} + c_3 p^{i+1} + c_R$$

$$= c_3 \sum_{j=1}^{i+1} p^j + c_2 \frac{k - C^1}{\beta} + c_R.$$

In the case of $s^j = (C^j, L^j)$ for $j = 1, 2, \ldots$

$$\underline{U}_R(s^1, \ldots) = c_3 \sum_{j=1}^{\infty} p^j + \lim_{j \to \infty} \frac{C^j - C^1}{\beta}.$$

Admitting $-\infty$ as a payoff then $\underline{U}_R(s^1, \ldots)$ is well defined because of $C^i \leqslant C_P$.
The same argument gives

$$\underline{U}_P(\pi) = c_4 \frac{k - C^1}{\beta} + c_P,$$

$$\underline{U}_P(\pi) = c_4 \lim \frac{C^j - C^1}{\beta};$$

and

$$\underline{U}_I(\pi) = c_5 \frac{k - C^1}{\beta} + c_6 \sum_{j=1}^{i+1} p^j + c_I,$$

$$\underline{U}_I(\pi) = c_5 \lim \frac{C^i - C^1}{\beta} + c_6 \sum_{j=1}^{\infty} p^i;$$

respectively.

The game is now completely described except for the definition of strategies. For simplification we admit only stationary strategies where the choices depend only on the last state and last measures of the other players.

Definition: A strategy σ_R of the regulator is a map:

$$\sigma_R : S \to \mathbf{R}$$

such that

$$\sigma_R\ (C,\ L) \in M_R\ (C,\ L) = \{l\ |\ 0 \leqslant l \leqslant L\},$$

$$\sigma_R\ (k) = 0.$$

A strategy σ_P of the producer is a map

$$\sigma_P: \{(s,\ l)\ |\ s \in S,\ l \in M_R\ (s)\} \rightarrow R,$$

such that

$$\sigma_P\ (C,\ L,\ l) \in M_P\ (C,\ L,\ l) = \left\{a\ |\ 0 \leqslant a \leqslant l,\ \frac{C_P - C}{\beta}\right\},$$

$$\sigma_P\ (k,\ 0) = 0.$$

A strategy σ_I of the impactee is a map

$$\sigma_I: \{(s,\ l,\ a)\ |\ s \in S,\ l \in M_P\ (s),\ a \in M_P\ (s,\ l)\} \rightarrow [0,\ 1],$$

such that

$$\sigma_I\ (C,\ L,\ l,\ a) \in [0,\ 1],$$

$$\sigma_I\ (k,\ 0,\ 0) = 0.$$

The sets of strategies are denoted by Σ_j $(j = R,\ P,\ I)$.

Due to the list of transition probabilities defined above infinitely many plays can occur. The appropriate σ-algebra over the set Π of all possible plays is defined as the minimal σ-algebra containing all cylinders with finite bases [see *Loève*, 8.3]. Due to the theorem of Tulcea there exist probability measures $P_j\ (\cdot\ |\ \sigma_R,\ \sigma_P,\ \sigma_I)$ on this σ-algebra where $P_j\ (\cdot\ |\ \sigma_R,\ \sigma_P,\ \sigma_I)$ stems from the iteration of given subjective probabilities.

The payoff function to player j is defined as his high subjective expected utility

$$V_j\ (\sigma_R,\ \sigma_P,\ \sigma_I) = \int \underline{U}_j\ (\pi)\ dP_j\ (\pi\ |\ \sigma_R,\ \sigma_P,\ \sigma_I)\ (j = R,\ P,\ I).$$

The formalism allows to derive a sharp upper bound for $V_j\ (\sigma_R,\ \sigma_P,\ \sigma_I)$. Due to the definition of the transition probability P_R the set of plays with a component state $s^m = (C^m,\ L^m)$, such that $C^m > C_R$ has probability $P_R\ (\cdot\ |\ \sigma_R,\ \sigma_P,\ \sigma_1) = 0$.

Hence only plays $\pi = (s^1,\ l^1,\ a^1,\ p^1;\ \dots)$ have to be considered where a component state s^m either equals $(C^m,\ L^m)$ such that $C^m < C_R$ or C_R. Hence

$$\underline{U}_R\ (\pi) = c_3 \sum_{j=1}^{i+1} p^j + c_2\ \frac{C_R - C^1}{\beta} + c_R\ \ \text{if } C^i \leqslant C_R < C^i + \beta a^i,$$

or

$$\underline{U}_R\ (\pi) = c_3 \sum_{j=1}^{i+1} p^j + \lim c_2\ \frac{C^j - C^1}{\beta}\ \ \text{if } C^j \leqslant C_R\ \ (j = 1, \dots).$$

In both cases $\underline{U}_R\ (\pi) \leqslant c_2\ \dfrac{C_R - C^1}{\beta}$ is obvious. Hence

$$V_R\,(\sigma_R,\sigma_P,\sigma_I)\leqslant c_2\,\frac{C_R-C^1}{\beta}\,.$$

The analogous argument yields $V_I\,(\sigma_R,\sigma_P,\sigma_I)\leqslant c_5\,\dfrac{C_I-C^1}{\beta}$ whereas

$\underline{U}_P\,(\pi)\leqslant c_4\,\dfrac{C_P-C^1}{\beta}$ immediately implicates

$$V_P\,(\sigma_R,\sigma_P,\sigma_I)\leqslant c_4\,\frac{C_P-C^1}{\beta}\,.$$

The bounds are sharp in the sense that strategy triples exist yielding the bounds as payoffs.

Let $\sigma_R\,(C,L)=L$, $\sigma_P\,(C,L,l)=\min\,(l,\,(C_P-C)\,/\,\beta)$, $\sigma_I\,(C,L,l,a)=0$. Then

$$V_R\,(\sigma_R,\sigma_P,\sigma_I)=c_4\,\frac{C_P-C^1}{\beta}\,.$$

We give examples for V_R and V_P below. If the establishment of the payoff as expected payoffs over Π were more elaborated [see e.g. *Kindler*] it would be obvious that we arrive at the same payoffs $V_j\colon \Sigma_R\times\Sigma_P\times\Sigma_I\to\mathbf{R}$ if we replace the component utility U_I by $U_{I,r}$:

$$U_{I,r}\,(C,L;l,a,p;C+\beta a,M)=\begin{cases}c_5a & \text{if }M=L;\\[2ex] c_5a+c_6/v & \text{if }M=L/2;\end{cases}$$

$$U_{I,r}\,(C,L;l,a,p;k)\quad=\quad U_I\,(C,L;l,a,p;k);$$

$$U_{I,r}\,(k;0,0,0;k)\quad=\quad U_I\,(k;0,0,0,k).$$

This remark permits to shorten proofs in the next section.

The Game-Theoretic Solution

Except for two-person zero-sum games or equivalent games, there is no unanimous solution concept. Instead there are a variety. Therefore we shall first give brief definitions of the solution concepts [for a broader discussion see *Avenhaus/Höpfinger*], and later on describe strategy three-tuples satisfying them.

Definition: A three-tuple $(\sigma_R^+,\sigma_P^+,\sigma_I^+)\in\Sigma_R\times\Sigma_P\times\Sigma_I$ of strategies is called a (weak) *equilibrium point* if

$$V_R\,(\sigma_R^+,\sigma_P^+,\sigma_I^+)\geqslant V_R\,(\sigma_R,\sigma_P^+,\sigma_I^+)\qquad(\sigma_R\in\Sigma_R);$$
$$V_P\,(\sigma_R^+,\sigma_P^+,\sigma_I^+)\geqslant V_P\,(\sigma_R^+,\sigma_P,\sigma_I^+)\qquad(\sigma_P\in\Sigma_P);$$
$$V_I\,(\sigma_R^+,\sigma_P^+,\sigma_I^+)\geqslant V_I\,(\sigma_R^+,\sigma_P^+,\sigma_I)\qquad(\sigma_I\in\Sigma_I).$$

Definition: The payoff vector $(V_j (\sigma_R, \sigma_P, \sigma_I))$ $j = R, P, I$ is called *Pareto-optimal* if there is no other payoff vector $(V_j (\tau_R, \tau_P, \tau_I)$ where $\tau_j \in \Sigma_j$ $(j = R, P, I)$, such that

$$V_j (\sigma_R, \sigma_P, \sigma_I) \geqslant V_j (\tau_R, \tau_P, \tau_I) \ (j = R, P, I),$$

and at least one inequality strictly holding.

Definition: Let $(W_R, W_P, W_i) \in \mathbf{R}^3$ denote the point of maximal possible payoffs which is called *bliss point*, i.e. $W_j = \max (V_j (\sigma_R, \sigma_P, \sigma_I) \mid \sigma_i \in \Sigma_i$ $(i = R, P, I))$. The payoff vector (v_R, v_P, v_I) is called bliss-optimal if

$$\sum_{j=R,P,I} (v_j - W_j)^2 = \min (\sum_j (V_j (\sigma_R, \sigma_P, \sigma_I) - W_j)^2 \mid (\sigma_R, \sigma_P, \sigma_I) \in \Sigma_R \times \Sigma_P \times \Sigma_I).$$

Definition: Let (d_R, d_P, d_I) be a triple of payoffs the players obtain in case they cannot reach an unanimous agreement on the choice of a payoff vector. Then the Nash *solution* is the point (W_R, W_P, W_I) which maximizes the term $(u_R - d_R) (u_P - d_P)$ $(u_I - d_I)$ subject to the requirements $u_j = V_j (\sigma_R, \sigma_P, \sigma_I)$ $(j = R, P, I)$ for some strategy three-tuple $u_j \geqslant d_j$ $(j = R, P, I)$.

Definition: A *hierarchic solution* is a triple (τ_R, τ_P, τ_I) consistent of a strategy $\tau_R \in \Sigma_R$ and two maps

$$\tau_P : \Sigma_R \to \Sigma_P,$$

$$\tau_I : \Sigma_R \times \Sigma_P \to \Sigma_I,$$

such that

$$V_I (\sigma_R, \sigma_P, \tau_I (\sigma_R, \sigma_P)) = \max_{\sigma_I \in \Sigma_I} V_I (\sigma_R, \sigma_P, \sigma_I);$$

$$V_P (\sigma_R, \tau_P (\sigma_R), \tau_I (\sigma_R, \tau_P (\sigma_R))) = \max_{\sigma_P \in \Sigma_P} V_P (\sigma_R, \sigma_P, \tau_I (\sigma_R, \sigma_P));$$

$$V_R (\tau_R, \tau_P (\tau_R), \tau_I (\tau_R, \tau_P (\tau_R))) = \max_{\sigma_R \in \Sigma_R} V_R (\sigma_R, \tau_P (\sigma_R), \tau_I (\sigma_R, \tau_P (\sigma_R))).$$

The game has a huge variety of equilibrium points. In the following we give three equilibrium points, the first two of which have Pareto-optimal payoffs, whereas the third is only given as an indicator of the variety of equilibrium points.

Theorem: The tuples of strategies given below are equilibrium points:

1. $\sigma_R^1 (C, L) := \min (L, \max (0, ((C_R - C)/\beta));$

$\sigma_P^1 (C, L, l) := l;$

$\sigma_I^1 (C, L, l, a) := 0.$

The inherent utilities are

$$V_R \, (\sigma_R^1, \sigma_P^1, \sigma_I^1) = c_2 \, \frac{C_R - C^1}{\beta};$$

$$V_P \, (\sigma_R^1, \sigma_P^1, \sigma_I^1) = c_4 \, \frac{C_R - C^1}{\beta};$$

$$V_I \, (\sigma_R^1, \sigma_P^1, \sigma_I^1) = \begin{cases} c_5 \, \dfrac{C_R - C^1}{\beta} & \text{if } C_R \leqslant C_I, \\[2mm] c_5 \, \dfrac{C_I - C^1}{\beta} + c_I & \text{if } C_R > C_I. \end{cases}$$

2. $\sigma_R^2 \, (C, L) := \min \, (L, \max \, (0, \frac{C_I - C}{\beta}))$;

$\sigma_P^2 \, (C, L, l) := l;$

$$\sigma_I^2 \, (C, L, l, a) := \begin{cases} 0 & \text{if } l = \min \, (L, (C_I - C)/\beta) \text{ and } C \leqslant C_I, \\[2mm] 1 & \text{if } l \neq \min \, (L, (C_I - C)/\beta) \text{ or } C > C_I. \end{cases}$$

The inherent utilities are

$$V_R \, (\sigma_R^2, \sigma_P^2, \sigma_I^2) = \begin{cases} c_2 \, \dfrac{C_I - C^1}{\beta} & \text{if } C_I \leqslant C_R, \\[2mm] c_2 \, \dfrac{C_I - C^1}{\beta} + c_R & \text{if } C_I > C_R. \end{cases}$$

$$V_P \, (\sigma_R^2, \sigma_P^2, \sigma_I^2) = c_4 \, \frac{C_I - C^1}{\beta};$$

$$V_I \, (\sigma_R^2, \sigma_P^2, \sigma_I^2) = c_5 \, \frac{C_I - C^1}{\beta}.$$

3. *Keep quiet point*

$\sigma_R^3 \, (C, L) = 0;$

$\sigma_P^3 \, (C, L, l) = 0;$

$$\sigma_I^3 \, (C, L, l, a) = \begin{cases} 0 & \text{if } l = 0 \text{ and } C = C^1, \\[2mm] 1 & \text{if } l > 0 \text{ or } C > C^1 \, ; \end{cases}$$

with utilities $V_j \, (\sigma_R^3, \sigma_P^3, \sigma_I^3) = 0 \quad (j = R, P, I).$

Proof: In order to avoid descriptions that are cumbersome but not illustrative we give sketches only.

1. Let $i_R \in \{1, 2, \ldots\}$ be defined by $C^1 + \beta\,(i_R - 1)\,L^1 \leqslant C_R < C^1 + \beta i_R L^1$. One can show by iteration on i that

$$C^{i+1} = C^1 + \beta i L^1 \ (i = 0, 1, \ldots, i_R - 1), \quad a^i = L^1 \ (i = 1, \ldots, i_R - 1),$$

$$a^{i_R} = \frac{C_R - C^{i_R}}{\beta},$$

$$C^{i+1} = C_R \ (i = i_R, i_R + 1, \ldots), \qquad a^i = 0 \ (i = i_R + 1, i_R + 2, \ldots),$$

due to the regulator's strategy. Hence

$$V_R\,(\sigma_R^1, \sigma_R^2, \sigma_R^3) = c_2 \sum_{i=1}^{i_R-1} L^1 + c_2 \frac{C_R - C^{i_R}}{\beta} = c_2 \frac{C_R - C^1}{\beta};$$

analogously

$$V_P\,(\sigma_R, \sigma_P, \sigma_I) = c_4 \frac{C_R - C^1}{\beta}.$$

In the case of $C_R \leqslant C_I$, (C_R, L^1) will be the state of the play for $i = i_R + 1$, $i_R + 2, \ldots$ also due to the subjective probability of the impactee. However, if $C_R > C_I$, catastrophe C_I will be the final state resulting in a payoff

$$c_5 \frac{C_I - C^1}{\beta} + c_I.$$

The regulator's condition for an equilibrium is obviously satisfied since the strategy triple gives him the maximal possible utility. Just as is obvious, there is no better payoff for the producer with another strategy, and this is also true for the impactee in the case of $C_R \leqslant C_I$.

Only $C_R > C_I$ requires more sophistication. Let σ_P' denote a different strategy of the impactee. Then a play π with $\lim C^i \leqslant C_I$ is only possible if the reduction of L^i to its half takes place an infinite number of times. But then $\underline{U}_{I,r}\,(\pi) = -\infty <$

$$c_5 \frac{C_I - C^1}{\beta} + c_I.$$ If the reduction of L^1 takes place only a finite number of times then $\underline{U}_{I,r}\,(\pi) \leqslant c_5 \dfrac{C_I - C_1}{\beta} + c_I$. Hence any other strategy cannot yield a better payoff.

2. In the case of $C_I < C_R$ the regulator can only get a better payoff if plays π with states (C^i, L^i) where $C^i > C_I$ occur with a subjective probability greater than zero. But then $\sigma_I^2\,(C^i, L^i, l, a) = 1$ infinitely often yielding the payoff $-\infty$ to the regulator. Thus he cannot get a better payoff with a different strategy. Obviously the producer cannot get a better payoff, whereas the impactee gets his maximal payoff.

In the case of $C_I = C_R$ regulator and impactee receive their maximal payoffs, whereas the producer has no better response. In the case of $C_I > C_R$ the regulator may want to escape catastrophe by applying a strategy like the one of the first equilibrium point. But then he is punished an infinite number of times by pressure from the impactee and gets a smaller payoff. Again it is obvious that producer and impactee cannot do better.

3. The impactee's ability to exert pressure infinitely often again makes the strategy triple $(\sigma_R^3, \sigma_P^3, \sigma_I^3)$ an equilibrium point.

The question arises: Which of these equilibrium points yield Pareto-optimal payoffs? The answer can immediately be deduced from the following:

Theorem: The set of payoffs

$$\{(V_R(\sigma_R, \sigma_P, \sigma_I), V_P(\sigma_R, \sigma_P, \sigma_I), V_I(\sigma_R, \sigma_P, \sigma_I)) \mid \sigma_j \in \Sigma_j \ (j = R, P, I)\}$$

is a subset of the following domain $D \subseteq \mathbb{R}^3$.

1. Let $C_R < C_I < C_P$. Then D consists of all $(x, y, z) \in \mathbb{R}^3$ such that a pair (p_R, p_I) of real numbers exists such that $0 \leqslant p_R, 0 \leqslant p_I, 0 \leqslant 1 - p_R - p_I$ and the following inequalities hold:

$$x \leqslant c_2 \frac{C_R - C^1}{\beta} + (1 - p_R) c_R ;$$

$$y \leqslant c_4 \left\{ p_R \frac{C_R - C^1}{\beta} + p_I \frac{C_I - C^1}{\beta} + (1 - p_R - p_I) \frac{C_p - C^1}{\beta} \right\} ;$$

$$z \leqslant c_5 p_R \frac{C_R - C^1}{\beta} + c_5 p_I \frac{C_I - C^1}{\beta} + (1 - p_R - p_I) \left(c_5 \frac{C_I - C^1}{\beta} + c_I \right).$$

2. Let $C_R = C_I < C_P$. Then D consists of all $(x, y, z) \in \mathbb{R}^3$ which are part of a solution $(x, y, z, p) \in \mathbb{R}^4$ of the following system of inequalities:

$$0 \leqslant p \leqslant 1;$$

$$x \leqslant c_2 \frac{C_R - C^1}{\beta} + (1 - p) c_R ;$$

$$y \leqslant c_4 \left\{ p \frac{C_R - C^1}{\beta} + (1 - p) \frac{C_P - C^1}{\beta} \right\} ;$$

$$z \leqslant c_5 p \frac{C_R - C^1}{\beta} + (1 - p) \ c_5 \frac{C_R - C^1}{\beta} + c_I.$$

3. Let $C_I < C_R < C_P$. Then D consists of all $(x, y, z) \in \mathbb{R}^3$ which are part of a solution $(x, y, z, p_I, p_R) \in \mathbb{R}^5$ of the following system of inequalities:

$$0 \leqslant p_I, 0 \leqslant p_R, 0 \leqslant 1 - p_I - p_R ;$$

$$x \leqslant c_2 p_I \frac{C_I - C^1}{\beta} + c_2 p_R \frac{C_R - C^1}{\beta} + (1 - p_I - p_R) \left(c_2 \frac{C_R - C^1}{\beta} + c_R \right) ;$$

$$y \leqslant c_4 \left\{ p_I \frac{C_I - C^1}{\beta} + p_R \frac{C_R - C^1}{\beta} + (1 - p_I - p_R) \frac{C_P - C^1}{\beta} \right\} ;$$

$$z \leqslant c_5 \, \frac{C_I - C^1}{\beta} + (1 - p_I) \, c_I \, .$$

Sketched proof: Let $(\sigma_R, \sigma_P, \sigma_I)$ denote a strategy triple. In the case of $C_R < C_I < C_P$ let p_R denote the probability $P_P \, (T_R \mid \sigma_R, \sigma_P, \sigma_I)$ that a play with states (C^i, L^i), $C^i \leqslant C_R$ will be realized, i.e. T_R is the set of all plays $(s^1, l^1, a^1, p^1, \dots)$ such that $C^i \leqslant C_R$ for all component states (C^i, L^i) $(i = 1, 2, \dots)$. Let $p_I = P_P \, (T_I \mid \sigma_R, \sigma_P, \sigma_I)$ denote the probability for the set of plays (s^i, l^i, a^i, p^i) $(i = 1, 2, \dots)$ such that $C^i \leqslant C_I$ for all i, where $s^i = (C^i, L^i)$ $(i = 1, 2, \dots)$ but $C^j > C_R$ for at least one j. Obviously

$$V_P \, (\sigma_R, \sigma_P, \sigma_I) \leqslant c_4 \left\{ p_R \, \frac{C_R - C^1}{\beta} + p_I \, \frac{C_I - C^1}{\beta} + (1 - p_R - p_I) \, \frac{C_P - C^1}{\beta} \right\} .$$

By definition of the regulator's transition probability, $P_R \, (T_R \mid \sigma_R, \sigma_P, \sigma_I) = p_R$, but with probability $1 - p_R$ the catastrophe will occur. Hence

$$V_R \, (\sigma_R, \sigma_P, \sigma_I) \leqslant c_2 \, \frac{C_R - C^1}{\beta} + (1 - p_R) \, c_R \, .$$

The impactee's probabilities for plays with only state components below C_R, and between C_R and C_I are p_R and p_I respectively. Therefore

$$V_I \, (\sigma_R, \sigma_P, \sigma_I) \leqslant c_5 p_R \, \frac{C_R - C^1}{\beta} + c_5 I \, \frac{C_I - C^1}{\beta} + (1 - p_R - p_I) \left(c_5 \, \frac{C_I - C^1}{\beta} + c_I \right) .$$

The proofs for the two remaining cases follow the same line of argumentation. One has only to consider that p is the producer's subjective probability that a play will occur where $C^1 \leqslant C_R = C_I$ for all component states C^1. In the last case p_I denotes the producer's probability for a play with component states not greater that C_I, p_R, the probability for a play with a component state greater than C_I, and all component states not greater than C_R.

Corollary: The first and the second equilibrium point of the last but one theorem have Pareto-optimal payoff vectors. In the case of $C_R > C^1$ and $C_I > C^1$ the keep-quiet point has no Pareto-optimal payoff vector.

Proof: Having chosen either $p_I = 1$ or $p_R = 1$ and $p = 1$, one immediately verifies that the payoff vectors of the first and second equilibrium points belong to the boundary plane given on the right-hand side of the inequalities of the last but one theorem. Hence the payoff vectors are Pareto-optimal.

Under the given conditions the keep-quiet point is dominated by the first or the second equilibrium point. The results are illustrated by Figures 1 and 2 showing the projection of subset D of the last theorem.

As can be seen from the figures even the combined solution concepts of equilibrium point and Pareto-optimality do not yield an unanimous solution. But what about the remaining solution concepts? In order to discuss them we give the boundary

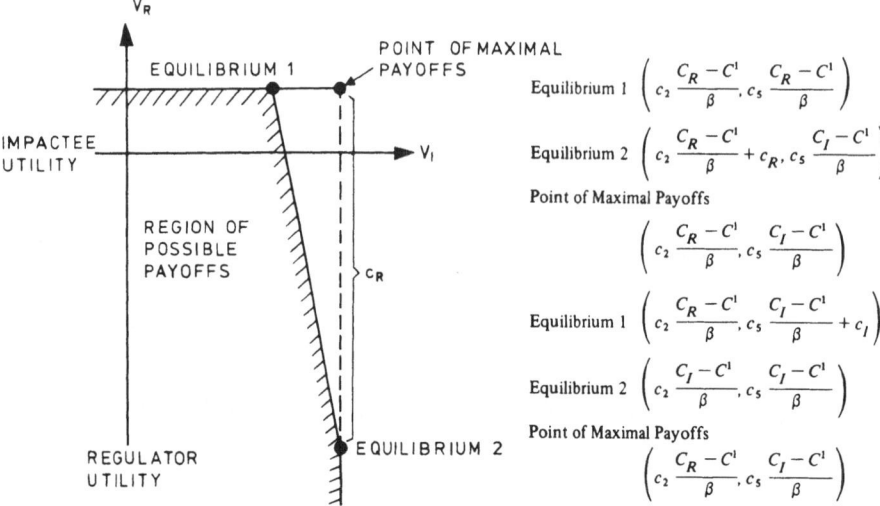

Fig. 1: Payoff diagram for regulator and impactee $(C_R < C_I)$

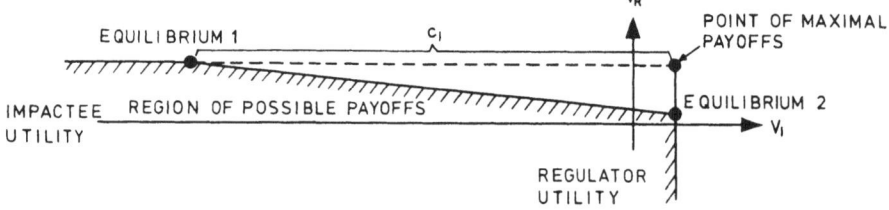

Fig. 2: Payoff diagram for regulator and impactee $(C_R > C_I)$

plane of the last theorem after elimination of the parameters for the case of $C_R < C_I < C_P$ by the following equation:

$$\frac{y}{c_4} + \frac{z}{c_I}\frac{C_P - C_I}{\beta} + \frac{x}{c_R}\left\{\frac{C_R - C_I}{\beta} + \frac{c_5}{c_I}\frac{C_R - C_I}{\beta}\frac{C_P - C_I}{\beta}\right\} = \text{constant}.$$

Since by assumption c_I and c_R are huge negative numbers the equation is dominated by the first term $\frac{y}{c_4}$. Hence the payoff vector $\left(c_2 \frac{C_R - C^1}{\beta}, c_4 \frac{C_R - C^1}{\beta}, c_5 \frac{C_R - C^1}{\beta}\right)$ is either bliss-optimal or very close to the bliss-optimal payoff vector. Hence we can regard it as approximately bliss-optimal.

The same holds for $C_R = C_I < C_P$, and in the case of $C_I < C_R < C_P$ for $\left(c_2 \frac{C_I - C^1}{\beta}, c_4 \frac{C_I - C^1}{\beta}, c_5 \frac{C_I - C^1}{\beta}\right)$.

Without proof we state the two approximate bliss-optimal points are Nash solutions for $d_j = 0$ $(j = R, P, I)$ as soon as the absolute values of c_I and c_R are large enough. This means that the bliss-point concept as well as the Nash solution favor a behavior based on the most pessimistic estimate $\min(C_R, C_I)$ of the critical value.

The hierarchic solution concept is much more complicated than the preceding ones since it involves maps from strategy spaces into strategy spaces. We circumvent the mathematical optimization problem specifying only the resulting strategies.

Theorem: Let $(\sigma_R^1, \sigma_P^1, \sigma_I^1)$ be the first equilibrium point of the last but one theorem, i.e.,

$$\sigma_R^1 \ (C, L) = \min (L, \max (0, (C_R - C)/\beta));$$

$$\sigma_P^1 \ (C, L, l) = l;$$

$$\sigma_I^1 \ (C, L, l, a) = 0.$$

Let $(\tau_R^1, \tau_P^1, \tau_I^1)$ denote a hierarchic solution. Then (τ_R, τ_A, τ_I) defined by

$$\tau_R = \sigma_R^1 ;$$

$$\tau_P (\sigma_R) = \tau_P^1 (\sigma_R) (\sigma_R \in \Sigma_R - \{\sigma_R^1\}), \tau_P (\sigma_R^1) = \sigma_P^1;$$

$$\tau_I (\sigma_R, \sigma_P) = \tau^1 (\sigma_R, \tau_P) (\sigma_j \in \Sigma_j - \{\sigma_j^1\}) (j = R, P);$$

$$\tau_I (\sigma_R^1, \sigma_P^1) = \sigma_I^1 ;$$

is also a hierarchic solution.

Proof: $V_I (\sigma_R^1, \sigma_P^1, \sigma_I^1) = \max_{\sigma_I} V_I (\sigma_R^1, \sigma_P^1, \sigma_I^1)$ since $(\sigma_R^1, \sigma_P^1, \sigma_I^1)$ is an equilibrium point. The next step is the verification of $V_P (\sigma_R^1, \sigma_P^1, \sigma_I^1) = $

$= \max_{\sigma_P} V_P (\sigma_R^1, \sigma_P, \tau_I (\sigma_R^1, \sigma_P))$. The regulator's strategy σ_R^1 prevents a larger amount than C_R of carbon dioxide in the atmosphere, whereas the producer's utility is the

larger the more dioxide is in the atmosphere. Therefore $V_P (\sigma_R^1, \sigma_P^1, \sigma_I^1) = $

$= c_4 \dfrac{C_R - C^1}{\beta} = \max_{\sigma_P, \sigma_I} V_P (\sigma_R^1, \sigma_P, \sigma_I)$, which is even stronger. The last condition

is trivially satisfied since $V_R (\sigma_R^1, \sigma_P^1, \sigma_I^1)$ gives the maximal possible utility

$c_2 \dfrac{C_R - C^1}{\beta}$ to the regulator.

It should be remarked that the theorem is independent of whether $C_R < C_I$ or not. It simply states that the regulator is strong enough to push through his standpoint.

The following example serves to illustrate the order of magnitude. Let $C^1 = 6 \cdot 10^{16}$ g, $C_I = 18 \cdot 10^{16}$ g, $L^1 = 0.2 \cdot 10^{16}$ g, $\beta = 0.3$, $c_2 = 0.002$ \$/g, $c_4 = 10^{-4} c_2$, $c_5 = 0.7 c_2$. C^1 is in the order of magnitude of the present amount of carbon dioxide in the atmosphere, and L^1 in the order of magnitude of the present release of carbon dioxide. \$3.6 $\cdot 10^{12}$ is an estimate of the gross world product of 1970. Then production is possible for 200 years and the payoff vector equals (\$8 $\cdot 10^{14}$, \$8 $\cdot 10^{10}$, \$5.6 $\cdot 10^{14}$).

Conclusion

The game has been analyzed for different solution concepts. It turns out that the Nash solution and the bliss-optimal concept yield solutions that are basically different from the hierarchic solution. In the case of $C_I < C_R$ where the impactee's view is more pessimistic than that of the regulator, the Nash solution and the bliss-optimum concept, by their tendency to fair bargains, favor the second equilibrium point based on the estimate C_I. Contrary to this the hierarchic solution yields the first equilibrium point which is based on the estimate C_R as critical value.

The results heavily depend on the fact that the summed up component payoffs are not discounted. Thus the impactee can principally push the regulator's payoff down to minus infinity. Actually he cannot exert pressure infinitely often since then he would also receive the payoff minus infinity. Hence this capability to punish or to exert pressure only yields a vastness of equilibrium points. It seems that the results may change substantially if discounting is included. Then the regulator may be able to resist pressure, and on the other side the impactee may be able to affort pressure. Another way would be to assume the game to be stopped as soon as the upper bound L is below a given limit, e.g., if L is less than ten percent of the carbon dioxide produced by the biosphere during one year. Again the question arises whether the impactee can enforce a total release that is less than $(C_I - C^1) / \beta$.

So far the impactee has been represented as a rational player with a utility function. Another possibility would be to represent him by a response function based on his perception of the regulator's and the producer's decision, i.e., to prescribe one strategy of the impactee. Then we would actually have a regulator-producer game, and as solution concept we might take the hierarchic solution. But which response should we use? Our analysis of the three-person game offers us two responses:

$$\sigma_I^1 (C, L, l, a) = \quad 0;$$

$$\sigma_I^2 (C, L, l, a) = \begin{cases} 0 & \text{if } l = \min (L, (C_I - C) / \beta) \text{ and } C \leqslant C_I, \\ 1 & \text{if } l \neq \min (L, (C_I - C) / \beta) \text{ or } C > C_I. \end{cases}$$

If we assume the first, then the impactee is actually a dummy player. Then equilibrium point one is part of the hierarchic solution. In the case of σ_I^2 however, the hierarchic solution yields the second equilibrium point as can be verified very easily. Thus, the three-person game can provide for ideas how to formalize a response function.

References

Augustsson, T., and *V. Ramanathan*: A Radiative Convective Model of the CO_2 Climate Problem. Journal of the Atmospheric Sciences **34**, 1977, 448–481.

Avenhaus, R., S. Fenyi, and *H. Frick*: Mathematical Treatment of Box Models for the CO_2-Cycle of the Earth. Carbon Dioxide, Climate and Society, IIASA Proceedings Series. Ed. by J. Williams, Oxford 1978.

Avenhaus, R., and *E. Höpfinger*: A Game-Theoretic Framework for Dynamic Standard Setting Procedures. RM-78-64, International Institute for Applied Systems Analysis, Laxenburg, Austria, 1978.

Kindler, J.: Definitheitskriterien für nichtstationäre stochastische Spiele. Doctoral thesis, University of Karlsruhe, FRG, 1971.

Loeve, M.: Probability Theory. Toronto 1955.

Markley, O.W., A.L. Webre, R.C. Carlson, and *B.R. Holt*: Sociopolitical Impacts of Carbon Dioxide Buildup in the Atmosphere Due to Fossil Fuel Combustion. EGU-6370, Stanford Research Institute, Menlo Park, Calif. 1977.

McKinsey, J.C.C.: Introduction of the Theory of Games. Rand Corporation, New York 1952.

Owen, G.: Game Theory. Philadelphia 1968.

Rotty, R.M.: The Atmospheric Consequences of Heavy Dependence on Coal. ORAU/IEA(M) 77-27, Institute for Energy Analysis, Oak Ridge, Tenn., 1977.

Williams, J.: Introduction to the Climate/Environment Aspects of CO_2 (A Pessimistic View). Carbon Dioxide, Climate and Society, IASA Proceedings Series. Ed. by J. Williams. Oxford 1978.

Applied Game Theory. 1979 ©Physica-Verlag, Wuerzburg/Germany

Coping with Deception

By *R. Axelrod*, Ann Arbor[1])

Abstract: This paper develops a game theoretic model for the optimal use of information which not only has some random error, but also has error which is under the control of the other player. Signal detection theory is applied in the context of a two-sided game to show how such partially reliable and partially deceptive information can best be used. Historical examples from international relations and military affairs are presented to demonstrate the applicability of the model to problems of strategic intelligence. Specific consideration is given to additional factors which would make the model more realistic, including the use of supervised learning, standard operating procedure, and cover stories.

How can we make sense out of the world when our information is not only randomly in error, but some of it is actually put there to deceive us?[2]) Coping with deception is a problem faced by players of poker and by Secretaries of State, by consumers in the marketplace and by cops on the beat, by voters in the precinct and by generals in the army.

The stakes are greatest in the arena of international politics. Therefore this paper will concentrate on deception between governments, although much of the analysis will also apply to other arenas.

The Prevalence of Surprise

One might think it is nearly impossible today for one major government to deceive another. After all, modern technology has advanced very far in the area of surveillance. For example, under ideal conditions a modern spy satellite has optical resolution approaching one foot [*Greenwood*, p. 6f.]. Methods of electronic eavesdropping are also very advanced. Moreover, the level of effort devoted to strategic intelligence is extremely high. For example, the United States today spends about 6.2 billion dollars per year on strategic intelligence (N.Y. Times, April 1, 1977) [*Marchetti/Marks*, p. 80]. This compares to about 10.6 billion on all aspects of strategic nuclear forces [*H. Brown*]. Over 150,000 people are employed by the United States working full time on strategic intelligence.

But premeditated surprise is still possible in the modern world. Consider the following examples taken just from the post-World War II era. In 1950 the United States

[1]) *R. Axelrod*, Dept. of Political Science and Institute of Public Policy Studies, University of Michigan, Ann Arbor, Michigan 48109, USA.

[2]) I would like to thank Brian Barry, Steven Brams, John Chamberlin, Steve Crocker, Alex George, Mark Granovetter, William R. Harris, and Zvi Lanir for their help. I would also like to thank the Center for Advanced Study in the Behavioral Sciences where this work was done.

was surprised when North Korea invaded South Korea. And then a few months later
the United States was surprised a second time when China intervened after the United
States had come to the aid of South Korea. In 1956, Egypt was surprised when Israel
attacked her. And then a week later, Egypt was again surprised when Britain and
France invaded the Suez Canal Zone. In 1967, the Arabs were again surprised by the
Israeli attack. In 1968, the United States was surprised by the Tet offensive in Viet-
nam. And then a few months later the United States was surprised when the Soviets
invaded Czechoslovakia. Most recently, in 1973, came one of the most interesting
cases where both the United States and Israel were surprised by the attack of Egypt
and Syria.

Not all attempts at premeditated surprise have succeeded, however. For example, in
1961 the American-sponsored Bay of Pigs attack was anticipated by Castro. And in
1962 the Soviet placement of missiles in Cuba was spotted and countered by the
United States.

It should be acknowledged at the outset of this analysis of deception that many of
these surprises are due as much or more to self-deception by the victim as to the efforts made by others to deceive. Nevertheless, deceptions in the strict sense of deliberate attempts to mislead are contributing causes to the success of most, if not all, of
these surprises.

Toward a Policy-Oriented Theory

My long-term goal is to develop a policy-oriented theory of deception and counter-
deception. This theory would have four uses. First it would help provide a realistic
view of the world. One can imagine a dimension with naiveté on the one hand, and
paranoia on the other. With a realistic view of the world we could keep away from the
two extremes of this dimension. The second purpose of the theory would be as a guide
to action. Thus a good theory of deception and counter-deception would help provide
appropriate methods for analysis of incoming information, would help in the provision of timely warnings, would help determine which kinds of surveillance techniques
were worth having, and would help in the development of verifiable arms control
agreements. The third purpose of the theory would be to evaluate the performance in
a given setting of a nation or of its intelligence establishment. The fourth purpose is
the most modest one, and the one that this paper will make most progress in: the development of a conceptual guide so that we can better understand the basic mechanisms of deception and how to cope with it.

Unfortunately we have few formal theories to work with in the area of deception.[3]
The primary reason probably is that natural scientists have given little attention to this
kind of problem. The weatherman, for example, never has to worry about someone
having put a dark cloud in the sky in order to make him think it's going to rain.

---

[3]) One exception is the work of *Brams* [1977] on two-person 2x2 games. He deals with the
question of how you should try to distort the other player's perception of your payoffs. Another
exception is *Axelrod* [1979] which treats timing in the use of resources for deception. For less
rigorous treatments of the principles of deception in international affairs, good places to start are
Jervis [1970] and *Whaley* [1969 and 1973].

Another reason is that secrecy surrounds the doctrines of deception used by governments. Many good histories, memoires, and exposés exist of secret operations, but not
much has come out in the way of theory. Indeed, as one former practitioner explained
it, "there is no doubt that strategic deception, whether in peace or in war, is the most
secret of secret operations. Accordingly, the less said about it the better" [*Felix*,
p. 154].

But I suspect that still another reason for the public lack of theory of deception is
that bureaucracies are the main practitioners of deception and counter-deception, and
bureaucracies are not very prone to theorizing.

Be that as it may, what we do have to work with are a great number of episodes of
deception described in memoirs and histories. We also have some abstract theories
which can be usefully applied to this subject, in particular game theory and signal detection theory. Eventually it would be good to bridge the gap from the empirical data
all the way to the abstract theory as applied to specific circumstances which are relevant to strategic deception in the international arena. The present paper begins the
process of bridging this gap, starting at the more theoretical end of the spectrum and
working back toward the more applied end.

An Illustrative Game

For the sake of illustrating concepts of deception, let us start a simple game which I
will call the mobilization game. There are two actors: an attacker and a defender. The
attacker can prepare a fake attack or a real attack, and the defender can decide to react or not to react. The attacker would prefer that if it is a real attack, the defender
not react, and therefore not be prepared for it. Or, if the attacker institutes a fake attack, he would prefer that the defender mobilize and thereby waste his resources.
From the defender's point of view, the object is to outguess the attacker so that the
defender will react only to real attack and not react to a fake attack.

This sort of interaction is common when mobilization is an important part of
meeting a military threat. There are many good historical illustrations — World War I,
for example. More recent illustrations come from the dynamics of the Arab-Israeli confrontation, especially in 1956 and 1973. In the near future there are many potential
applications: NATO versus the Warsaw Pact in Central Europe, the Soviet Union versus
China, the Korean Peninsula, or of course the Middle East. All have this sort of interaction, because in each case it pays for one party to react only to a genuine attack and
not to react to a fake attack.

Figure 1 shows a representation of this as a zero-sum game. The vertical axis is the
payoff to the attacker. The payoff to the defender is just the negative of this, and thus
the attacker wants to get as high as possible and the defender wants to get as low as
possible. The horizontal axis represents the attacker's choice. He can have a fake attack or a real attack, or pick any probability mix in between. The defender's choices
are shown by the two lines. If the defender does not react he gets payoff A if there is a
fake attack and C if there is a real attack. C, of course, is very good for the attacker
because the defender did not mobilize to meet a genuine attack. The other choice of

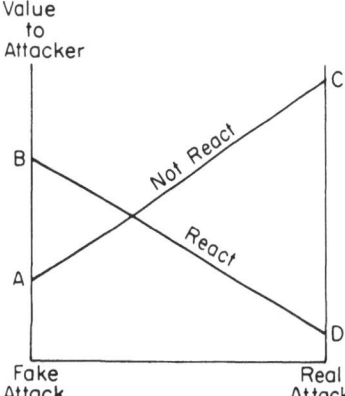

Fig. 1: The Mobilization Game

the defender is to react; the attacker gets B if there is a fake attack and D if there is a real attack. D, of course, is the worst payoff for the attacker because the attacker launches a real war but the defender is ready and waiting.

In the basic game theory analysis of this situation, there is no information for either side about what the other is doing, but there is perfect information about the payoff structures. The classical analysis says that the rational actor in such a situation should use the minimax strategy.

In the mobilization game the minimax strategy will involve a mixed strategy. For example, if the attacker uses the pure strategy of always having a fake attack, the defender might outguess it and never react. If the attacker always used a real attack, the defender might outguess it and always react. However, the attacker also has the option of using a probability mix. With any given probability mix of fake and real attack that the attacker uses, he can guarantee himself at least as much payoff as shown in the heavy line on Figure 2. This is because whether the defender reacts or not, the attacker will get at least as much as the heavy line. The attacker is trying to get the highest possible value, and the highest possible value that he can guarantee himself, then, is the peak of this heavy line shown in the point marked "minimax". In the illustrated case this means that the attacker will be using a mixed strategy with a real attack about one third of the time, and a fake attack about two-thirds of the time.[4] Some people balk at the idea of using a mixed strategy. But it should be remembered that Sadat undertook preparations for attacks that did not develop in December 1971, December 1972, and April/May 1973 as well as for the real attack in October 1973 [*Herzog*, p. 43].

The defender can also use a mixed strategy so that the attacker can not know for sure whether the defender will react or not in a given instance.[5]

---

[4]) The attacker's minimax strategy is given by launching a real attack P percent of the time, where $P = (D - C)/(A - B - C + D)$. See for example *Singleton/Tyndall* [1974, p. 62]. In the mobilization game $C > B > A > D$, and thus P is always less than $1/2$, meaning that fewer than half of the attacks should be real.

[5]) The defender's minimax strategy calls for a probability of not reacting of $Q = (D - B)/(A - B - C + D)$. If either or both sides use their minimax strategies the expected value of the game (to the attacker) is $(AD - BC)/(A - B - C + D)$. In the mobilization game this will always fall between the value of a fake attack which draws a mobilization (B) and a fake attack which does not draw a mobilization (A).

R. Axelrod

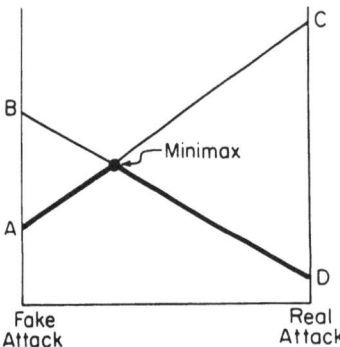

Fig. 2: The Mobilization Game. Heavy line shows the amount the attacker can guarantee himself.

The Use of a Warning System

The analysis so far has assumed that the defender has *no* information about what the attacker has decided to do. If the defender had *perfect* information the analysis would be very simple. When a fake attack is coming, the defender sees through it and does not react; when a real attack is coming, he detects it and reacts. In such a situation of perfect information, deception would be impossible. Alas, the world is not so benign.

The typical case is precisely the one in which deception is potentially a problem.[6] This is the situation in which one party, here the defender, has some *partially reliable* information about what the other party is about to do, and gets this information in time to undertake a suitable response, here in the form of reacting or not. Since the information is not perfectly reliable the attacker will sometimes be able to deceive the defender, but since the information is of some predictive value, the defender can try to make optimal use of what information he does have.[7] How can he do this?

Signal detection theory provides a helpful way to characterize the quality of partially reliable information [for example, *Lee*, pp. 193–244].[8] The information available to the defender can be thought of as providing a warning system which sometimes will issue false alarms and sometimes will issue accurate warnings.

To characterize such a warning system, we need two parameters. The first is the probability that a false alarm will be issued. This is the probability that a real attack is predicted given that a fake attack is actually undertaken. The other parameter is the

---

[6] Curiously enough, a recent article on the crisis warning system of the United States does not once mention the possibility of deliberate attempts at deception, although it does consider action-reaction problems and the possibility of miscalculation [*Belden*]. The author identifies himself as a member of the Intelligence Community Staff.

[7] This analysis will assume that the attacker does not have sufficient warning of his own to respond to the defender's response. A more realistic analysis would have to take this possibility into account.

[8] The language of signal and noise has become common in the discussion of surprise attacks ever since the landmark work on Pearl Harbor by *Wohlstetter* [1962]. This literature, however, does not develop the insight in any formal manner.

hit probability, that is to say the probability that when a real attack is being prepared it will be identified as a real attack.

For illustration, suppose that the false alarm probability of a given warning system is 40 percent and the hit probability is 80 percent. This gives the defender a third strategy in addition to his original two, i.e., of only reacting on warning. In this case, the attacker's guaranteed payoff takes the form of the heavy line in Figure 3. The best (in the sense of highest expected value) that the attacker can guarantee himself is the highest point on this heavy line. This point is indicated in Figure 3. Note that the presence of partially reliable information will always lower the payoff to the attacker, and thereby help the defender. This is true even though the attacker adjusts his strategy mix to take account of the warning system. In this illustration, for example, the attacker should adjust his strategy mix in such a way that he will more often use a fake attack and less often use a real attack compared to the situation in which the defender had no information about what the attacker had decided to do.

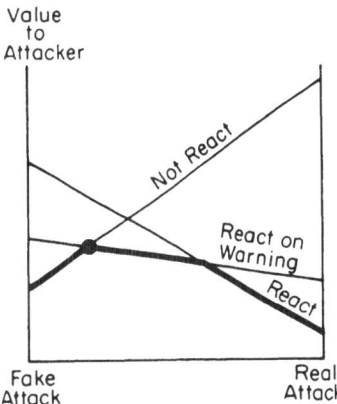

Fig. 3: The Mobilization Game With a Partially Reliable Warning System

Optimal Warning Systems

The preceding analysis was based on the idea of a fixed level of false alarm and hit probability. In actual practice for any given technology it would be possible for the defender to achieve a higher hit probability provided he was willing to accept a higher false alarm rate. In such a circumstance the warning system would be relatively "nervous." The defender, by accepting a higher false alarm rate would often be reacting unnecessarily, although he would be more confident of reacting when a real attack was in the offing. Alternatively, the defender could choose to have a relatively "phlegmatic" warning system with a low false alarm rate, but also a low hit rate.

These ideas can be illustrated with Figure 4. One can imagine that the defender has a wide variety of information available, some pointing to the probability of a real attack and some pointing to the probability of a fake attack. Let us assume for simplicity that this information can be aggregated into a single variable, to be called the discrimination variable. Thus a high value of the discriminable variable means that most of the information points toward a real attack, and a low value means that most of the information points toward a fake attack.

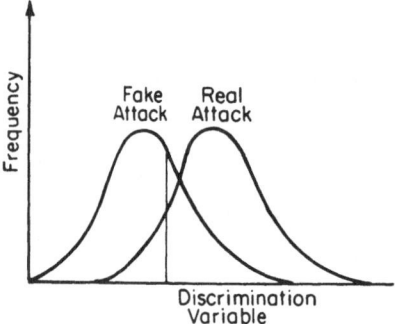

Fig. 4: Discrimination Using Critical Value of Discrimination Variable

The defender knows that if there is a real attack, the discrimination variable will often be high, pointing to a real attack, but it will sometimes be low because the information is not good enough to provide perfect discrimination between a real attack and a fake attack. Likewise if there is a false attack, the discrimination variable will probably be relatively low, but it will sometimes be high causing a false alarm. Thus the defender can represent his ability to discriminate between real and fake attacks by estimating two conditional probabilities distributions. The first is the conditional probability that a specific value of the discrimination variable will be observed if there is a fake attack. The second is the conditional probability that a specific value of the discrimination variable will be observed if there is a real attack. The better the discrimination ability of the defender, the less these two distributions will overlap (see Figure 4).

The defender can choose a critical value of the discrimination variable, and consider any higher value to indicate the need for reaction and any lower value to indicate no need for a reaction.[9])

Figure 4 shows the critical value which gives a false alarm rate of 40 percent and a hit rate of 80 percent. This is determined by the fact that 40 percent of the fake attack distribution is above the critical value, and 80 percent of the real attack distribution is above the critical value. If the critical value were lowered the system would be more nervous because the false attack rate as well as the hit rate would be higher. If the critical value were moved upwards, the system would be more phlegmatic because where would be a low false alarm rate although at the cost of a lower hit rate.

Thus the selection of a critical value for the observed variable gives a specific value of the hit rate and the false alarm rate. For any given technology this can be repres-

---

[9]) This can be shown to be the optimal if the two distributions are normal with equal variance. In the more general case, the defender can use likelihood ratios of the two conditional probability distributions to get an optimal decision rule for warning [Lee, pp. 201–203]. The likelihood method can be applied even when the discrimination variable is only measured on a nominal scale.

ented as a trade-off between the false alarm rate and the hit rate, as shown in Figure 5. The point indicating a 40 percent false alarm rate and an 80 percent hit rate is indicated.

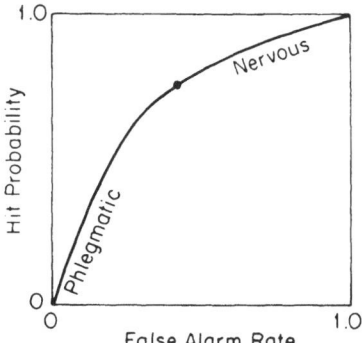

Fig. 5: Alternative Warning Systems for a Given Degree of Discrimination

The defender, therefore, has a range of options which can be plotted on the original game diagram shown in Figure 6. He can choose to have a relatively phlegmatic warning system which rarely issues false alarms but often misses real attacks, or he can choose to have a relatively nervous system which misses few real attacks but gives many false alarms. Obviously the best thing to do is to choose something in between, but how can the best choice be calculated?

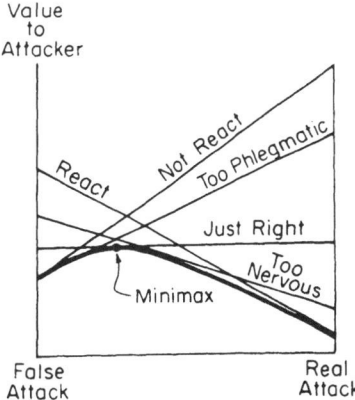

Fig. 6: The Mobilization Game Showing Alternative Warning Systems and Attacker's Guaranteed Payoff

This is where the analysis gets neat. To figure out what the defender should do, first consider the attacker's perspective. As before, the attacker wants to maximize the amount he can guarantee himself regardless of what choice the defender makes. This gives the heavy curved line shown in Figure 6. This takes into account the whole range of possible warning systems which the defender can use, from completely phlegmatic to overwhelmingly nervous. The best the attacker can guarantee himself is the point where this heavy line reaches its maximum, and this is where the warning system of the defender gives an exactly horizontal line.

Thus the best the defender can do is to select the warning system which corres- ponds to this horizontal line in Figure 6, since any other warning system can be abused by the attacker. If the warning system is more phlegmatic than this optimal one the attacker can always attack, but if the warning system is more nervous than the optimal one the attacker can always undertake a fake attack. But if the warning system yields the same expected payoff for a real attack as it does for a fake attack, the defender can not be outsmarted. Thus the optimal warning system is the one which has the same expected value whether the attack is real or fake. This warning system is imple- mented by selecting the critical value of the discrimination variable which corresponds to the horizontal line in Figure 6.[10])

This analysis has the interesting property that the defender no longer has to use a mixed strategy to decide whether to react or not. Instead he can use the pure strategy of selecting the optimal value and letting his decision about whether to react or not depend upon whether the value of the discrimination variable is greater or less than the threshold value. (This assumes that the probability of a tie between the observed value and the threshold value is negligible.) What is happening in effect is that the noise in the observation system, no matter how small, can provide the randomization necessary to prevent the attacker from outguessing the defender. This has the interest- ing implication that the defender does not have to add any more randomization of his own. On the other hand, the attacker must still use a mixed strategy in order to decide whether to make a given attack fake or real.

---

[10]) The defender's calculations are as follows. The condition that the warning system has the same expected value for a fake attack as for a real attack is $fB + (1-f)A = hD + (1-h)C$ where f is the false alarm rate and h is the hit rate. This gives a linear equation in h and f which can be plotted with the curve of the feasible trade-offs between h and f to give the ideal value of h among all those that are feasible, and the corresponding best feasible value of f. (See Figure 7.) Then the ideal critical value of the discrimination variable can be determined from Figure 4, picking the unique value which gives these values of h and f.

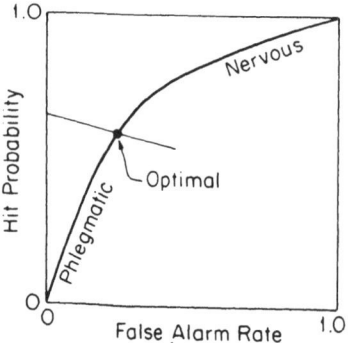

Fig. 7: Optimal Warning System for a Given Degree of Discrimination

Some Applications of the Model

The previous section showed how the defender can determine his optimal reaction strategy using warning and how the attacker can determine his optimal strategy of mixing real attacks and fake attacks. The same model can be used to answer a number of specific questions of interest about action-reaction situations.

1. *Intelligence budget*. Typically the defender has the choice of spending money to improve his intelligence capabilities. In this model improvement of intelligence capabilities is reflected in a greater ability of a defender to discriminate between the two contingent distributions shown in Figure 4, namely the *a priori* probability of a fake attack given a specific observed value of the discrimination variable, and the probability of a real attack given the same specific observed value. The extent to which these distributions are spread apart so that they do not overlap is a measure of a defender's ability to discriminate real from fake attacks. If the distributions are normal and have common standard deviation, then the ability of a defender to discriminate these two hypotheses, real and fake attack, is measured by the number of standard deviations which separate the means of the two distributions.

A defender can achieve a better degree of discrimination either by reducing the standard deviations of the distributions or making their means further apart. To achieve such discrimination requires added expenditures in the intelligence budget for such things as new observational capabilities. This suggests the question of how much the defender should be willing to pay to achieve a given increase in his ability to discriminate real and fake attacks.

This budgetary analysis is fairly straightforward. The first step is to determine for a given increase in discrimination the new optimal false alarm and hit rate probabilities. The method of doing this is shown in Figure 8. The straight line represents those combinations of false alarm rate and hit probability which make the defender indifferent to a real and fake attack. This line is determined only by the initial pay-off parameters of the game and therefore does not change with increased discrimination. The new optimal combination of false alarm rate and hit probability is the intersection of the original straight line with the new curve of false alarm and hit rate trade-off possibilities, given the new discrimination level. Once these optimal levels are determined the next step is to plot in Figure 9 the new optimal pay-off which corresponds to this particular false alarm and hit probabilities. Typically, with better discrimination this new line will be lower than before, indicating a better situation for the defender. The extent to which this horizontal line falls is a measure of the value, to the defender, of this increased ability to discriminate. The defender should be willing to pay this much, but no more, for this increase in his ability to discriminate between real and fake attacks.

2. *The value of cover and deception*. A similar analysis could be done for the attacker in terms of the amount he would be willing to spend in order to reduce the discrimination ability of the defender. The budgetary analysis would work in the opposite direction. If the attacker invests in such things as camouflage or deceptive maneuvers, the defender's ability to discriminate is lessened. The defender's optimal false

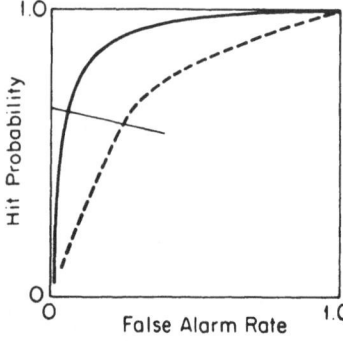

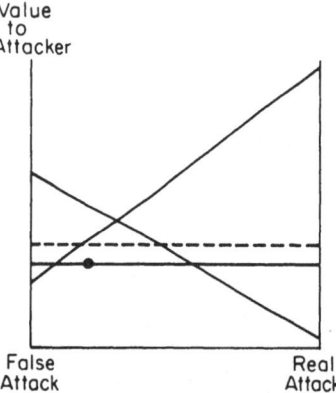

Fig. 8: Optimal Warning System with
         Higher Degree of Discrimination.
         Lower discrimination: Dotted line;
         Higher discrimination: Solid line.

Fig. 9: The Mobilization Game With Higher
         Degree of Discrimination. Lower
         discrimination: Dotted line; Higher
         discrimination: Solid line.

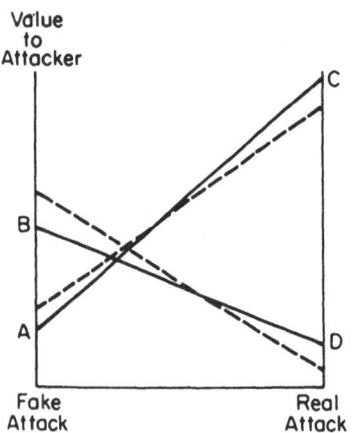

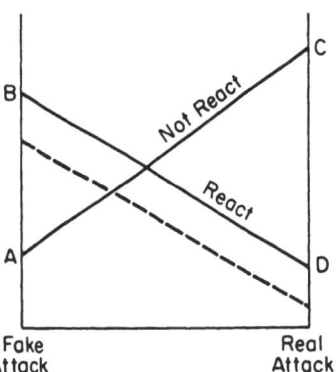

Fig. 10: The Mobilization Game With
         Conspicuous Preparations (Solid
         Lines)

Fig. 11: The Mobilization Game With
         High Reaction Costs (Solid Lines)

alarm and hit probability would then yield a horizontal warning line higher than if the
attacker had not spent such effort in reducing the discrimination ability of the defender. This budgetary analysis assumes that the attacker and defender have common
shared knowledge of the level of discrimination actually achievable by the defender.
This assumption is approximately correct because the ability of the defender to observe the attacker is often roughly known to the attacker. The reason is that the capabilities of such observational means as reconnaissance satellites are fairly well known to
both sides. Likewise the attacker's ability to reduce the defender's discrimination by
such things as camouflage are roughly known to both sides as well.

3. *Conspicuous preparations.* Another interesting case to analyze is when the attacker undertakes preparations which are perfectly visible to the defender. The attacker might do this for the sake of bluffing or for the sake of increasing his ability to launch a real attack. Of course if a real attack is not launched, the attacker winds up paying for the costs of these conspicuous preparations. Thus all four entries of the original pay-off matrix are changed with the attacker benefiting if he launches a real attack, whether or not the defender reacts, and the attacker losing compared to the previous situation if he does not launch a real attack, whether or not the defender reacts. This situation is shown in Figure 10. The net effect of these changes due to conspicuous preparations depends on their exact magnitudes, since the value of the game can go up or down, and the minimax probability of an attack can increase or decrease. However, if the exact amount of benefit to the attacker from these preparations and the exact costs of the preparations can be determined, then the net effect can also be calculated.

4. *Increased reaction costs.* An important application of this kind of model is to the determination of what happens when there is a change in the cost to the defender of reacting to potential attack. For example, in May, 1973, the Israelis undertook a partial mobilization in response to what turned out to be a fake attack by the Egyptians. The Israelis estimated that this mobilization cost about $ 11 million [*Herzog*, p. 29]. If the Israelis thereafter regarded the cost of mobilization as higher than they had previously estimated it, two of the original payoff parameters would have changed. In other words the cost of reacting, given a fake attack or given a real attack, would be higher than previously estimated. This change is illustrated in Figure 11. Notice that with higher reaction costs, the minimax probability of the attacker's using a real attack in the basic game is greater than it was with the lower reaction costs. In other words, as reaction costs of the defender go up, the attacker is more likely to launch a real attack.

Adding Realism to the Model

The preceding analysis used two important assumptions. The first is that the interaction between the two sides is a zero-sum game. The other assumption is that those players are rational. These two assumptions allowed the use of minimax calculations of the optimal strategy choice for each player and thereby allowed the determination of a good deal about the structure of the interaction and the consequences of various changes in it. In actual practice in international relations neither of these two assumptions is strictly valid, and therefore a more realistic analysis of strategic deception would involve going beyond the two simple assumptions of zero-sum interaction and complete rationality. In fact the use of these two assumptions moves us toward the paranoid side of the naive-to-paranoid dimension. After all, assuming that your opponent is in a zero-sum situation is equivalent to the assumption that he is out to get you, and assuming that he is rational means that he is out to get you, and he is able to do it in the most efficient possible way. This clearly is a somewhat paranoid point of view. Therefore a more sophisticated analysis would go beyond the zero-sum rational actor situation.

The problem, of course, is that when dropping the zero-sum and rational actor assumptions the available models are much less tidy. But we can still think conceptually about what is happening. What follows are several complementary approaches which can be developed to deal with some of these problems.

1. *Supervised learning.* Rather than follow a minimax strategy, it seems common that governments use something which may be called supervised learning. The defender watches a series of fake or real attacks over the years and tries to learn from experience to estimate what might happen next. The attacker knows that the defender is going to be learning from experience, and therefore he can supervise or shape the experience the defender gets in such a way as to maximize his own values. This can lead to the familiar "cry wolf" phenomenon. The attacker can launch one fake attack after another until the defender begins to believe that all attacks are fake. At that point the attacker can launch a real attack and perhaps catch the defender without having reacted to meet it. To launch such a series of fake attacks of course costs the attacker something each time since fake attacks are expensive to mount.

It is interesting to note that when two Soviet military specialists analyzed a game analogous to the basic one being studied here, the form of their analysis did not use the minimax assumptions but instead looked at supervised learning [*Druzhinin/Kontorov*, pp. 102–106].[11])

2. *Indicator theory.* It would be useful to look at how the defender can take specific pieces of information and aggregate them into an overall judgment. The preceding analysis using the idea of discrimination assumed that the defender had already aggregated all of his information into a single measurable variable. This process of aggregation, however, is one of the most important aspects of coping with deception.

The standard way to treat such data aggregation is with Bayesian analysis, but people are sometimes not very good at making the subjective probability judgments required for the use of Bayesian analysis. See for example the experiments reported in *Slovic/Fischhoff/Lichtenstein* [1977, p. 19f.] and *Brown/Kahr/Peterson* [1974, pp. 429–432]. It would certainly be useful to have a formal theory that did not rely so heavily on subjective probabilities which specified how a given state of knowledge can be applied to a given piece of information to determine how valuable that piece of information is. In addition it would be useful to have a method of aggregating various specific pieces of information into a single variable that could be used as a discrimination variable.

The use of indicators is aided by two special features of large organizations: they use standard operating procedures, and they can not achieve perfect results in the implementation of their policies. Standard operating procedures are a great help to the observer because they allow the observation of one thing to indicate the occurrence of other things. The only drawback in relying on standard operating procedures to make inferences is that they might be changed just when the observer is really relying on them. But it is probably an enduring feature of large organizations that their standard

---

[11]) Related Soviet work on man-machine interaction is reported by *Lefebvre* [1972, p. 189f.]. It is obvious from *Druzhinin/Kontorov*'s authoritative book that Soviet attitudes toward game theory have changed a good deal from the sceptical attitudes reported by *Robinson* [1970].

operating procedures are more stable than they ought to be for their own good. For example, in World War II the Germans were so methodical about their camouflage of submarine construction that the British photo interpreters were able to follow the progress of the construction by following the progress of the camouflage [*Babington-Smith*, p. 112f.]. Similarly, the Russian standard operating procedures for constructing missile sites helped the United States interpret photographs of the construction of sites in Cuba at an early stage [*Allison*, p. 107f.].

The other special feature of a large organization that helps in the interpretation of its behavior is that it can not implement its policies with perfection. For example, in 1956 the Israeli attack against Egypt started with what was supposed to look like a limited raid on the Sinai. Several Israeli units, however, immediately exceeded their carefully formulated instructions and displayed their wider intent [*Dayan*, pp. 60–63 and 91–93]. Likewise in 1973, the Egyptian deception plan called for leaving Cairo airport open before the attack to deceive the Israelis, but some senior official of the Egyptian civil airline became frightened of the possibility of war and ordered all civil flights cancelled [*Heikal*, p. 34 and *Handel*, p. 37]. Both of these deceptions succeeded despite flaws in their implementation. Evidently God takes care of little children and big deceivers.

3. *Cover story theory.* The analysis so far has dealt with two possibilities that the defender has to worry about, namely, a fake attack or a real attack. In a more realistic situation there are several different possible explanations for the attacker's behavior. In fact the attacker may shape his behavior in order to point to one or another specific explanation different from the actual explanation. It is up to the defender to decide whether the other party is even acting like an attacker.

A good cover story is one that is so plausible that it is hard to discriminate from the truth, such as preparations for real attack disguised as a maneuver. An even better cover story is one that is so unobtrusive that it arouses no suspicion at all. Better than either of these would be a cover story which is likely to be believed and which if believed would lead to activities by the defender opposite of what he would do if he expected the truth. For example, as Hitler was moving three million men into positions to attack the Soviet Union, Stalin was busy warning his generals not to shoot at reconnaissance overflights lest they provoke the Germans [*Bialer*, pp. 179–269, especially 201 and 203, note 28. See also *Whaley*, 1973].

We do not yet have good theories for how to use past experience in dealing with deception, for how to aggregate current information in the making of a judgment, or for the development of cover stories and how to see through them. But all these areas need to be developed if our theoretical understanding of deception and counter-deception is to proceed beyond the simplistic level of using zero-sum games and putatively rational actors.

Handles on Deception

As we have seen, there are many opportunities to deceive. An attacker can undertake a sequence of behavior which leads to a reduction of vigilance on the part of the defender; the attacker can also reduce the validity of specific indicators if he realizes that the defender may be using them to draw inferences; and finally, the attacker can invent cover stories which might make his behavior difficult to discriminate from some alternative hypothesis or perhaps even lead the defender to undertake inappropriate action if the cover story is accepted.

All these possibilities for deception exist. Fortunately, there are also some reasons why the target of deception has a good chance to cope successfully with deception.

1. In situations that have some cooperative aspect, deception is taken as a hostile act by the victim. For example, when the Soviet Union misrepresented its intentions to the United States about placing offensive nuclear missiles in Cuba in 1962, the United States government reacted very strongly. Part of the reason why President Kennedy regarded the placement of missiles in Cuba so seriously is that the Soviets tried to deceive him by saying in public and private that they were not sending offensive weapons to Cuba [*Allison*, pp. 40–42]. Thus the very use of deception carries with it a substantial risk that its eventual discovery will worsen relations with the target of the deception. For this reason, large-scale deceptions are probably used less often than they might otherwise be. In a sense, the predictable wrath of the victim provides some assurance that he will not be the target of big deceptions for small purposes. This isn't much, but it's something.

2. The basic goals and perspectives of foreign governments can usually be estimated with some degree of reliability. This means that the intensions of other governments, in broad outline at least, should be determinable in most cases. To the extent that one government can understand how another government views the world, they should be able to appreciate at least the broad outlines of what the other government intends to do. This in turn should reduce the potential for deception. Needless to say, the last four decades have seen many startling cases of governments failing to successfully understand the way other governments view the world. A striking case is how the United States failed to appreciate that Japan in 1941 regarded itself as being in such a desperate situation that highly risky actions were called for. And, on the other hand, Japan failed to understand that the United States, once provoked, would not simply acquiesce in Japanese victories in the western Pacific. Of course even if one government understands the basic perspectives of another, it may not understand its detailed intentions. Thus Stalin understood that Hitler might some day attack, but he did not expect the attack when it took place in June 1941. All in all, however, the greatest room for improvement in dealing with deception is probably in the realm of understanding the world as seen by other governments and especially their leaders.

3. Even if one does not understand the basic perspectives of other governments, it is still possible to prevent deception. After all, major conventional war is hard to undertake without significant preparations, and these preparations are difficult to hide. While the attacker does have the option of attacking with minimal advanced preparation, this would reduce the strength of the attack. On the other hand, significant preparation raises the chance that the preparations will be seen, understood, and reacted to. While nuclear war can be launched with virtually no warning, it can also be

countered with a second strike even if total deception were achieved. Thus deception plays less of a role in a nuclear attack than it does in a conventional attack. And in general, whenever preparations are required for an efficient attack, these preparations are difficult to hide and difficult to do without.

4. A final method of dealing with deception is to take advantage of the possibility that the target of deception may be able to see some aspects of the deceiver's behavior without the deceiver realizing that these aspect of his behavior is being observed. While such indicators can never be completely reliable because of the possibility of the deceiver's manipulating them, there may well be some indicators for which the target has a justifiable degree of confidence.

The moral of the story is that the better we understand deception, the better we can cope with it. Developing theories of deception can help prevent us from being either too naive or too paranoid.

References

Allison, G.: Essence of Decision. Boston 1971.

Axelrod, R.: The Rational Timing of Surprise. World Politics **31**, January 1979, 228–246.

Babington-Smith, C.: Air Spy, the Story of Photo Intelligence in World War II. New York 1957.

Belden, Th.G.: Indications, Warning, and Crisis Operations. International Studies Quarterly **21**, 1977, 181–198.

Bialer, S. (ed.): Stalin and His Generals. New York 1969.

Brams, St.: Deception in 2 x 2 Games. Journal of Peace Science **2**, 1977, 171–203.

Brown, A.C.: Bodyguard of Lies. New York 1975.

Brown, H.: Statement of Secretary of Defense Harold Brown to the Congress on the Amendments to the FY 1978 Budget and FY 1979 Authorization Request. February 22, 1977.

Brown, R.V., A.S. Kahr and *C. Peterson*: Decision Analysis for the Manager. New York 1974.

Dayan, M.: Diary of the Sinai Campaign. Jerusalem 1965. (English tr. 1966).

Druzhinin, V.V., and *D.S. Kontorov*: Concept, Algorithm, Decision. Moscow 1972 (tr. 1975).

Felix, Ch.: A Short Course in the Secret War. New York 1963.

Greenwood, T.: Reconnaissance, Surveillance and Arms Control. London, Adelphi Paper number 88, 1972.

Handel, M.I.: Perception, Deception and Surprise: The Case of the Yom Kippur War. Jerusalem, number 19. 1976.

Heikal, M.: The Road to Ramadan. New York 1975.

Herzog, Ch.: The War of Atonement. London 1975.

Jervis, R.: Logic of Images in International Relations. Princeton 1970.

Lee, W.: Decision Theory and Human Behavior. New York 1971.

Lefebvre, V.A.: A Formal Method of Investigating Reflection Processes. General Systems **17**, 1972, 181–187.

Marchetti, V., and *J.D. Marks*: The CIA and the Cult of Intelligence. New York 1974.

Robinson, Th.: Game Theory and Politics: Recent Soviet Views. Studies in Soviet Thought **10**, 1970, 291–315.

Singleton, R., and *W. Tyndall*: Games and Programs. San Francisco 1974.

Slovic, P., B. Fischhoff and *S. Lichtenstein*: Behavioral Decision Theory. Annual Review of Psychology **28**, 1977, 1–39.

Whaley, B.: Codeword BARBAROSSA. Cambridge, Mass., 1973.

– : Stratagem: Deception and Surprise in War. Cambridge, Mass., 1969.

Wohlstetter, R.: Pearl Harbor, Warning and Decision. Stanford, Ca., 1962.

Applied Game Theory. 1979 ©Physica-Verlag, Wuerzburg/Germany

An Optimal Conviction Policy for Offenses that May Have Been Committed by Accident

By *A. Rubinstein*, Jerusalem[1])

Abstract: When the tax authorities discover that a taxpayer has failed to report a certain part of his income, they cannot tell whether this is the result of deliberate tax evasion or, perhaps, the result of an innocent oversight. Penalties must be designed so as to apply whenever misreporting is discovered, but society may very well wish to distinguish between a deliberate offense and an offense that has been committed by accident and to be more lenient in the latter case. However, leniency will encourage people to commit the offense deliberately. In a one-shot game, equilibrium consists of society picking severe penalties and innocent offenders being hit hard. However, in the repeated game, there exists an equilibrium in which the optimal penalties imposed by society on people with a "reasonable" record are lenient and the optimal strategy for the individuals is to refrain from deliberate offenses.

1. Introduction

In almost all criminal proceedings, some doubt remains as to the guilt of the accused. Often the factual element (actus reus) is not completely conclusive and even if it is, the mental element necessary in order for an act to become a crime in law (mens rea) is also frequently questionable.

Examples are numerous; here is a short list:

1. The windscreen of a car in a parking lot carries a parking permit with yesterday's date. The owner claims it has been backdated by mistake.
2. A source of income which does not appear in the appropriate income tax return is discovered. The declarer swears that he had omitted it out of forgetfulness, despite his efforts to give a true account of his earnings.
3. One of the headlights of a car is unlit. Conceivably it had just failed, and the driver who had checked the headlights before driving away, had found it to be in working order.
4. A person is caught leaving a supermarket with unpaidfor merchandise in his possession. Two doubts may arise; perhaps the person was confused or possibly someone else might have placed the merchandise where it was found without the person's knowledge.

In these cases, a complete dispelling of the doubt is almost impossible and the legal system is faced with a dilemma; conviction may risk injustice whereas acquittal may open the door to widespread breach of the law.

---

[1]) *Ariel Rubinstein*, Department of Economics, The Hebrew University of Jerusalem, Israel.
My thanks to Professor M.E. Yaari for his valuable comments and encouragement.

As far as once-and-for-all situations are concerned (as for example in cases of crimes like murder) the law has no alternative but to consider which is more unpalatable: A possible miscarriage of justive or a weakening of the deterrent effect of the penalty. However, for offenses which an individual might commit periodically, a wide range of conviction and punishment policies, built around the offender's record, are possible.

In this paper, I shall consider a simple situation which has at its core the social dilemma described above. It will be described as a game in section 2. In section 3 I shall discuss the corresponding repeated game in order to examine the possible policies which the legal system may adopt. The main result (presented in section 4) is that in the repeated game, there exists a pair of strategies, one for the penal system and one for the individual, which are jointly optimal and which have the following structure: In any period, if the individual is discovered as having committed the offense, then he is penalized only if his long-run record is "unreasonably" bad; as for the individual, his optimal strategy is to refrain from committing the offense deliberately.

2. The Isolated Game

Consider the following two-person game whose players are society (player 1) and a typical member of society (player 2).

The individual has to choose between two kinds of behaviour:

B — Committing a given act which is advantageous to himself but harmful to society at large.

G — Refraining from this act.

Even if the individual chooses G, the act may still be committed, by a cause outside his control. Let us think of this unintentional committing of the act as the result of a move by a fictitious *chance player* who picks the move "+" ("commit") with a given probability, α, and the move "−" ("do not commit") with probability $1 - \alpha$, where $0 < \alpha < 1$. The chance player's set of strategies, that is the set, $\{+, -\}$ will be denoted S_c.

Society has complete information about whether an act has or has not been committed, but it has no way of telling whether the act was committed wilfully or accidentally. Suppose that there are but two possible sentences which the court can pass when the act has been committed, namely conviction with a fixed penalty, or acquittal.

Society's set of strategies — S_1 — contains two elements:

P — Convict and punish the individual if the act has been committed.

NP — Do not convict the individual even though the act has been committed.

We assume the rules of the legal system are a matter of public knowledge, i.e., player 1 announces his strategy at the beginning of the game. The individual's strategy is therefore of the form: "I will do X if society chooses P, and Y otherwise", where $X, Y \in \{B, G\}$. Let this strategy be denoted simply XY. Then, the individual's set of strategies — S_2 — contains four elements: GG, GB, BG and BB.

The following outcomes correspond to the various choices that the players and the chance player can make:

Outcome 1: (*NP, G,* −). Society is lenient and the act is not committed.

Outcome 2: (*P, G,* −). Society is strict and the act is not committed.

Outcome 3: (*NP, G,* +). The act is committed unintentionally and the offender is not punished.

Outcome 4: (*P, G,* +). The act is committed unintentionally but the offender is punished.

Outcome 5: (*NP, B*). The act is committed deliberately, but the individual is not punished.

Outcome 6: (*P, B*). The act is committed deliberately, and the individual is punished.

Let the set consisting of these six outcomes be denoted S, and let u_1 and u_2 be respectively the utility functions of society and of the individual, with $u_i: S \rightarrow \mathbf{R}$ $i = 1, 2$.

Assume that u_1 and u_2 are given by:

$$u_1 \ (NP, G, -) = u_1 \ (P, G, -) = 4$$

$$u_1 \ (NP, G, +) = 3$$

$$u_1 \ (P, B) = a \ \text{with} \ 1 < a < 3$$

$$u_1 \ (NP, B) = 2$$

$$u_1 \ (P, G, +) = 1$$

and

$$u_2 \ (NP, B) = 4$$

$$u_2 \ (NP, G, +) = 4$$

$$u_2 \ (NP, G, -) = u_2 \ (P, G, -) = 3$$

$$u_2 \ (P, B) = 2$$

$$u_2 \ (P, G, +) = 1 .$$

The specific numbers used in these definitions of u_1 and u_2 have been picked merely for ease of exposition. The numbers as said, have no significance, and the analysis depends only on the ordinal relationships among them. As a final piece of notation, let the symbol ⊕ be used for probability mixtures. More precisely, if u and v are utility numbers and if α satisfies $0 \leqslant \alpha \leqslant 1$, then $\alpha \cdot u \oplus (1 - \alpha) \cdot v$ will stand for the lottery that yields u with probability α and v with probability $1 - \alpha$.

Now it is possible to represent the game being discussed here both in extended and in normal forms:

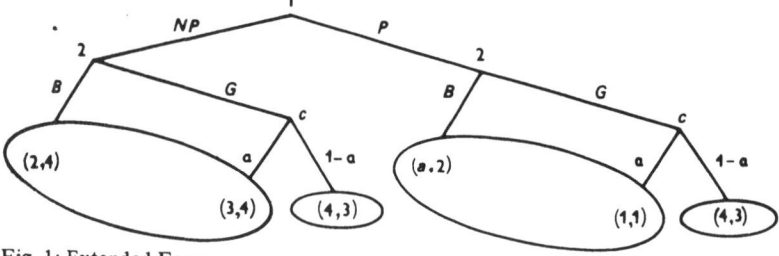

Fig. 1: Extended Form

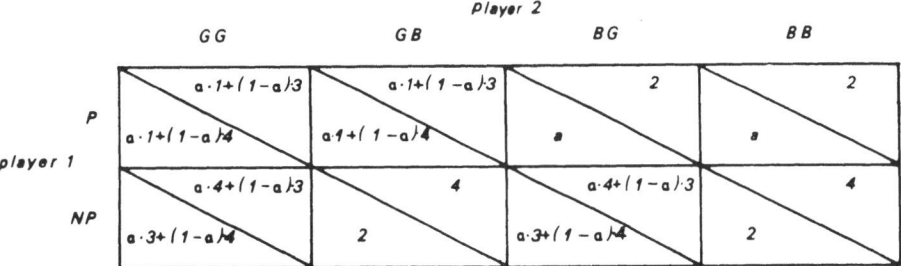

Fig. 2: Normal Form

If the individual prefers the utility outcome 2 over the utility lottery $\alpha \cdot 1 \oplus (1 - \alpha) \cdot 3$, then the strategy BB is the dominant strategy for the individual, and the pair (NP, BB) is an equilibrium point for $a < 2$ and the pair (P, BB) is an equilibrium point for $a > 2$. If the individual prefers the lottery $\alpha \cdot 1 \oplus (1 - \alpha) \cdot 3$ to the utility outcome 2, then punishment is a deterrent factor, and the strategy GB will be the individual's dominant strategy with the equilibrium point determined by society's preference between the lottery $\alpha \cdot 1 \oplus (1 - \alpha) \cdot 4$ and the utility outcome 2. If society is ready to take the risk of incurring injustice the act will be declared as a strict liability offense, and equilibrium will be (P, GB). If not, the act will be declared legal, and equilibrium point will be (NP, GB).

In all these cases the equilibrum points are also $\max_1 \max_2$ solutions. In other words, they are the outcomes of optimal legal strategies assuming the individual maximizes his utility subject to society's declared strategy. Notice that then the equilibrium points are (P, GB) or (P, BB) the pair (NP, GG) Pareto dominates the equilibrium points.

Society's most preferred pair is (NP, GG). However it is not an equilibrium point in the single game. The main message of this paper is that in the repeated game, society can "enforce" this pair of strategies.

3. The Repeated Game

Considering offenses that the individual may repeat many times, we assume that society and the individual both expect that after each play of the game there will be more plays of the same game. The suitable concept in game theory, for analysing this situation, is the repeated game [see for example, *Luce/Raiffa*], which consists of an infinite number of plays of the single game.

This structure seems to be unrealistic. But as *Aumann* [1959] says this notion is more suitable than a repeated game consisting of any fixed number of single games. "The fact that the players know when they have arrived at the last play becomes the decisive factor in the analysis overshadowing all other considerations . . . . *A.W. Tucker* has pointed out the condition that after each play the players expect that there will be more is mathematically equivalent to an infinite sequence of plays".

We assume that each game played at any stage of the repeated game, is identical with every other game, played at any other stage, irrespective of what had happend before.

The individual has complete information after each game. His information set is therefore

$$I_1 = \{\{(NP, B)\}, \{(NP, G, +,)\}, \{(P, G, -)\}, \{(NP, G, -)\}, \{(P, G, +)\}, \{(P, B)\}\}.$$

Since society cannot distinguish between deliberate and accidental acts, its information set at the end of every single game is given by

$$I_2 = \{\{(P, B), (P, G, +)\}, \{(NP, B), (NP, G, +)\}, \{(NP, G, -)\}, \{(P, G, -)\}\}.$$

Both society and the individual have perfect memories. For any natural number k let $I_i(k) = I_i$. Define $I_i^t = \underset{k=1}{\overset{t}{\times}} I_i(k)$.

In the repeated game, a strategy of a player is a sequence of the form $\{f^t\}_{t=1}^{\infty}$ where $f^1 \in S_1$ and $f^t: I_i^{t-1} \to S_i$. (If $\tau^1, \ldots, \tau^{t-1} \in I_i$ is the information possessed by i, then i will choose the strategy $f^t(\tau^1, \ldots, \tau^{t-1})$. F_i denotes the set of strategies open to i, and $F = F_1 \times F_2$.

Let $h_i^t(f, g)$ be the random variable of the utility of player i at time t, under the assumption that the players adopt strategies $f \in F_1$, $g \in F_2$. For a formal description of $h_i^t(f, g)$ see *Aumann* [1959].

As for the preferences of the players in the repeated game I assume that both society and the individual aim to maximize the limit of the long-run average utility.

If the limit $H_i(f, g) = \underset{T \to \infty}{\lim} \dfrac{1}{T} \overset{T}{\underset{t=1}{\Sigma}} h_i^t(f, g)$ exists a.s., the pair (f, g) will be called summable. Denote by $\widetilde{F}$ the set of summable pairs of strategies.

I shall write $(f, g) >_i (\bar{f}, \bar{g})$, if there exists $\epsilon > 0$ such that with positive probability there are an infinite number of T for which

$$\frac{1}{T} \overset{T}{\underset{t=1}{\Sigma}} h_i^t(f, g) > H_i(\bar{f}, \bar{g}) + \epsilon.$$

Definition: The pair $(f, g) \in \widetilde{F}$ will be called an *equilibrium point* of the repeated game if there is no $\bar{f} \in F_1$ or $\bar{g} \in F_2$ satisfying $(\bar{f}, g) >_1 (f, g)$ or $(f, \bar{g}) >_2 (f, g)$.

Let $(f, g) \in \widetilde{F}$. The strategy g is said to be $>_2$ maximal relative to f if there is no $\bar{g} \in F_2$ such that $(f, \bar{g}) >_2 (f, g)$.

Now we are ready for the main definiton:

Definition: The pair $(f, g) \in \widetilde{F}$ is a *max$_1$ max$_2$ solution* if g is $>_2$ maximal relative to f and there is no pair $(\bar{f}, \bar{g}) \in \widetilde{F}$ such that $\bar{g}$ is $>_2$ maximal relative to $\bar{f}$ and $(\bar{f}, \bar{g}) >_1 (f, g)$.

4. The Theorem

The main theorem asserts the optimality for society of the following legal policy: the individual will be convicted and punished at time $t + 1$ if and only if two conditions hold simultaneously: first the antisocial act is discovered to have been committed at time $t + 1$; and, second, the relative frequency of such acts having occured up to time t is greater than $\alpha + \alpha_t$ where $\{\alpha_t\}$ is a sequence of positive real number converging to 0 "sufficiently slowly". Formally, we have –

Theorem: Let $k > 1$ and $\alpha_t = \sqrt{2 k \alpha (1 - \alpha) \ln \ln t} / \sqrt{t}$.

Let

$\hat{f} = \{\hat{f}^t\}$ be the following social strategy:

$\hat{f}^1 = NP$

$$\hat{f}^{t+1}(\tau^1, \ldots, \tau^t) = \begin{cases} P \text{ if } & \left| \left\{ s \leq t \; \middle| \; \begin{array}{l} \tau^s = \{(P, B), (P, G, +)\} \text{ or} \\ \tau^s = \{(NP, B), (NP, G, +)\} \end{array} \right\} \right| \geq t\,(\alpha + \alpha_t) \\ NP \text{ otherwise} \end{cases}$$

Let $\hat{g} = \{\hat{g}^t\}$ be the strategy of the individual where he always adopts G. Then:

a) $H_1 (\hat{f}, \hat{g}) = 4 \cdot (1 - \alpha) + 3\alpha$

and

$\quad H_2 (\hat{f}, \hat{g}) = 3 \cdot (1 - \alpha) + 4\,\alpha.$

b) $(\hat{f}, \hat{g})$ is an equilibrium point of the repeated game.

c) $(\hat{f}, \hat{g})$ is a $\max_1 \max_2$ solution.

Proof:

a) Let D_t be the random variable which assumes the value 1 if the chance player causes the act to accur at time t, and assumes the value 0 otherwise.

From the law of the iterated logarithm [see, for example, *Lamperti*] we have that with probability 1 there exists T_0 such that for all $T \geq T_0$

$$\frac{1}{T} \sum_{t=1}^{T} D_t - \alpha < \alpha_T.$$

Therefore almost surely society will choose strategy P only a finite numbers of times and with probability 1

$$\frac{1}{T} \sum_{t=1}^{T} h_1^t (\hat{f}, \hat{g}) \to 4 \cdot (1 - \alpha) + 3\alpha$$

and

$$\frac{1}{T} \sum_{t=1}^{T} h_2^t (\hat{f}, \hat{g}) \to 3 \cdot (1 - \alpha) + 4\,\alpha.$$

b)–c) The ideal combination of strategies from society's point of view is the pair (NP, GG) to be repeated forever. But even this pair will only yield a utility for society of $4 \cdot (1 - \alpha) + 3\,\alpha$. Therefore in order to prove that $(\hat{f}, \hat{g})$ is an equilibrium point in the repeated game and also max$_1$ max$_2$ solution, it suffices to show that there does not exist a $g \in F_2$ with $(\hat{f}, g) >_2 (\hat{f}, \hat{g})$. Let $g \in F_2$ and let $\epsilon > 0$. We now show the emptiness of the event "for infinitely many T,

$$\frac{1}{T} \sum_{t=1}^{T} h_2^t\, (\hat{f}, g) > H_2\, (\hat{f}, \hat{g}) + \epsilon".$$

Let $\{w_t\}$ be a sequence of the individual's utilities obtained from $(\hat{f}, g)$. Denote $\bar{w}_T = \frac{1}{T} \sum_{t=1}^{T} w_t$. Let N be the minimal natural number for which $\alpha_N + 4/N < \epsilon$.

Our claim is that only for a finite number of times T, $\bar{w}_T > 4\,\alpha + 3\,(1 - \alpha) + \epsilon$. The claim is applied easily from the following three assertions:

For any $t > N$:

1. if $\bar{w}_t > 4\,(\alpha + \alpha_N) + 3\,(1 - \alpha - \alpha_N)$ then $\bar{w}_{t+1} < \bar{w}_t - \alpha/(t + 1)$.
2. if $\bar{w}_t \leqslant 4\,(\alpha + \alpha_N) + 3\,(1 - \alpha - \alpha_N)$ then $\bar{w}_{t+1} < 4\,\alpha + 3\,(1 - \alpha) + \epsilon$.
3. there is $T > N$ such that $\bar{w}_T \leqslant 4\,(\alpha + \alpha_N) + 3\,(1 - \alpha - \alpha_N)$.

Proof of assertions 1:
For any $t > N$ if $\bar{w}_t > 4\,(\alpha + \alpha_N) + 3\,(1 - \alpha - \alpha_N)$ then also
$$\bar{w}_t > 4\,(\alpha + \alpha_t) + 3\,(1 - \alpha - \alpha_t).$$

The relative frequency of the forbidden acts is greater than $\alpha + \alpha_t$ and therefore society's strategy at time $t + 1$ is P.
Now $w_{t+1} \leqslant 3$ and

$$\bar{w}_{t+1} \leqslant \frac{t \cdot \bar{w}_t + 3}{t + 1} \leqslant \bar{w}_t + \frac{3 - \bar{w}_t}{t + 1} \leqslant \bar{w}_t - \frac{\alpha + \alpha_t}{t + 1} < \bar{w}_t - \frac{\alpha}{t + 1}$$

Proof of assertion 2:
If $\bar{w}_t \leqslant 4\,(\alpha + \alpha_N) + 3\,(1 - \alpha - \alpha_N)$ then

$$\bar{w}_{t+1} \leqslant \frac{[4\,(\alpha + \alpha_N) + 3\,(1 - \alpha - \alpha_N)]\,t + 4}{t + 1} = [4\,\alpha + 3\,(1 - \alpha)]\frac{t}{t + 1} +$$
$$+ \frac{t \cdot \alpha_N + 4}{t + 1} < 4\,\alpha + 3\,(1 - \alpha) + \alpha_N + \frac{4}{N} < 4\,\alpha + 3\,(1 - \alpha) + \epsilon.$$

Proof of assertion 3: The harmonie series diverges and therefore assertion 1) implies 3.

Remark: If we replace the sequence $\{\alpha_t\}$ with the sequence $\alpha_t \equiv \epsilon > 0$, the previously optimal strategy is no longer optimal since the individual may deviate with relative frequency of say $\epsilon/2$, which would result almost surely in being punished only a finite number of times.

If society convicts the accused whenever the frequency of the harmful act exceeds α (i.e. if we let $\alpha_t \equiv 0$) then the individual will be punished infinitely many times (almost surely) even if he consistently adopts the stragey G. The same result is true if we choose $k < 1$ in the definition of $\{\alpha_t\}$ [see *Lamperti*]. However any sequence that tends to 0 more slowly than our sequence is suitable for a definition of an optimal strategy.

Remark: Some of the assumptions made above may be dropped or relaxed. For example, we can remove our assumption that committal of a forbidden act is always found out by society if we introduce a chance player representing the chance of discovery. Society's information set is represented here by loops.

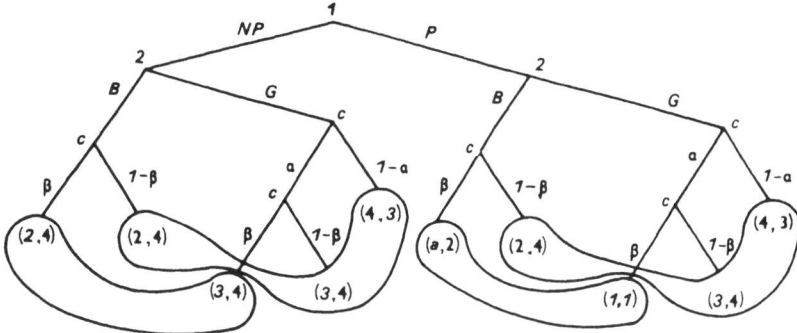

Fig. 3

In a way similar to the main theorem we can prove that there is a positive sequence $\beta_n \to 0$ and an optimal policy which punishes the individual if the frequency of his offences in the past is greater than $\alpha \cdot \beta + \beta_n$, where β is the probability of discovery.

References

Aumann, R.J.: Acceptable Points in General Cooperative n-person Game. Ed. by A.W. Tucker and R.C. Luce, Contributions to the Theory of Games IV. Annals of Math. Studies, **40**, Princeton, N.J., 1959, 287–324.

Lamperti, J.: Probability. New York 1966.

Luce, R.P., and H. Raiffa: Games and Decisions. New York 1957.

Applied Game Theory. 1979 ©Physica-Verlag, Wuerzburg/Germany

Game Theory Models of Intermale Combat
in Fiddler Crabs (Genus *Uca*)

By *G. W. Hyatt*, Chicago[1]), *St.D. Smith* and *T.E.S. Raghavan*, Chicago[2])

Abstract: Male fiddler crabs fight over real estate. The combatants play two roles, Resident or Wanderer, depending upon which one possesses a burrow. The burrow is postulated to be a resource which endows the owner with reproductive fitness. Conflicts observed in the wild consist of alternating acts by Resident and Wanderer, where each animal chooses from a repertoire of 12 ritualized motions. Analysis of over 400 conflicts shows that combat strategies vary intraspecificially according to status and relative animal size. We suggest that a conflict can be regarded as a two-person, alternating, multi-stage game. We hypothesize that the strategy types contributing to successful burrow defense or takeover are optimal strategies, and that they probably maximize individual reproductive fitness.

1. Introduction

Recent work relating game theory to animal aggressive behavior suggests that this tool may be useful for analyzing the structure and evolution of intraspecific combat [*Maynard Smith*, 1974, 1976; *Maynard Smith/Price; Parker; Maynard Smith/Parker*]. Field observations of fiddler crab (Genus *Uca*) conflicts show that these animals may be useful subjects for development of detailed aggression models, which may provide insight into the biological evolution of the fine structure of aggressive behavior, and which may be adaptable for use with other taxonomic groups [*Hyatt/Salmon*, 1978].

The fiddler crabs are semi-terrestrial and highly social crustaceans inhabiting beaches, mud flats, mangroves and salt marshes in both the tropical and temperate zones of both hemispheres. Male crabs focus their activities around a burrow which they protect by active defense from intrusions by conspecifics. Burrow-owning crabs are termed Residents; males without burrows are called Wanderers. The burrow serves as a shelter, especially when predators are in the vicinity, when climatic conditions are extreme, and during high tide when the animals remain underground. The burrow is also a nuptial chamber where copulation occurs (in most species).

---

[1]) Prof. *Gary W. Hyatt*, Dept. of Biological Sciences, University of Illinois at Chicago Circle, Box 4348, Chicago, Ill. 60680, USA.

[2]) *Stephen D. Smith* and *T.E.S. Raghavan*, Dept. of Mathematics, University of Illinois at Chicago Circle, Box 4348, Chicago, Ill. 60680, USA.

Territorial defense by males is particularly dramatic. Males possess an enlarged claw which may be as much as a third of their body weight. The claw is used to signal to females ("waving" or "beckoning" display), and to threaten or to actually fight with other males over burrow possession [*Hyatt/Salmon*, 1978; In Press]. Male-male combat is prevalent during the breeding season, and probably serves both to space out males and to determine which males will possess choice territories from which to court females.

In this paper, we present models in which an intermale conflict is formalized as a two-person, zero-sum alternating game. Analysis of empirical data suggests a specific set of rules for the game. Under such rules, we determine optimal strategies for the players. The underlying assumption is that if natural selection pressures cause animal behavior to converge toward optimal strategies, then game theory models should realistically represent the nature of their combat. The merit of a given model is judged by how well the independently calculated "best" strategies show up in the field data.

In preliminary analyses, we have constructed several models which are variations or generalizations of models already in the biological literature [*Maynard Smith*, 1974; *Maynard Smith/Parker*]. These models are simplistic in our context, since they basically involve a one-stage conflict: each player chooses a strategy, and the outcome is immediately determined. Such models are probably more useful in explaining the qualitative aspects of conflict patterns. In the real conflicts observed between unrestrained fiddler crabs, there is also a fine structure of act-to-act sequences; more complicated multi-stage models are required to address these finer details.

This work is part of an effort to understand the relationship of aggression to the behavioral biology and mating system of fiddler crabs. Our goal is to derive quantitative models to predict the direction of male-male aggressive encounters with regard to immediate winning and losing and with regard to the employment of combat strategies between conspecific males. Questions which come immediately to mind are: 1. What are the physical and/or behavioral characteristics of combatants which are most strongly predictive of victory or defeat? 2. Is there an optimal way to meet an opponent in combat? 3. Are there conflict variables which are convertible to a quantitative index of each combatant's individual reproductive fitness? The first question has been answered elsewhere [*Hyatt/Salmon*, 1978]. The last two are posed here.

2. Data Base, Combat Ethogram, and Parameter Definitions

2.1 Database

As described by *Hyatt/Salmon* [1978], the data base for this study consists of 426 agressive interactions between male fiddler crabs, *Uca pugilator*. The fights were observed and recorded in the field from June – August, 1975 near the Duke University Marine Laboratory, Beaufort, North Carolina. The animals were unrestrained and in their natural habitat. These conflicts now exist as computer datafiles, recorded in sequential format such that a conflict record appears as AP1, EX2, MP2, MP1, DS1, MP2, DS1, HR2, HR1, FI1, WD2. The letters stand for the combat acts (see below), and the numbers distinguish Resident (#1) from Wanderer (#2).

2.2 Combat Ethogram

The aggressive activities of *Uca* have long been known, and the names of several species (e.g., *pugilator, pugnax, bellator*) attest to the impression their combat must have made on early observers and taxonomists. Observations made by *Crane* [1966, 1975] on *U. rapax* led to the identification of 21 combat acts, 15 of which were classified as "ritualized" in the evolutionary sense of having taken on a signal function. Field experience and, indeed, qualitative and quantitative analysis, suggested that not all this detail need be incorporated into recorded descriptions of the fights. After a series of pilot observations, the following combat repertoire was chosen for recording. Many of these acts have been described and illustrated by *Crane* [1967; 1975, p. 582].

APPROACH (AP) — One animal approaches the other.

EXTEND (EX) — A sweeping, 90° positioning of the major cheliped, usually performed as initial "warding off" or terminal "move along" gesture by Residents.

MANUS PUSH (MP) — A low-intensity component of combat where contact is confined to the outer surfaces of the mani. The motion is usually perpendicular to the dorsal manus surfaces.

DACTYL SLIDE (DS) — A low-intensity component of combat where one crab rubs his dactyl teeth along the edge of the opponent's dactyl.

HEEL AND RIDGE (HR) — A high-intensity component of combat in which the actor places his dactyl outside the manus of the opponent, while the pollex passes to the inner or palm side. Oscillations of the interlaced claws result in rubbing of palmar and heel tubercles by the dactyl and pollex.

TAP (TP) — Usually occurs during the Heel and Ridge. The tips of the dactyl and/or pollex tap against the opponent's dactyl and/or pollex or manus.

FORCEFUL INTERLACE (FI) — Comprised of a grip, fling and/or upset of the opponent. The crab's chela is gripped and, through a sharp motion, the animal can be turned upside down, levered into the air, or tossed several centimeters away.

NULL MOVE (NM) — Artificial responding "act" inserted into the combat sequence when no overt behavior was observed for the recipient crab.

DOWNPUSH (DP) — Performed by the Wanderer after the Resident retreats into his burrow. Downpush shows three forms: 1. the Wanderer dashes 1—2 cm into the mouth of the burrow and may make contact with the chela or legs of the Resident, 2. the Wanderer enters the burrow a short distance and begins excavating with his walking legs, 3. the Wanderer remains on the surface and reaches into the burrow with his major chela.

TEMPORARY   — The distance between the combatants is increased.
WITHDRAW (TW)

BURROW   — The Resident retreats into his burrow.
WITHDRAW (BW)

TERMINAL   — The last move of a conflict. The Resident or Wanderer
WITHDRAW (TW)   leaves the area.

Crane [1975] categorizes these aggressive acts into "low-intensity" (ritualized) and "high-intensity" (forceful) components of combat. Although we cannot make statements regarding the evolutionary history of the acts, we can support the contention that a subjective intensity dichotomy exists, and does, indeed, show up in our models as a "decision level" in escalated encounters.

As originally transcribed, the sequential records (see above) allow an act to be followed by an act by the same combatant. The data set was modified by the addition of a "Null Move" when no overt response was detectable by the opponent. Thus, the sequences now follow a strict alternation of acts, and they assume a format in keeping with the two-individual game pattern.

2.3 Parameter Definitions

The alternating convention gives us a measure of time elapsing during a conflict, though it is only roughly related to real time. The actual duration of each conflict was measured at the time of data acquisition, but we do not consider it here. Since conflicts are initiated by the Wanderer, it is convenient to divide the sequences into dyads of the form (act by Wanderer:response by Resident). We use a time variable, t, to indicate the dyad number, measured from $t = 1$ at the first dyad. The longest observed conflict ends at $t = 89$, but the majority of conflicts are complete by $t = 15$. A basic property we expect of our model is to explain the distribution of conflict lengths, as measured by t. This is less ambitious than accounting for the fine structure of act sequencing, but has, nevertheless, been recently addressed in the biological literature [*Bishop/Cannings*].

Another important variable is relative size of animals. After each conflict, the animals were captured and the carapace width (CW) of each was measured to the nearest 0.5 mm. Conflict patterns depend consistently (though not necessarily in a simple way) on the size advantage, if any, enjoyed by one of the animals. We introduce a size index $d = $ (CW Resident $-$ CW Wanderer), which divides the conflicts into bins for 21 values of d. The observed conflicts fall into the range $d = -5.0$ mm to $d = +5.0$ mm. In Fig. 1, notice that conflicts are rare in the extreme size bins. This could indicate that burrow possession can be settled without conflict in the case of a large size mismatch, or that, for whatever motivational reason, fighting does not occur between crabs of these size classes. Also, Wanderers initiate conflicts when they have a slight size advantage, which suggests that Resident burrow possession is an advantage to be offset. A number of bins contain a reasonably rich collection of conflicts. We try to model the conflicts in each bin as a separate game, where all games have essentially the same rules, but, as will be seen, the values of certain parameters are determined by d. Thus, there should be some demonstration of size-related behavior as a continuous variation in terms of d.

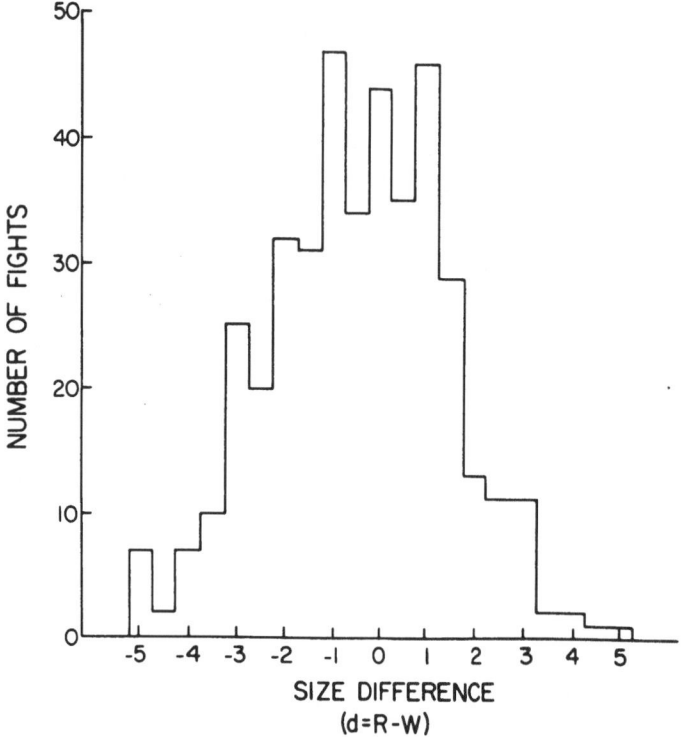

Fig. 1: Distribution of *Uca pugilator* conflicts according to size relationship of combatants. $d = $ (CW Resident − CW Wanderer). $N = 426$ fights.

Other aspects of the conflicts suggest that format modifications might be convenient. For instance, conflicts invariably begin with the dyad AP2, EX1 (Approach by Wanderer, Extend by Resident). It is convenient to regard this dyad not as a part of the conflict, but just as a marker indicating that a conflict has begun. However, the same sequence can reappear when a conflict is briefly interrupted and then resumed. Likewise, a more complicated phenomenon suggests the further subdivision of the withdrawal acts. In about 10 % of the conflicts, the following pattern is eventually observed: Resident withdraws down into the burrow, and the Wanderer performs Downpush near the opening of the burrow, usually to the response Null Move by the Resident. This continues until one animal terminates the conflict. The patterns of activity that are observed before and after such a turning point are so different that it seems unreasonable to treat them as the same game. The pattern afterwards is therefore regarded as a distinctive subgame which is initiated by the Resident, and which we analyze separately. Therefore we adopt the convention that a Burrow Withdraw by Resident, which is followed by a Downpush by Wanderer will be classified as a marker indicating the transition to the "late game". Sequences before any occurrence of Burrow

Withdraw are referred to as the "early game". Most early games never move into a late game phase [*Hyatt/Salmon*, 1978, Fig. 13]. The late game exhibits a much simpler structure, with a smaller effective repertoire of acts. It is probably more suitable for description by a simple model like the "war of attrition" of *Maynard Smith* [1974] and *Bishop/Cannings* [1978]. We are more interested in the early game, though some of our analysis can be applied to the late game as well.

3. Results

3.1 A Model with Simple Mixed Strategies

3.1.1 Motivation – *Maynard Smith* [1974] analyzes a simple conflict model in which a payoff, V, to the victor and zero to the loser is reduced by a factor, T, measuring the length of the conflict. Here the units are normalized by assuming unit loss in unit elapsed time. A pure strategy, m, is the choice to withdraw from the contest if it reaches time m. *Maynard Smith* shows that an equilibrium strategy is a mixed strategy. We consider two versions of the *Maynard Smith* model designed to investigate the possible presence of mixed strategies of this sort in *Uca* conflicts.

Under the assumptions of the model, we should be able to estimate which strategies are being used by the animals by considering the distribution of lengths of conflicts ended by Residents and by Wanderers. Since variation is observed to be related to size difference, d, we separate the empirical data into separate bins for fixed values of d. This distribution is not unequivocally unimodal. For instance, the distribution of terminal withdraw by Wanderer at $d = -1.0$ shown in Fig. 2 can easily be interpreted as having three or four peaks. Thus, we ask whether this distribution (in the early game) represents a mixture of a small number of pure strategies.

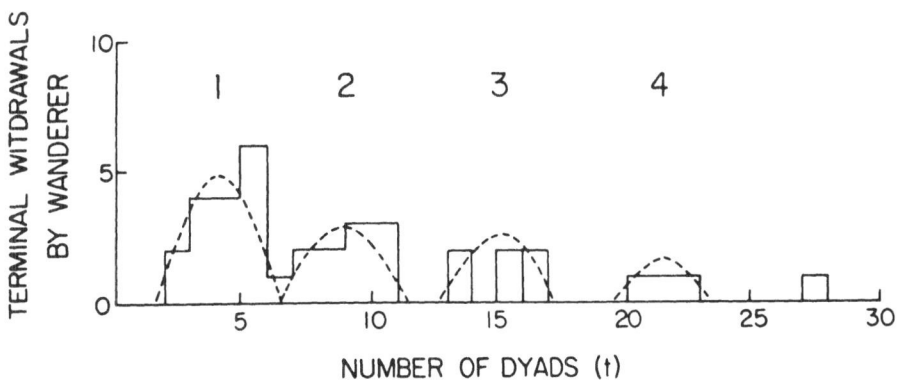

Fig. 2: Distribution of Terminal Withdraw by Wanderer for size difference $d = -1.0$ mm. Curves are fitted by eye, and are included to illustrate the division of the Wanderer persistence function into at least 4 peaks. Compare to Fig. 3. See text for further details.

There are several drawbacks to the earlier models. These models are symmetric, whereas empirically there is an initial asymmetrical advantage to the Resident. Indeed, it is hard to argue for the existence of simple patterns in the distribution of quit times for the Resident. Furthermore, the possible multimodal structure of the Wanderer distributions is absent from the aggregate data for all size differences, as can be seen in Fig. 3. This appears to be in agreement with an "Evolutionarily Stable Strategy" (ESS) of the type determined by *Maynard Smith* [1974]. To take account of this, we develop an alternative model for more complicated mixed strategies in section 3.2. We also mention at the end of that section a third possibility, which may reconcile the apparently contradictory patterns in the individual and aggregate distributions. Finally, we remark that this model is a single-stage game, and may not relect adequately the complexity of multi-stage fiddler crab conflicts.

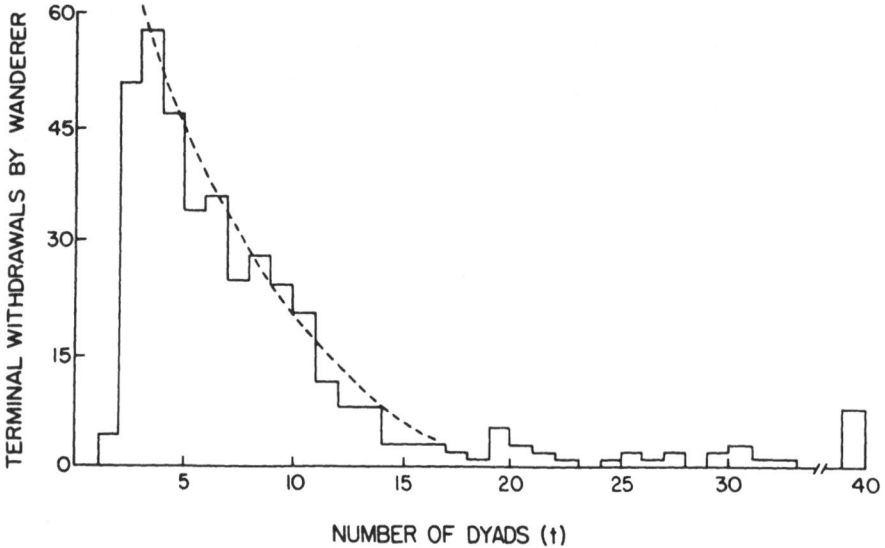

Fig. 3: Distribution of Terminal Withdraw times by Wanderer over all values of d. The curve represents the best fit exponential equation $y = 58\,(.85)^{t-4}$. Compare with Fig. 2.

3.2.2 The Model — In this model we seek only to determine what simple mixtures of strategies might be present in a distribution like that of Fig. 2. We do not attempt to estimate payoff, nor do we predict the best strategy. In particular, we are dealing only with lengths of conflicts, and not with actual act sequences within conflicts.

In a distribution like that in Fig. 2, it seems reasonable to replace a pure strategy of "quit at time m" by a normal distribution with mean m and suitable variance σ. A single such distribution is then represented by the graph of $p\,(t)$ the probability density for quit at time t, given by:

$$p\,(t) = \frac{1}{\sigma\sqrt{2\pi}}\ e^{-(t-m)^2/\sigma}$$

The given distribution $q\,(t)$ of quit times for an animal can then be considerd to be a sum of a small number, n, of such normal distributions. We may express this as follows:

Normalize so that:

$$\int_0^\infty q(t)\, dt = 1$$

Then:

$$q(t) = c_1 p_1(t) + c_2 p_2(t) + \ldots + c_n p_n(t)$$

with non-negative coefficients c_i such that $c_1 + c_2 + \ldots + c_n = 1$. Here, each $p_i(t)$ is a normal distribution with mean m_i and variance σ_i.

Given a distribution and these assumptions, an algorithm provided by *Sclove* [1977] allows us to optimize the parameters c_i, m_i and σ_i. As input we give, 1. a distribution, 2. the number n of strategies assumed to be present, and 3. a first approximation to the means $m_1, \ldots, m_n$. As output we receive the optimized values c_i, m_i and σ_i. Since our time scale, t, is discrete with an effective maximum of 40 units, it is not realistic to perform this optimization to a very small tolerance.

In the output, for values of n which result in a good fit to the initial distributions, the parameters show an approximately linear dependence on size difference, d. For instance, Fig. 4 displays the location of the second mean, m_2, for Wanderer when $n = 4$ strategies, for all values of d. This may be interpreted as saying that the Wanderer's second strategy results in terminating this conflict with $6 \leqslant t \leqslant 11$; but his persistence is lower when his size advantage is smaller.

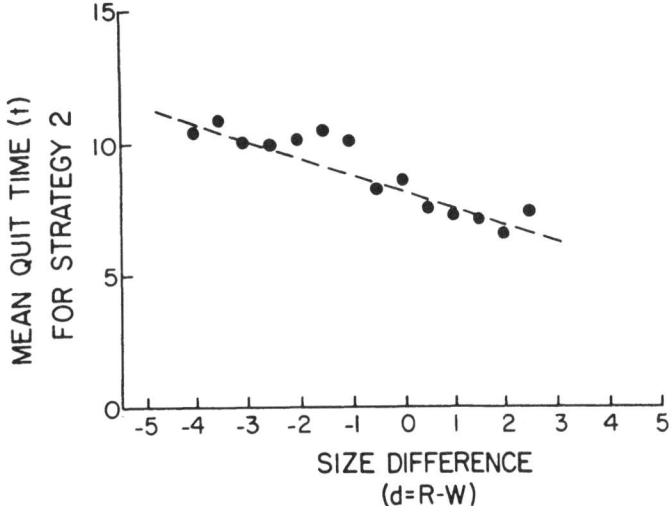

Fig. 4: Distribution over d of means of second Wanderer strategy assuming $n = 4$ strategies. Line is the best fit to linear dependence on d ($y = -.47d + 8.2$).

For this reason we also have an option to take the above parameters (c_i, m_i, σ_i), regard them as functions of d, and fit them to a straight line (with suitable renormalization to guarantee $\sum_{i=1}^n c_i(d) = 1$ for any d). This allows us to determine a function

$r(d, t)$ which gives the optimized instantaneous quit probability at time t of an animal with size advantage d, under the assumption of a mixture of n strategies. This pre-dicted distribution can then be compared with the actual distribution.

3.3.3 Testing the Model

3.3.3 Testing the Model — We simulate conflicts by assigning to each animal a func-tion $r(d, t)$ as above for suitable (not necessarily equal) values of n. A list of pseudo-random numbers is generated. At time t in a simulated conflict, for an animal with size advantage d, the next random number is compared with the value of $r(d, t)$ to deter-mine whether the animal continues or terminates the conflict. In this way, the desired number of conflicts in each size category may be generated, and a distribution of quit times simulated to compare with the original distribution. Thus, we are comparing ar-rays of size 21 (size difference bins) by 40 (effective conflict lengths). Significance is tested by chi-square using $r(d, t)$ for the predicted (expected) values.

As noted earlier, distributions for Resident do not fit this model well. Instead, we simulate conflicts using only Wanderer, independent of any activity by the Resident. This generates a distribution of quit times to compare with the actual data for those conflicts which were ended by the Wanderer. Here the fit is better. Simulations were made trying values of n between 1 and 10, and using a further range of first approxi-mations to $m_1, \ldots, m_n$. The best resulting fits occured for $n = 3$ and $n = 4$, with smallest standard deviation being 29.1. Lowest chi-square was 538.2, becoming worse as n moved away from 3 or 4. For the better values of n, The parameters m_i and σ_i varied roughly linearly with d. The variation of c_i, while not turbulent, may not be so simply described.

The assumption of a mixture of a small number of pure strategies (3 or 4) fits the data for Wanderers up to a certain limit of accuracy. Although not a conclusive de-monstration, the model allows an explanation of Wanderer behavior as a mixture of several more and more persistent strategies, where the actual length of conflict persist-ence depends on size advantage.

3.2 A Model with Optimal Strategy Mixtures

3.2.1 Motivation

3.2.1 Motivation — We now turn to the theoretical consequences of the rules of the game discussed above, with no particular reference yet to possible mixed strategies in the data. *Maynard Smith* [1974] determines that the best strategy is a mixed strategy assuming continuous variation of the time parameter, x. Here the probability density for withdrawing from the conflict at time x is given by a function $p(x)$. The optimal strategy is provided by:

$$p(x) = \frac{1}{V} e^{-x/V} \quad [\textit{Maynard Smith}, 1974; \text{eq. (4)}].$$

Thus, the distribution of conflict lengths for an animal playing a mixed strategy (con-sidering only conflicts which he terminates) will exhibit exponential decay. For a de-tailed proof see *Hines* [1977].

Clearly the region $3 \leqslant t \leqslant 14$ of Fig. 3 (derived from field data) suggests exponential decay. Consequently we will ask how well the data fit the assumption of some payoff value, V, assuming the animals adopt the corresponding optimal strategy.

A useful alternative to the probability density $p(x)$ is the "instantaneous probability of quit at x" represented in an actual distribution by:

$$r(x) = \frac{\text{number of conflicts terminated exactly at } x}{\text{number of conflicts terminated at } x, \text{ or any time afterwards}}$$

In terms of density $p(x)$, this may be given by:

$$r(x) = \frac{p(x)}{\int_x^\infty p(t)\, dt} .$$

In the case of the optimal strategy above, it is easily calculated that:

$$r(x) = \frac{1}{V} .$$

In this optimal strategy, the instantaneous probability that an animal will quit at time t is independent of t. It follows that the mean conflict duration is V. In our model we introduce variations of *Maynard Smith*'s model to study this situation.

This model involves the drawbacks of symmetry and single-stage structure mentioned for the model of section 3.1. It has the advantage of a definite payoff value, with predicted optimal strategy, but it still fails to account suitably for variation according to size relationship.

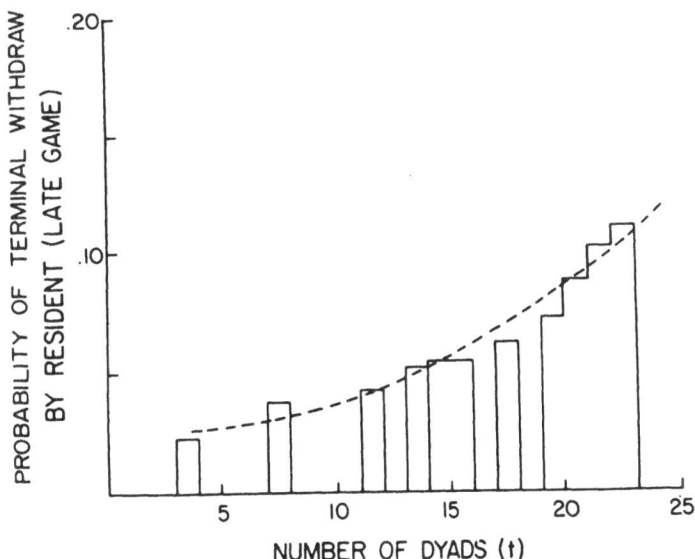

Fig. 5: Instantaneous probability of Terminal Withdraw by Resident in the late game ($t = 0$ at occurrence of Burrow Withdraw). See text for computation details. The curve represents the best fit exponential equation $y = .03(2)^{(t-8)/8}$.

3.2.2 The Model — For this model, we concentrate on the instantaneous probability $r(x)$ described above. The case "$r(x) = $ constant" is that of the optimal strategy in *Maynard Smith* [1974], and the reciprocal of this constant represents the value of the burrow to the winner. A value for V is easily obtained from the distribution in Fig. 3. If we concentrate on the region $3 \leqslant t \leqslant 14$, take logarithms of the values, and fit these points to the best straight line, we obtain a value for $1/V$ of 0.15. Thus, $V = 6.66$.

To account for the distribution outside the well-behaved region $3 \leqslant t \leqslant 14$, we introduce some *ad hoc* variations of the simple "optimal" strategy. We investigate $r(x)$ as it appears in the data, determined by counting numbers of conflicts terminated at a given time, t. Fig. 5 gives the calculated function, r, for Resident, in the late game (t is measured from the occurrence of Submissive Withdraw by Resident). In fact, $r(t)$ does not seem to be constant, but exhibits a gradual increase with t. For this reason we allow $r(t)$ to take the form

$$r(t) = r(0) \, e^{Kt},$$

where $r(0)$ is some initial value, and K represents a strategic term that increases or decreases the instantaneous quit probability with time. Thus, the simplified optimal strategy has $r(0) = 1/V$ and $K = $ zero. Finally, we allow the conflict time ($1 \leqslant t \leqslant 40$) to be partitioned into n regions, within each of which a value of $r(0)$ and K may be specified. This is motivated by observed discontinuities in the calculated $r(t)$, as in Fig. 6. We emphasize that the more general functions $r(t)$ of the form above do not correspond to obvious probability densities, as is the case for $r(0) = 1/V$ and $K = $ zero.

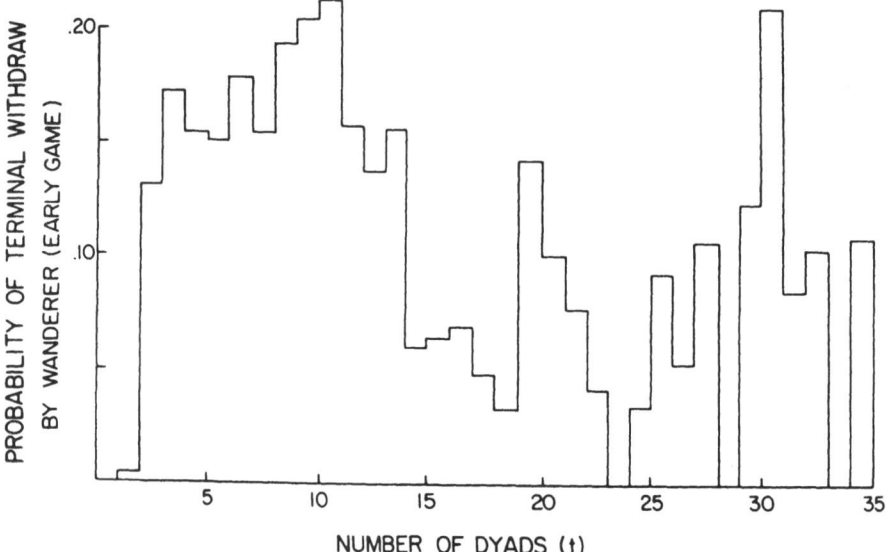

Fig. 6: Instantaneous probability of Terminal Withdraw by Wanderer in the early game. See text for details.

3.2.3 Testing the Model — We constructed a conflict simulator as for the previous model. Here the instantaneous quit probability $r(t)$ was compared to a pseudo-random number to determine whether an animal terminates the conflict at time t. As in section 3.1, this approach does not work well for the Resident in the early game. For the Wanderer in the early game, a value of $r(0)$ about 0.15 and K = zero is an optimal strategy (in the sense of *Maynard Smith* [1974]) that fits the data (ignoring variation in size) slightly better than the model of section 3.1. The fit can be slightly improved by using these values in the region $1 \leqslant t \leqslant 40$ with $n = 2$. Tests were run using values of n up to 10, and with various choices of $r(0)$ and K in each region. The fit was best in the region of $n = 1$ or $n = 2$ and $r(0) \leqslant 0.15$ with K small. The smallest of standard deviation and chi-square for these "better" values were 28.5 and 490.5 respectively.

The model can also be applied in the late game, with some success for both Resident and Wanderer. Here, because of the smaller sample (about 50 conflicts), it is inaccurate to look at distributions in a single size category. The best simulations occurred for $n = 2$ and $r(0) \leqslant 0.10$, with K small, but non-zero. Typical "better" fits had standard deviations about 120 and chi-square about 30.

This model fits the aggregate data better than that of section 3.1 and it agrees with the quitting time distribution suggested by *Maynard Smith* [1974]. However, the quitting time distributions found in individual size bins, which sum to the aggregate distribution, are forced mathematically to be different from the exponential. Our data verify this as follows. The distributions for the various values of d sum to an aggregate distribution of exponential type $e^{-\alpha t}$ where $\alpha \cong 0.15$. Although the separate distributions do not show simple exponential decay, they will do so in the aggregate if the d-th distribution has the form

$$P_d(t) \cong \frac{\alpha^{(\alpha E_d)}}{\Gamma_{(\alpha E_d)}} e^{-\alpha t} t^{\alpha E_d - 1}$$

where E_d is the mean of the d-th distribution. Such a curve is compared to the empirical distribution in Fig. 7.

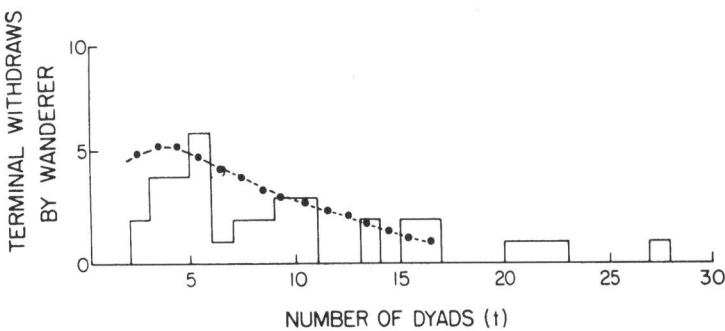

Fig. 7: Distribution of Terminal Withdraw by Wanderer for size difference $d = -1.0$ compared to graph of $e^{-\alpha t}$ equation × 37 conflicts. $\alpha = 0.15$ and $E_d = 8.51$ (displaced to start at $t = 2$).

3.3 A Differential Game Model in Discrete Approximation

3.3.1 Motivation — We have noted how the single stage format of the models in section 3.1 and 3.2 sacrifice much of the structure observed in *Uca* conflict. For a multi-stage, complex game, the normal form version is too unwieldly to allow a solution by means of a game matrix. It is more reasonable to view the game as a continuous process, which we observe only in a discrete approximation. Therefore, for our last model we use differential-equation techniques to solve for an optimum strategy, and see whether our data provide a discrete version of such a strategy.

Some aspects of *Uca* conflict suggest modeling by means of the "war of attrition and attack" discussed by *Isaacs* [1965]. There are similarities also to the "tactical air war" of *Berkowitz/Dresher* [1959]. In *Isaacs'* game two opponents can divide their forces between two tasks, "attrition" and "attack". Of these, attrition actually lowers the capacity of the enemy to continue. *Uca* conflicts seem to be divided between "low" and "high" intensity acts, and of these the high-intensity acts, such as HR, seem to be the only ones which reduce the willingness of the opponent to continue. Furthermore, it is often observed that a conflict will begin with low-intensity acts, such as MP or DS then one animal, especially the one which appears to be weaker, may begin to "escalate". Shortly thereafter both have escalated, and the fight continues at high-intensity until the end. This is generally consistent with the form of the solution of the game of *Isaacs*, which provides for a period of attack or mixed attrition and attack, followed by a period of all-out attrition directly preceding the end of the game.

3.3.2 The Model – The two participants, Resident (R) and Wanderer (W), are assumed to have an initial supply of a resource "currency", which we may think of as "willingness to continue fighting", measured by X_R and X_W. Each allocates his resource in acts of a varying level of intensity between "low" (*Isaacs'* "attrition") and "high" (*Isaacs'* "attack"). A fraction ϕ_R or ϕ_W measures the proportion of high intensity. We determine from the empirical data where the various combat acts fall in this intensity spectrum. We assume each animal has a susceptibility factor c_R or c_W, that measures how effectively his resources are reduced by the activity of his opponent. Finally, we make the (realistic) assumption, a simplification over the general *Isaacs* model, that neither animal can renew his resources during a conflict.

Since $\phi_R X_R$ measures the amount of Resident's resources committed to high intensity, the number $c_W (\phi_R X_R)$ measures the effect of this commitment. Thus, the differential equation for the development of X_W is

$$\frac{d}{dt} X_W = -c_W \phi_R X_R .$$

Here c_W is a constant, and X_W, ϕ_W and X_R are assumed to vary with t. Similarly, we have

$$\frac{d}{dt} X_R = -c_R \phi_W X_W .$$

The payoff to Wanderer is the difference between his "willingness" resource remaining at the end of a conflict, and that left for the Resident. If the conflict lasts a fixed time, T, then a reasonable estimate of payoff to Wanderer can be given by

$$\int_0^T (1 - \phi_W) X_W - (1 - \phi_R) X_R \, dt .$$

It is convenient to measure time from the end of a conflict, so that we can also regard T as a variable with $dT/dt = -1$.

For expository purposes assume that $c_R > c_W$; we can reverse the notation in the other case. Indeed, it seems in the data that this relation holds when the Wanderer has a large size advantage.

3.3.3 Testing the Model

Testing the Model — This work is in progress, but the results of modeling are sufficiently interesting to suggest further data collection and manipulative experimentation. Solution of the differential equations for the game shows that the following strategies are optimal:

i) The less susceptible animal, W, with $c_W < c_R$ should use low intensity ("attrition") acts up to a time $1/c_R$ units before the anticipated end of the conflict; then escalate to high-intensity ("attack") acts.

ii) The more susceptible animal, R, should follow the same act deployment pattern, but escalate to high-intensity at an earlier time $(1/c_R)\sqrt{2(c_R/c_W)} - 1$ units before the end of the conflict.

It is possible to look at actual conflict sequences, observing the pattern of high and low intensity acts, and to make an estimate of c_R and c_W, under the assumption that a conflict ends when X_R or X_W has been reduced to 0. We can similarly estimate the intensity factor (ϕ_R or ϕ_W) to be assigned to the various possible acts. For a first approximation, it is convenient to assign $\phi = $ zero to acts that seem intuitively to be frequent and "passive" and $\phi = 1$ to acts which appear to be rare and "aggressive" (i.e. decisive). An initial value for X_R and X_W is suggested by the normalized value $V \cong 6.66$ obtained from the quit time distributions mentioned earlier. The problem of suitable determination of intensity assignment ϕ is clearly critical, but additional field work may provide other empirical measures for the payoff equation, such as search cost for an alternative burrow, or an index of mating success. As for length of conflict, we can use a distribution derived from empirical data. This allows us to insert a term into the payoff equation so that T is not ridgidly fixed. Thus, empirically, we would expect to find the escalation points (however these are determined by the animals) to be distributed about $1/c_R$ or $1/c_R \sqrt{2(c_R/c_W)} - 1$ before the terminations of conflicts.

4. Discussion

The idea of applying quantitative predictive models, based upon game theoretic approaches, toward the analysis of biological data is not new. *Von Neumann/Morgenstern* [1953] suggested that, "Social events can best be described by models taken from suitable games of strategy. These games in turn are amenable to thorough mathematical analysis." *Kalmus* [1969] felt that games and psycholinguistic models should be useful for analyzing sequences of behavioral elements in and between species. *Trivers* [1971], in a review of the evolution of reciprocal altruism, suggested that the relationship between animals exposed to repeated reciprocal situations could be described in game form.

Maynard Smith's [1972] original intent was, " . . . to see whether an explanation of conventional fighting can be given in terms of selection at the individual level." To do this he applied the theory of games to the evolution of animal conflict behavior. *Maynard Smith/Price* [1973] expanded the animal conflict model and used game theory and computer simulation to show that the "limited war" strategy benefitted individual animals. Two models were examined, viz., combat in species possessing heavy weapons capable of inflicting injury, and combat in species where serious injury was impossible, thus favoring the extended conflict ("war of attrition"). For each case they sought a strategy that was stable under multiple iterations (i.e., "evolution") of the model and coined the term "Evolutionarily Stable Strategy", or ESS. An ESS was defined, roughly as " . . . a strategy such that, if most of the members of a population adopt it, there is no 'mutant' strategy that would give higher reproductive fitness."

Parker [1974] accepted *Maynard Smith's* argument that "limited war" was an ESS over "total war" or "total peace", and went on to show that any mutant able to assess its own resource holding power (RHP) compared to its opponent would have a selective advantage, since it could retreat if the RHP of the opponent was significantly greater, or persist without damage if it is significantly less.

Finally, *Maynard Smith/Parker* [1976] considered the sources of asymmetry in contests. Heretofore contests were assumed to be "equal" (i.e., zero-sum) in payoff and/or in contestant prowess. This is seldom the case in nature, since size, age, status, and other intrinsic and extrinsic factors can contribute to biaš outcome or payoff.

These theoretical guidelines are valuable templates. They have been supported ocassionally with examples from the behavioral literature which "fit the models". Similarly, we have supplied real data, and modified the models to meet the specific demands of the *Uca* behavorial repertoire. In the case of fiddler crabs, we are dealing with animals of unequal strength and with varying strategies depending upon Resident or Wanderer status. For this reason we cannot adopt a symmetric payoff. Furthermore, the game is truly sequential and multimove games are better captured by a continuous version. The model we propose closely mimics the zero-sum differential air war game of *Isaacs* [1965], and it allows for size and status asymmetry, as well as for variable contest duration. Indeed, a recent paper by *Mirmirani/Oster* [1978] also involves a continuous rather than a discrete model. Its solution is of the type found in our model of Section 3.3 above.

Acknowledgements

This work was supported by NSF BNS 76-08524 to G.W.H., by the Research Board of the University of Illinois at Chicago Circle, and by a travel grant from the Joyce Foundation. Computing services were provided by the Computer Center of the University of Illinois at Chicago Circle. We would like to recognize Yair Daon and Chris Krubel for their excellent computer programming skills. We appreciate the insight gained from discussions with John Maynard Smith, Robert Axelrod, Dennis Nyberg and Richard Videbeck. This paper is a contribution of the Duke Marine Laboratory and the Illinois Marine Biological Association.

References

Berkovitz, L.D., and *M. Dresher:* A game-theory analysis of tactical air war. Operations Res. 7, 1959, 599–620.

Bishop, D.T., and *C. Cannings:* A generalized war of attrition. J. theoret. Biol. 70, 1978, 85–124.

Crane, J.: Combat, display & ritualization in fiddler crabs (Ocypodidae, Genus *Uca*) Phil. Trans. Roy. Soc. London B. **251**, 1966, 459–472.

– : Combat and its ritualization in fiddler crabs (Ocypodidae) with special reference to *Uca rapax* (Smith). Zoologica **52**, 1967, 49–76.

– : Fiddler Crabs of the World. Princeton 1975.

Hines, W.G.S.: Competition with an evolutionary stable strategy. J. theoret. Biol. 67, 1977, 141–153.

Hyatt, G.W., and *M. Salmon:* Combat in the fiddler crabs *Uca pugilator* and *U. pugnax:* A quantitative analysis. Behaviour **65** (3–4), 1978, 182–211.

– : Comparative statistical and information analysis of combat in the fiddler crabs. *Uca pugilator* and *U. pugnax.* Behaviour. In Press.

Isaacs, R.: Differential Games. New York 1965.

Kalmus, H.: Animal behaviour and theories of games and of language. Anim. Behav. **17**, 1969, 607–617.

Maynard Smith, J.: On Evolution. Edinburgh 1972.

– : The theory of games and the evolution of animal conflicts. J. theoret. Biol. 47, 1974, 209–221.

– : Evolution and the theory of games. Am. Scientist. **64**, 1972, 41–45.

Maynard Smith, J., and *G.A. Parker:* The logic of asymmetric contests. Anim. Behav. **24**, 1976, 159–175.

Maynard Smith, J., and *G.R. Price:* The logic of animal conflict. Nature **246**, 1973, 15–18.

Mirmirani, M., and *G. Oster:* Competition, kin selection and evolutionarily stable strategies. Theor. Pop. Biol. 13, 1978, 304–339.

Neumann, J. von, and *O. Morgenstern:* Theory of Games and Economic Behavior. Third edition, Princeton 1953.

Parker, G.A.: Assessment strategy and the evolution of fighting behaviour. J. theoret. Biol. **47**, 1974, 223–243.

Sclove, S.L.: Population mixture and clustering algorithms. Commun. Statist.-Theor. Math. **46**, 1977, 417–434.

Trivers, R.: The evolution of reciprocal altruism. Quart. Rev. Biol. **46**, 1971, 35–57.

Applied Game Theory. 1979 © Physica-Verlag, Wuerzburg/Germany

Faith Versus Rationality in the Bible: Game-Theoretic Interpretations of Sacrifice in the Old Testament

By *S.J. Brams,* New York[1])

Abstract: Abraham's attempted sacrifice of his son, Isaac (as given in Genesis), and Jephthah's actual sacrifice of his daughter (as given in Judges), are viewed as two-person, nonzero-sum games played by the fathers with God. Different interpretations of these games suggest that there is a trade-off between "faith" and "rationality": the more sophisticated rationality calculations biblical characters make, the less need for them to have blind faith in God to achieve their goals. Thus, as the faith of a character wavers, his rationality may sustain him — but not necessarily if he deviates seriously and God is unsympathetic.

Implications of treating God as a game player are discussed, and comparisons are made with other interpretations of the biblical stories analyzed.

Introduction

The Bible is a sacred document to millions of people. It expresses supernatural elements of faith that do not admit of any natural explanations. At the same time, however, some of the great narratives in the Bible do not seem implausible reconstructions of events. Indeed, biblical characters exhibit on occasion common human failings in their behavior toward one another.

Is it possible to reconcile natural and supernatural elements in the Bible? This would not seem an easy task since God, in some manifestation, makes His presence felt in practically all biblical stories. A naturalistic interpretation of the Bible immediately confronts the fact of His commanding presence as well as His uniqueness.

In any biblical analysis or interpretation, then, God must be given His proper due. He *is* the central character in the Bible. Accordingly, I propose to treat Him as such, but my treatment assumes more than His omnipresence. I also assume that God is motivated to do certain things — that He has goals He would like to achieve.

I do not assume that God is omnipotent. To be sure, He can perform miracles and even endow others with great powers. But the Bible is clear on one thing: human beings *do* have free will and can exercise it, even if it invokes God's wrath. (Reasons why God chose not to make man a puppet are discussed in *Brams* [1980, chapt. 2]). Consequently, God, powerful as He is, is sometimes thwarted in His desires. When this happens, the Bible tells us that God may be angry, jealous, or vengeful — and His actions tend to reflect these emotions.

---

[1]) Professor *Steven J. Brams,* Department of Politics, New York University, New York, N.Y. 10003, USA.

Seen in this light, God is a very human character, despite His unique presence and awe-inspiring powers. Is it not sensible, then, to view Him as a player in a game, who chooses among different courses of action to try to achieve certain goals? Similarly, is it not reasonable that more ordinary characters in the Bible, knowing of God's presence, make choices to further their own ends in light of possible consequences they perceive can occur?

Although the idea that God and a human cast of characters play games may seem bizarre, I shall try to show in the context of two biblical stories that it is not only a reasonable interpretation but also that the players in these games (including God) — for plausible interpretations of their preferences — acted rationally. That is, given their knowledge of each others' preferences, they made strategy choices that would lead to better rather than worse outcomes. I shall offer a more precise definition of rationality later.

The main conclusion I draw from these stories is that "rationality" interpretations of biblical actions are not less bizarre than "faith" interpretations. In fact, I would argue that a more mundane, rationalistic explanation, precisely because it does not assume of biblical characters superhuman righteousness in the face of adversity, is more credible.

Apart from credibility considerations, the game-theoretic analysis demonstrates a trade-off between faith and rationality: the more sophisticated rationality calculations biblical characters make, the less need for them to have blind faith in God to achieve their goals. Thus, as the faith of a character wavers, his rationality may sustain him — but not necessarily if he deviates seriously and God is unsympathetic.

One lesson in all this, I believe, is that God can brook some deviance but not complete rejection. Knowing the failings of His subjects, He is capable of — but not always willing to demonstrate — mercy. This attitude of reasonable toleration, in my opinion, is consistent with conceptualizing God as a game player, omnipresent but not omnipotent, rational but not emotionless. With vastly scaled-down presence and power, He is thus not unlike most of us.

This is not, of course, the way most religions view God. Indeed, the Bible continually portrays Him as not only awesome but unknowable — utterly beyond our comprehension. The clear implication is that He is someone incapable of being calculating and conniving, stooping to the level of "playing games."

Yet, this is not the impression many of His actions convey. In fact, since God Himself often provides explicit reasons for acting in a particular way, it is hard to maintain that His motivations and design are unfathomable.

In the two stories of human sacrifice from the Old Testament that I shall analyze, God's motivations seem quite plain. Nevertheless, I shall consider alternative interpretations of His preferences, as well as alternative interpretations of the preferences of His protagonists, in trying to model their behavior as that of players in a game. Since the two stories have different outcomes, they nicely show up the dependence of game outcomes on player preferences. I shall conclude by comparing the game-theoretic explanations of player choices in these game with nonrational interpretations of these stories.

Abraham's Sacrifice

With characteristic economy of language, chapter 22 of Genesis begins, "The time came when God put Abraham to the test" [Gen. 22:1][2]). Then, in just eighteen verses, one of the greatest and most poignant stories from the Bible is told. The significance of this story, and its interlocking themes of faithfulness, sacrifice, and murder, have been subjected to prodigious analysis and interpretation, some of which I shall briefly discuss later.

In the story, God commands Abraham:

> Take your son Isaac, your only son, whom you love, and go to the land Moriah. There you shall offer him as a sacrifice on one of the hills which I will show you [Gen. 22:2].

Faithful servant of God that he is, Abraham sets out on his ass with Isaac, accompanied by two of his men with firewood for the sacrifice.

On the third day of the journey, Abraham sees the place for the sacrifice and leaves his ass and two men behind. He gives Isaac the firewood to carry and he himself carries the fire and knife. When Isaac asks, "Here are the fire and wood, but where is the young beast for the sacrifice" [Gen. 22:7]? Abraham answers, "God will provide himself with a young beast for a sacrifice, my son" [Gen. 22:8].

Abraham builds an altar and arranges the wood, after which he binds Isaac and lays him on the altar on top of the wood. As he stretches out his hand, his knife poised, to kill his son,

> the angel of the LORD called to him from heaven, "Abraham, Abraham." He answered, "Here I am." The angel of the LORD said, "Do not raise your hand against the boy; do not touch him. Now I know that you are a God-fearing man. You have not withheld from me your son, your only son." Abraham looked up, and there he saw a ram caught by its horns in a thicket. So he went and took the ram and offered it as a sacrifice instead of his son [Gen. 22:11–13].

Abraham is then rewarded for his faithfulness when the angel calls from heaven a second time:

> This is the word of the LORD: By my own self I swear: inasmuch as you have done this and have not withheld your son, your only son, I will bless you abundantly and greatly multiply your descendants until they are as numerous as the stars in the sky and the grains of sand on the sea-shore. Your descendants shall possess the cities of their enemies. All nations on earth shall pray to be blessed as your descendants are blessed, and this because you have obeyed me [Gen. 22:16–18].

If this game is viewed as played between Abraham and God, Abraham has two strategy choices:

1. Offer Isaac: O
2. Don't offer Isaac: $\bar{O}$.

God, in turn, also has two choices:

1. Renege (if Isaac offered)/relent (if not): R
2. Don't renege/relent: $\bar{R}$.

---

[2]) All biblical quotations are from New English Bible with the Apocrypha [1976].

For God, his first choice is a cooperative response, implying – whatever Abraham does – that He intended just to test him. On the other hand, God's second choice would indicate that He was deadly serious about His command to sacrifice Isaac. The consequences of these strategy choices for both players are summarized verbally in the outcome matrix shown in Figure 1. (The pairs of numbers associated with the different outcomes will be described shortly.)

God

		Renege (if Isaac offered)/ relent (if not): R	Don't renege/relent: $\bar{R}$
Offer Isaac:	O	Abraham faithful, a) (4,4) God merciful, b) (4,4) Isaac saved c) (4,4)	Abraham faithful, a) (3,3) God adamant, b) (2,3) Isaac sacrificed c) (1,3)
Abraham			
Don't offer Isaac:	$\bar{O}$	Abraham resistant, a) (2,1) God merciful, b) (3,1) Isaac saved c) (3,1)	Abraham resistant, a) (1,2) God adamant, b) (1,2) Isaac's fate c) (2,2) uncertain

Key: $(x, y) =$ (Abraham, God).
   "4" best; "3" next best; "2" next worst; "1" worst.
   a) **Abraham faithful regardless**: prefers "offer" over "don't offer"
   b) **Abraham wavers somewhat**: prefers God "renege/relent" over "don't renege/relent"
   c) **Abraham wavers seriously**: Isaac's life paramount – same as b) above except if God adamant, would prefer "don't offer"

Fig. 1: Outcome Matrix for Abraham's Sacrifice

In fact, however, God did not have the choice of just an unconditionally cooperative (R) or noncooperative $(\bar{R})$ response. His moves occurred after Abraham's, in full knowledge of the strategy choice (0 or $\bar{O}$) that Abraham made.

The representation of this *sequence* of moves is shown by the game tree in Figure 2 (read from top to bottom): Abraham first chooses to offer or not offer to sacrifice Isaac; only then does God choose to be merciful or not. The fact that Abraham's move precedes God's, and God is aware of Abraham's prior choice, means that the game cannot properly be presented as a 2 × 2 game (two players, each with two strategy choices), as illustrated by the outcome matrix in Figure 1. Its proper repesentation in normal, or matrix, from is a a 2 × 4 game (Abraham has two strategies, God has four), which I shall discuss presently. For the purpose of evaluating the four possible outcomes that can occur, however, the 2 × 2 form will be used.

In my evaluation, I shall attempt only to rank the outcomes for each player from best to worst, without attaching any specific values, or cardinal utilities, to these ranks. In the representation shown in Figures 1 and 2, "4" is considered a player's best outcome, "3" next best, "2" next worst, and "1" worst. Thus, the higher the number, the better the outcome.

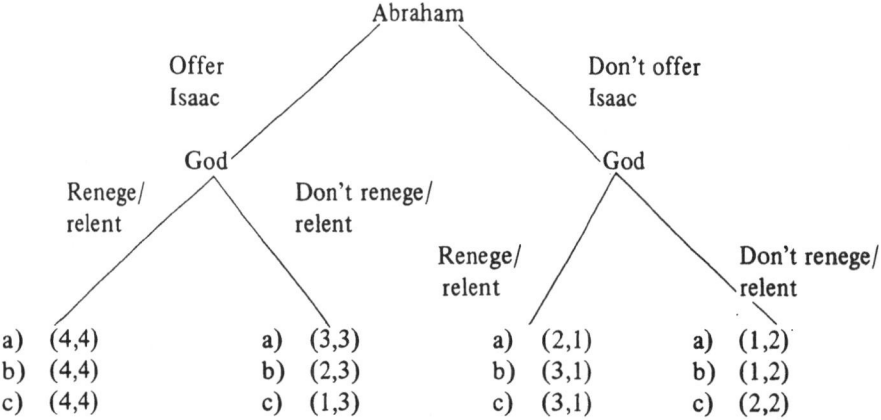

Key: $(x, y) = $ (Abraham, God)
       "4" best; "3" next best; "2" next worst; "1" worst

Fig. 2: Game Tree for Abraham's sacrifice

The first number in each pair is assumed to be Abraham's preference ranking (row player in Figure 1), the second number God's preference ranking (column player in Figure 1). Thus, for example, the payoff (3,1) means the next-best outcome for Abraham, the worst outcome for God. In general, the games analyzed here are not games of pure opposition (or zero-sum if cardinal utilities were assigned) — what is best for one player is not necessarily worst for the other player.

Now consider God's ranking of the four possible outcomes shown in Figure 1. (While varying Abraham's ranks in three different ways to be described shortly, I assume that God's ranks in the three cases shown for each outcome in Figure 1 remain fixed.) First, I assume God prefers that Abraham show his faith by offering Isaac, as OR and $O\bar{R}$ in the first row of the outcome matrix of Figure 1 are His two most-preferred outcomes. Given Abraham chooses O, I assume that God would prefer only to put Abraham to the test (R) — as the Bible says He intends — and not allow Isaac actually to be sacrificed $(\bar{R})$, so for God $OR = 4$ and $O\bar{R} = 3$. If Abraham should not offer Isaac $(\bar{O})$, however, I assume that God would prefer not to relent, so $\overline{OR} = 2$ and $\bar{O}R = 1$. Given these assumptions, I now want to show what consequences they have for the rational play of the game when Abraham's faith wavers.

In the key to Figure 1, and in Figure 3, I have briefly characterized three different assumptions about Abraham's preference rankings. Also in Figure 3, I have given the proper 2×4 normal-form representation of the game, derived from the Figure 2 game tree. It reflects the fact that, since Abraham has the first move, he can choose either to offer or not offer Isaac. God, on the other hand, whose moves occur only after Abraham has made a choice, has four possible choices, depending on what Abraham chooses. Thus, God has four strategies, or complete plans that describe His possible choices contingent upon Abraham's prior choices:

1. R/R – Be merciful regardless: Renege if Isaac offered, relent if not.
2. $\bar{R}/\bar{R}$ – Be adamant regardless: Don't renege if Isaac offered, don't relent if not.
3. $R/\bar{R}$ – Tit-for-tat: Renege if Isaac offered, don't relent if not.
4. $\bar{R}/R$ – Tat-for-tit: Don't renege if Isaac offered, relent if not.

a) **Abraham faithful regardless**: whatever God chooses subsequently – renege/relent (R) or not $(\bar{R})$ – Abraham prefers to offer (O).

	God		God			
	R	$\bar{R}$	R/R	$\bar{R}/\bar{R}$	$R/\bar{R}$	$\bar{R}/R$
O	(4,4)	(3,3)	(4,4)	(3,3)	**(4,4)**	(3,3) ← Offer dominant
$\bar{O}$	(2,1)	(1,2)	(2,1)	(1,2)	(1,2)	(2,1)

Abraham

↑
Tit-for-tat dominant

b) **Abraham wavers somewhat**: whatever Abraham chooses – offer (O) or don't offer $(\bar{O})$ – he prefers that God subsequently renege/relent (R).

	God		God			
	R	$\bar{R}$	R/R	$\bar{R}/\bar{R}$	$R/\bar{R}$	$\bar{R}/R$
O	(4,4)	(2,3)	(4,4)	(2,3)	**(4,4)**	(2,3)
$\bar{O}$	(3,1)	(1,2)	(3,1)	(1,2)	(1,2)	(3,1)

Abraham

Neither strategy dominant – must anticipate God's choice

↑
Tit-for-tat dominant

c) **Abraham wavers seriously**: same as b) above, except given God is adamant $(\bar{R})$, Abraham prefers not to offer $(\bar{O})$.

	God		God			
	R	$\bar{R}$	R/R	$\bar{R}/\bar{R}$	$R/\bar{R}$	$\bar{R}/R$
O	(4,4)	(1,3)	(4,4)	(1,3)	**(4,4)**	(1,3)
$\bar{O}$	(3,1)	(2,2)	(3,1)	(2,2)	(2,2)	(3,1)

Abraham

Neither strategy dominant – must anticipate God's choice

↑
Tit-for-tat dominant

Key: $(x, y) = $ (Abraham, God).
"4" best; "3" next best; "2" next worst; "1" worst.

Fig. 3: Payoff Matrices for Abraham's Sacrifice

For the three different assumptions I have made about Abraham's preference rankings, the 2 × 4 payoff matrices in Figure 3 give the payoffs each player receives for every pair of strategy choices (two for Abraham as row player, four for God as column player). For example, given assumption a) about Abraham's preference ranking of outcomes, if Abraham chooses to offer Isaac (O), and God chooses tat-for-tit ($\bar{R}/R$), $O\bar{R}$ is the resultant outcome since the choice of O by Abraham implies the choice of $\bar{R}$ by God under tat-for-tit. As can be seen from Figure 1, this yields a payoff of (3,3), the next best outcome for both players. By comparison, under assumption b) the choice of $O\bar{R}$ yields a payoff of (2,3), and under assumption c) a payoff of (1,3).

What meaning can be attached to these different preference assumptions for Abraham? As I indicate in Figure 3, it seems reasonable to assume that Abraham may a) be faithful regardless, b) waver somewhat, or c) waver seriously. Operationally, these assumptions have the following interpretations:

a) Whatever God chooses subsequently, Abraham prefers to offer Isaac (O). Of course, Abraham would prefer that God renege on His demand that Isaac be sacrificed (R), so $OR = 4$ and $O\bar{R} = 3$, which makes God and Abraham's first-row preferences identical. If Abraham does not offer Isaac ($\bar{O}$), he would prefer to do so when God relents (R) and Isaac is thereby saved, so $\bar{O}R = 2$ and $\overline{O}\overline{R} = 1$ for Abraham.

b) Whatever Abraham chooses, he prefers that God subsequently renege/relent (R). Since Abraham would prefer to show his faith by offering Isaac (O), $OR = 4$ and $\bar{O}R = 3$. If God is adamant ($\bar{R}$), however, Abraham would prefer to offer Isaac (O) as a show of his faith, so $O\bar{R} = 2$ and $\overline{O}\overline{R} = 1$.

c) Same as b) above, except now, given God is adamant ($\bar{R}$), Abraham would prefer not to offer Isaac ($\bar{O}$) – perhaps hoping that this will save his son's life, even if he himself is punished – so $\overline{O}\overline{R} = 2$ and $O\bar{R} = 1$. This assumption says, in effect, that Isaac's life is paramount – Abraham's worst outcome occurs when Isaac is offered and God does not arrest his sacrifice.

What are the game-theoretic implications of these different preference assumptions I have posited for Abraham? Notice, first, that in the case of all three 2 × 4 payoff matrices in Figure 3, God's tit-for-tat strategy is *dominant*: God's payoffs associated with this strategy are at least as good as, and sometimes better than, His payoffs associated with any of His three other strategies, whatever strategy Abraham chooses. For example, under assumption a), if Abraham chooses O, tit-for-tat yields God a payoff of 4, which obviously cannot be improved upon; if Abraham chooses $\bar{O}$, tit-for-tat yields God a payoff of 2, which is as good as R/R yields and better than $\bar{R}/R$ and $R/\bar{R}$ yield (payoffs of 1 in each of the latter cases). Thus, tit-for-tat is God's unconditionally best strategy – not dependent on which strategy (O or $\bar{O}$) Abraham chooses – and presumably the choice a rational player would make in this game. In fact, I define a *rational player* to be one who chooses a dominant strategy if he has one[3]).

---

[3]) It is easy to show that the expansion of every 2 × 2 ordinal game in which preferences are strict (i.e., whose four outcomes can be ranked from best to worst without ties) to a 2 × 4 ordinal game always results in the second-moving player's having a dominant strategy, whether he had one or not in the 2 × 2 game.

Under assumption a), Abraham also has a dominant strategy – to offer Isaac. The intersection of Abraham's dominant strategy (O) and God's dominant strategy ($R/\bar{R}$) yields the best outcome [(4,4)] for both players. Of course, this was the outcome that was actually chosen in the game.

The best outcomes for both players under assumptions b) and c) are also (4,4), but a wavering Abraham in these cases no longer has a straightforward choice. Rather, since he does not have a dominant strategy under assumptions b) and c), Abraham must anticipate the strategy God will choose in order to determine his own best choice.

To illustrate this point, consider Abraham's choices under assumption b) in Figure 3. If God should choose either R/R, $\bar{R}/\bar{R}$, or $R/\bar{R}$, Abraham would obtain a higher payoff by choosing O rather than $\bar{O}$. However, if God should chooses $\bar{R}/R$, O would yield Abraham a payoff of only 2 whereas $\bar{O}$ would yield Abraham a payoff of 3. Hence, neither O nor $\bar{O}$ is an unconditionally best choice for Abraham, for "best" depends on what God subsequently does.

What strategy, then, will God as a rational player choose? Since God's tit-for-tat strategy of $R/\bar{R}$ remains dominant under assumptions b) and c), I assume He will choose it. Anticipating God's choice of tit-for-tat, Abraham in both cases would obtain a higher payoff (4) by choosing O rather than $\bar{O}$, which would yield him payoffs of 1 and 2, respectively, under assumptions b) and c).

If rational, then, Abraham will presumably choose O. The choice of O by Abraham, and $R/\bar{R}$ by God, once again results in both players' obtaining their best outcome of (4, 4); now, however, the process by which it is arrived at is different. Specifically, a wavering Abraham under preference assumptions b) and c) must first anticipate God's rational (dominant) strategy choice before he can decide what is best for him. (In general, a player without a dominant strategy is *rational* if, for the dominant strategy choice of the other player, he chooses the strategy which yields for him the highest-ranked outcome of those associated with the other player's dominant strategy.) Under assumption a), by contrast, Abraham has no need to make such a calculation since he himself has a dominant strategy – a best choice not conditional on what God might subsequently choose – and hence can act rationally without knowing anything about God's preferences and the choices they might entail.

The conclusion I draw from this analysis of Abraham's sacrifice is that faith in God may ease the often difficult choices biblical characters face. Faith, at least in Abraham's case, meant that he did not have to consider God's possible reactions to his own course of action, which was to obey God.

To obey God blindly is, in fact, to act *as if* one has a dominant strategy – an unconditionally best choice – that requires no detailed preference information about the other player, much less an anticipation of what strategy he might choose. On the other hand, when a character's faith in God is not blind, he needs to make more sophisticated calculations to ascertain how to act rationally. Although his strategy choice may be the same in either case, the logical process needed to arrive at it may be more demanding in terms of both the preference information required and the sophistication needed to process this information. Elsewhere I have argued that such demands were well within Abraham's capabilities, based on his earlier behavior [*Brams*, 1980, chapt. 3].

The three games I have posited all offer, in my view, plausible game-theoretic explanations for Abraham's sacrificial offering of Isaac to God. One is based on the blind faith of an unswerving Abraham, the other two on the more sophisticated calculations of a concerned father. Since all games dictate the same rational choice for Abraham (O), they do not really test the "blindness" of Abraham's faith.

Thus, although God's harrowing test of Abraham succeeds in establishing that Abraham will obey His command — however ghastly — Abraham may well have done so for reasons other than faith. Hence, God's test does not assuredly dispel doubts about Abraham's faith, given Abraham knows God's preferences and is rational. I shall next show that, in another sacrificial game with a different outcome, faith *is* distinguishable from a more sophisticated game-theoretic rationality for one wavering assumption but not the other.

Jephthah's Sacrifice

As told in chapter 11 of Judges, Jephthah, a great warrior and the son of a prostitute, was driven from his home in Gilead by the legitimate sons of his father. In the land of Tob where he settled, he became the leader of a band of adventurers. The elders of Gilead, however, faced by an Ammonite invasion, recalled him and sought his aid, which he consented to give on condition that they appoint him chief of Gilead after his victory.

The elders agreed, and Jephthah then tried to negotiate with the Ammonites, but the negotiations broke down. Forced into battle, Jephthah made the following vow to God:

> If thou wilt deliver the Ammonities into my hands, then the first creature that comes out of the door of my house to meet me when I return from them in peace shall be the LORD's [Judg. 11:30–31].

Having made this fateful vow, Jephthah routed the Ammonites and slaughtered great numbers of them. Upon returning to his house, Jephthah, to his utter dismay, was greeted by his daughter and only child "with tambourines and dances" [Judg. 11:34]. His heart broken, Jephthah told his daughter, "I have made a vow to the LORD and I cannot go back" [Judg. 11:35].

Resigned to her fate, Jephthah's daughter asked only that her sacrifice be postponed for two months so that "I may roam the hills with my companions and mourn that I must die a virgin" [Judg. 11:37]. Jephthah granted her this wish, but at the end of this period he grimly fulfilled his vow. The Bible reports that it henceforth became a tradition that women of Israel commemorate this tragic event by an annual four-day mourning period.

In Jephthah's case, God clearly was not in a mood for testing but instead played for keeps. But what does this say about His preferences, and how does this translate into choices He and Jephthah made in the game they played?

I would suggest that there are at least two interpretations that can be made of God's preferences in this game. One I call *show-of-faith*, the other *vindictive*. In the case of Abraham's sacrifice, it will be recalled, God preferred that Abraham show his faith by offering to sacrifice Isaac, but given that he did, God preferred to renege on His demand. Now, in the show-of-faith interpretation, I assume that God prefers that Jephthah offer his daughter and thereby fulfill his vow; but, given that Jephthah offers his daughter, I assume that God prefers to allow the sacrifice to be consummated (see Figure 4). Note that the only difference between these games and those shown in Figure 3 is that "4" and "3" for God are interchanged in the first rows of the payoff matrices.

a) **Jephthah faithful regardless:** whatever God chooses subsequently — renege/relent (R) or not ($\bar{R}$) — Jephthah prefers to offer (O).

		God			God			
		R	$\bar{R}$		R/R	$\bar{R}/\bar{R}$	$R/\bar{R}$	$\bar{R}/R$
	O	(4,3)	(3,4)		(4,3)	**(3,4)**	(4,3)	(3,4) ← Offer dominant
Jephthah								
	$\bar{O}$	(2,1)	(1,2)		(2,1)	(1,2)	(1,2)	(2,1)

↑
Don't renege/relent regardless dominant

b) **Jephthah wavers somewhat:** whatever Jephthah chooses — offer (O) or don't offer ($\bar{O}$) — he prefers that God subsequently renege/relent (R).

		God			God			
		R	$\bar{R}$		R/R	$\bar{R}/\bar{R}$	$R/\bar{R}$	$\bar{R}/R$
	O	(4,3)	(2,4)		(4,3)	**(2,4)**	(4,3)	(2,4) ⎫ Neither strategy dom-
Jephthah								⎬ inant — must anti-
	$\bar{O}$	(3,1)	(1,2)		(3,1)	(1,2)	(1,2)	(3,1) ⎭ cipate God's choice

↑
Don't renege/relent regardless dominant

c) **Jephthah wavers seriously:** same as b) above, except given God is adamant ($\bar{R}$), Jephthah prefers not to offer ($\bar{O}$).

		God			God			
		R	$\bar{R}$		R/R	$\bar{R}/\bar{R}$	$R/\bar{R}$	$\bar{R}/R$
	O	(4,3)	(1,4)		(4,3)	(1,4)	(4,3)	(1,4) ⎫ Neither strategy dom-
Jephthah								⎬ inant — must anti-
	$\bar{O}$	(3,1)	(2,2)		(3,1)	**(2,2)**	(2,2)	(3,1) ⎭ cipate God's choice

↑
Don't renege/relent regardless dominat

Key: (x, y) = (Jephthah, God).
  "4" best; "3" next best; "2" next worst; "1" worst.

Fig. 4: Payoff Matrices for Jephthah's Sacrifice: Show-Of-Faith Interpretation of God's Preferences

Why this change in God's mood? Since Abraham's time, the Israelites had caused God much grief [*Brams*, 1980, chapts. 5 and 7], so He was not inclined to be sympathetic with people like Jephthah who were too quick to make solemn vows. Whether Jephthah could anticipate this problem, or remembered only the test of Abraham, is hard to say. To carry out the subsequent analysis, however, I shall assume that Jephthah anticipated that God might be less sympathetic than He was with Abraham since I have already analyzed the consequences that flow from this more benign view of God's intent.

I assume three different preference orderings for Jephthah that duplicate those of Abraham given earlier: a) faithful regardless; b) wavers somewhat; and c) wavers seriously. These different interpretations of Jephthah's preferences are shown in Figure 4 for the show-of-faith interpretation of God's preferences described previously.

Under the three different interpretations of Jephthah's preferences, God has a dominant strategy of don't renege/relent regardless. Observe that by making the outcome in which Jephthah offers and God accepts God's best (4) – rather than His next best (3) – God's rational strategy choice is changed from conditional cooperation (tit-for-tat) to unconditional noncooperation (don't renege/relent regardless).

When Jephthah is faithful regardless under interpretation a), he, like Abraham, has a dominant strategy of offering his child, obtaining his next-best outcome (3) when God accepts his offer. And like Abraham, Jephthah does worse if he wavers somewhat and God does not renege (2); to avoid his worst outcome of 1 under interpretation b), he must anticipate God's dominant strategy choice since he himself does not have one. Thus, if faith does not sustain Jephthah, game-theoretic rationality will if he does not waver too seriously. Yet, if Jephthah wavers more – under interpretation c) – his rational choice is not to offer his daughter. Neither God nor Jephthah will be happy with the resulting (2,2) outcome, however, especially since both could do better if they chose strategies associated with one of the (4,3) outcomes.

The problem with this game, like the classic Prisoners' Dilemma game, is that the rational choices of both players lead to an outcome [(2,2)] inferior for both players to some other outcome(s) [(4,3)][4]. Unlike Prisoners' Dilemma, however, one but not both players have dominant strategies, so Jephthah has to anticipate God's dominant strategy choice in order to avoid his worst outcome that would result if he offered to sacrifice his daughter.

Yet, Jephthah did not refuse to consummate his vow, so interpretation c) does not provide an explanation of the actual choices of the players in this game. Interpretations a) and b) work, on the other hand, which means that Jephthah's sacrifice has both a "faith" and a "rationality" explanation when God prefers a show of faith – with sacrifice preferred to nonsacrifice. This show-of-faith interpretation of God's preferences would not work in Abraham's case, however, because God's rational strategy choice under all three interpretations in Figure 4 is to be unrelenting regardless, which He of course was not when Isaac was about to be sacrificed.

If God is more vindictive, He does better when Jephthah wavers seriously, as shown under interpretation c) in Figure 5 [compare the boldface outcome (2,3) in this figure

---

[4]) For a description of Prisoners' Dilemma and examples of its occurrence in politics, see *Brams* [1976, chapt. 4].

with boldface outcome (2,2) in Figure 4 for this interpretation]. Under the vindictive interpretation of God's preferences, I no longer assume that He most prefers that Jephthah offer to sacrifice his child, as I did in the Figure 4 interpretation. Instead, I assume that God most prefers not to renege/relent ($\bar{R}$), with his best outcome (4) occurring when Jephthah offers (O), his next-best outcome (3) when Jephthah does not offer ($\bar{O}$). Note that the only difference between the Figure 4 and Figure 5 games is that preference rankings "3" and "2" for God are interchanged in the payoff matrices.

a) **Jephthah faithful regardless:** whatever God chooses subsequently – renege/relent (R) or not ($\bar{R}$) – Jephthah prefers to offer (O).

	God: R	God: $\bar{R}$	God: R/R	God: $\bar{R}/\bar{R}$	God: $R/\bar{R}$	God: $\bar{R}/R$	
Jephthah O	(4,2)	(3,4)	(4,2)	**(3,4)**	(4,2)	(3,4)	← Offer dominant
Jephthah $\bar{O}$	(2,1)	(1,3)	(2,1)	(1,3)	(1,3)	(2,1)	
				↑ Don't renege/relent regardless dominant			

b) **Jephthah wavers somewhat:** whatever Jephthah chooses – offer (O) or don't offer ($\bar{O}$) – he prefers that God subsequently renege/relent (R).

	God: R	God: $\bar{R}$	God: R/R	God: $\bar{R}/\bar{R}$	God: $R/\bar{R}$	God: $\bar{R}/R$	
Jephthah O	(4,2)	(2,4)	(4,2)	**(2,4)**	(4,2)	(2,4)	Neither strategy dominant – must anticipate God's choice
Jephthah $\bar{O}$	(3,1)	(1,3)	(3,1)	(1,3)	(1,3)	(3,1)	
				↑ Don't renege/relent regardless dominant			

c) **Jephthah wavers seriously:** same as b) above, except given God is adamant ($\bar{R}$), Jephthah prefers not to offer ($\bar{O}$).

	God: R	God: $\bar{R}$	God: R/R	God: $\bar{R}/\bar{R}$	God: $R/\bar{R}$	God: $\bar{R}/R$	
Jephthah O	(4,2)	(1,4)	(4,2)	(1,4)	(4,2)	(1,4)	Neither strategy dominant – must anticipate God's choice
Jephthah $\bar{O}$	(3,1)	(2,3)	(3,1)	**(2,3)**	(2,3)	(3,1)	
				↑ Don't renege/relent regardless dominant			

Key: (x, y) = (Jephthah, God).
"4" best; "3" next best; "2" next worst; "1" worst.

Fig. 5: Payoff Matrices for Jephthah's Sacrifice: Vindictive Interpretation of God's Preferecnes

As with the show-of-faith interpretation in Figure 4, God always has a dominant strategy of being unrelenting regardless $(\bar{R}/\bar{R})$. A blindly faithful and somewhat wavering Jephthah will offer to sacrifice his daughter — confirming the choices reported in the Bible — but if Jephthah is somewhat wavering he must first anticipate God's dominant strategy choice in order to make his own rational choice. As with the show-of-faith interpretation of God's preferences, a seriously wavering Jephthah playing against a vindictive God would not offer to sacrifice his daughter, so interpretation c) for both the show-of-faith and vindictive interpretations does not rationalize the biblical outcome of this game.

In summary, a less merciful God than in Abraham's case must be posited to explain the choices of Jephthah and God in the game played over the sacrifice of Jephthah's daughter. It is not necessary, however, to assume God to be totally vindictive — that His two best outcomes are associated with the consummation of Jephthah's vow — in order to rationalize the players' choices in this game. Indeed, the two months' respite granted Jephthah's daughter supports the proposition that God could tolerate some delay.

As with Abraham's sacrifice, both "faith" and "rationality" constitute alternative explanations of Jephthah's sacrifice. In Jephthah's case, however, the rationality explanation works only if Jephthah is somewhat wavering — he prefers to offer his daughter if God is adamant, otherwise not. Thus, while a sympathetic God allows a biblical character (Abraham) more free rein, even an unsympathetic God affords a wavering character (Jephthah) the opportunity to make rational calculations that prevent his worst outcome from occurring.

Other Interpretations and Conclusions

It hardly needs to be pointed out that not all analysts see a calculating rationality at work in the biblical stories I have analyzed. *Kierkegaard* [1954], for example, in a stunningly prolix and self-conscious treatment of the Isaac story, argues that Abraham, having no authoritative basis on which to choose between obeying God and refusing to sacrifice his son's life, was stricken with dread that made his supreme act of faith all the more magnificent and courageous. *Kolakowski* [1972, p. 18], on the other hand, disparages this interpretation and says that Abraham was simply following God's orders:

> Superiors are not in the habit of explaining orders to subordinates. The essence of an order is that it must be executed because it is an order and not because it is reasonable, promising of success, or well thought out. It is by no means required that the executant understand its meaning; were it so, it would inevitably lead to anarchy and chaos. A subordinate who asks about the meaning of an order is a sower of disorder, a sterile argumentative person. At bottom, he is a smart aleck, an enemy of authority, of the social order, and of the Establishment.

Kolakowski [1972], in his usual irreverent way, goes on to embellish the Isaac story by adding that Abraham botched the sacrifice. After Isaac discovered his father's intent, *Kolakowski* [1972, p. 19] insists, he "never got over his shock."

Kierkegaard [1954, p. 27] also offers a scenario in which Abraham does not conceal his intent. Indeed, the Jewish comentator *Rashi* [1970, 51–52] says unequivocally that when Abraham answered Isaac's question about the whereabouts of the sacrificial beast by responding that "God will provide himself with a young beast for a sacrifice, my son," Isaac understood he was to be murdered[5]).

If Isaac were aware of his impending fate, he might perhaps be included with Abraham and God as a player in a more complex three-person game. In my opinion, however, even if Isaac were aware of his intended sacrifice, the game is adequately – and more parsimoniously – modeled as one played between its two principal players, Abraham and God.

I would also argue that Jephthah's sacrifice was essentially a two-person game, though in this story there is no question that Jephthah's daughter was not only aware of her fate but, because of this, was also able to obtain a temporary reprieve. Yet, her foreknowledge had no effect on the final outcome.

What light does game theory cast on the meaning of these stories of sacrifice in the Old Testament and the nature of their players? First, I believe, the game-theoretic interpretations I have offered provide plausible *alternative* explanations for the choices made by the human biblical characters, based on their own preferences and the perceived preferences of the other player – namely, God. In particular, if Abraham and Jephthah are completely faithful servants of God, their preferences suffice to direct their actions[6]). But when Abraham and Jephthah waver, they no longer have dominant strategies and must, instead, anticipate God's choice of a dominant strategy, based on His preferences.

If God is sympathetic, as in Abraham's case, Abraham can waver seriously and still rationalize the offering of his son. Some wavering is possible if God is less sympathetic, as in Jephthah's case, but if Jephthah prizes his daughter's life above everything else, it is not rational for him to offer to sacrifice her, given the two sets of preferences I have postulated for God, and Jephthah's awareness of these. Since Jephthah did in fact consummate his vow, the game-theoretic analysis suggests he had greater faith in – or fear of – God.

As with Abraham, however, given the different human preferences I have postulated, it is impossible to ascertain whether biblical characters were blindly faithful or wavering. Thus, a major alternative explanation of their actions might be that they did indeed waver, but anticipating God's rational strategy, they were compelled by their own rationality to demonstrate their faith by offering to sacrifice their children. In one case the outcome was clearly salutary for the father, but in the other it was not.

Even for Jephthah, however, this is not to say that his action was irrational. Rather, he seems to have been caught up in a truly gruesome game. Perhaps the story of Jephthah is "exceptional and cannot be treated as indicative of the norm of human sacrifice in Israel" ["*Jephthah*", p. 1342]. Nonetheless, if Jephthah had not made his parlous vow, he presumably would have been killed in battle with the Ammonites. More-

---

[5]) Isaac is, in fact, slain by Abraham in the anti-war poem by Wilfred Owen, "The Parable of the Old Man and the Young," in which the old man also slew "half the seed of Europe, one by one" [Collected Poems of Wilfred Owen, p. 42].

[6]) Their dominant strategies, of course, may indicate fear of God as much as faith, but the Bible provides insufficient information to say whether Abraham and Jephthah's possible blind faith was fear-induced.

over, he not only lived for several years after the sacrifice but also won other military victories and ruled as judge until his death. In the parallel Greek legend related in Euripides's play, *Iphigenia in Aulis,* Agamemnon, too, survived the sacrifice of his daughter[7]).

God, it appears from the stories I have analyzed, may be merciful, vindictive, or even ambivalent. Whatever his mood or preferences, however, it seems not beyond the pale to interpret His choices as if He were a player in a game. In fact, when confronted by a human player who is told, or gains some understanding of, His preferences, the conditions for modeling their choices in a game are met. Since God, as I argued earlier, is not omnipotent, He cannot dictate the game's outcome; rather, the outcome depends on the choices of *all* players, which is the kind of interdependent decision situation that game theory can help illuminate.

Most of the interpretations I have offered of God and His protagonists as rational actors are consistent with the outcomes of the stories of sacrifice I have analyzed. Only when Jephthah wavers seriously does game theory predict an outcome different from that which occurred, whether God's preferences are viewed as show-of-faith or vindictive. This means that a seriously wavering Jephthah must be rejected, but not necessarily a seriously wavering Abraham, given the preferences of God I have postulated and the awareness of these by God's human protagonists. In either case, both biblical characters need not be blindly faithful to explain the choices they made under extremely difficult personal circumstances, which provides, I believe, an alternative explanation for the outcomes of these games.

The interpretation of God as a game player may strike some as odd, to say the least. It suggests, among other things, a more human portrait of God than is conjured up in most religions. God, in this view, has His own problems He is trying to cope with; moreover, He does not always get His way. This being the case, is it so strange that His moods fluctuate and He sometimes appears inconsistent – being merciful and sympathetic on one occasion, angry and vindictive on another?

The fact that God does not always appear cool and even-tempered, I submit, has no bearing on His rationality. The question is, *given* His moods and preferences – however erratic they may appear – do He and His protagonists make choices consistent with their preferences in a game?

I believe the evidence from the two game I have analyzed, backed up by both "faith" and "rationality" arguments, is that they do. The difference between these arguments, however, is that faith, though rational, is not game-theoretic in the calculating sense: a faithful character, by virtue of having a dominant strategy, need not consider what other players do but need only consult his own preferences. Whether faith expresses a character's trust in, or fear of, God, it allows him to act blindly, thereby lifting from him the burden of "playing games" (in the everyday sense).

---

[7]) Differences between biblical narratives and Greek mythology, as exemplified in the writings of Homer, are discussed in *Auerbach* [1953, chapt. 1]. Needless to say, I disagree with *Auerbach*'s opinion that "a rationalistic interpretation [of the Bible] seems psychologically absurd" (p. 14), especially since the life of its characters is "permeated with the stuff of conflict" (p. 22), which I regard as grist for the game-theoretic mill.

References

Auerbach, E.: Mimesis: The Representation of Reality in Western Literature. Tr. by W.R. Trask. Princeton 1953.

Brams, S.J.: Paradoxes in Politics: An Introduction to the Nonobvious in Political Science. New York 1976.

—: Biblical Games: A Strategic Analysis of Stories in the Old Testament, Cambridge 1980.

Collected Poems of Wilfred Owen. Ed. by C.D. Lewis. Norfolk 1964.

"Jephthah," in Encyclopedia Judaica. Vol. 9. Jerusalem 1971.

Kolakowski, L.: The Key to Heaven. Tr. C. Wieniewska and S. Attanasio. New York 1972.

Kierkegaard, S.: Fear and Trembling. Tr. by W. Lowrie. Princeton 1954.

New English Bible with the Apocrypha, Oxford Study Edition. New York 1976.

Rashi: Commentaries on the Pentateuch. Tr. by Ch. Pearl. New York 1970.

List of Contributors and Participants

Professor *John H. Aldrich*, Michigan State University, Department of Political Science, East Lansing, Michigan 48824, USA.

Professor *Deborah L. Allen*, Department of Economics, VPI, Blacksburg, Virginia 24061, USA.

Professor *Robert Axelrod*, Department of Political Science and Institute of Public Policy Studies, The University of Michigan, 1516 Rackham Building, Ann Arbor, Michigan 48109, USA.

Dr. *Gerhard Bonelli*, Institute of Sociology, University of Vienna, Alserstraße 33, 1080 Vienna, Austria.

Professor *Steven J. Brams*, New York University, Department of Politics, 25 Waverly Place, New York, N.Y.10003, USA.

Professor *A. Charnes*, The University of Texas at Austin, Center for Cybernetic Studies, Business Economic Building, Austin, Texas 78712, USA.

Mr. *Szaniszlo Fenyi*, Kernforschungszentrum Karlsruhe, Weberstraße 5, 7500 Karlsruhe, BRD.

Dr. *Hans Frick*, Kernforschungszentrum Karlsruhe, Weberstraße 5, 7500 Karlsruhe, BRD.

Professor *James W. Friedman*, The University of Rochester, College of Arts and Science, Department of Economics, Rochester, New York 14627, USA.

Professor *Dermot Gately*, New York University, Department of Economics, 40 W 4th, 10003 New York, USA.

Professor *Joseph Greenberg*, Virginia Polytechnic Institute and State University, Blacksburgh, VA 24061, USA.

Professor *Bernhard Grofman*, University of California, School of Social Sciences, Irvine, California 92717, USA.

Professor *Koichi Hamada*, University of Tokyo, Department of Economics, Bunkyo-ku, Tokyo, Japan.

Professor *Roger I.C. Hansell*, University of Toronto, Department of Zoology, 25 Harbord Street, 101 Toronto, Ontario, Canada.

Professor *Sergiu Hart*, Stanford University, Institute for Mathematical Studies in the Social Sciences, Fourth Floor, Encina Hall, Stanford, California 94305, USA.

Professor *James P. Heaney*, University of Florida, Department of Environmental Engineering Sciences, A.P. Black Hall, Gainesville, Florida 32611, USA.

Professor *Eckhard Höpfinger*, University of Karlsruhe, Institut für Wirtschaftstheorie und Operations Research, Kaiserstraße 12, 7500 Karlsruhe 1, BRD.

Professor *Gary W. Hyatt*, University of Illinois, Department of Biological Sciences at Chicago Circle Circle, Box 4348, Chicago, Illinois 60680, USA

Professor *James P. Kahan*, University of Southern California, Psychological Research and Service Center, Los Angeles, California 90007, USA.

Professor *Ehud Kalai*, Northwestern University, Graduate School of Management, Nathaniel Leverone Hall, Evanston, Illinois 60201, USA.

Dr. *Eckehart Köhler*, University of Nebraska, Lincoln, Nebr. 68508, USA

Mr. *Mark Lake*, 400 Argyle Road, Brooklyn, N.Y. 11218, USA.

Professor *Jean Lemaire*, Université Libre de Bruxelles, Campus de la plaine, Institut de Statistique, C.P. 210, 50, boulevard du triomphe, 1050 Bruxelles, Belgium.

Professor *S.C. Littlechild*, University of Birmingham, Department of Industrial Economic and Social Science, North Ring Road, Birmingham, B15 2TT, England.

Professor *William Lucas*, Cornell University, School of Operations Research and Industrial Engineering, Upson Hall, Ithaca, N.Y. 14853, USA.

Dr. *Christoph Mandl*, Institute for Advanced Studies, Department of Mathematics, Stumpergasse 56, 1060 Vienna, Austria.

Professor *Michael Maschler*, The Hebrew University of Jerusalem, Jerusalem, Israel.

Professor *Manuel Lopez Mateos*, Departamento de Matematicas, Facultad de Ciencias, UNAM., Mexico 20, D.F., Mexico.

Professor *Richard D. McKelvey*, Carnegie Mellon University, School of Urban and Public Affairs, Schenley Park, Pittsburgh, Pennsylvania 15213, USA.

Professor *Ishaq M. Nadiri*, National Bureau of Economic Research, 261 Madison Avenue, New York 10016, USA.

Professor *Christopher H. Nevison*, Colgate University, Department of Mathematics, Hamilton, New York 13346, USA.

Professor *Peter C. Ordeshook*, Carnegie Mellon University, School of Urban and Public Affairs, Schenley Park, Pittsburgh, Pennsylvania 15213, USA.

Professor *Guillermo Owen*, Instituto SER de Investigaciones, Calle 26, # 25–50, Bogota 1, D.E., Colombia.

Professor *Bezalel Peleg*, The Hebrew University, Institute of Mathematics, Jerusalem, Israel.

Professor *Jean Pierre Ponssard*, Ecole Polytechnique, Centre de Recherche en Gestion, 17, Rue Descartes, 75005 Paris, France.

Professor *Amnon Rapoport*, University of North Carolina at Chapel Hill.

Professor *Ariel Rubinstein*, Nuffield College, Oxford OX1 1NF, England.

Dr. *Markku Sääksjärvi*, Helsinki School of Economics, Runeberginkatu 14–16, 00100 Helsinki 10, Finland.

Professor *Heinz Schleicher*, Université de Paris, 58, Avenue Didier, 94210 La Varenne St. Hilaire, France.

Professor *Andrew Schotter*, New York University, Department of Economics, 538 Tisch Hall, Washington Square, New York, USA.

Professor *Gerhard Schwödiauer*, Universität Bielefeld, Fakultät für Wirtschaftswissenschaften, Postfach 8640, 4800 Bielefeld, BRD.

Professor *Mark Winer*, Carnegie Mellon University, School of Urban and Public Affairs, Schenley Park, Pittsburgh, Pennsylvania 15213, USA.

Professor *H. Peyton Young*, IIASA, International Institute for Applied Systems Analysis, Schloß Laxenburg, 2361 Laxenburg, Austria.

Professor *Rolf Ziegler*, Institute for Sociology, University of Vienna, Alserstraße 33, 1080 Vienna, Austria.

COMPSTAT 1974

Proceedings in Computational Statistics
1st Symposium held in Vienna
Edited by G. Bruckmann, F. Ferschl, L. Schmetterer
1974. 539 pages. Paperbound DM 48.–
ISBN 3 7908 0148 8

COMPSTAT 1976

Proceedings in Computational Statistics
2nd Symposium held in Berlin
Edited by J. Gordesch, P. Naeve
1976. 496 pages. Paperbound DM 56.–
ISBN 3 7908 0172 0

COMPSTAT 1978

Proceedings in Computational Statistics
3rd Symposium held in Leiden
Edited by L.C.A. Corsten, J. Hermans
1978. 540 pages. Paperbound DM 59.–
ISBN 3 7908 0196 8

COMPSTAT LECTURES I

Lectures in Computational Statistics
Edited by H. Skarabis, P.P. Sint
1978. 132 pages. Paperbound DM 30,–
ISBN 3 7908 0197 6